BIOLOGY OF BREEDING POULTRY

Poultry Science Symposium Series

Executive Editor (Volumes 1–18): B.M. Freeman

1 Physiology of the Domestic Fowl*
2 Protein Utilization by Poultry*
3 Environmental Control in Poultry Production*
4 Egg Quality – a Study of the Hen's Egg*
5 The Fertility and Hatchability of the Hen's Egg*
6 i. Factors Affecting Egg Grading*
ii. Aspects of Poultry Behaviour*
7 Poultry Disease and World Economy
8 Egg Formation and Production
9 Energy Requirements of Poultry*
10 Economic Factors Affecting Egg Production*
11 Digestion in the Fowl*
12 Growth and Poultry Meat Production*
13 Avian Coccidiosis*
14 Food Intake Regulation in Poultry*
15 Meat Quality in Poultry and Game Birds
16 Avian Immunology
17 Reproductive Biology of Poultry
18 Poultry Genetics and Breeding
19 Nutrient Requirements of Poultry and Nutritional Research*
20 Egg Quality – Current Problems and Recent Advances*
21 Recent Advances in Turkey Science
22 Avian Incubation
23 Bone Biology and Skeletal Disorders
24 Poultry Immunology*
25 Poultry Meat Science
26 Poultry Feedstuffs
27 Welfare of the Laying Hen
28 Avian Gut Function in Health and Disease
29 Biology of Breeding Poultry

*Out of print
Volumes 1–24 were not published by CAB International. Those still in print may be ordered from:

Carfax Publishing Company
PO Box 25, Abingdon, Oxfordshire OX14 3UE, UK

Biology of Breeding Poultry

Poultry Science Symposium Series
Volume Twenty-nine

Edited by

P.M. Hocking
Division of Genetics and Genomics, The Roslin Institute and Royal (Dick) School of Veterinary Studies, University of Edinburgh, UK

www.cabi.org

CABI is a trading name of CAB International

CABI Head Office
Nosworthy Way
Wallingford
Oxfordshire OX10 8DE
UK
Tel: +44 (0)1491 832111
Fax: +44 (0)1491 833508
E-mail: cabi@cabi.org
Website: www.cabi.org

CABI North American Office
875 Massachusetts Avenue
7th Floor
Cambridge, MA 02139
USA
Tel: +1 617 395 4056
Fax: +1 617 354 6875
E-mail: cabi-nao@cabi.org

A catalogue record for this book is available from the British Library, London, UK.

Library of Congress Cataloging-in-Publication Data

Biology of breeding poultry / edited by P.M. Hocking.
p. cm. -- (Poultry science symposium series ; v. 29)
Includes bibliographical references and index.
ISBN 978-1-84593-375-3 (alk. paper)
1. Poultry--Breeding--Congresses. 2. Poultry--Genetics--Congresses. I. Hocking, P. M. (Paul M.) II. Title. III. Series: Poultry science symposium ; no. 29.

SF492.B56 2009
636.5'082--dc22

2008033138

ISBN-13: 978 1 84593 375 3

Typeset by Columns Design Ltd, Reading.
Printed and bound in the UK by the MPG Books Group, Bodmin.

The paper used for the text pages in this book is FSC certified.
The FSC (Forest Stewardship Council) is an international network to promote responsible management of the world's forests.

CONTENTS

Contributors

S.L. Al-Musawi, *Department of Veterinary Basic Sciences, The Royal Veterinary College, Camden, London NW1 0TU; email: salmusawi@rvc.ac.uk*

P. Bijma, *Animal Breeding and Genomics Centre, Wageningen University, Marijkeweg 40, 6709PG Wageningen, The Netherlands; email: piter.bijma@wur.nl*

T.R. Birkhead, *Department of Animal and Plant Sciences, University of Sheffield, Sheffield, S10 2TN, UK; email T.R.Birkhead@sheffield.ac.uk*

H. Bovenhuis, *Animal Breeding and Genomics Centre, Wageningen University, Marijkeweg 40, 6709PG Wageningen, The Netherlands; email: henk.bovenhuis@wur.nl*

V. Bruggeman, *K.U.Leuven, Faculty of Bioscience Engineering, Department of Biosystems, Laboratory of Livestock Genetics, Immunology and Physiology, Department Animal Production, Kasteelpark Arenberg 30, B-3001 Leuven, Belgium; email: Veerle.Bruggeman@biw.kuleuven.be*

C. Butter, *Division of Immunology, Institute for Animal Health, Compton, Newbury, Berks, RG20 7NN, UK; email: Colin.Butter@bbsrc.ac.uk*

N.A. Ciccone, *Department of Veterinary Basic Sciences, The Royal Veterinary College, Camden, London NW1 0TU*

M. Clinton, *The Roslin Institute and Royal (Dick) School of Veterinary Studies, Roslin, Midlothian, EH25 9PS, UK; email: mike.clinton@bbsrc.ac.uk*

S.R. Collett, *The University of Georgia, College of Veterinary Medicine, Poultry Diagnostic and Research Centre, 953 College Station Road, Athens, Georgia, 30602-4875, USA; email: colletts@uga.edu*

T. Cserep, *Intervet UK, Walton Manor, Walton, Milton Keynes, MK7 7AJ, UK; email: tibor.cserep@sp.intervet.com*

E. Decuypere, *K.U.Leuven, Faculty of Bioscience Engineering, Department of Biosystems, Laboratory of Livestock Genetics, Immunology and Physiology, Department Animal Production, Kasteelpark Arenberg 30, B-3001 Leuven, Belgium; email: eddy.decuypere@biw.kuleuven.be*

D.C. Deeming, *Department of Biological Sciences, University of Lincoln, Riseholme Park, Lincoln, LN2 2LG, UK; email: cdeeming@lincoln.ac.uk*

I.J.H. Duncan, *Department of Animal and Poultry Science, University of Guelph, Guelph, Ontario, Canada N1G 2W1; email: iduncan@uoguelph.ca*

I.C. Dunn, *Department of Genetics and Genomics, The Roslin Institute and Royal (Dick) School of Veterinary Studies, Roslin, Midlothian, EH25 9PS, UK; email: ian.dunn@roslin.ed.ac.uk*

I. Estevez, *Neiker-Tecnalia, Arkaute's Agrifood Campus, PO Box 46, E-01080 Vitoria-Gasteiz, Araba, Spain; email: iestevez@neiker.net*

C. Fisher, *20 Longstaff Court, Hebden Bridge, West Yorkshire, HX7 6AB, UK; email: colinfisher1@btconnect.com*

N.A. French, *Aviagen Turkeys Ltd, Chowley Five, Chowley Oak Business Park, Tattenhall, Cheshire, CH3 9GA, UK; email: nfrench@aviagen.com*

R.T. Gladwell, *School of Biological Sciences, The University of Reading, Whiteknights, Reading, RG6 6UB, UK; email: r.t.gladwell@reading.ac.uk*

R.M. Gous, *University of KwaZulu-Natal, Pietermaritzburg, South Africa; email: gous@ukzn.ac.za*

P.M. Hocking, *Division of Genetics and Genomics, The Roslin Institute and Royal (Dick) School of Veterinary Studies, Roslin, Midlothian, EH25 9PS, UK; email: paul.hocking@roslin.ed.ac.uk*

N.T. Joseph, *Department of Genetics and Genomics, The Roslin Institute and Royal (Dick) School of Veterinary Studies, Roslin, Midlothian, EH25 9PS, UK; email: nerine.joseph@roslin.ed.ac.uk*

M.T. Kidd, *Mississippi State University, Department of Poultry Science, Mississippi State, Mississippi 39762, USA; email: mkidd@poultry.msstate.edu*

P.G. Knight, *School of Biological Sciences, The University of Reading, Whiteknights, Reading, RG6 6UB, UK; email: p.g.knight@reading.ac.uk*

K.F. Laughlin, *Aviagen, 11 Lochend Road, Newbridge, Midlothian, EH28 8SZ, UK; email: klaughlin@aviagen.com*

P.D. Lewis, *Northcot, Cowdon Lane, Goodworth Clatford, Andover, SP11 7HG, UK; email: pdlewis@dsl.pipex.com*

T.M. Lovell, *School of Biological Sciences, The University of Reading, Whiteknights, Reading, RG6 6UB, UK; email: t.m.lovell@reading.ac.uk*

J.C. McKay, *EW Group, Lochend Road, Newbridge, Midlothian, EH28 9SZ, UK; email: jim.mckay@ew-group.de*

S. Nandi, *The Roslin Institute and Royal (Dick) School of Veterinary Studies, Roslin, Midlothian, EH25 9PS, UK; email: sunil.nandi@bbsrc.ac.uk*

O.M. Onagbesan, *K.U.Leuven, Faculty of Bioscience Engineering, Department of Biosystems, Laboratory of Livestock Genetics, Immunology and Physiology, Department Animal Production, Kasteelpark Arenberg 30, B-3001 Leuven, Belgium; email: onagbesanok@yahoo.com*

T. Pizzari, *Edward Grey Institute of Ornithology, Department of Zoology, University of Oxford, South Parks Road, Oxford, OX1 3PS, UK; email: tommaso.pizzari@zoology.oxford.ac.uk*

H.M. Sang, *The Roslin Institute and Royal (Dick) School of Veterinary Sciences, Roslin, Midlothian, EH25 9PS, UK; email: helen.sang@roslin.ed.ac.uk*

P.J. Sharp, *Department of Genetics and Genomics, The Roslin Institute and*

Royal (Dick) School of Veterinary Studies, University of Edinburgh, Roslin, Midlothian, EH25 9PS, UK; email: peter.sharp@bbsrc.ac.uk

K. Tona, *University of Lomé, Faculty of Sciences, Department of Animal Physiology, BP 1515 Lomé, Togo*

H.J. Walter, *Division of Immunology, Institute for Animal Health, Compton, Newbury, Berks, RG20 7NN, UK; email: helenjskinner78@btinternet.com*

G.J. Wishart, *Division of Biotechnology, University of Abertay, Bell Street, Dundee, BB1 1HG, UK*

PREFACE

Commercial broilers, turkeys and ducks are largely the products of 50 years of organized genetic selection for growth, feed efficiency and carcass yields in North America and Western Europe. This process has revolutionized the poultry industry and resulted in the efficient worldwide production of nutritious and healthy meat for the consumer. The intensive production of poultry meat continues to expand in many parts of the world, particularly in the emerging economies of Brazil, China and India.

Chicks, poults and ducklings necessarily require adult male and female birds that are also required to reproduce efficiently. Adults of current meat breeding lines are so radically changed from traditional lines that gave rise to them that an essentially new class of farm livestock has been produced. The high growth rates of these birds lead inevitably to high adult body weights, which have also affected the reproductive systems of females and the mating efficiency of males. The management and husbandry systems for breeding birds have also developed in parallel with the genetic changes, and a review of the current scientific knowledge of these birds is both timely and opportune.

This book contains reviews of the literature pertaining to breeding poultry of the three main poultry species (broiler, turkey and duck) and a chapter on minor species for which there is some information (Chapter 16). Typically these birds are fed on cereal-based diets and are housed on deep litter with various standards of environmental control, depending on the climate and region. The broiler chicken is probably more advanced genetically than any other species, and in general the husbandry and management of the other species are based on the broiler chicken model. There is relatively little published information on the duck and even less on geese, both of which are kept in some countries with access to water for swimming and green plant material as a feed. Geese are not commonly kept in large intensive operations or indoors, and Romanov (1999) has reviewed the available literature. Ducks are also kept for the production of foie gras in France, and further information is available in the report by Guemene and Guy (2004).

An overview of genetic selection and developments in the management of breeding birds is given in Part I, followed by three chapters in Part II that summarize current developments in genetic knowledge that may be useful in the future; Parts III, IV and V review current knowledge on reproduction, mating, fertility and incubation. The rest of the book covers the management of breeding birds: lighting and environmental enrichment (Part VI), nutrition (Part VII) and health (Part VIII).

The symposium, the 29th in the Poultry Science Symposium series, was held on 23–25 July 2007 at Surgeons' Hall, Edinburgh, and consisted of short overviews of the material by each of the authors. Unfortunately Dr John Kirby and Dr Rob Renema were unable to produce a manuscript for the book.

I am greatly indebted to the organizing committee, who generously provided their expertise in the disparate fields encompassed by this book, including initial technical editing of the draft manuscripts. I am grateful for the support and advice of John Parsons and Kelvin McCracken, respectively secretary and treasurer of the UK Branch of WPSA, and to Liz Archibald for her sterling administrative support in preparation for the symposium. The organizing committee consisted of P.M. Hocking (Chairman), J.A. Parsons, K.J. McCracken, J.A. Ball, J.S. Bentley, T.F. Davison, K.J. Laughlin, P.J. Sharp and the late G.C. Perry.

Dr Graham Perry was regrettably taken terminally ill just before the symposium and Dr Peter Lake kindly chaired the session on Mating Behaviour and Fertility. Dr Perry organized the two very successful preceding symposia and provided a great deal of helpful advice and encouragement in the planning of this meeting. His enthusiasm, commitment and cheerful personality will be greatly missed.

Finally, I wish to thank the sponsors for their generous support for the symposium, without which it would not have been possible to meet and discourse over 3 days in the genial surroundings of Surgeons' Hall.

P.M. Hocking
Edinburgh
June 2008

REFERENCES

Guemene, D. and Guy, G. (2004) The past, present and future of force-feeding and 'foie gras' production. *Worlds Poultry Science Journal* 60, 210–222.

Romanov, M.N. (1999) Goose production efficiency as influenced by genotype, nutrition and production systems. *Worlds Poultry Science Journal* 55, 281–294.

ACKNOWLEDGEMENTS

Support for this symposium is gratefully acknowledged from the following organizations:

Principal Sponsors:
British United Turkeys (now Aviagen Turkeys)
Moypark

Session Sponsors:
Aviagen
Genesis Faraday
Intervet Ltd
P.D. Hook (Hatcheries) Ltd

Sponsors:
British Poultry Science
DSM Nutrition
Hubbard S.A.S.
Janssen-Cilag Ltd
Schering-Plough Ltd
Taylor and Frances Group

Part I
Introduction

CHAPTER 1
The Genetics of Modern Commercial Poultry

J.C. McKay
EW Group, Newbridge, Midlothian, UK

ABSTRACT

Genetically improved strains of poultry have been a major contribution to the success of the poultry industry, which is a major source of animal protein for the human population in most countries of the world. Improvements in health, nutrition and environmental management have also contributed to improved performance, but the majority of the change has been attributed to genetic improvement.

Egg production has been improved consistently since the late 1930s, and the industry continues to improve the efficiency of laying hen production by at least 1% per year. This requires the simultaneous improvement of multiple traits, including egg number, egg size, liveability, persistency and mature body weight. There is also continuing progress in uniformity of egg size and colour and freedom from defects. In broilers, combined selection for growth, body composition, feed efficiency and liveability continues to deliver 2–3% improvement per year in the efficiency of meat production. Other traits such as robustness, specific and general disease resistance, and absence of metabolic defects have also contributed to this progress.

INTRODUCTION

Poultry have been domesticated for thousands of years, and man has made many genetic changes during the process of domestication and since then by establishing local varieties and selecting for various traits. The genetic progress made since the late 1950s has been the foundation of a modern poultry industry that is a major source of animal protein in most countries of the world. The history of poultry domestication and the development of a modern poultry industry are well reviewed (Crawford, 1990). Recent developments in knowledge and technology have changed the dynamics of poultry breeding.

The most important influence on commercial success has been the improvement in feed conversion ratio (FCR). This chapter reviews the historical improvements in the efficiency of production of meat and eggs, the current performance of commercial strains of poultry and the selection criteria that are likely to be important in the future. In discussing improvements in FCR it is essential to relate them to improvements in welfare. These improvements make the chickens fitter to perform well in a broad range of environments, production systems and disease challenges with high welfare standards.

THE MODERN POULTRY INDUSTRY

The production of poultry meat and eggs is a worldwide industry which supplies at least one-third of the animal-derived food for the 6 billion people on earth. The statistical service of the Food and Agriculture Organization records that in 1961 the world produced less than 10 million tonnes of poultry meat. By 2006 the world's production of poultry meat was 81 million t. This represents a compound annual growth rate of more than 5%. Poultry meat production has increased every year since FAO records began. In 1965 the world produced less than 5 kg of poultry meat per capita, and 45 years later we produce more than 13 kg per capita. The vast majority of the meat (70 million t) is produced by broiler chickens, with the remainder produced by turkeys (5 million t), ducks (3.5 million t), and geese and others (2.5 million t). The production of 71 million t of chicken meat requires an annual crop of at least 40 billion broilers.

The world's egg production has also increased steadily throughout this period, growing from 15 million t in 1961 to 60 million t in 2006, an annual compound growth rate of 3%. This represents an annual production of at least 1 trillion eggs (1×10^{12}), and these are produced by a population of approximately 6 billion layer hens (1×10^{9}). There are as many layer hens as there are people in the world today. In 1965 we produced 5 kg of eggs per capita and today we produce more than 10 kg per capita. Ninety-two per cent of the world's eggs are produced by layer chickens, with ducks, geese and other species making up the rest.

The development of such an industry has required coordinated improvements of technologies in a number of areas. The most significant improvements have been in the following four areas.

1. Environmental control. Controlled-environment housing has ensured safety from predation, more predictable production and improved biosecurity.
2. Nutrition. The nutritional requirements have changed as birds have been selected for efficient production.
3. Poultry health. The development of effective vaccines and therapeutics, improved biosecurity and better nutrition have all contributed to improved health. The emergence of breeding companies which are able to supply stock reliably free of the major vertically transmitted pathogens means that replacement stock can always be of a high health status.

4. Genetics. There has been consistent selection for improved productivity and quality.

THE CONTRIBUTION OF GENETICS

Improvements in health, nutrition and environmental management have contributed to improved performance, but the majority of the change has been attributed to genetic improvement. Havenstein *et al.* (2003a,b) compared the performance of contemporary broilers and a line random bred since 1957. They estimate that at least 85% of the improvement in performance is attributable to genetic changes. In broilers, combined selection for growth, body composition, feed efficiency, reproduction, health and welfare continues to deliver 2–3% improvement per year in the efficiency of meat production. Other traits such as robustness, specific and general disease resistance, and absence of metabolic defects have also contributed to this progress (Aviagen data).

In production environments the data also show clear genetic trends. For example, in the United States their Industry Reporting Service, which records the performance of the majority of the broilers produced there, shows that over the last 5 years growth rates have improved by 0.74 days per year for broilers grown to 2.27 kg. Breast meat yields have improved by 0.5% per year and FCR is decreasing by 0.025 per year. The combined improvements in growth, yield and efficiency mean that the overall efficiency of meat production is improving by more than 3% per year. Even with such improvements in growth and efficiency, the liveability of broilers is improving by 0.22% per year and the condemnation rates have fallen by 0.7% per year over this period. This outcome requires combined selection for many traits and full recognition of the importance of the welfare of the birds.

Egg production has been improved consistently since the late 1930s, and the industry continues to improve the efficiency of production by at least 1% per year (Hy-Line and industry data). This requires the simultaneous improvement of multiple traits, including egg number, egg size, liveability, persistency and mature body weight. United States industry estimates are that egg number to 60 weeks has improved by more than one egg per year and the feed conversion ratio (FCR) is improving by 0.01 per year. A major component of this progress has been selection for improved robustness and disease resistance. Liveability to 60 weeks of age is 0.12% better each year and 0.18% better to 80 weeks of age. There is also continuing progress in uniformity of egg size and colour and freedom from defects. Again the most important feature of layer breeding programmes is the ability to improve multiple traits simultaneously even though some of the traits have adverse genetic correlations.

Although these rates of change cannot be entirely due to genetics, as discussed above, there are clear indications that the main driver for improved performance is genetic selection. However, many producers cannot or choose not to use the full genetic potential of the stock and set performance standards at locally acceptable levels.

THE IMPORTANCE OF FEED CONVERSION RATIO (FCR)

The most important influence of genetics on the development of the poultry industry has been the improvement in FCR. Sustained improvements in FCR have an impact on the industry through the requirement for less feed per unit weight of product. This affects the demand for animal feed resources (mainly grains) and ultimately the cost of production. There are also positive effects on the environmental impact of poultry production. Less water is required, less waste is produced and the environmental impact is reduced. All of these factors have an effect on the sustainability of the poultry industry. In discussing improvements in FCR it is essential to relate them to improvements in welfare. The objective of selection is to make the chickens fitter to perform well in a broad range of environments, production systems and disease challenges. These factors all contribute to overall bird welfare.

Comparing modern egg layers with those available 30 years ago shows that in 1975 it took 2.4 t of feed to produce each tonne of eggs whereas today it takes 1.9 t of feed to produce 1 t of eggs (Hy-Line and FAO: http://faostat.fao.org). Today at least 115 million t of feed is used to produce eggs. Using the 1975 genotypes to produce all of today's eggs would require 144 million t of feed, an increase of 26%. The genetic improvements in efficiency are cumulative and permanent, and this has made the products of the industry available to a higher proportion of the world's population.

The improvements in broiler efficiency are even more dramatic. Between 1975 and today the combined effects of selection for growth, efficiency, yield and liveability have reduced the feed requirement for meat production from 20 million t of feed per million tonnes of meat to 8.5 million t of feed per million tonnes of meat (Aviagen and FAO). The genetic potential of birds is even better but is not realized in all production environments. It took approximately 700 million t of feed to produce the 81 million t of poultry meat in 2005. Using a 1970s genotype would have required 1600 million t, an increase of 128%. The annual improvement of 2–3% in efficiency of meat production has made a huge cumulative impact on our ability to supply affordable animal protein to a growing proportion of the world's population.

A recent study in Australia has examined the sustainability of animal production industries in light of the growing concern about the environmental impact of various production systems (Foran *et al.*, 2005). By taking account of all inputs and outputs they compare the greenhouse gas emissions of beef, lamb and pork production with that of poultry meat and eggs. Beef production in Australia produces 26 kg of carbon dioxide equivalent per unit value. Poultry meat or eggs produce less than one-tenth of this (2.5 kg carbon dioxide equivalent per unit value). Poultry meat and eggs also have 20% less impact than pork production (3.2 kg carbon dioxide equivalent per unit value) and 60% less than lamb production (6.4 kg carbon dioxide equivalent per unit value).

Thus the modern poultry industry has used genetic improvements in the birds that they care for to establish a very efficient and sustainable industry.

Continued improvements in poultry should be faster than in other species because poultry breeders have advantages of large population size, short generation interval and considerable genetic variation available to them.

THE FUTURE OF GENETICS IN COMMERCIAL POULTRY

Breeding companies have the responsibility to manage their genetic resources to deliver stock of predictable performance at high health standards. Population sizes must be sufficient to avoid inbreeding and ensure that the genetic variation is maintained to sustain long-term selection responses. The most important developments in genetics since the late 1980s have been in the ability of breeding programmes to deliver predictable and coordinated changes in multiple traits. Thus, selection for improved skeletal quality and heart and lung function has allowed simultaneous improvements in growth and feed efficiency and decreasing incidences of skeletal defects and ascites. Major investments are now being made to further improve the relevance and accuracy of the measurements made. This will allow more efficient and accurate selection to make further progress in many traits.

Welfare traits

Successful breeding programmes must recognize that they should place appropriate emphasis on the welfare of their pure lines and the crosses that will constitute their commercial products. In layers, for instance, this has required the application of group selection methodology to improve the liveability of layers when housed in large populations (see Bijma and Bovenhuis, Chapter 3, this volume). By reducing intra-group aggression, welfare and productivity have been improved together. In broilers and turkeys, great emphasis is placed on improvements in skeletal quality, and heart and lung function to improve welfare in a broad range of production environments. All successful breeding programmes will ensure that welfare standards continue to improve to ensure that poultry production is a sustainable industry.

Robustness

Poultry production globally involves a broad range of environments, which represent many different environmental, nutritional and disease challenges. Selection programmes are now selecting to ensure that their products are robust and thus have predictable performance across this range of environments. The most important variable worldwide is disease challenge, and breeding programmes have incorporated selection for specific or general disease resistance. Production systems are changing in response to the needs of the birds or the preferences of public opinion, retailers and consumers. For instance, more layer birds are being housed in non-cage systems, and breeding programmes must ensure that birds will perform predictably in a range of alternative production systems. Nutritional variation has many components but the major divide in the world's industry is between maize/soy diets and wheat-based diets. Wheat-based diets offer a particular challenge for the predictable uptake of minerals for skeletal development. Besides selecting for birds that can perform in a broad range of environments, the breeding companies will continue to cooperate with universities, research centres and producers to improve the advice given for the technical management of the stock.

Genomics

The publication of the chicken genome sequence (Hillier *et al.*, 2004) and a description of the variation between individuals (Wong *et al.*, 2004) have quickly changed the structure and operation of commercial breeding programmes. More than three million single nucleotide polymorphisms (SNPs) are now available throughout the genome and the technology for large-scale genotyping is readily accessible. This means that associations can be established between marker SNPs and traits, allowing more accurate selection for multiple traits. However, genomics is not an alternative to traditional selection methods but a means of more fully describing the variation available within populations and of using the same phenotypic measurements to make more accurate selection decisions (Andreescu *et al.*, 2007). This will involve considerable investments in bioinformatics and an integration of traditional and new technologies. The benefits are likely to be greatest for traits that are difficult to measure (especially disease resistance and welfare traits) or traits of low heritability (e.g. some reproductive traits).

Ethics

Breeding companies have a major influence on food safety, animal health, animal welfare and the security of the food supply. They also have a responsibility to ensure that their programmes are sustainable. This requires careful management and conservation of genetic resources. The number of products available continues to increase to meet many different production systems and environments and the demand for a wide range of products. Successful breeding companies must have a long-term strategy for the management of their genetic resources for sustainable genetic progress in multiple traits over future decades. It is therefore important that they operate within an agreed ethical framework. Their products must be fit for purpose and support sustainable production. This requires that animal health and welfare are given full recognition by the selection strategies and that sufficient emphasis is given to traits affecting efficiency of resource utilization. The target is to deliver balanced, rapid genetic progress.

CONCLUSION

Breeding companies have worked with producers to revolutionize the production of poultry meat and eggs, especially since the late 1950s. Genetic change continues and it is focused on the health and welfare of the animals as well as producer, retailer and consumer requirements. The investments required in research, development, production facilities and distribution systems means that there has been a decreasing number of breeding companies able to maintain a competitive position in the international market. Genetic change will continue to be a major contributor to the future development of the industry. The successful breeding companies will be those that make effective use of the feedback from producers, retailers and consumers in guiding their genetic programmes. This will produce maximum benefits for food safety, animal health and welfare, and efficient utilization of natural resources and will reduce the environmental impact of animal production.

REFERENCES

Andreescu, C., Avendano, S., Brown, S.R., Hassen, A., Lamont, S.J. and Dekkers, J.C.M. (2007) Linkage disequilibrium in related breeding lines of chickens. *Genetics* 177, 2161–2169.

Crawford, R.D. (1990) *Poultry Breeding and Genetics*. Elsevier, Amsterdam.

Foran, B., Lenzen, M. and Dey, C. (2005) Balancing act: a triple bottom line analysis of the 135 sectors of the Australian economy. CSIRO Technical Report, Sydney.

Havenstein, G.B., Ferket, P.R. and Qureshi, M.A. (2003a) Carcass composition and yield of 1957 versus 2001 broilers when fed representative 1957 and 2001 broiler diets. *Poultry Science* 82, 1509–1518.

Havenstein, G.B., Ferket, P.R. and Qureshi, M.A. (2003b) Growth, livability, and feed conversion of 1957 versus 2001 broilers when fed representative 1957 and 2001 broiler diets. *Poultry Science* 82, 1500–1508.

Hillier, L.W., Miller, W., Birney, E., Warren, W., Hardison, R.C., Ponting, C. *et al.* (2004) Sequence and comparative analysis of the chicken genome provide unique perspectives on vertebrate evolution. *Nature* 432, 695–716.

Wong, G.K.S., Liu, B., Wang, J., Zhang, Y., Yang, X., Zhang, Z.J. *et al.* (2004) A genetic variation map for chicken with 2.8 million single-nucleotide polymorphisms. *Nature* 432, 717–722.

CHAPTER 2
Breeder Management: How Did We Get Here?

K.F. Laughlin
Aviagen, Newbridge, Midlothian, UK

ABSTRACT

Broiler breeders are managed to maximize the output of day-old chicks. This involves control of body weight, feed intake and day length during both the rearing and production periods. The relative importance of these factors has changed over time and with the commercial production strategies of different countries. Several novel concepts of managing broiler breeders have been developed in scientific studies. However, examination of the breeder management advice from commercial breeding companies shows that the precise methods have then evolved in practical use within the production industry.

Some ideas, such as dwarf breeders, caged housing and artificial insemination, have not gained industry acceptance to the extent which was originally assumed. Genetic improvements in breeder performance have been steady and consistent since the late 1970s but there remains a significant variation about the average performance, which is an indication of the 'environmental' effect. Reducing this variation represents an opportunity for further improvement in flock performance.

INTRODUCTION

The relationship between the scientific studies of meat breeder management and the practical recommendations provided by breeding companies for management in a commercial situation will be described in this chapter. Several later chapters in these proceedings will provide reviews of specific areas of biology and/or management: broodiness (Sharp, Chapter 11, this volume), incubation and hatching (French, Chapter 12, this volume), photoperiod and control of breeding (Lewis, Chapter 14, this volume), feed restriction and nutrition (Hocking, Chapter 17 and Fisher and Gous, Chapter 18, this volume), which will consider in more detail the development of these aspects of breeder management. In addition to developing management systems for broiler breeders raised in large groups on the floor, there were investigations relating to the separate management of males and females, either during rearing or throughout the productive life. The dwarf (*dw*) gene was incorporated into broiler breeds to create a parent female which was itself dwarfed but when mated to a normal male gave rise to normal broiler progeny. The raising of broiler breeders in cages, both normal and dwarf, with and without the use of artificial insemination (AI) has also been a topic for many studies.

Regardless of the state of scientific knowledge and its inherent uncertainties, at any point in time, breeding companies must provide reliable advice in the form of management guides to their customers and each of the flock managers that are directly responsible for approximately 18,000 broiler breeder flocks per year and who must take daily actions to manage 35 million broiler breeder males and 350 million females producing 44 billion broilers.

Since the late 1950s the poultry science literature has included many hundreds of papers reporting studies on breeder management. Primarily these examined feed programmes and nutritional modification, initially to control growth and then to enhance performance. Additionally the effects of light and lighting programmes were studied and various novel strategies such as the use of the dwarf gene (*dw*) and separate feeding of males. More recently the focus has changed to the welfare aspects of various management strategies. Over this same period, primary breeding companies supplying the genetic stocks to the industry worldwide have provided management advice to assist producers in optimizing the output of their breeding stock. Some relationship can be seen between the scientific studies and the advice produced by the breeding companies, including in some cases use of the industry standard management recommendations as the 'control' against which to test novel ideas.

A major difference between the opportunities for scientific experiments and the industry approach is that by its nature the scientific method controls as many variables as possible while changing one or two whereas the real world situation must deal with the simultaneous variation of multiple inputs. Breeder management guides must also address situations where there are significant differences in the major inputs such as feed raw materials, environment (climate and housing) and management strategies.

The ultimate output of a breeding operation is a 1-day-old commercial chick to grow for meat production, and this is therefore the best measure of the success of a management strategy, ideally supplemented with an assessment of the cost of each chick. This measure of output begins with the total egg production, reduced to the number of hatching eggs by factors which determine an egg unsuitable for incubation and finally by the hatching success. The latter is determined by both the fertility of the eggs and the hatchability of fertile eggs. Finally, from a commercial point of view, basing the chick production per breeder alive at transfer to the production house or capitalization age takes account of the liveability of the breeder flock. Of course to get a true chick cost the costs incurred in rearing the birds, including the males, must be included. Thus translation of scientific study data, which may provide measures only at certain points in this process, into a commercial strategy can be difficult.

It is important to distinguish between breeder management techniques and strategies. The former are the day-to-day interactions with a flock – feed amount, weight profile, lighting, etc., – which achieve the desired strategy. The strategy itself is dependent on the economic (and cultural) system in which breeders are being farmed. A particular driver on strategy is the investment in the farm, the equipment and the flock itself. The annual returns to a breeding enterprise are largely influenced by whether farms are contracted or belong to an integrated production chain: a contract can be adjusted to match the needs of the farmer, who may also have another source of income apart from the breeder flock, which effectively subsidizes the latter. The sophistication of the farm construction and its equipment would add to the investment in either situation. This need for return on investment may then influence whether the aim is to have 5% production at 23 weeks or 25 weeks, i.e. earlier returns that require an earlier age at photostimulation. Finally the target production period, to 60 weeks of age or 40 weeks of production (or more), is also driven by local economics. There is no universally right system, but once the strategy is determined (and understood) then a management process to best achieve this can be developed.

FEED MANAGEMENT

In one of the earliest papers on breeder management Waldroup *et al.* (1966) noted 'Poultrymen have long been seeking a method to alter the growth pattern of both egg-type and meat-type pullets in an attempt to improve performance during the laying period.' In support of this statement they cite the work of Vondell (1943) and Novikoff and Byerly (1945), which would appear to be the oldest references to the need to control growth. The aim of the studies reported during the 1950s and 1960s was variously to delay sexual maturity, control early egg size and generally improve performance during the laying period.

The 'Broiler Breeder Paradox', as recently defined by Decuypere *et al.* (2006), which requires reconciliation of the production requirements of the breeder and satisfactory growth performance of the offspring without recourse to severe feed restriction, has driven much of the research since the late 1950s. The use of high-fibre grower feeds was proposed by Couch *et al.* (1957) and Issacks *et al.* (1960), but their reports did not agree as to the effect on subsequent performance. Alternatively manipulation of protein levels as a means to delay sexual maturity was examined by Howes and Cottier (1964), who fed a low-protein ration (100 g crude protein (CP)/kg) but found no delay in maturity. Conversely, Singsen *et al.* (1964) proposed a grower diet deficient in lysine and achieved a delay of 18.4 days in age at 25% production with 50% production delayed 12.4 days. However, following this, production from 26 to 54 weeks on the restricted group equalled or exceeded production of those on the normal diet. Waldroup *et al.* (1966) examined low-protein (100 g CP/kg), high-fibre (15%) and control (160 g CP/kg protein) diets. The low-protein and high-fibre diets delayed age at 50% production by 23.6 and 8.7 days compared with the control, and this was accompanied by a reduction in body weight.

Performance during the laying phase was not adversely affected by any of the grower treatments. No differences existed in total hen day production, eggs per hen, feed consumed, feed utilization, fertility, hatchability or mortality. Delaying sexual maturity in breeders resulted in more extra-large eggs than were produced on the control diet. These studies are some of the first attempts to describe methods to control sexual maturity and breeder performance, but a notable feature was that treatments began around 8 weeks, until that age birds were fed *ad libitum* with no weight control, and natural light was provided until 2 weeks before mating. Since that time it has become increasingly obvious in the commercial production situation that the real challenge for managing breeder performance is a complex relationship involving the control of bird weight, feed quality and quantity, and light (day length) at all stages from day old to flock depletion. Further it is necessary in the field situation to make real-time modifications to the rearing and production programmes when unexpected deviations occur.

Siegel and Dunnington (1985) summarized the problem of broiler breeders as 'selection for increased body weight is negatively correlated with the onset of sexual maturity', and this concept has remained in the literature in spite of all the success of management systems and genetic selection to achieve increases in breeder performance at the same time as maintaining the improvements in broiler characteristics. In both the scientific literature and breeder management guides the target and terminology of feed restriction supplanted the real aim of the exercise, which was to control growth of the breeder during rearing. There was a focus on feed restriction *per se* as a way to delay sexual maturity, which was also a feature of work on growing pullets (Lee *et al.*, 1971).

The industry focus on feed restriction led to the development of a system of skip-a-day feeding, in which the allocation of feed for 2 days was given on alternate days. This was occasionally varied to give the weekly feed allocation equally on 4 alternate days in each week or on 5 of the 7 days in a week. This system evolved because the equipment available was not capable of delivering the small amounts of feed required for 1 day to all birds in the flock in a way in which all could have equal access. In areas where this practice became unacceptable, systems were eventually developed to distribute the feed more quickly ('high-speed feeders'), and these were sometimes combined with covers over the feeding track which could be quickly released after feed was distributed throughout the house.

One solution to the problem of feed distribution associated with limited feed allowance during rearing of broiler breeders has been to broadcast pellets on to the floor. Such a system maximizes feed distribution in a minimum time and theoretically gives all birds in the flock almost instant equal access to feed. It is critical for the effective operation of this system that pellets can be made hard enough to withstand the spreading mechanism and that nipple drinkers are provided. Bell drinkers will catch a certain proportion of the feed and reduce the effectiveness of this method. Litter material, litter depth and management of litter condition are important to the successful operation of this system, along with a relative absence of other equipment in the house. Thus the technique may not be suitable for use in all housing and management systems. This practice itself, though it addresses some of the bird welfare issues of feeding broiler breeders, does raise other issues related to floor feeding and beak treatment. Birds feeding from a track or metal pan routinely smooth the beak tip, as would occur through natural foraging, whereas on litter systems with pellet feeding this control of beak condition does not always occur. Birds can therefore retain sharp beaks, which are a problem if other management features stimulate feather pecking or cannibalism.

LIGHT MANAGEMENT

For a long period, in the literature, the effects of light were mainly ignored. This latter situation was probably driven by the fact that in the majority of the world's growing areas houses were open sided or had curtains which did not isolate the birds from changes in the daily light period. Only after the 're-cognition' of the importance of light did commercial housing recommendations include first 'brownout', then 'darkout' and finally true 'blackout' to ensure that natural light was excluded from the birds during the rearing phase to achieve the dissipation of juvenile photorefractoriness (see Lewis, Chapter 14, this volume). In most of these systems, the birds are then transferred at the end of the rearing phase to houses which are not light tight and therefore the light stimulation occurs instantaneously at the time of transfer.

THE *dw* ALTERNATIVE

The sex-linked dwarf *dw* gene has been known for many years (Hutt, 1959) and described in a variety of chicken lines. The gene was introduced into commercial products, most notably the ISA Vedette, first sold in 1968, and Ross PM3, which has been available for more than 30 years. When crossed with a normal (*DW*) male dwarf hens produce offspring with similar growth rates to normal broilers. In a 1976 review Guillaume summarized the potential advantages of the commercial use of the *dw* gene as follows. 'With its reduced body size, the adult dwarf chicken needs less room in the house. Related expenses of heating, ventilation, etc.... appear reduced by the same amount. Since basal metabolism per unit weight is also reduced food energy expenses are even more clearly decreased.' Guillaume concluded his review by stating, 'Calet's 1972 statement that in the near future all broiler female strains will be dwarf seems well founded.' There was some initial commercial enthusiasm for dwarf broiler breeders in countries where cage breeder systems were used or where financial incentives were provided to invest in new house construction. However, the best current estimate is that no more than 5% of the world's 360 million broiler parent stock are of a dwarf type. Notwithstanding this a recent report (Decuypere *et al.*, 2006) remains optimistic about the potential of the *dw* gene to contribute to resolving some of the welfare issues associated with broiler breeder management.

CAGES AND ARTIFICIAL INSEMINATION

At various times and in various countries since the late 1950s the use of cages and artificial insemination have been suggested and tried in commercial practice as a way to overcome the perceived difficulties of managing and improving the performance of broiler breeders. Both of these techniques require novel investment in equipment and skills plus the availability of a reliable labour force. History has shown that these two approaches to broiler breeder management have gained relatively little use. In developed countries the labour to perform artificial insemination would not be readily available at the right price and concerns about animal welfare would not encourage producers to invest in cages for breeders. In many developing countries the skills and investment capital have not been available. Thus with the exception of a few countries, e.g. Russia, China and India, these opportunities have not received any attention, and even in the exceptions have not been widely adopted. The remaining suppliers of breeding stock have not developed lines or promoted management systems that involve cages and artificial insemination. Brillard (2001) has argued that optimizing AI in broiler breeder flocks would provide geneticists with a large potential for selection of even heavier male lines. This, however, is precisely what breeding companies have avoided, preferring to maintain a balance between the sexes in order to avoid potential welfare issues.

SEPARATE SEX FEEDING DURING THE REARING PERIOD

Breeding company management guides have included separate male and female target weight curves for many years but in many cases the two sexes were housed and fed together. This raised the question of which sex should be 'controlled' by feed allocation. Raising the males and females in separate houses or pens allowed separate control of body weight until the point of mating.

The techniques for feeding the sexes separately after mixing were introduced by McDaniel (1986). This underlying concept could be stated quite simply as 'It is beneficial to provide feed separately to males and females in order that the physical amounts and potentially feed type can be varied between the sexes. This allows separate control of male body weight during the production period and female feed amount to ensure continued production.' Since that time there has been extensive work on the precise nature of the control and the methods to effect it. Essentially a system was devised in which the females were provided with feed in feeders from which the males were physically excluded and the males were fed in feeders which were positioned at a height which the smaller females could not access. Differences in relative heights and head sizes of males and females, age at mating and feeder types have led to much practical refinement of the original concept. Such a system offered the possibility to provide a separate feed type to the males and females. A view developed that males did not require as much calcium for reproduction as females and indeed that excess calcium might be detrimental. Similarly males might benefit from feed with lower protein content than that required for females. The lack of convincing data on these issues and the difficulty of making, storing and delivering an additional feed type in a commercial environment means that little practical use is now made of separate feed types for males. Field evidence indicates that by far the major benefit for the male performance is controlling the feed amounts and hence the body weight gained during the production period (see Hocking, Chapter 17, this volume). The majority of scientific studies relating to the management of broiler breeders refer to feed restriction and are designed on the basis of feed allowance defined as amount or quality. However, the main aim of the rearing process is to grow a bird according to a desired weight profile, and the most common way to do this is to control feed intake, quantitatively or qualitatively. Thus, although growth control is achieved by controlling the feed intake, all of the practical management strategies that have evolved are based on control of body weight. In the absence of equipment, time or skills necessary to practise routine weighing and hence manage the growth of breeding stock, management guides have usually provided an expected feed allocation. This feed allocation assumes certain parameters about the feed, largely concerning its energy content (MJ/kg). The guides also always caution that feed quality varies geographically and temporally and therefore monitoring of bird weights is essential to achieve the desired weight profile. The nature of a scientific experiment is to define the inputs *a priori* (feed amount and quality) whereas the management guides define the objectives (weekly target weight). In an environment in which feed can be produced to a reliable standard on a regular basis and where feed raw materials are simple and subject to little variation it is possible to determine with some confidence the amount of daily feed allocation which will result in a desired body weight profile. This would generally be the case in the USA but is not the situation in many other countries. For this reason two distinct strategies for managing broiler breeders have evolved. These are based either on controlling daily feed amount, as in the USA, or on monitoring bird weight and adjusting feed allocation accordingly, as in much of the rest of the world.

If we examine the developments in the latter strategy we see that the initial focus in management guides was on following a target weight at an age rather than managing growth rate to a particular profile. Later the significance of this age–weight profile was better understood and management was soon refined to include an assessment of uniformity of the flock around the average weight. Initially this was expressed as a percentage of the birds in the flock that were within ±10% of the average weight (NB not the target weight). Initially such an evaluation was carried out by plotting the actual weights on squared paper and counting the number within ±10% of the average weight. The flock was therefore assessed according to how close the average weight was to the target and how uniform the birds were around that average. With the advent of electronic calculators and computers, weight data could be analysed statistically and the coefficient of variation (CV) calculated. This has a direct relationship with the old uniformity calculation and is much simpler to determine. Managing flocks to a target weight and uniformity was therefore possible and in some cases flock managers were financially rewarded for achieving a desired CV. However, it has long been recognized that two birds of equal weight may not have the same reproductive potential (the 'long lean' bird and the 'short fat' bird). In fact Zelenka *et al.* (1984) hypothesized that multiple thresholds, including chronological age, body weight and body composition, influence the onset of sexual maturity in female quail. The inclusion of a routine assessment of body condition using both a 'fleshing score' (breast muscle depth) and 'fat score' (abdominal fat pad size or fat cover on the pubic bone) was introduced in the late 1990s in some production operations in South America. The uniformity of these body condition scores measured at 16–20 weeks was seen as critical to good breeder performance. Such an evaluation can be important in assessing the reproductive potential for breeders of the recently developed high-yield broiler lines. Field evidence in South America, where these techniques are common practice, indicates a benefit in breeder performance of adequate body condition on breeder productivity.

WEIGHT OR LIGHT?

Throughout the history and evolution of practical breeder management there have been periods when weight control was seen as being of paramount importance and others when feed amounts were regarded as most critical. Light and lighting programmes were always mentioned but quite frequently misunderstood, particularly with respect to the importance of juvenile photorefractoriness and the need for a period of short days during early rearing. As part of an elegant study Ciacciariello and Gous (2004) (see Chapter 14) presented results which confirm that broiler breeders do not require a lighting stimulus in order to initiate ovarian activity and that, where no lighting stimulus is given, body weight or feeding level plays a critical role in stimulating the birds to attain sexual maturity. However, when a lighting stimulus is given, factors such as body weight and body composition become relatively less important in regulating the age at sexual maturity. This could explain the apparent changes of emphasis on these three important parameters over time.

LIGHT AND LIGHTING PROGRAMMES

Lewis (Chapter 14, this volume) describes results of an exhaustive and elegant study carried out at the University of Natal which has revisited many aspects of lighting and photoperiodic control of breeding in broiler parents, which extended work on commercial layers and allowed significant comparisons to be made. An important part of this work was the elucidation of photorefractoriness, both juvenile and adult, in broiler breeders. Commercial management guides have for many years provided for 'in season' and 'out of season' flocks, i.e. those which hatch in winter and experience natural spring and those which hatch in summer and mature during naturally declining daylengths in the northern hemisphere. The alternative to this advice was to persuade producers that they must control day length in 'blackout' housing during the rearing period in order to control sexual maturity. This blackout housing prevents any ingress of natural light by ensuring that all cracks in the house structure are sealed and by adding light traps to the ventilation system. Without this system the adventitious light of long summer days prevents the dissipation of juvenile photorefractoriness. Thus the true cause of poor production in out-of-season flocks is addressed rather than the previously perceived problem of the reducing natural daylength at the time of achieving sexual maturity, which was assumed to be overcome by providing artificial light of adequate intensity to effectively extend the natural day.

RECENT 'INCLUSIVE' RESEARCH

The difference between scientific experiments, which tend to fix most parameters and vary only one or two, and the commercial requirements to manage and respond to variations in several parameters through the life of a flock has already been mentioned. Several recent papers report a more 'inclusive' approach to broiler breeder research. Renema *et al.* (2001a,b) studied interactions between the growth curve and the timing of light stimulation, as have the many recent papers from Gous, Lewis and others from the University of Natal. Carcass composition or 'bird condition' was considered in the work of Reddish and Lilburn (2004), Renema *et al.* (1999), Dixson and Teeter (2001) and Ciacciarello and Gous (2005). Finally a modelling approach to describing the interaction of various parameters in determining broiler breeder performance has been taken by Alvarez and Hocking (2007) and Johnston and Gous (2007) in commercial layers. The latter approach may eventually be able to make use of the vast amounts of data available from commercial practice.

COMMERCIAL MANAGEMENT ADVICE

In order to describe and hence understand the process of the evolution of industrial management advice, management guides for three years, 1976, 2001 and 2007, were examined. In Fig. 2.1A the suggested female body weight profile was plotted for each of these three years. The shape of the body weight curve was significantly altered between 1976 and 2001 but only the absolute values changed in 2007. Target weights from 15 weeks to 30 weeks increased significantly and continue to do so. A simple examination of the target weights does not, however, tell the complete story in a way which is relevant to the biological needs and behaviour of the birds.

If the rate of weight gain is examined (Fig. 2.1B) we see that there is an initial sharp reduction in the weekly percentage increase in weight to about 10 weeks, followed by a level period to 20 weeks, after which there is a further reduction in the rate of gain to 30 weeks. It is noticeable that in the 1976 data the changes are somewhat erratic – a feature which would not be evident from simple examination of the weight profile – and is not a natural biological progression. Data from the 2001 and 2007 guides plotted in the same way

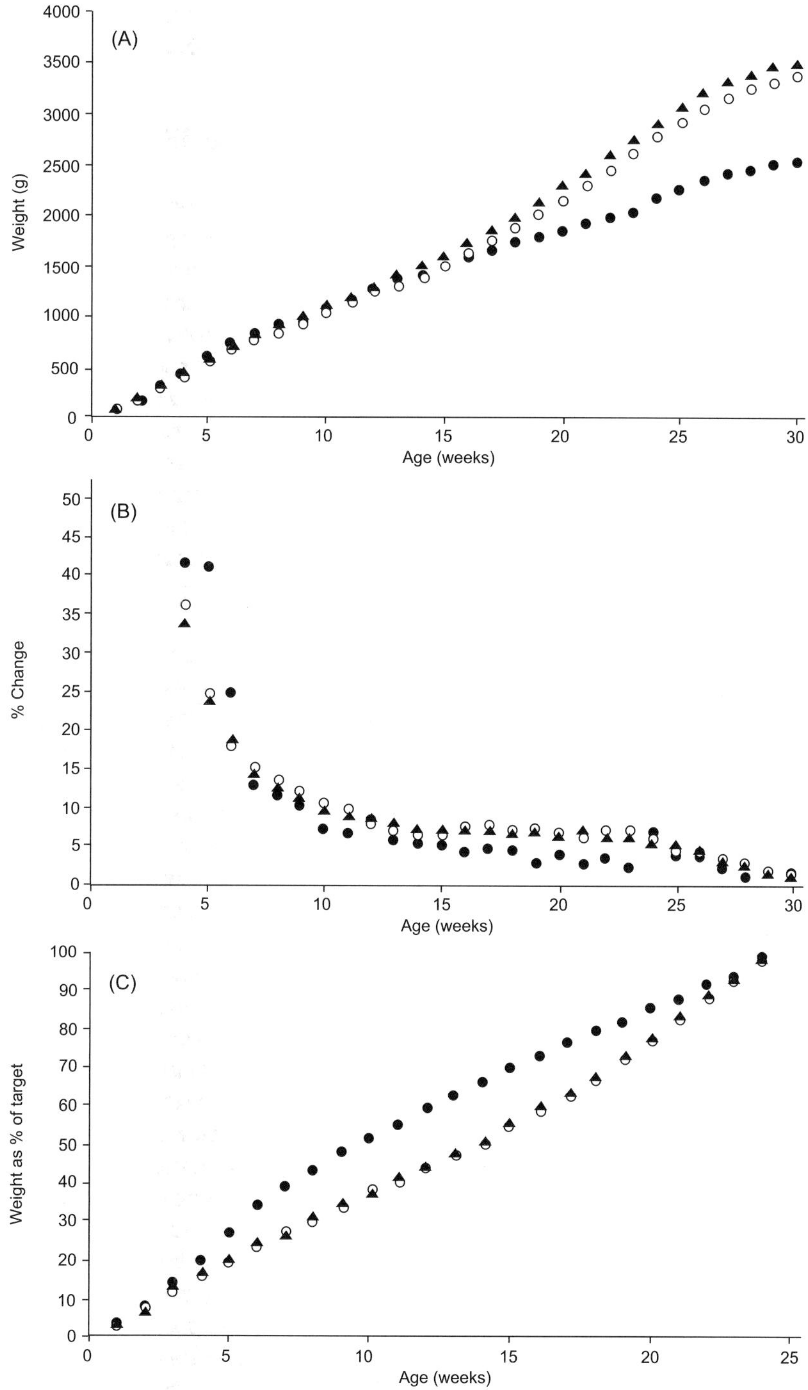

Fig. 2.1. Female broiler breeder target body weight profile to 30 weeks for 1976 •, 2001 ○ and 2007 ▲: (A) body weight; (B) percentage weekly increase; and (C) body weight as a % of 24-week weight.

show a much smoother progression, which it could be argued more closely resembles a biological process. Furthermore, the weekly gains between 10 and 24 weeks are significantly greater and contribute to the desired sigmoid growth curve. If the absolute weight at 24 weeks is ignored and changes are plotted as a percentage of the 24-week target (Fig. 2.1C) then we can see more clearly the change in the programme of weight control. In the recent guides the birds achieve 50% of their 24-week weight at around 14 weeks as opposed to 9 weeks in 1976. The result of this is that the weight gain of the birds can accelerate as they approach sexual maturity rather than decrease. Once again a much more natural biological process is demonstrated. While only three guides (years) were taken for clarity, the process described above has been gradual and continuous since the late 1960s. For reference the expected production performance is provided (Table 2.1), based on the actual data for the top 25% of flocks. These data show an increase in broiler breeder reproductive performance at all stages of production of 12% in total eggs, 15% in hatching eggs and 6% in hatchability, giving an overall improvement of 20% in total chicks per hen housed. Clearly the genetic and management changes over this period of time have shown significant improvement in breeder performance over the same time frame, in which there were considerable advances in broiler performance. Thus the field and commercial experience would now be at variance with the view of Siegel and Dunnington (1985) gained from experimental data on selection experiments. The general observations in this section would apply to each of the major broiler breeds though absolute values and timings and the output parameters may not be identical for all genotypes.

CHANGES IN PRODUCTIVITY

French (Chapter 12, this volume) has reviewed recent advances in incubation research and practice, and the improvement in expected hatchability over time has been noted above. Hodgetts (1991) provided data on hatchability improvements in the UK from 1976 to 1988. This has been extended by reference to data held by the major breeding companies supplying the UK market and are reproduced in Fig. 2.2. The most significant improvements in hatchability of broiler breeders occurred prior to 1990, with relatively little

Table 2.1. Expected performance taken from breeder guides which are intended to be representative of the top 25% performance of current flocks.

	Total eggs/hen housed	Hatching eggs/hen housed	Cumulative hatchability	Total chicks/hen housed
1976	154.6	144.8	80.4*	116.4†
2001	167.7	159.1	85.6	136.1
2007	173.2	166.6	85.2	141.7

* From Hodgetts (1991); † Calculated.

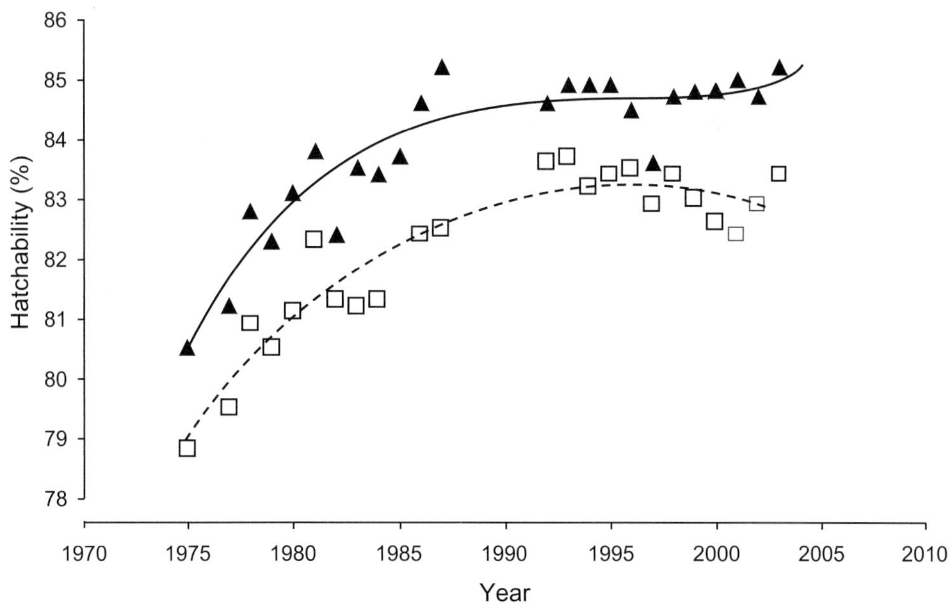

Fig. 2.2. UK hatchability changes with time (top 25% ▲, average □).

improvement thereafter, with average hatchabilities ranging between 82 and 84%. This situation in the UK is similar to that reported for the period 1985 to 2005 for the industry in the USA (Schaal and Cherian, 2007). Hatchability during this period ranged from 79 to 82% for broiler eggs. The authors concluded that advances in nutrition, genetic selection and management of broiler flocks during this time period for the top quartile of flock performance, which ranges from 84 to 85%, might suggest that, if allowance is made for fertility (of around 95% in these best flocks), and data are expressed as hatch of fertile eggs set, then we are close to the biological potential for hatchability. Further advances in hatchery practice may be limited to improving chick viability and subsequent broiler performance. Applying this approach to the overall performance of broiler breeders, as measured by chick numbers per breeder female, and using field data from different countries and seasons indicate that there is a remarkably similar variation about the average performance, with a CV of about 6 in all situations (Table 2.2). This would indicate that there is opportunity to improve breeder performance by reducing this variation specifically by improving the poorer performing flocks. In almost all data sets there is a skew towards the poorer performing flocks (Fig. 2.3), which provides an opportunity for overall industry improvement. However, an indication of the technical challenge this presents can be gained from the fact that from Table 2.1 we can see an improvement of 25.3 chicks in 31 years – an average of 0.82 chicks per year. If a production operation wished to advance more rapidly, by say one more chick, then 20% of the 175 flocks represented in Fig. 2.3 would require an improvement of five chicks per breeder. Clearly this is a significant task for any management or advisory organization.

Table 2.2. Field data on the number of chicks per broiler breeder female housed (2004).

Variable	UK	Japan	USA Summer	USA Winter
Number of flocks	175	232	58	55
Mean	131.4	145.5	130.7	131.0
SD	8.7	8.6	8.7	8.7
CV	6.6	5.9	6.7	6.6

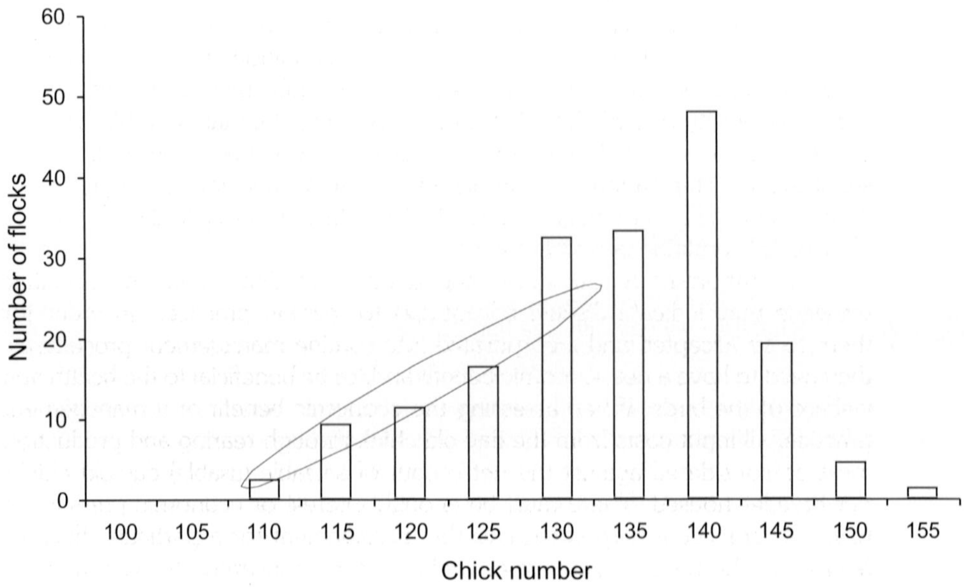

Fig. 2.3. UK data on chicks per broiler breeder female housed, indicating distribution of flocks (n = 175) and area of opportunity for improvement (circled).

DISCUSSION

It is very significant that Waldroup *et al.* (1966) introduced their paper with the statement 'Poultrymen have long been seeking a method to alter the growth pattern of both egg-type and meat-type pullets in an attempt to improve performance during the laying period.' This is precisely the challenge for effectively managing broiler breeding stock, a fact which has often been lost in work which focused on feed restriction. This focus produced a part of the solution but also a major part of the problems of management. Having decided that the solution lay in restricting the birds' feed intake, the focus in many studies then moved from understanding the optimal growth pattern to a relationship between feed restriction and production and health parameters.

There has been an evolution of management strategies from a process which recognized the need for weight control initially beginning from 8 to 10 weeks of age and progressing over time to have target weekly body weights from hatch. The recommended body weight programmes were successful in improving breeder performance but probably suffered from a broad generalization that the problem of broiler breeders was that they tended to overeat and become heavy and/or fat. Therefore the focus was on lowering weights and restricting feed to control obesity. In practice it soon became clear that variations in performance between flocks could often be explained by the uniformity of individuals in the flock. Assuming there was an ideal target weight to achieve sexual maturity then the more similar the individuals were in the flock the easier it was to determine the time for photostimulation and the better the resulting productivity. Initially uniformity and ultimately CV were included along with average weight as a measure of flock condition. This approach gave significant success, but it was then realized that uniformity of weight did not mean uniformity of bird condition. A 2 kg bird could be tall and thin or short and fat and was unlikely therefore to have a similar body composition and suitability for stimulation into production. Fleshing and fat pad scores were then introduced into management strategies, which also required uniformity of the desired condition across the flock.

It is important to recognize and understand that, while the scientific evidence may indicate distinct advantages for certain practices, in order for them to be accepted and incorporated into routine management procedures, they need to have a net economic benefit and/or be beneficial to the health and welfare of the birds. When assessing the economic benefit of a management practice, all input costs from the day-old chick through rearing and production must be considered against the net output of saleable (usable) day-old chicks per breeder housed. There must be enough interest or economic pressure to have the practice incorporated into the management for a particular flock or flocks and the result must be measurable and demonstrated to be beneficial. Breeder flocks have a life of more than 1 year, and if the ultimate assessment is on total egg or chick production or hatchability then there is essentially a lag time of more than 1 year before a practice can be accepted. Thereafter practices which are beneficial can gain rapid implementation within companies and from there across countries. However, there are numerous examples of practices which do not find expression in all companies or regions of the world due to differences in fundamental breeder strategy or simple practicalities. Some of the more labour-intensive practices, such as routine bird weighing and uniformity assessment, were slow to be incorporated into the management of breeder flocks in the USA, where almost all breeder farms are privately owned and the time taken for these activities may not show a benefit to the individual concerned. Equally in this example the consistency of feed manufacture and composition in the USA is such that a reasonable standard of body weight control can be achieved by controlling feed amount. Finally the practice of delaying flock maturity (first light increase after 21 weeks) with the aim to have 5% production at 25 weeks of age provides more opportunity for flock uniformity at the time of photostimulation. The breeder farming system is based on the historical availability of many farms and supply contracts which take account of this management strategy. In Europe and many other areas of the world where most broiler breeder farms are company owned and/or farming space is in short supply (and stocking densities are higher), the strategy would be to have 5% production at 23 weeks of age, with first light stimulation before 21 weeks.

Breeder management strategies and techniques have evolved since the late 1950s, with some ideas originating in scientific studies gaining wide acceptance and many being modified through field experience. Some ideas, although theoretically promising, have failed to gain widespread use due to cost or practicality or the management strategy being followed in a specific region or time period. Nevertheless there has been a significant improvement in the normal measures of reproductive performance of broiler breeders at the same time as production, health and welfare aspects of their broiler offspring have also increased.

REFERENCES

Alvarez, R. and Hocking, P.M. (2007) Stochastic model of egg production in broiler breeders. *Poultry Science* 86, 1445–1452.

Brillard, J.P. (2001) Future strategies for broiler breeders: an international perspective. *World's Poultry Science Journal* 57, 243–250.

Calet, C. (1972) Introduction to the symposium on the dwarf (dw) gene in chickens. Tours 1972 (unpublished).

Ciacciariello, M. and Gous, R.M. (2004) A comparison of the effects of feeding treatments and lighting on age at first egg and subsequent laying performance and carcase composition of broiler breeder hens. *British Poultry Science* 46, 246–259.

Couch, J.R., Issacks, R.E., Reid, B.L., Crawford, W.P. and Quisenberry, J.H. (1957) Feeding and management of replacement broiler stock. *Feed Age* 7, 34.

Decuypere, E., Hocking, P.M., Tona, K., Onagbesan, O., Bruggeman, V., Jones, E.K.M., Cassy, S., Rideau, N., Metayer, S., Yego, Y., Putterflam, J., Tesseraud, S., Collin, A., Duclos, M., Trevidy, J.J. and Williams, J. (2006) Broiler breeder paradox: a project report. *World's Poultry Science Journal* 62, 443–453.

Dixson, S.J. and Teeter, R.G. (2001) Fasted heat production and body composition of broiler breeder females ranging from 5 to 50 weeks of age. Animal Science Research Report P986, Oklahoma Agricultural Experiment Station, Oklahoma State University, Oklahoma.

Guillaume, J. (1976) The dwarfing gene dw: its effects on anatomy, physiology, nutrition, management. Its application in poultry industry. *World's Poultry Science Journal* 27, 285–303.

Hodgetts, B. (1991) Current hatchabilities in species of domestic importance and the scope for improvement. In: Tullett, S.G. (ed.) *Avian Incubation*. Poultry Science Symposium Number 22, pp. 139–144.

Howes, J.R and Cottier, G.J. (1964) A comparison of various methods of delaying maturity in heavy-type breeders. *Poultry Science* 43, 1330.

Hutt, F.B. (1959) Sex linked dwarfism in the fowl. *Journal of Heredity* 50, 209–221.

Issacks, R.E., Reid, B.L., Davies, R.E., Quisenberry, J.H. and Couch, J.R. (1960) Restricted feeding of broiler type replacement stock. *Poultry Science* 39, 339–346.

Johnston, S.A. and Gous, R.M. (2007) A mechanistic, stochastic, population model of egg production. *British Poultry Science* 48, 224–232.

Lee, P.J.W., Gulliver, A.L. and Morris, T.R. (1971) A quantitative analysis of the literature concerning the restricted feeding of growing pullets. *British Poultry Science* 12, 413–437.

McDaniel, G.R. (1986) Sex separate feeding of broiler parent stock. *Zootechnica International*, November 1986, pp. 54–58.

Novikoff, M. and Byerly, J. (1945) Observations on two methods of feeding chickens from one day to twelve months of age. *Poultry Science* 24, 245–251.

Reddish, J.M. and Lilburn, M.S. (2004) A comparison of growth and development patterns in diverse genotypes of broilers. 2. Pullet growth. *Poultry Science* 83, 1072–1076.

Renema, R.A., Robinson, F.E., Newcombe, M. and McKay, R.C. (1999) Effects of body weight and feed allocation during sexual maturation in broiler breeder hens. 1. Growth and carcase characteristics. *Poultry Science* 78, 619–628.

Renema, R.A., Robinson, F.E. and Goerzen, P.R. (2001a) Effects of altering growth curve and age at photostimulation in female broiler breeders. 1. Reproductive development. *Canadian Journal of Animal Science* 81, 467–476.

Renema, R.A., Robinson, F.E., Goerzen, P.R. and Zuidhof, M.J. (2001b) Effects of altering growth curve and age at photostimulation in female broiler breeders. 2. Egg production parameters. *Canadian Journal of Animal Science* 81, 477–486.

Schaal, T. and Cherian, G. (2007) A survey of the hatchability of broiler and turkey eggs in the United States from 1985 through 2005. *Poultry Science* 86, 598–600.

Siegel, P.B. and Dunnington, E.A. (1985) Reproductive complications associated with selection for broiler growth. In: Hill, W.G., Manson, J.M. and Hewitt, D. (eds) *Poultry Genetics and Breeding*. British Poultry Science Ltd, Harlow, UK, pp. 59–72.

Singsen, E.P., Nagel, J., Patrick, S.G. and Matterson, L.D. (1964) The effect of lysine deficiency on growth characteristics, age at sexual maturity and reproductive performance of meat-type pullets. *Poultry Science* 43, 1362.

Vondell, J.H. (1943) Range utilization by growing chickens. *U.S. Egg and Poultry Magazine* 49, 464.

Waldroup, P.W., Damron, B.L. and Harms, R.H. (1966) The effect of low protein and high fibre grower diets on the performance of broiler pullets. *Poultry Science* 45, 393–402.

Zelenka, D.J., Cherry, J.A., Nir, I. and Siegel, P.B. (1984) Body weight and composition of Japanese quail (*Coturnix coturnix japonica*) at sexual maturity. *Growth* 48, 16–28.

Part II

Genetic Improvement

CHAPTER 3
Developments in Quantitative Genetics and Genomics Relevant for Poultry Breeding

P. Bijma and H. Bovenhuis

Animal Breeding and Genomics Centre, Wageningen University, The Netherlands

ABSTRACT

Developments in genetics and genomics relevant for poultry breeding are reviewed, focusing on the use of molecular information in breeding programmes and the potential for improving traits affected by social interactions among individuals. QTL mapping studies have resulted in almost 700 QTL for a wide variety of traits, showing that they have been successful. However, few applications of marker-assisted selection in commercial poultry breeding exist, mainly because evidence for the QTL is often not conclusive and confidence intervals for QTL locations are large, making utilization in practice difficult. Moreover, moving from QTL to causal mutation has been successful only in a few cases. Thus, neither MAS nor selection for known genes will greatly increase responses to selection in the near future. Genome-wide selection (GWS), however, may offer a solution. Basically GWS is a method to estimate breeding values. Simulation studies suggest that accuracies of EBVs in the range of 0.7–0.8 are feasible. Tests of GWS on real data are currently ongoing, and initial results are promising. The second section of this chapter shows that social interactions create considerable heritable variance that is hidden in classical analyses. Results show that social effects contribute more than half of the heritable variance in mortality due to cannibalism in laying hens, and in growth rate and feed intake in growing pigs. We present statistical models and breeding strategies to utilize this extra heritable variation for genetic improvement. Though social genetic effects may not always be important or may be difficult to use in some cases, the promising results observed in laying hens and pigs should be sufficient incentive for further research in this area.

INTRODUCTION

Poultry breeding has been very successful in both broilers and layers as outlined in Chapter 1. Improvements have been large in production traits but smaller in traits related to fitness. Fitness-related traits pose a difficulty because they have lower heritability, are often more affected by environmental circumstances and are more sensitive to genotype × environmental interaction between the breeding and commercial environments. Two recently developed tools have the potential to contribute significantly to the improvement of fitness-related traits: first, the use of whole-genome marker-assisted breeding value estimation ('genomic selection') and, second, the inclusion of social genetic effects in breeding programmes.

Though molecular genetics has long been advocated, its contribution to genetic improvement has been limited, partly because research has focused on identifying quantitative trait loci (QTL). QTL are difficult to use operationally because they have often been detected in experimental crosses; QTL frequency and linkage phase change over time; and the number of essential QTL soon becomes impractically large. Recently, emphasis has shifted to marker-assisted breeding value estimation using dense maps covering the entire genome ('genomic selection'). In this approach, the sole objective is to maximize the accuracy of estimated breeding values (EBVs). This approach has high potential for breeding practice because marker information translates directly into EBVs for selection candidates, data from commercial environments can be utilized for estimating breeding values on candidates in the nucleus and application to multiple traits is straightforward.

Social interactions among individuals have often been treated as environmental noise. However, compared with the physical environment, social interactions are more interesting because they may have a heritable component, similar to maternal effects. Our results in layers and pigs show that social interactions create considerable heritable variation that is hidden in classical analyses. We present statistical models and breeding strategies to utilize this extra heritable variation for genetic improvement.

UTILIZING MOLECULAR INFORMATION FOR GENETIC IMPROVEMENT

In the last decade several genome-wide QTL mapping studies have been performed in poultry. The aim of these studies was to identify chromosomal regions associated with traits of interest. Recently Hocking (2005) and Abasht *et al.* (2006) reviewed the findings of these studies. Abasht *et al.* (2006) reported almost 700 QTL, of which approximately one-third was significant at the 5% genome-wide level. QTL have been identified for a wide variety of traits which are related to growth, egg production, disease resistance, behaviour and metabolism. These reviews show that QTL mapping studies have been successful in identifying chromosomal regions affecting quantitative traits in poultry. In a number of cases, crosses between divergent breeds have been used, such as the cross between red jungle fowl and White Leghorn (Crittenden *et al.* 1993) or the Sasumadorri and White Plymouth Rock cross (Tatsuda and Fujinaka, 2001). In other cases the lines used were less divergent, such as a cross between lines divergently selected from a common base (Yonash *et al.*, 2001; Siwek *et al.*, 2003), or a cross between commercial breeds (e.g. Van Kaam *et al.*, 1998; Rabie *et al.*, 2005).

Analyses contrasting chromosomal segments originating from different lines are expected to detect genes that explain line differences. There has been debate on whether QTL explaining between-line variation are the same as those explaining genetic variation within lines. Because commercial lines have been under intense selection for many generations, it has been argued that QTL with important effects must have been fixed within lines. However, recently de Koning *et al.* (2003, 2004) showed that several QTL identified in experimental crosses of chickens are still segregating in a commercial population of broilers. Pleiotropic effects on other traits might be an explanation for the fact that these loci are still segregating in commercial broiler populations.

From QTL detection to utilization

It can be concluded that the QTL harvest after 10 years of mapping studies is rich (Abasht *et al.* 2006). An obvious next step would be to utilize these findings in commercial breeding schemes. However, to our knowledge very few (if any) applications of marker-assisted selection (MAS) in commercial poultry breeding exist. A number of reasons can be identified. First, most QTL mapping experiments are not very large and will only detect genes of large effect; for most QTL the evidence will not be conclusive. These suggestive QTL will need to be verified in further experiments or by using the findings of other studies. Abasht *et al.* (2006) report that in several cases QTL have been detected for similar traits in similar chromosomal locations. Those confirmed QTL are ready to be used in breeding schemes. However, traditional QTL mapping experiments will give inaccurate information on the location of the causative mutation. Depending upon the size of the experiment and the size of the mapping population, confidence intervals are in the order of 20–50 cM. Therefore, utilizing this information directly in a commercial breeding scheme is far from trivial, because the linkage phase between markers and QTL needs to be established within families. Further, the established linkage information is subject to 'inflation' as a result of recombination between the marker(s) and the functional mutation. Arthur and Albers (2003) concluded, therefore, that commercial application of DNA genotyping will largely be through direct genotyping for critical genes rather than for MAS. However, moving from QTL to causal mutation (the quantitative trait nucleotide – QTN) is difficult and has only been successful in a few cases, such as for IGF2 in pigs and DGAT1 in cattle (Andersson and Georges, 2004). New developments like the chicken genome sequence (Hillier *et al.*, 2004) and the availability of the SNP map (Wong *et al.*, 2004), which makes it possible to genotype individuals using genome-wide SNP assays containing over 20,000 SNPs per assay, will greatly contribute to identifying QTNs. However, some recent examples show that genetic mechanisms underlying these traits might be complicated, e.g. the callipyge mutation described by Takeda *et al.* (2006). This illustrates that even with the availability of the chicken genome sequence and genome-wide SNP assays the identification of QTNs is difficult. Further, the extent of linkage disequilibrium in poultry seems to vary considerably between regions and between lines (Aerts *et al.*, 2007), which will affect the resolution of association studies in specific lines or chromosomal regions.

We conclude, therefore, that neither MAS nor selection for known genes will in the short term have a great impact on commercial poultry breeding. Qualitative traits may be an exception. For example, there exists an association between the FMO3 gene and fishy taint of eggs that can be used to identify carriers and the defect subsequently eradicated from the population (Honkatukia *et al.*, 2005). The contribution of MAS and known genes to improvement of true quantitative traits, however, will be limited for the near future.

There are, however, interesting opportunities for genome-wide selection (GWS) as proposed by Meuwissen *et al.* (2001). GWS has become within reach with the availability of genome-wide SNP assays. Basically GWS exploits linkage disequilibrium between markers and QTL, so as to estimate breeding values of selection candidates. By typing individuals for a dense set of markers covering the whole genome and linking that information to phenotypic measurements, the total genetic merit of related individuals without phenotypic records can be predicted with accuracies in the range of 0.7–0.8 (Meuwissen *et al.*, 2001). Note that GWS does not attempt to identify genes or QTL regions but simply estimates associations between chromosomal regions of interest and uses these associations to predict breeding values of related individuals. In that sense GWS selection can be characterized as an improved Black-Box approach. So far, GWS has only been tested using simulated data, and several questions still need to be addressed, such as the number of markers required and the need for re-estimation of associations in ongoing breeding schemes. Tests of GWS on real data are currently ongoing and initial results suggest that GWS may revolutionize poultry breeding.

Potential contribution of genomic information

Several studies have indicated that use of genomic information is expected to contribute to traits that are difficult or impossible to measure on the selection candidate, including sex-limited traits, which can only be measured late in the animal's life, or are of low heritability (e.g. Dekkers and Hospital, 2002). These findings are general and apply to MAS, selection for known genes or GWS, and are based on the general principle that genomic information is an information source that does not rely on the availability of phenotypic observations. For some traits, traditional selection using phenotypic information does an excellent job and adding genomic information is not expected to make an important contribution. An example is body weight in broilers, which can be measured on both sexes, has a reasonable heritability and can be measured early in life. Consequently, traditional selection is very effective to improve body weight. In contrast, a trait like resistance to ascites is much more difficult to improve using traditional selection. Selection for ascites resistance can be done by challenging relatives of selection candidates under cold conditions or at high altitude, and recording the fraction that survive. Alternatively, indicator traits like the ratio of right ventricular over total ventricular weight can be used. However, this also requires that the individual is sacrificed and therefore traits need to be recorded on relatives of the selection candidates rather than on the selection candidate itself. Selection for ascites resistance is thus complicated and expensive. New opportunities to reduce ascites by means of selection will become available if genomic information is used. Pakdel *et al.* (2005) evaluated the consequences of alternative selection strategies for body weight and resistance to ascites in broilers using deterministic simulation. Alternative selection strategies that used information on different ascites-related traits and the benefits of using genomic information were quantified. Pakdel *et al.* (2005) concluded that, by using information on QTL that explain only 5% of the genetic variance, ascites could be controlled while still permitting considerable selection response for body weight. An important advantage of having genomic information is that the information is available on all the selection candidates, which makes it possible to differentiate full sib individuals and thus facilitate selection within families. Moreover, no birds need to be euthanized to measure ascites-related traits. Similar arguments apply for the expected contribution of genomic information in selecting, for example, for carcass traits, disease resistance or egg production. We believe that, with the availability of genome-wide SNP assays and by using genomic selection, new opportunities have become available to selectively improve these types of traits.

SOCIAL INTERACTIONS AMONG INDIVIDUALS

Commercial chickens are usually kept in groups, ranging from groups of four laying hens kept in cages, to groups of thousands of birds kept on floor systems. The performance of birds may, therefore, depend not only on the properties of the bird itself but also on (social) properties of other birds present in the same group. A well-known example is mortality due to cannibalistic behaviour in non-beak-trimmed laying hens (Muir, 2003). It is unlikely, however, that effects of social interactions are restricted to cannibalism. Our first results in pigs, for example, indicate that growth rate in pigs can be substantially affected by social interactions, even in *ad libitum* feeding (Bergsma *et al.*, 2008). In general, other individuals kept in the same group are an important component of the environment experienced by an individual. Social interactions may therefore affect most traits of interest in commercial chickens.

From a breeder's perspective, a social environment is fundamentally different from a physical environment, because it may have a heritable component (Wolf *et al.*, 1998). In contrast to a physical environment, the social environment experienced by birds can be improved by means of selection. Thus, when social effects have a heritable component, the performance of birds can be improved by breeding birds that provide a better social environment to their group members. With the exception of maternal effect models, however, classical breeding theory does not incorporate heritable social effects. Thus there is a need for extension of quantitative genetic theory. Generalization of quantitative genetic theory is required to understand how interactions among individuals affect performance and which selection methods are suited to improve traits affected by social interactions. This involves both the development of basic theory of inheritance and response to selection, and the development of methodology to estimate the genetic parameters of socially affected traits in real populations.

Here we summarize the basic theory presented in Muir (2005), Bijma *et al.* (2007a,b) and Ellen *et al.* (2007), and discuss implications and extensions relevant for poultry breeding. The focus will be on translating the basic results into terms familiar to animal breeders, such as accuracy of selection and additive genetic variance in trait value.

Theory of the genetics of social interactions

Model

Consider a population consisting of many groups of n birds each. Social interactions occur among birds within a group and affect phenotypic values of birds. The phenotypic value of a bird will be the sum of a direct effect due to the individual itself and the social effects of the $n - 1$ other birds kept in the same group. Both the direct effect and the social effect may be decomposed into a heritable component, A, and a non-heritable component, E. The observed phenotypic value of each individual, therefore, is the sum of the direct breeding value and direct environmental effect of the individual itself, and the summed social breeding values and social environmental effects of its $n - 1$ group members:

$$P_i = A_{D,i} + E_{D,i} + \sum_{i \neq j}^{n} A_{S,j} + \sum_{i \neq j}^{n} E_{S,j} \quad (3.1)$$

In Equation 3.1, $A_{D,i}$ is the direct breeding value (DBV) of individual i, and $A_{S,j}$ the social breeding value (SBV) of its group member j. The DBV is equivalent to the classical breeding value, whereas the SBV is a generalization of a breeding value for maternal effect. The $E_{D,i}$ and $E_{S,j}$ are the corresponding non-heritable effects ('environment').

Response to selection

The DBV and SBV represent the heritable components underlying phenotypic trait values. Thus response to selection will depend on improvement in both DBVs and SBVs. The relative impact of DBVs and SBVs on response to selection follows from decomposing the overall population mean into a mean direct effect and a mean social effect, $\overline{P} = \frac{1}{m}\sum_{m} P_i = \overline{P}_D + (n-1)\overline{P}_S$, where averages are taken over *m* individuals, *m* denoting population size. Total genetic response per generation, therefore, equals the increase in the DBV (ΔA_D) plus (*n* – 1) times the increase in the SBV (ΔA_S):

$$\Delta \bar{P} = \Delta A_D + (n-1)\Delta A_S \tag{3.2}$$

This result shows that social effects should be weighted by group size minus one. The $(n-1)\Delta A_S$ in Equation 3.2 represents response to selection in the social environment that individuals experience.

In classical breeding theory, response to selection equals the change in the (direct) breeding value per generation (Lynch and Walsh, 1998). With interaction among individuals, however, response contains a component due to SBVs (Equation 3.2). We therefore generalize the definition of breeding value to incorporate both direct and social effects, and define a total breeding value:

$$\text{TBV}_i = \text{DBV}_i + (n-1)\text{SBV}_i = A_{D,i} + (n-1)A_{S,i} \tag{3.3}$$

Analogous to classical theory, response equals the change in TBV per generation, and the TBV replaces the usual breeding value. This result shows that, with social interactions among individuals, the total breeding value of a bird will depend on group size (*n*). Note that the social component of the TBV of an individual is expressed not in the individual itself or in its offspring, but in the phenotypes of their group members.

In addition to a total breeding value, we can also identify the total heritable variation in the trait. Due to social interactions, the total heritable variation that is available to generate response to selection will differ from the usual (direct) additive genetic variance. Analogous to classical theory, the total heritable variation is the variance of TBVs among individuals:

$$\sigma^2_{TBV} = Var[A_{D,i} + (n-1)A_{S,i}] =$$

$$\sigma^2_{TBV} = \sigma^2_{A_D} + 2(n-1)\sigma_{A_{DS}} + (n-1)^2\sigma^2_{A_S} \tag{3.4}$$

Analogous to the usual σ^2_A, the σ^2_{TBV} measures the potential of a trait to respond to selection. For example, an efficient breeding scheme can generate a response to selection of approximately one genetic standard deviation. Hence, for traits affected by social interactions, the standard deviation of TBVs, σ_{TBV}, is a measure of the amount of response that can be generated. The term $\sigma^2_{A_D}$ in Equation 3.4 corresponds to the additive genetic variance in classical theory. The term $(n-1)^2\sigma^2_{A_S}$ shows that the presence of heritable social interactions can substantially increase total heritable variation. This may explain the rapid responses observed with group selection (e.g. Muir, 1996). Essentially, the term $(n-1)^2\sigma^2_{A_S}$ represents the heritable variation present in the social environment. The term $2(n-1)\sigma_{A_{DS}}$ in Equation 3.4 shows that a negative genetic covariance between DBVs and SBVs ($\sigma_{A_{DS}}$) reduces the total heritable variation. With negative $\sigma_{A_{DS}}$, individuals with positive breeding values for their own phenotype (DBV) have, on average, negative social effects on the phenotypes of their group members (SBV). Thus negative $\sigma_{A_{DS}}$ may be interpreted as 'heritable competition'. Heritable competition, therefore, reduces the total heritable variation and thus the potential of the trait to respond to selection. In contrast, heritable cooperation ($\sigma_{A_{DS}} > 0$) increases the total heritable variation.

In stark contrast to classical theory, the total heritable variation in a trait can be larger than the phenotypic variance in the trait, i.e. Var(TBV) > Var(P). It follows from Equation 3.1 that, when groups consist of *unrelated* individuals, the phenotypic variance equals $\mathrm{Var}(P) = \sigma^2_{A_D} + \sigma^2_{E_D} + (n-1)(\sigma^2_{A_S} + \sigma^2_{E_S})$. In that case, a population contains a heritable variation which is greater than the observed phenotypic variance. Social interactions, therefore, result in hidden heritable variation. The reason that the total heritable variation is not observed in the phenotypic variance is because the TBV of an individual is distributed over multiple phenotypes. The total social effect of individual i, $(n-1)A_{S,i}$ is not expressed in a single individual, but each of the $n-1$ group members of individual i receives an amount $A_{S,i}$. The total effect, therefore, does not surface in the phenotypic variance but is hidden to direct observation. In other words, in the classical case $P_i = A_i + E_i$, so $\mathrm{Var}(A) \leq \mathrm{Var}(P)$ and $h^2 \leq 1$, whereas with social interactions $P_i \neq TBV_i + E_i$, so there is no need for $\mathrm{Var}(TBV) \leq \mathrm{Var}(P)$. Our preliminary results in pigs indicate that, in a real population, the heritable variance may indeed exceed the phenotypic variance. As a consequence, response in traits affected by social interactions can be very large when compared with the phenotypic standard deviation of the trait. This agrees with the very large responses that have been observed in some selection experiments (Muir, 1996). Prospects for genetic improvement of traits affected by social interactions may, therefore, be substantially better than currently perceived.

Selection methods

The analysis above has shown that social interactions create additional heritable variation. However, this additional heritable variation may not be utilized to the full by classical selection strategies, such as mass selection or selection based on sib information. Improvement of traits affected by social interactions, therefore, may require modification of our selection methods. Analogous to classical theory, the response for each selection strategy may be expressed as the product of intensity, accuracy and genetic standard deviation:

$$\Delta G = i\rho\sigma_{TBV} \tag{3.5}$$

in which ρ is the 'accuracy of selection', which is the correlation between the selection criterion and the TBV of the selection candidate. Below we present accuracies for mass selection, group selection and sib selection schemes. In all cases, the individuals that provide the information for selection are kept in groups of n members but the selection criterion may differ.

Mass selection

With mass selection, individuals are selected by truncation based on their own phenotypic record only. For traits affected by social interactions, accuracy of mass selection equals (Ellen *et al.*, 2007):

$$\rho_{mass} = \frac{\sigma^2_{A_D} + (n-1)\sigma_{A_{DS}} + r(n-1)[\sigma_{A_{DS}} + (n-1)\sigma^2_{A_S}]}{\sigma_{TBV}\sigma_P}. \quad (3.6)$$

This result shows that, in contrast to the classical case, accuracy of mass selection depends on relatedness, r, between birds in a group. When groups consist of unrelated birds, $r = 0$, Equation 3.6 reduces to $\rho_{mass} = \frac{\sigma^2_{A_D} + (n-1)\sigma_{A_{DS}}}{\sigma_{TBV}\sigma_P}$. The numerator of this expression, $\sigma^2_{A_D} + (n-1)\sigma_{A_{DS}}$, can take negative values when direct and social effects have a negative genetic covariance, $\sigma_{A_{DS}} < 0$. (A negative $\sigma_{A_{DS}}$ indicates that individuals that have positive direct breeding values have, on average, negative social breeding values.) As a consequence, the accuracy of mass selection can be negative, meaning that the population will respond in the opposite direction to that of selection. What happens in this case is that mass selection yields a negative correlated response in social effects, which exceeds the direct response in absolute value, leading to a negative net result, $|(n-1)\Delta\bar{A}_S| > \Delta A_D$. In other words, due to negative correlated response in the social effect, mass selection may lead to increased competition, annulling the increase in the direct effect. This theoretical result is in line with empirical results of mass selection against mortality due to cannibalism in laying hens (Muir, 1996).

Equation 3.6 shows that an increase in relatedness between group members moves the accuracy in a positive direction. For example, with full relatedness, $r = 1$, the accuracy equals $\rho_{mass} = \frac{\sigma^2_{A_D} + 2(n-1)\sigma_{A_{DS}} + (n-1)^2\sigma^2_{A_S}}{\sigma_{TBV}\sigma_P} = \frac{\sigma^2_{TBV}}{\sigma_{TBV}\sigma_P} = \frac{\sigma_{TBV}}{\sigma_P}$. This result is an analogy of the usual accuracy of mass selection, which equals the square root of heritability, $h = \sigma_A/\sigma_P$, and is always positive. Thus, a breeding strategy to counteract an increase in competition is to keep individuals in groups consisting of relatives, e.g. full sibs, which was originally proposed by Griffing (1976).

Group selection

In group selection, the entire group of individuals is either selected or rejected as a whole, based on mean performance of individuals in the group. Thus selection is by truncation among groups, and the selection criterion is mean group performance, $\bar{P}_{grp}$. The accuracy of group selection (Ellen *et al.*, 2007) equals:

$$\rho_{grp} = \frac{[(n-1)r + 1]\sigma_{TBV}}{n\sigma_{\bar{P}_{grp}}} \quad (3.7)$$

This result shows that the accuracy of group selection is always positive, because both the numerator and denominator of Equation 3.7 are positive. Moreover, the accuracy increases with increasing relatedness among group members, because the numerator of ρ_{grp} is higher, although it also increases $\sigma_{\bar{P}_{grp}}$ but to a lesser extent. Thus selecting between groups and using groups composed of relatives are breeding solutions to prevent negative response due to increased competition.

Sib selection

The use of sib information is common in poultry breeding. This section investigates response to selection based on sib information, using two different strategies. First, selection based on mean performance of sibs ($\bar{P}_{sibs}$) kept in groups of *unrelated* individuals. For example, selection based on mean performance of 20 full sibs kept in 20 distinct groups, in which the group members of a particular full sib are unrelated to the full sib. Second, selection based on mean performance of sibs kept in *family* groups. For example, selection based on mean performance of 20 full sibs kept in five distinct groups, each consisting of four full sibs.

The accuracy of selection when using sib information coming from groups of unrelated individuals equals:

$$\rho_{sibs} = \frac{r[\sigma^2_{A_D} + (n-1)\sigma_{A_{DS}}]}{\sigma_{TBV}\sigma_{\bar{P}_{sibs}}} \tag{3.8}$$

in which r is the additive genetic relatedness between the candidate and its sibs. The numerator of this expression is proportional to that of mass selection with unrelated groups (Equation 3.6), and thus the response shows the same pattern. The accuracy takes negative values when $\sigma^2_{A_D} + (n-1)\sigma_{A_{DS}} < 0$, which may occur when direct and social effects have a negative genetic covariance, $\sigma_{A_{DS}} < 0$. This result shows that selection based on sib information, coming from groups of unrelated individuals, may lead to an increase in competition that exceeds the direct response, yielding a negative net response.

The accuracy of selection when using sib information coming from groups composed of related individuals is an analogy of the classical accuracy for traits not affected by interactions among individuals. Without interactions among individuals, the accuracy of sib selection is given by $\rho = rh/\sqrt{t + (1-t)/N}$, in which r is relatedness between the candidate and its sibs, h the square root of heritability, t the intraclass correlations between the sibs, $t = r_{br}h^2$ with r_{br} denoting mutual relatedness between the sibs, and N is the total number of observations (Falconer and Mackay, 1996). For example, when using a group of full sibs that are half-sibs of the candidate, i.e. a group of descendants of the sire of the candidate but of a single other dam, r = ¼ and r_{br} = ½. With social interactions and groups composed of related individuals, accuracy is an analogy of this classical expression (Ellen *et al.*, 2007):

$$\rho = \frac{r\eta}{\sqrt{\tau + (1-\tau)/N}} \tag{3.9}$$

In Equation 3.9, η is an analogy of the square root of heritability, and τ is an analogy of the intraclass correlations between sibs. The $\eta^2 = \sigma^2_{TBV}/\sigma^2_{TPV}$ and $\tau = r_{br}\eta^2$, in which σ^2_{TBV} is the total heritable variation in trait value (Equation 3.4) and σ^2_{TPV} is the variance in total phenotypic values of individuals, $TPV_i = (A_{D,i} + E_{D,i}) + (n-1)(A_{S,i} + E_{S,i})$. The TPV_i represents the total phenotypic effect of a bird on the trait values of its group (including itself), of which an amount $(A_{D,i} + E_{D,i})$ is expressed in its own phenotype and an amount $(A_{S,i} + E_{S,i})$ in the phenotypes of each of its n – 1 group members. (Note that the TPV differs from the observed phenotypic value of an individual.) Analogous to Equation 3.4, the variance of total phenotypic values is given by $\sigma^2_{TPV} = (\sigma^2_{A_D} + \sigma^2_{E_D}) + 2(n-1)(\sigma_{A_{DS}} + \sigma_{E_{DS}}) + (n-1)^2(\sigma^2_{A_S} + \sigma^2_{E_S})$.

Investigation of Equation 3.9 shows that the limiting accuracy when using information of a large number of sib groups is identical to that in the classical case. For large N, the accuracy approaches $r/\sqrt{r_{br}}$, which equals 0.5 for half-sib information, 0.71 for full sib information and 1 for progeny information. These values are the same as for classical selection based on sibs in the absence of social interactions. Equation 3.9, therefore, shows that accuracies for traits affected by social interactions can be similar to accuracies for classical traits. The key issue is that the sibs need to be kept in family groups. For example, when 16 half-sibs are available and cages consist of four birds, four cages of four birds each should be used; the 16 half-sibs should not be distributed over 16 cages each, in addition, containing three unrelated birds.

Parameter estimation

The selection methods described above can be applied without knowledge of the genetic parameters. Without this information, however, it is unclear whether those selection strategies are needed at all and what the expected response to selection is. Moreover, most applied breeding programmes make use of BLUP for estimating breeding values, which requires knowledge of the genetic parameters. Thus there is a need to estimate the following genetic parameters: the direct genetic variance, $\sigma^2_{A_D}$; the covariance between direct and social effects, $\sigma_{A_{DS}}$; and the additive genetic variance in social effects, $\sigma^2_{A_S}$. These parameters can be estimated by extending the usual mixed model equations, in a manner similar to estimating maternal genetic effects. This results in the following model (Muir, 2005; Bijma *et al.*, 2007b):

$$\mathbf{y} = \mathbf{Xb} + \mathbf{Z}_D\mathbf{a}_D + \mathbf{Z}_S\mathbf{a}_S + \mathbf{e}, \tag{3.10}$$

in which $\mathbf{y}$ is the usual vector of observations, $\mathbf{Xb}$ the fixed effects, $\mathbf{Z}_D\mathbf{a}_D$ the (direct) additive genetic effects and $\mathbf{a}_D$ is a vector of DBVs, with incidence matrix $\mathbf{Z}_D$ linking phenotypic values of individuals to their DBV. The $\mathbf{Z}_S\mathbf{a}_S$ represent the social additive genetic effects: $\mathbf{a}_S$ is a vector of SBVs, with incidence matrix $\mathbf{Z}_S$ linking phenotypic values of individuals to the SBVs of their associates. When considering four bird cages, for example, the $\mathbf{Z}_D$ has a one at the position of the individual producing the record, and the $\mathbf{Z}_S$ has three ones, one for each cage member of the individual producing the record. The covariance structure of the random genetic terms is: $Var\begin{bmatrix}\mathbf{a}_D\\ \mathbf{a}_S\end{bmatrix} = \mathbf{C}\otimes\mathbf{A}$, where $\mathbf{C} = \begin{bmatrix}\sigma^2_{A_D} & \sigma_{A_{DS}}\\ \sigma_{A_{DS}} & \sigma^2_{A_S}\end{bmatrix}$, **A** is the relationship matrix, and ⊗ denotes the Kronecker product of matrices. (Note that the element y_i of **y** in Equation 3.10 corresponds to P_i in Equation 3.1.) Thus, analogous to a maternal effects model, Equation 3.10 is a bivariate model involving the estimation of two variances, $\sigma^2_{A_D}$ and $\sigma^2_{A_S}$, and a single covariance, $\sigma_{A_{DS}}$.

The residual term of Equation 3.10 requires special attention. Because social interactions may have a non-heritable component, denoted E_S in Equation 3.1, records of group members will be correlated for non-genetic reasons. Such non-genetic correlations between group members need to be accommodated in the model, to avoid biased estimates of the genetic parameters. It follows from Equation 3.1 that the non-genetic covariance between phenotypic values of group members equals $2\sigma_{E_{DS}} + (n-2)\sigma^2_{E_S}$, and that the residual variance is $\sigma^2_e = \sigma^2_{E_D} + (n-1)\sigma^2_{E_S}$. Residuals of group members, therefore, have a non-genetic correlation equal to $\rho_e = \left[2\sigma_{E_{DS}} + (n-2)\sigma^2_{E_S}\right] / \left[\sigma^2_{E_D} + (n-1)\sigma^2_{E_S}\right]$. Depending on $\sigma_{E_{DS}}$, $\sigma^2_{E_S}$ and n, this correlation may take either positive or negative values. The within-group non-genetic correlation creates an overall residual covariance structure that is block diagonal:

$$Var(\mathbf{e}) = \mathbf{R}\sigma^2_e, \tag{3.11}$$

with $R_{ii} = 1$, $R_{ij} = \rho_e$ when i and j are in the same group, and $R_{ij} = 0$ when i and j are in different groups.

This structure can be fitted in ASREML using the CORU statement in the description of the R-structure (Gilmour *et al.*, 2002). Assuming independent residuals instead of correlated residuals may severely bias the estimated genetic variance components. Analysis of both simulated and real data show that using $Var(\mathbf{e}) = \mathbf{I}\sigma^2_e$, instead of Equation 3.11, may yield genetic variance components that are overestimated by as much as 200–300% (Bijma *et al.*, 2007b).

With large group size (n), the residual correlation is likely to be positive due to the termσ $(n-2)\sigma^2_{E_S}$ in the residual covariance. When the residual correlation is positive, an equivalent but simpler solution is to fit a random group effect instead of a correlated residual. It follows from $\text{Cov}_{\text{within}} = \text{Var}_{\text{between}}$ that the between-group variance, i.e. the variance of the random group effect, equals $\sigma^2_{group} = 2\sigma_{E_{DS}} + (n-2)\sigma^2_{E_S}$. When the residual covariance is positive, both models are equivalent and their relationship is given by $\rho_e = \sigma^2_{group} / \sigma^2_e$. Analysis of both simulated and real data shows that both models yield identical estimated genetic parameters and an identical likelihood (unpublished results). When groups are of equal size there is no information in the data to separate $\sigma_{E_{DS}}$ from $\sigma^2_{E_S}$. A random group effect or a residual correlation can be estimated, but their causal components cannot be identified. In most cases this will not be important because the major interest is in estimating the genetic variances without bias, not in the non-genetic parameters.

Varying group sizes

In our model, each individual has a social effect on each of its group members. Hence, the total social effect originating from individual i equals $(n-1)A_{S,i}$, and the total heritable variation has a term $(n-1)^2\sigma^2_{A_S}$. If the magnitude of $A_{S,i}$ is independent of group size, then the total heritable variation is infinite for large n, which is clearly not realistic. It is likely that the social breeding value, $A_{S,i}$, and the social genetic variance, $\sigma^2_{A_S}$, will become smaller when group size becomes larger. As a consequence, the σ^2_{TBV} as given by Equation 3.4 applies to a specific group size. For other group sizes, the values of $\sigma_{A_{DS}}$ and $\sigma^2_{A_S}$ may be different. A more general formulation of Equation 3.4 would therefore be $\sigma^2_{TBV} = \sigma^2_{A_D} + 2(n-1)\sigma_{A_{DS}(n)} + (n-1)^2\sigma^2_{A_S(n)}$, which makes explicit that $\sigma_{A_{DS}}$ and $\sigma^2_{A_S}$ are functions of group size. When $\sigma_{A_{DS}}$ and $\sigma^2_{A_S}$ are decreasing functions of group size, the σ^2_{TBV} increases less with group size than suggested by Equation 3.4.

The dependency of $\sigma_{A_{DS}}$, $\sigma^2_{A_S}$ and σ^2_{TBV} on group size may be interpreted as a kind of genotype × environment (G × E) interaction, meaning that the ranking of individuals will depend on group size. Extrapolation of estimates obtained from a specific group size to populations with different group sizes will require knowledge of the relationship between $\sigma_{A_{DS}}$, $\sigma^2_{A_S}$ and n. That relationship may be non-linear, or differ among traits, and needs to be established empirically. For example, when the trait of interest is mortality due to cannibalism in four bird cages versus eight bird cages, there may be little dependency of $\sigma_{A_{DS}}$ and $\sigma^2_{A_S}$ on n. This is because a single individual has sufficient opportunity to kill all group members since their number is limited. The situation is different, however, when comparing four bird cages with floor systems with thousands of birds. In that case, a single bird can never interact with all others. Thus, for mortality due to cannibalism, $\sigma^2_{A_S}$ may be fairly independent of group size as long as groups are not too large and then decrease with group size as groups become larger.

When data contain a single group size, there is no need to consider the relationship between $\sigma^2_{A_S}$ and group size when analysing data. The estimated $\sigma^2_{A_S}$ simply applies to the group size considered. When data contain groups of different size, the relationship between $\sigma^2_{A_S}$ and group size can be investigated using random regression models (Hadfield and Wilson, 2007). In such models, the social effect of an individual can be modelled as a polynomial function of group size, e.g. using an intercept and a term linear in group size ('slope'). In this case, however, the use of random regression models is a complex solution to a simple problem. A random regression model allows the social effect of each individual to have a different sensitivity to group size, whereas the main requirement is for $\sigma^2_{A_S}$ to depend on group size. A much simpler solution is to weight the A_S in the model by a function of group size, and find the function of group size that best fits the data (e.g. Arango *et al.*, 2005). This involves estimating a single additional fixed effect only. For example, when individuals compete for a fixed total amount of resources, one may evaluate a model in which social effects are inversely proportional to $n - 1$;

$P_i = A_{D,i} + E_{D,i} + \frac{1}{n_i - 1}\left[\sum_{i \neq j}^{n_i} A_{S,j} + \sum_{i \neq j}^{n_i} E_{S,j}\right]$, in which n_i is the size of the group of individual i. In this case the total heritable variation is independent of group size; $\sigma^2_{TBV} = \sigma^2_{A_D} + 2\sigma_{A_{DS}} + \sigma^2_{A_S}$ for any value of n.

The variance components for social effects, $\sigma_{A_{DS}}$ and $\sigma^2_{A_S}$, may not be estimable in all data structures occurring in practice. The $\sigma_{A_{DS}}$ and $\sigma^2_{A_S}$ are estimable when groups are composed at random with respect to family but not when groups are entirely composed of full sibs or half-sibs. In poultry breeding, family cages are often used to record egg production traits in recurrent tests. Unfortunately, such data cannot be used to estimate $\sigma_{A_{DS}}$ and $\sigma^2_{A_S}$, and there may be a need for additional data collection in order to estimate them. Note that the data structure required for variance component estimation differs from that required for efficient selection without knowledge of the genetic parameters. When genetic parameters are unknown, socially affected traits can be improved efficiently by using groups composed of family members (see section Selection methods). However, data on family groups cannot be used to estimate genetic parameters.

DISCUSSION

This summary of theoretical analyses and early results suggests that social effects are potentially very relevant for poultry genetic improvement. This is most clear from Equation 3.4, which shows that social interactions may contribute substantially to the heritable variation in trait values. Whether or not social interactions are of interest to poultry breeders will depend on the trait considered and on the environment in which birds are kept. Results on mortality in non-beak-trimmed layers kept in four-bird cages illustrate the relevance of social genetic effects in that situation (Bijma *et al.*, 2007b). In that study, two-thirds of the total heritable variation in survival days originated from social effects. Our preliminary results in pigs indicate that social genetic effects are important for growth rate and feed intake. In contrast, results of Arango *et al.* (2005, 2006) indicate that social genetic effects may be difficult to estimate and not deviate significantly from zero. At present there is little knowledge of the importance of social genetic effects in livestock genetic improvement. In cases where groups are large, such as in broilers, it will be difficult to estimate genetic parameters. One solution for such cases would be to invest in data collection on a large number of smaller groups. This would, however, require substantial investment, and social effects in smaller groups may not correspond accurately to those in larger groups. A small-scale selection experiment, in which small groups are used for selection and results are evaluated in larger groups, might be more promising. Though social genetic effects may not always be important and may be difficult to use in some cases, the promising results observed in layer chickens should be sufficient incentive for further research in this area.

REFERENCES

Abasht, B., Dekkers, J.C.M. and Lamont, S.J. (2006) Review of quantitative trait loci identified in the chicken. *Poultry Science* 85, 2079–2096.

Aerts, J., Megens, H.J., Veenendaal, T., Ovcharenko, I., Crooijmans, R., Gordon, L., Stubbs, L. and Groenen, M. (2007) Extent of linkage disequilibrium in chicken. *Cytogenetic and Genome Research* 117, 338–345.

Andersson, L. and Georges, M. (2004) Domestic animal genomics: deciphering the genetics of complex traits. *Nature Reviews Genetics* 5, 202–212.

Arango, J., Misztal, I., Tsuruta, S., Culbertson, M. and Herring, W.O. (2005) Estimation of variance components including competitive effects of Large White growing gilts. *Journal of Animal Science* 83, 1241–1246.

Arango, J., Misztal, I., Tsuruta, S., Culbertson, M., Herring, W.O. and Long, T. (2006) Competitive effects in Large White growing gilts. Proceedings of the 8th World Congress on Genetics Applied to Livestock Production, Belo Horizonte, MG, Brazil. Communication 06-24.

Arthur, J.A. and Albers, G.A.A. (2003) Industrial perspective on problems and issues associated with poultry breeding. In: Muir, W.M. and Aggrey, S.E. (eds) *Poultry Genetics, Breeding and Biotechnology*. CAB International, Wallingford, UK, pp. 1–12.

Bergsma, R., Kanis, E., Knol, E.F. and Bijma, P. (2008) The contribution of social effects to heritable variation in finishing traits of domestic pigs (Sus scrofa). *Genetics* 178, 1559–1570.

Bijma, P., Muir, W.M. and. Van Arendonk, J.A.M (2007a) Multilevel selection 1: quantitative genetics of inheritance and response to selection. *Genetics* 175, 277–288.

Bijma, P., Muir, W.M., Ellen, E.D., Wolf, J.B. and Van Arendonk, J.A.M. (2007b) Multilevel selection 2: estimating the genetic parameters determining inheritance and response to selection. *Genetics* 175, 289–299.

Crittenden, L.B., Provencher, L., Sanatangello, L., Levin, I., Abplanalp, H., Briles, R.W., Briles, W.E. and Dodgson, J.B. (1993) Characterization of a red jungle fowl by White Leghorn backcross reference population for molecular mapping of the chicken genome. *Poultry Science* 72, 334–348.

Dekkers, J.C.M. and Hospital, F. (2002) The use of molecular genetics in the improvement of agricultural populations. *Nature Reviews Genetics* 3, 22–32.

de Koning, D.J., Windsor, D., Hocking, P.M., Burt, D.W., Law, A., Haley, C.S., Morris, A., Vincent, J. and Griffin, H. (2003) Quantitative trait locus detection in commercial broiler lines using candidate regions. *Journal of Animal Science* 81, 1158–1165.

de Koning, D.J., Haley, C.S., Windsor, D., Hocking, P.M., Griffin, H., Morris, A., Vincent, J. and Burt, D.W. (2004) Segregation of QTL for production traits in commercial meat-type chickens. *Genetical Research* 83, 211–220.

Ellen, E.D., Muir, W.M., Teuscher, F. and Bijma, P. (2007) Genetic improvement of traits affected by interactions among individuals: sib selection schemes. *Genetics* 176, 489–499.

Falconer, D.S. and Mackay, T.F.C. (1996) *Introduction to Quantitative Genetics*. Pearson, Harlow, UK.

Gilmour, A.R., Gogel, B.J., Cullis, B.R., Welham, S.J. and Thompson, R. (2002) ASReml User Guide Release 1.0. http://www.vsn-intl.com/, VSN International Ltd, Hemel Hempstead, UK.

Griffing, B. (1976) Selection in reference to biological groups. VI. Use of extreme forms of non-random groups to increase selection efficiency. *Genetics* 82, 476–490.

Hadfield, J.D. and Wilson, A.J. (2007) Multilevel selection 3: modeling the effects of interacting individuals as a function of group size. *Genetics* 177, 667–668.

Hillier, L.W., Miller, W., Birney, E., Warren, W., Hardison, R.C., Ponting, C. *et al.* (2004) Sequence and comparative analysis of the chicken genome provide unique perspectives on vertebrate evolution. *Nature* 432, 695–716.

Hocking, P.M. (2005) Review of QTL results in chicken. *World's Poultry Science Journal* 61, 215–226.

Honkatukia, M., Reese, K., Preisinger, R., Tuiskula-Haavisto, M., Weigend, S., Roito, J., Mäki-Tanila, A. and Vilkkia, J. (2005) Fishy taint in chicken eggs is associated with a substitution within a conserved motif of the FMO3 gene. *Genomics* 86, 225–232.

Lynch, M. and Walsh, B. (1998) *Genetics and Analysis of Quantitative Traits*. Sinauer Associates Inc., Sunderland, MA, USA.

Meuwissen, T.H.E., Hayes, B.J. and Goddard, M.E. (2001) Prediction of total genetic value using genome-wide dense marker maps. *Genetics* 157, 1819–1829.

Muir, W.M. (1996) Group selection for adaptation to multiple-hen cages: selection program and direct responses. *Poultry Science* 75, 447–458.

Muir, W.M. (2003) Indirect selection for improvement of animal well-being. In: Muir, W.M. and Aggrey, S. (eds) *Poultry Genetics, Breeding and Biotechnology*. CABI, Cambridge, MA, USA, pp. 247–256.

Muir, W.M. (2005) Incorporation of competitive effects in forest tree or animal breeding programs. *Genetics* 170, 1247–1259.

Pakdel, A., Bijma, P., Ducro, B.J. and Bovenhuis, H. (2005) Selection strategies for body weight and reduced ascites susceptibility in broilers. *Poultry Science* 84, 528–535.

Rabie, T.S.K.M., Crooijmans, R.P.M.A., Bovenhuis, H., Vereijken, A.L.J., Veenendaal, T., van der Poel, J.J., Van Arendonk, J.A.M., Pakdel, A. and Groenen, M.A.M. (2005) Genetic mapping of quantitative trait loci affecting susceptibility in chicken to develop pulmonary hypertension syndrome. *Animal Genetics* 36, 468–476.

Siwek, M, Cornelissen, S.J.B., Nieuwland, M.G.B., Buitenhuis, A.J., Bovenhuis, H., Crooijmans, R.P.M.A., Groenen, M.A.M., Vries-Reilingh, G., Parmentier, H.K. and van der Poel, J.J. (2003) Detection of QTL for immune response to sheep red blood cells in laying hens. *Animal Genetics* 34, 422–428.

Takeda, H., Caiment, F., Smit, M., Hiard, S., Tordoir, X., Cockett, N., Georges, M. and Charlier, C. (2006) The callipyge mutation enhances bidirectional long-range DLK1-GTL2 intergenic transcription in cis. *Proceedings of the National Academy of Sciences of the USA* 103, 8119–8124.

Tatsuda, K. and Fujinaka, K. (2001) Genetic mapping of the QTL affecting body weight in chickens using a F2 family. *British Poultry Science* 42, 333–337.

Van Kaam, J.B.C.H.M., van Arendonk, J.A.M., Groenen, M.A.M., Bovenhuis, H., Vereijken, A.L.J., Crooijmans, R.P.M.A., van der Poel, J.J. and Veenendaal, A. (1998) Whole genome scan for quantitative trait loci affecting body weight in chickens using a three generation design. *Livestock Production Science* 54, 133–150.

Wolf, J.B., Brodie, E.D. III, Cheverud, J.M., Moore, A.J. and Wade, M.J. (1998) Evolutionary consequences of indirect genetic effects. *Trends in Ecology and Evolution* 13, 64–69.

Wong, G.K.S., Liu, B., Wang, J., Zhang, Y., Yang, X., Zhang, Z.J. *et al.* (2004) A genetic variation map for chicken with 2.8 million single nucleotide polymorphisms. *Nature* 432, 717–722.

Yonash, N., Cheng, H.H., Hillel, J., Heller, D.E. and Cahaner, A. (2001) DNA microsatellites linked to quantitative trait loci affecting antibody response and survival rate in meat-type chicken. *Poultry Science* 80, 22–28.

CHAPTER 4
Genetic Modification of the Chicken: New Technologies with Potential Applications in Poultry Production

Helen M. Sang

The Roslin Institute and Royal (Dick) School of Veterinary Sciences, University of Edinburgh, UK

ABSTRACT

The ability to genetically modify any organism provides a powerful tool with many applications both in contributing to basic science and with the potential for practical applications. The development of methods to genetically modify the chicken has only recently made significant advances. We have shown that gene transfer vectors derived from lentiviruses (for example HIV) can be used to introduce transgenes with high efficiency. The efficiency of the production of founder transgenic birds relates to the titre of the vector that is produced. We have efficiencies ranging from 100% to approximately 10%, with transmission through the germline from the founder transgenic cockerels ranging from 45 to 0.2%. Transgenes introduced using lentiviral vectors are stable through several generations and transgene expression is also maintained, in contrast to transgenic birds generated using retroviral vectors. A second new method for production of transgenic chickens, involving long-term *in vitro* culture of primordial germ cells (the precursors of the gametes), has recently been described. This approach has significant potential as the basis for sophisticated methods for genetic modification of chickens.

We are using this technology in a range of applications. Transgenesis can be used to develop very useful tools for the study of development in chick embryos and to investigate the function of genes thought to regulate development. Gene function is often investigated by knocking out a gene and we are developing the use of lentiviral vectors to deliver microRNA(miRNA)-expression constructs to knockdown the expression of specific target genes. One application of this technology is to express miRNA constructs to knockdown specific mRNAs of viral genes to interfere with viral replication or investigate function of viral genes during the course of infection. The potential to utilize the protein synthetic capability of the laying hen's oviduct to synthesize human therapeutic proteins was proposed many years ago. We have shown that regulatory sequences from the ovalbumin gene can be used to direct expression of three candidate human therapeutic proteins in a tightly tissue-specific manner.

INTRODUCTION

A significant tool has been missing from the armoury available to understand the genetic mechanisms involved in development and growth of poultry. This has been the ability to genetically modify the chicken. Transgenesis has been shown in many studies of the mammalian paradigm, the mouse, to be a very powerful tool in analysis of gene function. Now that the chicken genome sequence is available (Hillier *et al.*, 2004) we have many possibilities for applying transgenesis to the study of gene function in the chicken, in basic biological studies but also in understanding the function of genes involved in production traits, for example the genes involved in innate and acquired immunity. Transgenesis is not only a useful tool for studying gene function but it can also be used to introduce novel and/or artificial genes into the genome. This technology has been exploited by the plant breeding industry to generate new traits, for example herbicide resistance or for the expression of proteins that are toxic to insect pests. The lack of a robust technology for the genetic modification of the chicken has meant that the investigation of gene function has been limited and that novel genetic alterations to the chicken genome with possible applications for the breeding industry have not been possible. Significant advances towards developing transgenesis in the chick have been made by a small number of research groups (reviewed in Zajchowski and Etches, 2000; Sang, 2004). These results have underpinned the recent successful development of two methods for genetic modification of the chicken: the use of lentiviral vectors and the *in vitro* culture and genetic manipulation of primordial germ cells. These methods will greatly facilitate the study of production traits in poultry and also, more controversially, enable the generation of birds carrying novel traits generated by genetic modification.

LENTIVIRAL VECTOR TRANSGENESIS

The earliest reports of generation of transgenic chickens involved the use of vectors derived from avian retroviruses (Salter *et al.*, 1987; Bosselman *et al.*, 1989). These studies were of limited success because the frequency of production of transgenic birds was low and the transgenes introduced using these vectors were not reliably expressed. These vectors have continued to be used to a limited extent for the production of transgenic birds (Harvey *et al.*, 2002) and in the very useful RCAS (replication-competent ASLV long terminal repeat with a splice acceptor) vector system for analysis of short-term transgene expression in embryos (Hughes *et al.*, 1987). A more recent advance in the use of viruses as gene transfer vectors has been the development of vectors from the lentivirus class of retroviruses. Transgenic mammals have been produced with high efficiency using vectors derived from human immunodeficiency virus (Hofmann *et al.*, 2003; Whitelaw *et al.*, 2004) and reliable tissue-specific expression of transgenes seen after germline transmission (Lois *et al.*, 2002).

We tested the efficiency of production of germline transgenic birds using vectors derived from the lentivirus equine infectious anaemia virus (EIAV) (McGrew *et al.*, 2004). Reporter transgenes were cloned into an EIAV vector genome and concentrated virus produced by transfection of tissue culture cells with the vector and plasmids expressing the viral proteins required for assembly of viral particles (Fig. 4.1). The EIAV vectors are self-inactivating, minimal vectors that have been developed for human gene therapy and which are incapable of generation of viral particles once integrated into the host genome. The vectors we used were pseudotyped with vesicular stomatitis virus glycoprotein (VSV-G), which facilitates infection of cells of a wide range of types from most species. The packaged virus was concentrated to a titre of 10^7–10^8 TU/ml and 1–2 μl of viral suspension injected into the subgerminal cavity below the embryonic disc of embryos in newly laid eggs. The eggs were cultured using the method for chick embryo culture that we had described previously (Perry, 1988), resulting in 27% of the manipulated embryos developing to give healthy hatched chicks. These experiments generated 12 cockerels, ten of which were shown to be germline transgenic by breeding with stock hens. These cockerels transmitted the vector to between 4 and 45% of their offspring. We showed stable transmission of the integrated vectors from the G_1 to the G_2 generation, establishing that we could set up transgenic lines from the G_1 generation. Examination of expression reporter gene constructs, carried by the EIAV vectors, showed a conserved expression profile between individuals and maintenance of expression from one generation to the next. This study demonstrated that lentiviral vectors can be used to generate transgenic chickens with high efficiency and with no detectable silencing of the expression of transgenes introduced into the genome using these vectors.

The report described above involved the use of transgenes whose expression was driven by a viral enhancer/promoter, resulting in transgene expression at differing levels in some tissues and no detectable expression in others. Ubiquitous transgene expression driven by the mouse phosphoglycerol kinase promoter was described in transgenic chickens generated using an HIV vector (Chapman *et al.*, 2005). For many applications of transgenesis it will be important to direct expression of transgenes to specific tissues. This has been demonstrated in quail (Scott and Lois, 2005), where an HIV vector was used to generate transgenic birds with a transgene consisting of the human synapsin1 promoter driving expression of green fluorescent protein (GFP). GFP was detected only in neurons and expression was consistent from one generation to the next.

A potential application of transgenesis in the chicken that has been discussed is the production of therapeutic proteins in the eggs of transgenic hens (Lillico *et al.*, 2005). There are an increasing number of therapeutic proteins in development, which are produced in industrial bioreactors. These systems are very expensive to establish and run, and the potential for using transgenic animals as bioreactors, with the possibility of producing high-quality, post-translationally modified proteins, is already advanced, for example in the use of transgenic goats to produce therapeutic proteins in the mammary gland. The

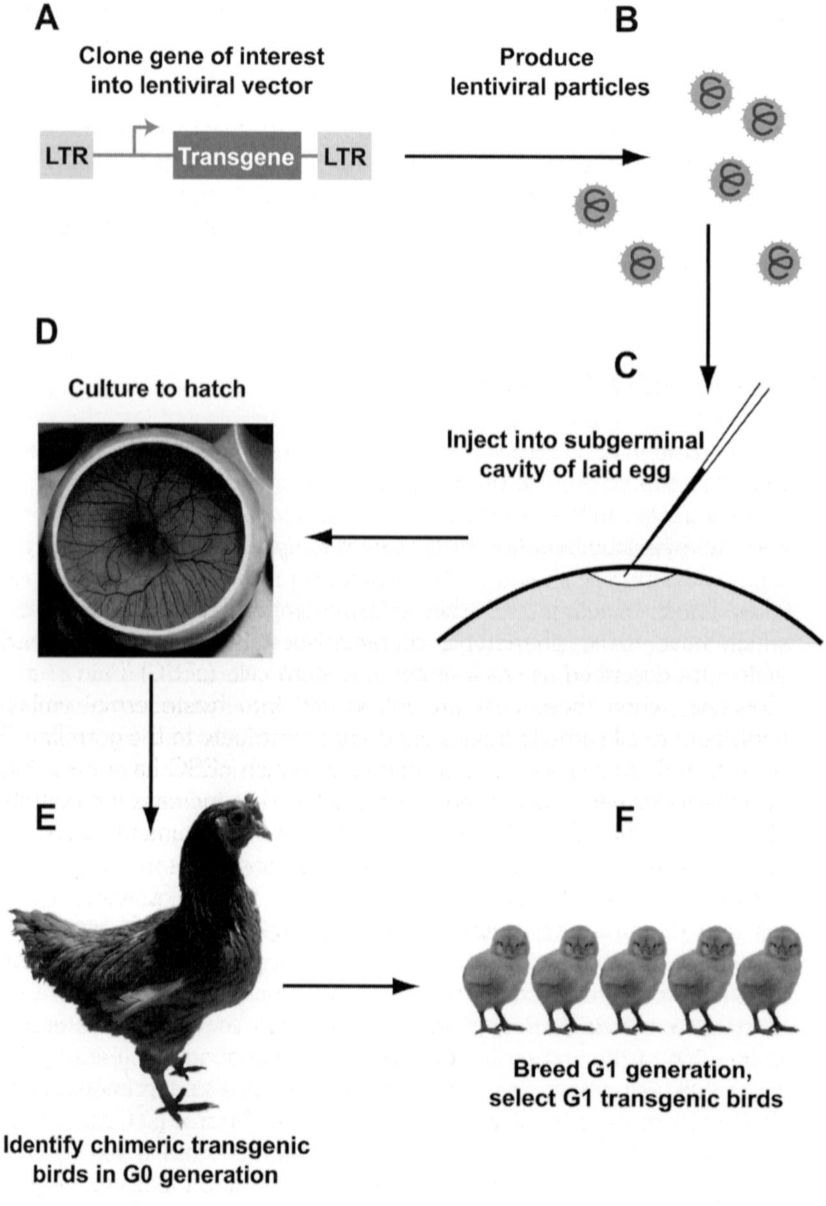

Fig. 4.1. Production of transgenic chickens using lentiviral vectors. A. A transgene is inserted into a minimal virus vector genome. B. The vector is cotransfected with plasmids encoding essential viral proteins into tissue culture cells, the culture medium harvested and packaged viral vector particles concentrated by centrifugation. C. The concentrated viral vector suspension is injected into the embryos of newly laid eggs. D. The eggs are cultured using an open shell culture sytem to hatch. E. Hatched chicks are screened to identify chimeric transgenic birds and at sexual maturity are bred to generate transgenic and non-transgenic chicks. F. The G1 chicks are screened to identify transgenic birds, which are then bred to establish transgenic lines for analysis.

basic strategy that is proposed is to target expression of a therapeutic protein to either the yolk or white of the egg, from which it will be extracted to produce a purified drug. We have demonstrated that this approach is possible by using regulatory sequences of the ovalbumin gene, the major egg white protein, to direct expression of two model therapeutic proteins to the oviduct of laying hens, using EIAV vectors to generate the transgenic hens (Lillico *et al.*, 2007). The expression of the proteins is tightly regulated to the magnum of the oviduct, confirming that tissue-specific expression can be achieved using lentiviral vector transgenesis.

CELL-MEDIATED TRANSGENESIS

The derivation of mouse embryonic stem cells (mESC) and the demonstration that they can be genetically modified *in vitro*, including modification of endogenous genes, followed by generation of transgenic mice from the modified cells by their incorporation into chimeras, has been an enormously powerful tool in the study of gene function in mice. Significant advances have been made in the culture of cells from the blastodermal stage of chick embryo development, which have many phenotypic characteristics in common with mESC, and which are described as chick embryonic stem cells (cESC) (Pain *et al.*, 1996). However, when these cells are introduced into blastodermal embryos they contribute to all somatic tissues but do not contribute to the germline. This has been tested extensively in experiments in which cESC have been introduced into blastodermal embryos that were irradiated to increase the contribution to chimeras of the cESC. Even though many of the chimeras were over 50% derived from the cESC, none showed germline transmission from gametes derived from the cESC (van de Lavoir *et al.*, 2006a). This restricts their use as a route to transgenesis, using genetically modified cESC, to the study of chimeric birds. This can be useful, as shown by the expression of a monoclonal antibody as a component of egg white in chimeric hens generated using cESC carrying a transgene incorporating ovalbumin regulatory sequences driving expression of the two chains of a monoclonal antibody (Zhu *et al.*, 2005).

Significant efforts have been focused at targeting primordial germ cells (PGC) as a route to modification of the chicken genome (D'Costa *et al.*, 2001). At about 2–3 days of incubation, PGC migrate from a region outside the developing embryo through the circulatory system to the developing gonads. It has been shown that PGC can be transferred from one embryo to another and generate functional gametes in the recipient bird. It has proved difficult to isolate primary PGC, genetically modify them *in vitro* and generate transgenic birds after their transfer to recipient embryos. Recently a method has been described for long-term culture of proliferating PGC, isolated from this migratory phase of their development (van de Lavoir *et al.*, 2006b). This method facilitates the genetic modification of PGC in culture and allows the selection of cells carrying desired changes prior to their recovery through the germline by transfer to recipient embryos. This exciting development opens up the possibility of making complex genetic modifications to the chicken, including gene knockout and other forms of targeting of changes to the chicken genome.

FUTURE APPLICATIONS OF TRANSGENIC TECHNOLOGIES

The two methods outlined above, lentiviral vector transgenesis and transgenesis using cultured PGC, are much more efficient than previously described methods for the generation of transgenic birds. Both methods have particular strengths and weaknesses but their availability makes consideration and testing of many applications of transgenesis in poultry possible. Lentiviral vectors are limited in the size of transgene they can deliver and currently cannot be used to target endogenous genes. They are easy to develop and a single construct can be used to generate transgenic birds in any genotype by injection of fertile eggs from a chosen breed. The use of primordial germ cells requires more sophisticated cell culture expertise and would require significant efforts to establish cell lines from different genotypes, but they can be transfected with transgenes of possibly unlimited size and the technology and expertise will be developed to allow gene knockout by homologous recombination and further sophisticated modifications of the endogenous genome.

These technologies will have many applications in basic biology, probably limited mainly by the cost of such experiments. They also allow development of genetic modification strategies that could be introduced into commercial poultry breeding programmes. These could involve simple overexpression of growth factors, for example to increase growth rate, but this sort of approach has not been successful in other farm animal species as there are associated negative effects on the phenotype (Pursel *et al.*, 1997). More sophisticated approaches are likely to be more successful. For example, knockdown of expression of myostatin, a negative regulator of muscle growth, may result in increased muscle and lower fat, as seen in natural mutations in myostatin in cattle (Grobet *et al.*, 1997). This knockdown could be achieved either by reducing expression of the myostatin gene by a targeted mutation or by expression of a short interfering RNA expressed from a transgene based, for example, on the microRNA (miRNA) knockdown approach (Das *et al.*, 2006).

The use of RNA interference to deplete a target mRNA has also been proposed as a novel method for introducing resistance to viral diseases. The suggested strategy is to express short interfering RNAs (siRNAs) that will block expression of target viral genes using a transgenic approach (for example by introduction of transgenes that encode miRNA genes) designed to target conserved sequences within the viral mRNA (or genomic RNA) sequences. This will lead to degradation of the targeted viral mRNA, and, if the targeted mRNA encodes a gene required for viral replication, viral replication and therefore productive infection will be reduced or blocked. The potential of this approach for blocking avian influenza virus replication has been demonstrated by introduction of siRNAs into embryonated eggs and reduction in influenza virus replication (Ge *et al.*, 2003). A further approach to blocking viral infection is the possibility of using intracellular interference by expression of a protein that will act as a dominant negative inhibitor of viral infection. This approach has been modelled by González *et al.* (2005) for replication of infectious bursal disease virus (IBDV) in tissue culture cells. They demonstrated that expression of a mutant version of the VP3 viral structural peptide effectively interfered with the IBDV replication cycle. This approach could easily be adopted by making transgenic birds that express the interfering protein and subsequently testing the effectiveness of blocking IBDV infection in commercial poultry.

The above examples illustrate that there is significant potential for production of disease-resistant poultry using genetic modification. As robust technologies are now in place these approaches can be evaluated. The remaining challenge will be to present these opportunities to the public and determine if the use of genetic modification to confer disease resistance, which will be of benefit to both the birds and production companies, and also potentially to human health in the case of avian influenza, will be acceptable in commercial production.

REFERENCES

Bosselman, R.A., Hsu, R.-Y., Boggs, T., Hu, S., Bruszewski, J., Ou, S., Kozar, L., Martin, F., Green, C., Jacobsen, F., Nicolson, M., Schultz, J.A., Semon, K.M., Rishell, W. and Stewart, R.G. (1989) Germline transmission of exogenous genes in the chicken. *Science* 243, 533–535.

Chapman, S.C., Lawson, A., Macarthur, W.C., Wiese, R.J., Loechel, R.H., Burgos-Trinidad, M., Wakefield, J.K., Ramabhadran, R., Mauch, T.J. and Schoenwolf, G.C. (2005). Ubiquitous GFP expression in transgenic chickens using a lentiviral vector. *Development* 132, 935–940.

Das, R.M., Van Hateren, N.J., Howell, G.R., Farrell, E.R., Bangs, F.K., Porteus, V.C., Manning, E.M., McGrew, M.J., Ohyama, K., Sacco, M.A., Halley, P.A., Sang, H.M., Storey, K.G., Placzek, M., Tickle, C., Nair, V.K. and Wilson, S.A. (2006) A robust system for RNA interference in the chicken using a modified microRNA operon. *Developmental Biology* 294, 554–563.

D'Costa, S., Pardue, S.L. and Petitte, J.N. (2001) Comparative development of avian primordial germ cells and production of germ line chimeras. *Avian and Poultry Biology Reviews* 12, 151–168.

Ge, Q., McManus, M.T., Nguyen, T., Shen, C.H., Sharp, P.A., Eisen, H.N. and Chen, J. (2003) RNA interference of influenza virus production by directly targeting mRNA for degradation and indirectly inhibiting all viral RNA transcription. *Proceedings of the National Academy of Sciences of the USA* 100, 2718–2723.

González, D., Rodriguez, J.F. and Abaitua, F. (2005) Intracellular interference of infectious bursal disease virus. *Journal of Virology* 79, 14437–14441.

Grobet, L., Martin, L.J., Poncelet, D., Pirottin, D., Brouwers, B., Riquet, J., Scheberlein, A., Dunner, S., Ménissier, F., Massabanda, J., Fries, R., Hanset, R. and Georges, M. (1997) A deletion in the bovine myostatin gene causes the double-muscled phenotype in cattle. *Nature Genetics* 17, 71–74.

Harvey, A.J., Speksnijder, G., Baugh, L.R., Morris, J.A. and Ivarie, R. (2002) Consistent production of transgenic chickens using replication-deficient retroviral vectors and high-throughput screening procedures. *Poultry Science* 81, 202–212.

Hillier, L.W., Miller, W., Birney, E., Warren, W., Hardison, R.C., Ponting, C. *et al.* (2004) Sequence and comparative analysis of the chicken genome provide unique perspectives on vertebrate evolution. *Nature* 432, 695–716.

Hofmann, A., Kessler, B., Ewerling, S., Weppert, M., Vogg, B., Ludwig, H., Stojkovic, M., Boelhauve, M., Brem, G., Wolf, E. and Pfeifer, A. (2003) Efficient transgenesis in farm animals by lentiviral vectors. *EMBO Reports* 4, 1054–1060.

Hughes, S.H., Greenhouse, J.J., Petropoulos, C.J. and Sutrave, P. (1987) Adaptor plasmids simplify the insertion of foreign DNA into helper-independent retroviral vectors. *Journal of Virology* 61, 3004–3012.

Lillico, S., McGrew, M., Sherman, A. and Sang, H.M. (2005) Transgenic chickens as bioreactors for protein based drugs. *Drug Discovery Today* 10, 191–196.

Lillico, S.G., Sherman, A., McGrew, M.J., Robertson, C.D., Smith, J., Haslam, C., Barnard, P., Radcliffe, P.A., Mitrophanous, K.A., Elliot, E.A. and Sang, H.M. (2007) Oviduct-specific expression of two therapeutic proteins in transgenic hens. *Proceedings of the National Academy of Sciences of the USA* 104, 1771–1776.

Lois, C., Hong, E.J., Pease, S., Brown, E.J. and Baltimore, D. (2002) Germline transmission and tissue-specific expression of transgenes delivered by lentiviral vectors. *Science* 295, 868–872.

McGrew, M.J., Sherman, A., Ellard, F.M., Lillico, S.G., Gilhooley, H.J., Kingsman, A.J., Mitrophanous, K.A. and Sang, H. (2004) Efficient production of germline transgenic chickens using lentiviral vectors. *EMBO Reports* 5, 728–733.

Pain, B., Clark, M.E., Shen, M., Nakazawa, H., Sakurai, M., Samarut, J. and Etches, R.J. (1996) Long-term *in vitro* culture and characterisation of avian embryonic stem cells with multiple morphogenetic potentialities. *Development* 122, 2339–2348.

Perry, M.M. (1988) A complete culture system for the chick-embryo. *Nature* 331, 70–72.

Pursel, V.G., Wall, R.J., Soloman, M.B., Bolt, D.J., Murray, J.E. and Ward, K.A. (1997) Transfer of an ovine metallothionein-ovine growth hormone fusion into swine. *Journal of Animal Science* 75, 2208–2214.

Salter, D.W., Smith, E.J., Hughes, S.H., Wright, S.E. and Crittenden, L.B. (1987) Transgenic chickens: insertion of retroviral genes into the chicken germ line. *Virology* 157, 236–240.

Sang, H.M. (2004). Transgenics, chickens and therapeutic proteins. *Vox Sanguinis* 87, 164–166.

Scott, B.B. and Lois, C. (2005) Generation of tissue-specific transgenic birds with lentiviral vectors. *Proceedings of the National Academy of Sciences of the USA* 102, 16443–16447.

van de Lavoir M.C., Mather-Love, C., Leighton, P., Diamond, J.H., Heyer, B.S., Roberts, R., Zhu, L., Winters-Digiacinto, P., Kerchner, A., Gessaro, T., Swanberg, S., Delany, M.E. and Etches, R.J. (2006a) High-grade transgenic somatic chimeras from chicken embryonic stem cells. *Mechanisms of Development* 123, 31–41.

van de Lavoir, M., Diamond, H.H., Leighton, P.A., Mather-Love, C., Heyer, B.S., Bradshaw, R., Kerchner, A., Hooi, L.T., Gessaro, T.M., Swanberg, S.E., Delany, M.E. and Etches, R.J. (2006b) Germline transmission of genetically modified primordial germ cells. *Nature* 441, 766–769.

Whitelaw, C.B., Radcliffe, P.A., Ritchie, W.A., Carlisle, A., Ellard, F.M., Pena, R.N., Rowe, J., Clark, A.J., King, T.J. and Mitrophanous, K.A. (2004) Efficient generation of transgenic pigs using equine infectious anaemia virus (EIAV) derived vector. *FEBS Letters* 571, 233–236.

Zajchowski, L.D. and Etches, R.J. (2000) Transgenic chickens: past, present and future. *Avian and Poultry Biology Reviews* 11, 63–80.

Zhu, L., van de Lavoir, M.C., Albanese, J., Beenhouwer, D.O., Cardarelli, P.M., Cuison, S., Deng, D.F., Deshpande, S., Diamond, J.H., Green, L., Halk, E.L., Heyer, B.S., Kay, R.M.,

Kerchner, A., Leighton, P.A., Mather, C.M., Morrison, S.L., Nikolov, Z.L., Passmore, D.B., Pradas-Monne, A., Preston, B.T., Rangan, V.S., Shi, M.X., Srinivasan, M., White, S.G., Winters-Digiacinto, P., Wong, S., Zhou, W. and Etches, R.J. (2005) Production of human monoclonal antibody in eggs of chimeric chickens. *Nature Biotechnology* 23, 1159–1169.

CHAPTER 5
Prospects for Sex Determination in Poultry

S. Nandi and M. Clinton

The Roslin Institute and Royal (Dick) School of Veterinary Studies, University of Edinburgh, UK

ABSTRACT

In developmental biology terms, sex determination describes the differentiation of the gonadal primordia into testes or ovaries. In mammals, testis differentiation is initiated by expression of the male-specific Sry gene, and many of the genes that regulate the subsequent tissue remodelling have been identified. In birds, although the process of gonadal development appears morphologically similar to that in mammals and a number of differentially expressed genes have been identified, the molecular regulation of sex determination is not conserved in mammals and birds. To the poultry industry, sex determination is also of interest, although this usually equates to finding a means of altering sex ratios: layer breeders would prefer to raise only female offspring whereas broiler and turkey breeders would prefer males. This short chapter will briefly review how our basic research findings can impinge on efforts to manipulate sex ratios in the poultry industry.

INTRODUCTION

In the UK 28 million layer chickens are raised each year and the drive for more efficient production has led to the generation of layer breeds that can lay more than 300 eggs per year. It is uneconomic to raise male layer birds for meat production and, as a consequence, these birds are culled at hatch. Similarly, in some production systems female turkeys are culled at hatch because males are significantly more profitable than females.

The annual UK disposal of such a large number of hatchlings is both a significant welfare problem and, in terms of waste generation, a growing environmental issue. One possible strategy to address these problems would be to identify the sex of developing embryos prior to hatch and to dispose of the unwanted eggs at that stage. The following gives a brief overview of the early stages of chick development and how knowledge of the underlying molecular biology might be used to hatch chicks of one sex only.

SCIENTIFIC BACKGROUND

After ovulation the egg is fertilized and begins the process of cell cleavage. As the oocyte progresses down the oviduct it is surrounded by a vitelline membrane, layers of albumen and, finally, shell. Twenty-four hours after ovulation, the egg is laid, and at this stage the early embryo comprises a disc of around 60,000 cells on the surface of the yolk. Over the following 3 days, the vascular system develops to the extent that a beating heart can be clearly observed. The amniotic membranes start to develop at day 3 and grow to eventually form a protective sac around the embryo. Subsequently, the allantois develops into a richly vascularized structure that surrounds the amnion and embryo and enables the exchange of gases with the environment (Wolpert *et al.*, 2001).

The first evidence of embryonic gonad development can be seen at day 4 after lay. A genital ridge forms on the ventral surface of the embryonic kidney, the mesonephros, and develops into an elongated structure that extends for approximately half the length of the embryo. At day 5 the gonads are morphologically identical in males and females and have not yet developed the specialized structures in which the male and female germ cells will mature. Over the subsequent 4 days the development of male and female gonads diverges: germ cells migrate through the bloodstream to the gonads and become associated with granulosa cells in the cortex of the female ovary or with Sertoli cells in the medullary sex cords of the male testis (Smith and Sinclair, 2004; Bellairs and Osmond, 2005). There is a programmed sequence of gene expression correlated with the morphological changes of the gonads. Many of the genes that are differentially expressed in ovary and testis (such as Sox9, AMH, WT1 and aromatase) appear to have conserved roles during gonadogenesis in mammals and birds (Clinton and Haines, 2001; Smith and Sinclair, 2004).

At the beginning of gonadal development, right and left gonads are symmetrical. In males this symmetry is maintained as the gonads thicken and form testes while in females the right ovary regresses so that by day 9 it is clearly smaller than the left. This gonadal asymmetry is the first clear morphological difference between males and females.

The full period of development from laying to hatch is 21 days. The major events of organogenesis, including the development of ovaries and testes, take place in the first half of this period. Although the developmental signals that direct tissue development toward testes in males and ovaries in females are unknown, this process is clearly dependent on the sex chromosome composition. In chickens, males have two Z chromosomes and females have one Z and one W chromosome. The W chromosome is much smaller than the Z and has only a small region that pairs with the Z chromosome at meiosis. Genomic analysis reveals that the W chromosome has a relatively poor complement of genes compared with the Z and that much of its length comprises short repeat sequences that are not transcribed.

SEX SELECTION IN THE POULTRY INDUSTRY

For most breeds, day-old male and female chicks are virtually indistinguishable. There are slight differences in the appearance of the cloacal vent, but only highly trained specialists are able to reliably sex day-old chicks on this basis (Martin, 1994). In certain breeds feather growth rate and colour are sex-linked traits that can be used to identify sex; however, this does not avoid the problem of culling one or other sex at hatch.

The ability to identify the sex of individuals at an early stage of embryonic development would begin to address the welfare and environmental issues associated with layer production. To be practical, this would require molecular methods that can distinguish sex on the basis of a small number of cells and the means to make such a procedure automated and high-throughput. In reality, the molecular methods and the necessary automated equipment are already available; however, these would have to be adapted to fit in with current industrial practices.

MOLECULAR METHODS OF SEX IDENTIFICATION

Research laboratories already make use of sex chromosome differences to identify the sex of early embryos. Only a very small amount of material is required, for example a few μl of blood or a few cells from a feather pulp suspension. These methods utilize nucleic acid probes or primers that hybridize to specific regions of the Z and W chromosomes.

The following are three examples of some of the many methods available that require only minimal quantities of starting material.

Fluorescent *in situ* hybridization (FISH)

Metaphase spreads of somatic cells allow chromosomes in individual cells to be visualized under the microscope. The spreads are examined following hybridization with W and Z chromosome probes that have been labelled with different colours – for example, red for a W-specific sequence and blue for a Z-specific sequence (Tone *et al.*, 1982). Male cells would display chromosome spreads with two blue spots while female cells would show one blue and one red spot.

Polymerase chain reaction (PCR)

In PCR reactions, DNA primers can be used to amplify targeted regions of the Z and W chromosomes. In standard PCR reactions, these products are then visualized following agarose gel electrophoresis, while in quantitative PCR these products are monitored by the accumulation of fluorescence. The W chromosome contains multiple copies of a short, non-coding DNA fragment known as the XhoI repeat, and a number of PCR sexing assays exploit this to generate a strong female-specific product of a particular size (Clinton *et al.*, 2001). Variations of this principle can be applied to other sequence differences between the Z and W chromosomes, to provide several options for PCR assays in sex diagnosis (e.g. Petitte and Kegelmeyer, 1995; Griffiths *et al.*, 1998).

Invader® assay

DNA is incubated with two primers designed to hybridize in an overlapping fashion to the same target region (see Third Wave Technologies, http://www.twt.com). When the target is present in a sample, the probes generate a structure that is cleaved by a specific enzyme. This reaction releases a short 'invader' oligonucleotide that interacts with a special chromophore that is in turn cleaved, releasing a fluorescent signal. The probe is present in excess so continues to produce cleaved products and amplify the fluorescent signal when its target DNA sequence is present in the reaction mix. Z and W chromosome differences can be exploited to generate 'male' and 'female' signals.

For any sexing assay to be of practical interest to the poultry industry it must be robust, rapid, simple and inexpensive. In addition, it must be applicable during the existing short windows when eggs are removed from incubation, e.g. for 'candling' or vaccination. In reality, the Invader®-based assay is the only existing method that comes close to 'fitting the bill'. FISH analysis is technically demanding, time-consuming and limited to small numbers of samples. PCR is also technically demanding, prone to contamination, and even the most rapid versions require around 1 h to generate results. On the other hand, Invader® assays are not PCR-based, require only a water-bath and plate-reader and can sex chick embryos in 5–10 min.

The technology to take blood samples from young embryos in the egg is already available. The Embrex company has developed robots that can inject thousands of eggs per hour with vaccine (see http://www.embrex.com). The needle enters the egg, delivers vaccine and withdraws. The puncture in the shell is sealed when the needle is withdrawn. Although the trace amounts of blood or amniotic fluid that are retained in the needle could potentially be used to identify the sex of embryos, the equipment can also be adjusted to specifically collect samples from individual eggs. For example, the method has been successfully used to collect samples to determine the levels of circulating hormones in embryos between 13 and 18 days (Phelps *et al.*, 2003). The Invader® assay could be applied to similar samples and it is sufficiently sensitive to determine sex from 1 μl of blood.

CONCLUSIONS

Separate breeds of poultry are used to increase the efficiency of egg production in layer hens or weight gain and muscle yields in broiler and turkey males. To reduce the costs of rearing, birds of the unwanted sex may be culled at hatch, resulting in the mass slaughter of one-day chicks and poults. This represents a huge welfare issue, and the disposal of large amounts of animal waste also creates an environmental problem. A potential solution is to identify the sex of the embryos at an early stage and dispose of the unwanted eggs before the embryo is fully formed. This would require a rapid and inexpensive assay that could be applied in large numbers. The most promising candidate at the moment is the Invader® assay, which is a non-PCR-based molecular assay. With the correct probes it can be used to identify the sex of embryos within 10 min using 1 µl of blood. The technology for rapid sampling of eggs is available now and there is potential to combine these technologies to produce a rapid *in ovo* sexing assay.

REFERENCES

Bellairs, R. and Osmond, M. (2005) *Atlas of Chick Development*. Elsevier Academic Press, San Diego, CA.

Clinton, M. and Haines, L.C. (2001) An overview of factors influencing sex determination and gonadal development in birds. In: Scherer, G. and Schmid, M. (eds) *Genes and Mechanisms in Vertebrate Sex Determination*. Birkhaeuser, Basle, Switzerland, pp. 97–115.

Clinton, M., Haines, L., Belloir, B. and McBride, D. (2001) Sexing chick embryos: a rapid and simple protocol. *British Poultry Science* 42, 134–138.

Griffiths, R., Double, M.C., Orr, K. and Dawson, R.J. (1998) A DNA test to sex most birds. *Molecular Ecology* 7, 1071–1075.

Martin, R.D. (1994). *The Specialist Chick Sexer*. Bernal Publishing, Box Hill, UK.

Petitte, J.N. and Kegelmeyer, A.E. (1995). Rapid sex determination of chick embryos using the polymerase chain reaction. *Animal Biotechnology* 6, 119–130.

Phelps, P., Bhutada, A., Bryan, S., Chalker, A., Ferrell, B., Neuman, S., Ricks, C., Tran, H. and Butt, T. (2003) Automatic identification of male layer chicks prior to hatch. *World's Poultry Science Journal* 59, 33–38.

Smith, C.A. and Sinclair, A.H. (2004) Sex determination: insights from the chicken. *Bioessays* 26, 120–132.

Tone, M., Nakano, N., Takao, E., Narisawa, S. and Mizuno, S. (1982) Demonstration of W-chromosome-specific repetitive DNA-sequences in the domestic fowl, *Gallus g. domesticus*. *Chromosoma* 86, 551–569.

Wolpert, L., Beddington, R., Jessell, T.M., Lawrence, P., Meyerowitz, E.M. and Smith, J. (2001) *Principles of Development*. Oxford University Press, Oxford, UK.

Part III
Physiology of Reproduction

CHAPTER 6
Endocrinology and Genetics of the Hypothalamic–Pituitary–Gonadal Axis

I.C. Dunn,[1] N.A. Ciccone[2] and N.T. Joseph[1]

[1] *The Roslin Institute and Royal (Dick) School of Veterinary Sciences;*

[2] *Department of Veterinary Basic Sciences, The Royal Veterinary College, London*

ABSTRACT

Hypothalamic control of the pituitary–gonadal axis in broad terms operates at two levels. One mediates longer-term responses to maturational signalling, to environmental and behavioural factors such as changes in photoperiod and availability of food, and to the expression of incubation behaviour. This level of control is principally at gonadotrophin-releasing-hormone (GnRH-I) cell bodies in the hypothalamus, where GnRH-I synthesis occurs. The second level of control mediating shorter-term reproductive responses, for example the pre-ovulatory releases of luteinizing hormone (LH) or the first-day hormone response to photostimulation, may be principally at the median eminence (ME), where GnRH-I release occurs. Dopamine is emerging as a potential key regulator at both levels of control, being stimulatory at the GnRH-I cell body and inhibitory at the ME. Glial cell plasticity may play a key role at both levels of control.

At the level of the pituitary much has been achieved in understanding the control of follicle-stimulating hormone (FSH) secretion and synthesis. While it is clear that FSH secretion responds to feedback signals from the ovary, the mechanism responsible for the dramatic increase in circulating levels of FSH around puberty is not fully understood and may involve changing pulsatile patterns of GnRH-I release. GnRH-I signals through two receptors in the chicken, but further work is required to understand their respective functions in the regulation of gonadotrophin synthesis and release. Particularly exciting is the emerging role of the recently discovered gonadotrophin inhibitory hormone (GnIH), which is stimulating further research to understand how FSH and LH synthesis and release are controlled. A particular focus is on the role of GnIH in the regulation of common gondadotrophin α-subunit expression and consequent gonadotrophin secretion. This may lead to an understanding of how the complex functions of the ovary are regulated using a limited repertoire of primary hypothalamic signalling neuropeptides. Compared with egg-laying-type hens the reproductive neuroendocrine system in meat-type hens is abnormal, resulting in the requirement for feed restriction for satisfactory reproductive performance. In meat-type birds, feed restriction suppresses the activity of GnRH-I neurones. It is now possible to identify genes critical to the control of the hypothalamic–pituitary–gonadal axis, and the gene variation critical to the differences in activity of the axis. This, combined with further dissection of the genetics of reproduction, should result in new approaches to improving broiler breeder reproduction and ameliorating the negative effects of selection for muscle growth on ovarian function.

INTRODUCTION

The hypothalamus integrates signals from the environment, gonads and body to synchronize the development and maturation of the reproductive system through appropriate activation of gonadotrophin output from the pituitary gland to stimulate growth and maintain the functions of the gonads (Fig. 6.1). The success of the poultry industry and the power of genetic selection for desirable reproductive traits rely on the correct functioning of the hypothalamic–pituitary–gonadal axis (HPGA) to control ovulation rate, ovulation, semen production, incubation behaviour, the development of photorefractoriness and sexual senescence. Since the last time the HPGA was reviewed at a Poultry Science Symposium in 1984 (Scanes, 1984; Wilson and Cunningham, 1984) some aspects of this topic have been reviewed (Johnson and Wang, 1993; Norgren, 1996; Dunn and Millam, 1998; Decuypere *et al.*, 2002; Pawson and McNeilly, 2005; Bedecarrats *et al.*, 2006; Tsutsui *et al.*, 2006). Other contributors to this volume, describing aspects of ovarian development, photoperiodism and incubation, also refer to functions of the HPGA. In this review, more recent contributions to understanding the functioning of the axis in poultry and to new or emerging areas in this field will be addressed while avoiding specific topics dealt with elsewhere in this volume (Knight *et al.*, Chapter 7; Sharp, Chapter 11; Lewis, Chapter 14, this volume).

HYPOTHALAMIC CONTROL

The gonadotrophin-releasing hormone neurone

The chicken gonadotrophin-releasing hormone-I (cGnRH-I) neurone is recognized as the point of integration of signals from the external and internal environment to regulate reproduction through the functions of the HPGA. It produces a pulsatile output of cGnRH-I whose tempo defines sexual activity. cGnRH-I cell bodies are located in the pre-optic hypothalamus, to which they migrate from the presumptive nasal regions of the embryo. Neurones containing cGnRH-I are detectable as early as embryonic day 3 or 4 (Norgren and Lehman, 1991). The source of the neurones was thought to be the olfactory placode

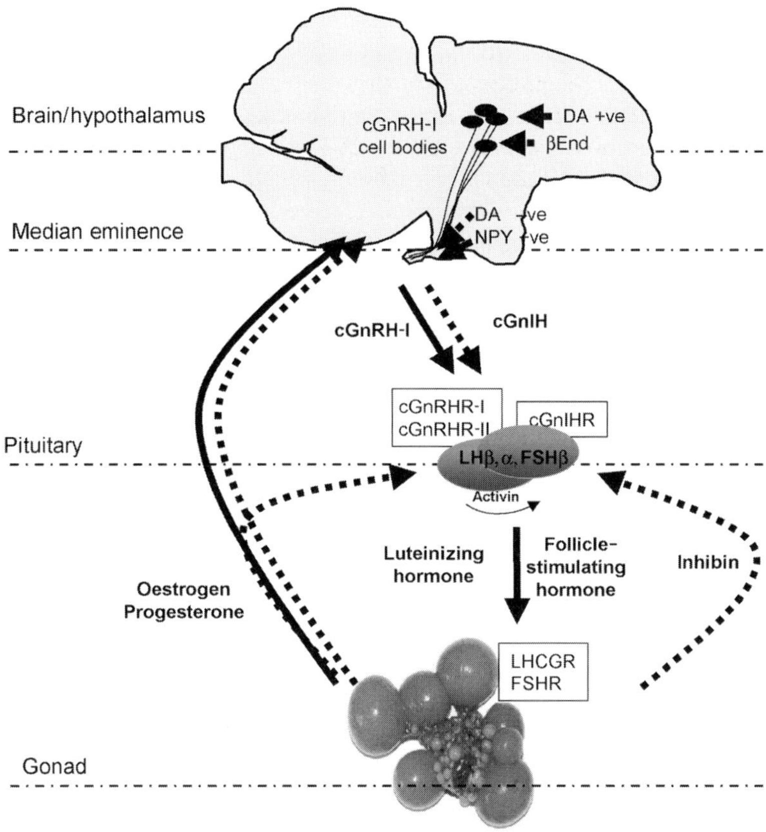

Fig. 6.1. Diagrammatic representation of the HPG axis showing the key components discussed in the chapter: hypothalamus, median eminence of the hypothalamus, pituitary and gonad. Key messenger molecules, neurotransmitters, endocrine hormones and paracrine factors are represented by their molecular abbreviation or full name and are unboxed. Secreted factors are in bold. Key receptors are shown in boxes. Factors with stimulatory effects are represented with solid lines and the factors with negative effects with a broken line. Abbreviations: cGnRH-I, chicken gonadotrophin-releasing hormone I; DA, dopamine; βEnd, beta endorphin; NPY, neuropeptide Y; cGnIH, chicken gonadotrophin inhibitory hormone; cGnRHR-I, chicken gonadotrophin-releasing hormone receptor I; cGnRHR-II, chicken gonadotrophin-releasing hormone receptor II; cGnIHR, chicken gonadotrophin inhibitory hormone receptor; LHβ, luteinizing hormone β subunit; α, common alpha subunit; FSHβ, follicle-stimulating hormone β subunit; LHCGR, luteinizing hormone receptor; FSHR, follicle-stimulating hormone receptor

itself, but the presumptive ectoderm of the nasal cavity was first suggested in the chicken (Elamraoui and Dubois, 1993), and this has more recently been supported by an *in situ* hybridization study of the development of cGnRH-I-expressing cells (Witkin *et al.*, 2003). Whatever their initial source, cGnRH-I neurones migrate from the olfactory epithelium along the olfactory nerve to reach the septal area of the hypothalamus by embryonic day 7 (Norgren,

1996), when cGnRH-I peptide concentrations peak; they then decrease and peak again at day 17 (Millam *et al.*, 1993). In the adult it has been observed that less than 12% of the cell bodies in the chicken reside in the pre-optic hypothalamus and therefore the majority of cGnRH-I cells are actually extra-hypothalamic (Kuenzel and Golden, 2006), suggesting that for birds the term HPGA may be technically incorrect.

Measurement of cGnRH-I mRNA as a marker of GnRH-I neuronal activity has made it possible to demonstrate that environmental changes or stimuli such as increased photoperiod (Dunn and Sharp, 1999; Kang *et al.*, 2006), the presence or removal of a nest (Dunn *et al.*, 1996) and an increase in food availability (Ciccone *et al.*, 2007) directly influence the cGnRH-I neurone. Both hypothalamic cGnRH-I peptide and mRNA are regulated by the inhibitory effects of gonadal steroids. In the pre-optic hypothalamus cGnRH-I peptide content is increased after castration in cockerels (Knight *et al.*, 1983; Sharp *et al.*, 1990; Sun *et al.*, 2001a), and this is associated with an increase in the release of cGnRH-I from the median eminence (Lal *et al.*, 1990). Castration increases the number of visible immunoreactive hypothalamic cGnRH-I neurones and the amount of cGnRH-I mRNA, and this is reversed after administration of oestrogen (Sharp *et al.*, 1994; Sun *et al.*, 2001a; Dunn *et al.*, 2003).

Role of cGnRH-I in key events in the reproductive cycle

Puberty

Knight and colleagues (Knight, 1983; Knight *et al.*, 1985) demonstrated that puberty in cockerels is accompanied by increases in hypothalamic cGnRH-I peptide content and capacity to release cGnRH-I from nerve terminals in the median eminence. This occurs in the face of increased inhibitory steroid feedback that reduces both cGnRH-I peptide content (Knight *et al.*, 1983; Lal *et al.*, 1990) and expression of cGnRH-I mRNA (Dunn and Sharp, 1999; Sun *et al.*, 2001a). In view of these inhibitory effects of steroids on hypothalamic cGnRH-I neurones, major stimulatory inputs appear to develop to increase the activity of cGnRH-I neurones in cockerels and hens at the onset of puberty. At this time, cGnRH-I neurones seem most sensitive to stimulatory inputs, and this may be a result of neural organizational changes resulting from increasing steroid levels. In the chicken, as observed in the ewe (Evans *et al.*, 1995), organizational changes in the hypothalamus correlated with sexual maturation may generate specific patterns of pulsatile cGnRH-I release that may be central to changes leading to sexual maturation. Increased cGnRH-I neuronal activity associated with precocious puberty in the chicken is correlated with an increase in dopamine neuronal activity, which may act directly or indirectly to increase the stimulatory tone on the cGnRH-I neurone (Fraley and Kuenzel, 1993). Rapid progress has been made in mammals to identify factors, such as kisspeptin, that stimulate the hypothalamic GnRH-I pulse generator to activate puberty (Fernandez-Fernandez *et al.*, 2006), but similar observations have not been reported in birds.

Photoperiod

The onset of sexual activity in the chicken is accelerated by increased photoperiod. It has long been recognized that changes in day length are perceived in the hypothalamus (Perera and Follett, 1992), and measurement of hypothalamic cGnRH-I mRNA content has demonstrated stimulation of cGnRH-I neurones after 7 days of exposure to long days in both somatically mature hens and cockerels (Dunn and Sharp, 1999) and within 3 days in quail (Baines *et al.*, 1999) and turkeys (Kang *et al.*, 2006). Similar results have been observed using cGnRH-I *in-situ* hybridization (Kuenzel and Golden, 2006) in photostimulated juvenile chickens treated with sulfamethazine, which induces premature puberty. However, an increase in luteinizing hormone (LH) secretion seen after photostimulation in juvenile cockerels was not associated with an increase in cGnRH-I mRNA, suggesting that there may be neural maturational changes required before an increase in mRNA can be detected. As demonstrated by Dunn and Sharp (1999), these changes may be induced by oestrogen. They may result in different patterns of release of cGnRH-I that favour FSH secretion and recruitment of follicles to the rapidly growing hierarchy (Dunn *et al.*, 2003). It seems likely that the reproductive effects of oestrogen are mediated through the alpha form (Table 6.1) of the oestrogen receptor (Ball *et al.*, 1999).

Ovulation

The pre-ovulatory release of LH is triggered by a pre-ovulatory release of cGnRH-I (Fraser and Sharp, 1978). This is induced by the positive feedback action of circulating progesterone derived from the dominant pre-ovulatory ovarian follicle (Wilson and Sharp, 1973). The amount of cGnRH-I in the hypothalamus and median eminence decreases after an exogenous progesterone-induced surge of LH (Johnson and Advis, 1985; Wilson *et al.*, 1990). However, during a spontaneous pre-ovulatory surge of LH, hypothalamic cGnRH-I peptide content increases (Johnson and Advis, 1985).

Incubation

When hens incubate eggs there is a rapid inhibition of the reproductive system. This inhibition is, in large part, mediated at the level of the cGnRH-I neurones, as cGnRH-I mRNA levels decrease dramatically at the onset of incubation. This depression in GnRH-I mRNA is reversed within 4 days of depriving an incubating hen of its eggs and nest box (Dunn *et al.*, 1996).

Food restriction

Food restriction is used as a tool in the broiler breeder industry to control reproductive function, specifically to produce an orderly pattern of ovarian hierarchical follicular growth. Food restriction does this is by altering the activity of cGnRH-I neurones. Chronic or short-term food restriction reduces the release of cGnRH-I from the median eminence (Lal *et al.*, 1990; Contijoch *et al.*, 1992), and cGnRH-I mRNA is lower in food-restricted hens than in hens fed *ad libitum* (Ciccone *et al.*, 2007).

Table 6.1. Names and standard acronyms for key genes in the hypothalamic–pituitary–gonadal axis showing the chromosome and position of the gene with accession number and a key reference. Data for position have been taken from the 2006 build of the chicken genome sequence (http://www.ensembl.org/Gallus_gallus/index.html or http://genome.ucsc.edu/cgi-bin/hgGateway).

Gene	Symbol	Chromosome	Position	Accession number	Reference
Vasoactive intestinal peptide	VIP	3	51,388,607–51,395,826	NM_205366	Talbot *et al.* (1995)
Neuropeptide Y	NPY	2	31,392,138–31,400,047	NM_205473	Blomqvist *et al.* (1992)
Gonadotrophin-releasing hormone I	GNRH-I	22	836,307–839,634	NM_001080877	Dunn *et al.* (1993)
Gonadotrophin-releasing hormone II	GNRH-II	4	92,044,252–92,047,790	AB194403	Ikemoto and Park (2006)
Gonadotrophin inhibitory hormone GnIH	NPVF	2	31,731,799–31,735,364	AB120325	Tsutsui *et al.* (2000)
Neuropeptide FF receptor 1 GnIH receptor	NPFFR-I	6	12,592,606–12,596,181	NM_204362	Ikemoto and Park (2005)
Gonadotrophin-releasing hormone receptor I	GnRH-RI	10	19,961,229–19,963,079	AJ304414	Sun *et al.* (2001b)
Gonadotrophin-releasing hormone receptor II	GnRH-RII	10	22,162,439–22,164,225	AY895154	Shimizu and Bedecarrats (2005)
Follicle-stimulating hormone beta polypeptide	FSHβ	5	4,580,528–4,583,901	NM_204257	Shen and Yu (2002)
Luteinizing hormone beta polypeptide	LHβ	NA	NA	S70834	Ando and Ishii (1994) (quail)
Glycoprotein hormone alpha polypeptide	GCA	3	79,311,472–79,320,910	BI395075	Predicted
Luteinizing hormone receptor	LHCGR	3	7,517,756–7,537,096	NM_204936	Johnson *et al.* (1996)
Follicle-stimulating hormone receptor	FSHR	3	7,353,248–7,430,911	U51097	You *et al.* (1996)
Oestrogen receptor 1 alpha	ESR1	3	50,936,183–51,041,281	NM_205183	Krust *et al.* (1986)
Oestrogen receptor 2 beta	ESR2	5	55,537,412–55,579,008	AB036415	Lakaye *et al.* (1998)
Progesterone receptor	PGR	1	187,457,391–187,496,309	NM_205262	Conneely *et al.* (1986)
Activin A receptor, type I	ACVR-I	7	37,894,083–37,907,942	NM_204560	Stern *et al.* (1995)
Activin A receptor, type IIA	ACVR-IIA	7	36,127,982–36,182,504	NM_205367	Stern *et al.* (1995)
Activin A receptor, type IIB	ACVR-IIB	2	5,662,067–5,680,476	NM_204317.1	Stern *et al.* (1995)
Inhibin, beta A	INHβA	2	50,699,071–50,711,410	NM_205396	Chen and Johnson (1996)
Inhibin, alpha	INHα	7	23,643,750–23,645,964	NM_001031257	Wang and Johnson (1993)
Inhibin beta B	INHβB	7	27,124,138–27,124,437	AF055478	Hecht *et al.* (2000)

Reproductive ageing

In ageing reproductively active domestic hens, circulating LH and ovarian function are diminished, but no decrease has been observed in hypothalamic cGnRH-I content (Sharp *et al.*, 1992). However, some studies have demonstrated less cGnRH-I in the median eminence (Contijoch *et al.*, 1992), and cGnRH-I levels in the hypothalamus of older hens are more susceptible to negative stimuli (Contijoch *et al.*, 1992). Compared with hens at peak of lay, hypothalamic cGnRH-I mRNA is not reduced in old laying or out-of-lay hens, despite large differences in follicle numbers and other markers of reproductive activity (Ciccone *et al.*, 2005). This suggests that changes in release, or the pattern of release, of GnRH-I are more critical to declining reproductive function as hens age than changes in hypothalamic cGnRH-I peptide or mRNA content.

Neurotransmitter and neuropeptide control of cGnRH-I

Many of the key changes in the avian reproductive cycle are associated with changes in the activity of cGnRH-I neurones, but progress in identifying the neurotransmitters and neuropeptides mediating these changes has been slow. Much of the evidence is from immunocytochemical observations on the anatomical relationships between cell bodies and fibres containing GnRH-I and neurotransmitters and neuropeptides, and is correlational rather than functional. Only the neural systems for which there is good evidence for an interaction with cGnRH-I neurones or fibres will be summarized.

Administration of the neurotransmitter *N*-methyl-D-aspartate (NMDA) stimulates LH release, showing that glutamate receptors are important activators of the HPGA. However, NMDA may act indirectly on cGnRH-I neurones since no increase in cFos, an indicator of cellular activation, was observed in cGnRH-I neurones after NMDA administration (Jozsa *et al.*, 1997).

Vasoactive intestinal polypeptide (VIP), which has a critical role in controlling prolactin release in birds, has been implicated in mediating photoperiodic effects on GnRH neurones in mammals. In quail, VIP neurones contact cGnRH-I cells; however, they are not the same VIP neurones that respond to changes in photoperiod (Teruyama and Beck, 2001). Nerve terminals containing cGnRH-I and VIP are in close proximity in the median eminence, suggesting the possibility of regulatory control. Opiates are thought to modulate cGnRH-I release in the chicken. The release of cGnRH-I from the median eminence is increased by the opiate antagonist, naloxone, and inhibited by β-endorphin at the time of, but not before, a spontaneous pre-ovulatory LH surge (Contijoch *et al.*, 1993b). There is evidence for synaptic contact between enkephalinergic-containing neurones and cGnRH-I cell bodies in the turkey (Millam *et al.*, 2002) and a sparse innervation by β-endorphin neurones of the internal zone of the median eminence (Contijoch *et al.*, 1993b).

Hypothalamic neuropeptide Y (NPY) gene expression is increased after food restriction (Boswell *et al.*, 1999) and may be involved in the reduction of cGnRH-I gene expression in food-restricted hens (Ciccone *et al.*, 2007).

Hypothalamic NPY may also be involved in the control of the ovulatory cycle at the level of the median eminence, which has a high density of NPY nerve terminals in the internal zone with some overlap with the external zone cGnRH-I terminals (Contijoch *et al.*, 1993a). NPY stimulates the release of cGnRH-I from the median eminence during the pre-ovulatory release of LH, and not at other times in the ovulatory cycle, and the NPY content of the median eminence decreases at the time of the pre-ovulatory LH surge (Contijoch *et al.*, 1993a). These observations on NPY and opiates in the regulation of ovulation suggest that they act at the level of the median eminence to stimulate the release of stored cGnRH-I peptide to generate the pre-ovulatory release of luteinizing hormone. It is suggested that progesterone from the mature pre-ovulatory ovarian follicle may stimulate NPY release from the median eminence, possibly by lifting an inhibitory action of opiates such as β-endorphin (Advis and Contijoch, 1993).

Dopamine may regulate GnRH-I release at the levels of both the hypothalamus and the pituitary. Pituitary LHβ mRNA increases when dopamine is infused into the third ventricle of the turkey brain, and this effect may be mediated by an increased release of GnRH-I (Bhatt *et al.*, 2003). A regulatory role for dopamine at the level of the cGnRH-I cell body is suggested by the observation that a group of dopamine neurones adjacent to cGnRH-I neurones are activated after photostimulation (Thayananuphat *et al.*, 2007). Immunocytochemical observations on relationships between dopaminergic and cGnRH-I nerve fibres in chickens in which early sexual maturation has been induced (Fraley and Kuenzel, 1993) or after acute food restriction (Contijoch *et al.*, 1992) suggest that GnRH-I release may also be mediated by dopamine in the median eminence.

The interaction between neurotransmitters or neuropeptides and cGnRH-I neurones may be modulated by glial cells. A role for glial cells in altering the input to cGnRH-I neurones, particularly after photostimulation, is suggested by experiments in quail, chickens and mammals. The expression of cFos, a marker of cellular activation, is upregulated after photostimulation in the quail basal hypothalamus, but interestingly also in glial cells in the median eminence (Meddle and Follett, 1997). It has been proposed in the quail that cGnRH-I is released after photostimulation in response to an increase in the local generation of triiodothyronine (T_3) in the basal hypothalamus. This increase in T_3 may induce withdrawal of glial ensheathment of GnRH-I terminals in the median eminence to allow GnRH-I to be released (Yamamura *et al.*, 2006). cDNA microarray studies of photoinduced changes in gene expression in the chicken hypothalamus suggest that heat shock protein 108 (HSP108), a marker of glial cells, is downregulated in the anterior hypothalamus after photostimulation (Graham *et al.*, 2003). This may imply a role for glial cells in mediating cGnRH-I cell function at the cell body, as has been suggested in mammals, where glial cells regulate the formation of synapses on GnRH neurones (Garcia-Segura and McCarthy, 2004).

PITUITARY

Gonadotrophes

The anterior pituitary contains gonadotrophes, the target cells for cGnRH-I released from the median eminence. The gonadotrophes synthesize and release luteinizing hormone (LH) and follicle-stimulating hormone (FSH), which are dimers of the hormone-specific β subunits (LHβ and/or FSHβ) non-covalently linked to a common α subunit (Table 6.1). In chickens, FSH and LH are synthesized in separate cell types (Proudman *et al.*, 1999; Puebla-Osorio *et al.*, 2002), providing a basis for an analysis of specific mechanisms controlling LH or FSH synthesis and release.

In the chicken, gonadotrophes producing FSH are identifiable from about embryonic day 7, while cells producing LH are observed from day 8 (Maseki *et al.*, 2004). They appear to be part of a functional axis which controls development of the gonads as oestrogen receptors are found in embryonic gonadotrophes (Liu and Cui, 2005) and embryonic FSH levels are reduced by oestrogen (Rombauts *et al.*, 1993). Functional removal of the embryonic pituitary results in a reduction in somatic, steroidogenic and germ cells in the ovary, and incubation of the embryonic ovary with FSH increases somatic cell proliferation (Peralta *et al.*, 2004; Sanchez-Bringas *et al.*, 2006). Gonadal development is therefore shaped at an early stage by the output of the embryonic gonadotrophes (Rombauts *et al.*, 1993).

Avian GnRH receptors

Fundamental to the regulation of reproduction is the interaction of the GnRH ligand with its corresponding receptor (cGnRH-R), which is assumed to be present on avian gonadotrophes. The binding and subsequent activation of the receptor result in the activation of second messenger pathways, which ultimately control the transcription of the gonadotrophin subunit genes. Two isoforms of GnRH (cGnRH-I and cGnRH-II) occur in the chicken hypothalamus (Dunn and Millam, 1998) and both stimulate the release of LH in the chicken *in vitro* (Millar *et al.*, 1986). Many *in vivo* studies have confirmed the ability of both cGnRH-I and -II to stimulate the release of LH in chickens (Proudman *et al.*, 2006). However, gonadotrophin synthesis is assumed to be a result of cGnRH-R activation by cGnRH-I, since GnRH-II is almost absent from the chicken median eminence (Sharp *et al.*, 1990).

The number of GnRH-Rs and GnRH ligands has decreased during the course of evolution, with multiple forms of GnRH and GnRH-Rs occurring in protochordates, teleost fish and amphibians, while mammals express only one or two GnRH ligands or receptors, and in some mammals there is specific silencing of either cGnRH-II or the type II GnRH receptor genes or both cGnRH-II and the type II GnRH receptor genes (Morgan and Millar, 2004; Sherwood *et al.*, 2006). cGnRH-RI is expressed relatively ubiquitously in the chicken (Sun *et al.*, 2001a), and pharmacological studies show that this receptor has a higher binding affinity and stimulates more inositol phosphate accumulation when activated by cGnRH-II than by cGnRH-I (Sun *et al.*, 2001b). This observation suggests that there may be another GnRH-receptor which has a higher binding affinity for GnRH-I. A second chicken G-protein-coupled receptor (cGnRH-RII) has been identified on chromosome 10 (Table 6.1), which is close to the location of cGnRH-RI (Table 6.1). The close proximity of the two receptors on chromosome 10 (approximately 0.6 Mb) suggests they may have been derived by gene duplication. The two cGnRH receptor isoforms share a sequence identity of 55%. Sequence comparison of the two cGnRH-Rs shows conservation of the aspartic acid and asparagine ligand-binding residues in transmembrane region 2. cGnRH-RI has a proline–glutamic acid–tyrosine (PEY) motif in the extracellular domain 3 whereas in cGnRH-RII this motif is proline–proline–serine (PPS). According to the scheme proposed by Millar *et al.* (2004), these motifs in extracellular loop 3 of each receptor type are receptor type 'classification motifs'.

The expression of cGnRH-RII mRNA in the pituitary correlates with the reproductive status of the domestic hen, suggesting that this receptor is likely to contribute to the regulation of reproduction at the level of the pituitary gland (Shimizu and Bedecarrats, 2006), while maximal levels of pituitary cGnRH-RI mRNA coincide with the LH surge prior to ovulation (Lovell *et al.*, 2005). Pituitary responsiveness to cGnRH-I and cGnRH-II is sexually differentiated in the domestic hen, with LH release being greater in the cockerel than in the hen. The potencies of cGnRH-I and -II are similar in the cockerel, while cGnRH-II is more potent than GnRH-I in releasing LH in the hen (Sharp *et al.*, 1987b). Differences in the control of cGnRH-RI and cGnRH-RII expression may result in the differential expression of the two receptors during the reproductive cycle and this may underlie the sexual differentiation of the LH response to cGnRH-I and cGnRH-II.

Avian GnRH receptor structure and function

Avian and other non-mammalian GnRH receptors, unlike mammalian receptors, have a cytoplasmic tail (C-terminal). The tail is involved in receptor-mediated internalization, ligand binding, receptor expression, phosphorylation and desensitization (Heding *et al.*, 1998). cGnRH-RII has a cytoplasmic C-terminal tail that is slightly longer than the C-terminal tail in cGnRH-RI. Regions in the cytoplasmic tail known to be critical for receptor internalization (Pawson *et al.*, 1998) are conserved in both receptors, as are regions in the cytoplasmic region in transmembrane region 3 and 7 (Arora *et al.*, 1996, 1997). The cGnRH-RI receptor internalizes through a dynamin-dependent mechanism (Pawson *et al.*, 2003). Desensitization of receptors is caused by sustained stimulation of the receptor and is a consequence of receptor phosphorylation which stabilizes association of β-arrestin with the receptor, resulting in the inhibition of G-protein binding and subsequent inactivation of effector proteins (McArdle *et al.*, 1999, 2002).

GnRH-Rs couple to G_q proteins causing an increase of phosphoinositide-specific phospholipase C (PLC) activity, resulting in the hydrolysis of phosphatidylinositol-4,5-biphosphate (PIP_2) and formation of inositol-1,4,5-trisphosphate and diacylglycerol with the subsequent mobilization of Ca^{2+} and phosphokinase C (PKC) activation (Ruf *et al.*, 2003). GnRH receptors can additionally couple to G_i and G_s proteins (Ruf *et al.*, 2003), which results in the inhibition and stimulation of adenylyl cyclase effector proteins, respectively. Activation of adenylyl cyclase results in production of the cAMP second messenger that activates protein kinase A (PKA), which phosphorylates cytoplasmic and nuclear proteins to regulate gene transcription through cAMP response elements. cGnRH-RI and cGnRH-RII have so far been demonstrated to signal through G_q proteins (Sun *et al.*, 2001b; Shimizu and Bedecarrats, 2005), but the activation of the other pathways has yet to be demonstrated for the avian receptors.

Follicle-stimulating hormone (FSH)

The recruitment and subsequent growth of small ovarian follicles to the rapidly growing hierarchy is dependent on their ability to acquire responsiveness to FSH (Palmer and Bahr, 1992; Hernandez and Bahr, 2003). However, the study of FSH has been relatively neglected in comparison with that of LH because of the lack of tools to measure the FSH protein or mRNA. An early chicken FSH radio immunoassay (Sakai and Ishii, 1983) appeared to cross-react with LH and therefore was not used extensively. More recently a more specific chicken FSH radioimmunoassay (Krishnan *et al.*, 1993) and competitive assay for FSHβ mRNA (Ciccone *et al.*, 2003) have been developed and used to elucidate how FSH synthesis and release are controlled.

A rapid increase in both pituitary (Bruggeman *et al.*, 1998a) and plasma FSH (Vanmontfort *et al.*, 1995; Bruggeman *et al.*, 1998a) coincides with the increase in ovarian follicle growth prior to the onset of lay. Similarly, the number of FSH cells visible using immunohistochemistry with an antibody to mammalian FSH increases in hens prior to the return to egg production after an induced pause (Chowdhury and Yoshimura, 2002). Oestrogen reduces pituitary FSH expression in chicken embryos (Rombauts *et al.*, 1993) and, in juvenile hens, oestrogen treatment decreases pituitary FSH content by 97% (Dunn *et al.*, 2003). This inhibitory effect of oestrogen on pituitary FSH could be mediated, at least in part, by suppression of hypothalamic cGnRH-I synthesis and release (Lal *et al.*, 1990; Dunn and Sharp, 1999), and/or by suppression of pituitary GnRH receptor expression (Sun *et al.*, 2001a). However, the depressive effect of oestrogen treatment on pituitary FSH but not pituitary LH content is ameliorated after photostimulation at 54 but not at 34 days of age (Dunn *et al.*, 2003). Further, photostimulation of 34-day-old hens treated with oestrogen increases plasma FSH but not plasma LH. This observation suggests that oestrogen may have organizational effects on the maturing hypothalamic–pituitary axis in the hen to regulate FSH secretory response to photostimulation.

In mammals, the synthesis of FSH is GnRH dependent (Padmanabhan *et al.*, 2002) and involves paracrine interactions between pituitary activin, a homodimer of inhibin βA (activin A), inhibin βB (activin B) the βAβB heterodimer (activin AB) and follistatin (Table 6.1). Activin stimulates FSH synthesis while follistatin blocks this action of activin by binding to its receptor (Padmanabhan *et al.*, 2002). Activin and follistatin mRNAs have been observed in the chicken pituitary by Ciccone *et al.* (2007) but were not detected by another group (Lovell *et al.*, 2005). Inhibin, a heterodimer of inhibin βA or inhibin βB and inhibin α (Table 6.1), inhibits FSH synthesis in mammals, and in birds is secreted from the ovary under the control of FSH (Davis *et al.*, 2001b), and is negatively correlated with plasma FSH levels (Lovell *et al.*, 2000b). In mammals, GnRH stimulates the synthesis of pituitary activin and follistatin, with follistatin synthesis being dependent on GnRH pulse frequency (Burger *et al.*, 2002). However, in the chicken, no experiments have been reported on the effects of activin and inhibin on FSH synthesis. In one study, treatment of cultured chicken pituitaries with pulses of GnRH-II increased follistatin mRNA levels, but this was not associated with changes in FSHβ expression (Ciccone *et al.*, 2007), while in another study *in vitro*, using more frequent pulses of cGnRH-I, FSHβ mRNA was increased (Shen and Yu, 2002). Active immunization against inhibin stimulates FSHβ and LHβ subunit gene expression in the turkey pituitary (Ahn *et al.*, 2001), and in the domestic hen stimulates gonadal development (Lovell *et al.*, 2000a, 2001) and increased reproductive performance (Satterlee *et al.*, 2002, 2006). In juvenile domestic hens active immunization against inhibin does not affect plasma FSH concentrations, and it was therefore concluded that the effect of active immunization against inhibin on ovarian function is mediated by interfering with the paracrine actions of inhibin in the ovary rather than at the level of the pituitary to regulate FSH secretion (Lovell *et al.*, 2000a, 2001). However, betaglycan (the accessory receptor for inhibin) and activin receptors (which also mediate inhibin's action) are present in the pituitary and co-localize with FSH cells (Lovell *et al.*, 2005; Sweeney and Johnson, 2005).

When hens naturally cease laying, circulating FSH and pituitary FSH*β* mRNA concentrations increase and circulating LH decreases (Ciccone *et al.*, 2005), as in hens after the onset of ovarian atrophy induced by feed deprivation (Vanmontfort *et al.*, 1994; Lovell *et al.*, 2000b). Similarly, surgical removal of the largest hierarchical follicles from laying hens also increases FSH secretion (Johnson *et al.*, 1993). This increase in FSH is ascribed to the lifting of the inhibitory ovarian feedback of oestrogen and/or inhibin from the developed gonad at the level of the hypothalamic–pituitary complex.

The likelihood that cGnRH-I regulates FSH secretion indirectly by controlling pituitary activin and follistatin may explain why GnRH injection does not increase plasma FSH (Krishnan *et al.*, 1993; Bruggeman *et al.*, 1998b; Dunn *et al.*, 2003). Further evidence that the mechanisms controlling FSH and LH secretion are different comes from the observation that pulses of LH and FSH secretion are not in phase (Vizcarra *et al.*, 2004).

FSH is clearly critical to the development and recruitment of follicles in the ovary, and the control of FSH release is almost certainly key to the onset of reproductive activity. As in mammals, it is not completely clear what factors control FSH synthesis and release in poultry. There is evidence that cGnRH-I is involved in modulating FSHβ expression, either directly or indirectly, and oestrogen certainly reduces FSHβ expression. However, it is still necessary to determine if and how the activin–inhibin system is involved, and there is still the possibility that other factors modulate the synthesis of the FSHβ subunit and/ or FSH secretion.

Luteinizing hormone (LH)

The control of LH secretion in the chicken is relatively well understood in comparison with FSH, with clear stimulatory effects of photostimulation (Dunn and Sharp, 1990) and cGnRH-I and -II (Sharp *et al.*, 1987a) and negative oestradiol (Dunn *et al.*, 2003) and positive progesterone feedback effects (Wilson and Sharp, 1975). Pituitary and plasma LH increase prior to the onset of laying (Bruggeman *et al.*, 1998a), as do visible LH cell numbers in the pituitary after an induced cessation in egg production (Chowdhury and Yoshimura, 2002). The pulsatile release of LH was described over 30 years ago in cockerels (Wilson and Sharp, 1973), and more recent studies have demonstrated that females also secrete LH, with pulses of about one per hour (Senthilkumaran *et al.*, 2006). Administration of cGnRH-I to turkeys showed that desensitization occurs with pulses at frequencies less than one per hour (Guemene and Williams, 1999).

The temporal relationships between the pre-ovulatory surges of plasma LH, progesterone and testosterone (Wilson and Cunningham, 1984) have been confirmed in the turkey, using a continuous sampling technique (Yang *et al.*, 1997). The functional significance of the pre-ovulatory release of testosterone has recently been demonstrated by showing that treatment of laying hens with a testosterone receptor antagonist blocks the pre-ovulatory releases of LH and progesterone (Rangel *et al.*, 2006).

Gonadotrophe common alpha subunit

The common alpha (α) subunit (Table 6.1) is generally ignored as a rate-limiting step in the secretion of gonadotrophins. However, it has been shown to be a good indicator of GnRH stimulation in man (Hayes *et al.*, 1999). In mammals, α-subunit expression is four- to fivefold in excess of the β-subunits (Papavasiliou *et al.*, 1986), but data collected in chickens (Ciccone *et al.*, 2004, 2005, 2007) show that the ratio is between 0.5 and 1.24 (Dunn *et al.*, 2006). In the chicken, α-subunit mRNA measurements reflect changes in the reproductive condition better than measurements of gonadotrophin β-subunit mRNAs. Thus, α-subunit mRNA levels change with age (Ciccone *et al.*, 2003), expression of incubation behaviour (Dunn *et al.*, 2006) and changes in food intake (Ciccone *et al.*, 2007) and, in *in vitro* experiments, α-subunit mRNA changes in response to pulsatile cGnRH-I and GnIH administration (Ciccone *et al.*, 2004). Finally, measurements of α, but not LHβ, mRNAs are directly correlated with plasma

LH concentrations and reflect changes in hypothalamic cGnRH-I mRNA content in ageing broiler breeders (Ciccone *et al.*, 2003). A similar observation was made in Japanese quail (Kobayashi and Ishii, 2002; Kobayashi *et al.*, 2002, 2004). It is therefore concluded that pituitary α-subunit mRNA is a better predictor of reproductive axis activity than the gonadotrophin β-subunit mRNAs and is correlated with plasma LH levels. It is not correlated with plasma FSH levels. The control of α-subunit expression may be important in birds in the control of circulating gonadotrophin levels, particularly LH.

Gonadotrophin inhibitory hormone (GnIH)

The observation that FSH and LH are differentially released has prompted a search for the regulatory factors that might be responsible for this difference. Gonadotrophin inhibitory hormone (GnIH) is one such recently discovered factor (Table 6.1). It is a RFamide (SIKPSAYLPLRF-NH_2) that is expressed in neurones in the hypothalamus (Tsutsui *et al.*, 2000) with nerve terminals in the median eminence (Ukena *et al.*, 2003). Functional receptors for the peptide have also been cloned, although they may be promiscuous for other RFamides (Ikemoto and Park, 2005; Yin *et al.*, 2005). GnIH depresses FSHβ and gonadotrophin α-subunit mRNAs in chicken pituitaries *in vitro*, associated with a depression in FSH and LH secretion (Ciccone *et al.*, 2005). LHβ subunit mRNA concentration was not affected by acute GnIH treatment (Ciccone *et al.*, 2005) but may be affected by chronic GnIH treatment (Ubuka *et al.*, 2006). GnIH neurones contain melatonin receptor, and melatonin treatment depresses GnIH neuronal function (Ubuka *et al.*, 2005). GnIH mRNA increases in incubating hens and may be responsible, in part, for the associated depression in plasma LH; however, GnIH mRNA is not increased in out-of-lay hens at the end of a laying year (Ciccone *et al.*, 2004). Clearly GnIH has the potential to play a role in the control of gonadotrophin secretion, but more research is needed before its role in the regulation of the HPGA of poultry is firmly established.

Other forms of GnRH

The question of a role for cGnRH-II in the control of reproduction has been unanswered since it was first discovered (Miyamoto *et al.*, 1984). Although evidence is lacking in poultry, the peptide has been implicated in diverse roles in mammals, including mediating signalling of nutritional status to the brain, paracrine roles in the ovary and effects on reproductive behaviour (Kauffman, 2004). However, it is not believed to control reproduction in birds (Sharp *et al.*, 1990) or mammals directly (Gault *et al.*, 2003). The molecular cloning of cGnRH-II (Table 6.1, Ikemoto and Park, 2006) opens up new possibilities for research on its function in poultry. The chicken GnRH-II gene is unusual amongst GnRH genes so far cloned in that it shares exon sequences with a ribosomal protein. There has also been speculation that a third form of GnRH, lamprey GnRH, may be present in the chicken and that it may be a specific

FSH-releasing factor, but this possiblity has been shown to be unlikely (Proudman *et al.*, 2006).

OVARY

Gonadotrophin receptors

Gonadotrophin receptors are located in the gonads, specifically the granulosa and theca cells in the ovary and the Sertoli cells in the testes. The expression of FSH receptor (Table 6.1) in the domestic hen is correlated with the develpment of ovarian follicular responsiveness to FSH, being highest in the granulosa cells of small yellow follicles (6–8 mm diameter), whilst expression in the theca remains relatively constant in all follicles (You *et al.*, 1996). The hierarchy of pre-ovulatory follicles is selected from the small yellow follicles and follicular selection may involve a decrease in epidermal growth factor in the selected follicle that lifts a suppressive effect on FSH-induced steroidogenesis in granulosa cells (Hernandez and Bahr, 2003).

LH increases the expression of the LH receptor (Table 6.1) but not the FSH receptor, and this effect is antagonized by activin A in follicles in the middle of the rapidly growing pre-ovulatory follicular hierarchy. Activin A may therefore act in a paracrine manner to control the responsiveness of these follicles to LH (Davis *et al.*, 2001a). Conversely, in prehierarchical follicles activin A increases the expression of LH receptor and, together with FSH, it induces FSH receptor expression (Johnson *et al.*, 2004). LH receptor is also induced in long-term granulosa cell cultures in the presence of FSH and LH (Johnson and Bridgham, 2001). *In vivo* LH receptor mRNA is expressed in the granulosa of hierarchical follicles, increasing as the follicles mature, and in theca cells from prehierarchical follicles, increasing until the F2 follicle (Johnson *et al.*, 1996). Increasing amounts of LH receptors in the granulosa cells of hierarchical follicles reflect an increased capacity to produce progesterone, which is greatest in the mature pre-ovulatory follicle.

GENETICS AND THE HYPOTHALAMIC–PITUITARY–GONADAL AXIS

Modern commercial poultry have been selected for the two key economic traits, meat or table egg production, which has resulted in meat-type and egg-type poultry. Both selection strategies have affected the reproductive system. In egg-type breeds, the age of puberty has been reduced, the persistency of egg production increased with almost no non-laying days (Preisinger and Flock, 2000), and the requirement for external cues, such as increased day length, to control reproductive activity is greatly reduced (Morris *et al.*, 1995). In meat-type birds the selection for growth and meat yield is associated with polyfollicular ovarian growth resulting in lost egg production (Hocking, 2004; Chapter 17, this volume). In egg-type poultry there are few, if any, problems reported in the reproductive system, but little is known of how selection has acted upon the endocrine function to improve egg production. In meat-type poultry considerable attention has been paid to understanding the causes of, and to improving, poor reproductive performance (Decuypere *et al.*, 2002; Siegel, 2002). The underlying problem may be in the ovary, where selection for somatic growth may have affected interactions between growth factors produced in the ovary that are required for the development of an orderly follicular hierarchy, or it may be at the level of the hypothalamic–pituitary axis, where selection for somatic growth may have affected the secretion of reproductive hormones and the feedback systems required to main endocrine homeostasis. It is likely that endocrine mechanisms in both the ovary and the hypothalamic–pituitary axis and the interaction between the two contribute to the poor reproductive performance of meat-type poultry. However, it is unlikely that the mechanism controlling ovulation is a contributory factor since the ovulatory cycle in meat-type turkey hens is similar to that in unselected hens (Buchanan *et al.*, 2002) and is not influenced by food restriction (Liu *et al.*, 2004).

The responsiveness of thecal tissue from meat-type hens to gonadotrophic stimuli is greater than that of thecal tissue from egg-type hens, indicating that selection for growth increases ovarian responsiveness to gonadotrophins and may be a contributory factor to the tendency for broiler breeders to have polyfollicular ovarian growth (Hocking and McCormack, 1995). However, concentrations of plasma gonadotrophins are not predictive of reproductive traits in broiler breeders (Lewis *et al.*, 2005). There is some evidence that selection has detrimentally altered the response to photoperiod in meat-type birds (Dunn and Sharp, 1990; Eitan and Soller, 2001) and is considered in more detail by Lewis (Chapter 14, this volume). Genetic selection has not affected hypothalamic NPY gene expression, which is thought to mediate the effect of food intake on reproduction, although NPY gene expression is affected by food restriction (Boswell *et al.*, 1999). Comparison of layers with dwarf broilers indicates higher LH levels in the dwarf broiler (Dunn and Sharp, 1990); however, in turkeys selected for a high growth rate with a high incidence of multiple ovulation, baseline circulating LH and oestrogen levels are lower than in control turkeys with normal, single ovulations (Buchanan *et al.*, 2002). This latter observation is surprising given that the mass of oestrogenic tissue was much greater in the turkey hens selected for high growth rate. This observation in turkeys has been confirmed in a comparison of broiler breeders from a fast-growing line and lines with better reproductive performance (Onagbesan *et al.*, 2006). In this experiment LH-stimulated progesterone production by cultured granulosa cells collected from the mature pre-ovulatory follicle was enhanced in birds subjected to food restriction and was increased further by the growth factors BMP-7 and IGF-I (Onagbesan *et al.*, 2006). A further growth factor, tumour necrosis factor α, originating from macrophages in the thecal layer of hierarchical follicles, is also implicated in modulating the actions of LH and IGF-I on granulosa cell progesterone production and proliferation (Onagbesan *et al.*, 2000).

Ultimately it is hoped that we can use genetic methods to identify the causative genes in broiler breeders that are responsible for polyfollicular ovaries.

This would allow precision selection against the trait whilst maintaining broiler traits, provided that identical gene polymorphisms are not responsible for both traits. Somewhat surprisingly indications are that the heritability of polyfollicular ovaries is close to zero (Hocking *et al.*, 2007), suggesting that the contribution of additive genetic variation is low. This leads us to assume that the genetic control is complex, perhaps comprising effects at interacting loci as well as dominance components. This is in line with the understanding from physiological studies which point to multiple aetiologies.

Although there are a number of quantitative trait loci (QTL) studies that have looked for QTL for reproduction (Hocking, 2005), there are none which suggest genes expressed in the HPGA, although we have demonstrated an association between markers in the cGnRH-RI gene and the production of double-yolked eggs, an indicator of polyfollicular ovaries, in a commercial broiler breeder line (Dunn *et al.*, 2004). However, in the same study, we did not confirm an association between markers in the growth hormone receptor and egg production, as shown in White Leghorns by Kuhnlein *et al.* (1997).

In contrast to the underlying aetiology of the problem of polyfollicular ovaries, it is becoming increasingly clear that food restriction ameliorates the deleterious effect of selection for meat-type traits on reproduction by decreasing the activity of the HPGA. This includes decreased expression of cGnRH-I mRNA in the hypothalamus and α-subunit mRNA in the pituitary (Ciccone *et al.*, 2007), and decreased cGnRH-I peptide in the median eminence and of LH and FSH in the pituitary.

SUMMARY AND CONCLUSIONS

Since the publication of the last review on the poultry HPGA in 1984, progress in understanding details of the axis has been facilitated in large part by cloning chicken genes expressed in the HPGA, and the sequencing of the chicken genome. We have become aware that, in broad terms, there appear to be two levels of control of the hypothalamus. One level of control may mediate longer-term or chronic changes in reproductive function, associated with, for example, the onset of puberty, the response to photostimulation and availability of food, or to the expression of incubation behaviour. This seems to act principally at the level of the cGnRH-I cell bodies in the hypothalamus. The second level of control mediates shorter-term or acute changes in reproductive function, such as the generation of the pre-ovulatory surge of LH and the initial hypothalamic response to photostimulation. The second level of control may act principally at the median eminence to regulate the release of GnRH-I. Dopamine is emerging as a potential key regulator of both mechanisms, being stimulatory at the GnRH-I cell body and inhibitory at the median eminence. Chronic changes in the functions of GnRH neurones involve changes in interactions with glial cells.

At the level of the pituitary it is clear that FSH synthesis and secretion are regulated by endocrine feedback signals from the ovary; however, it is far from clear what causes the dramatic increase in circulating levels of FSH around puberty. A change in patterns of GnRH-I pulsatile secretion is likely to be involved by altering pituitary paracrine interactions with gonadotrophes.

GnRH-I is now known to signal through two receptors in the chicken, and further work will be required to determine whether they are involved in the regulation of LH and FSH release from separate cell types and to identify the second messenger pathways they use. A particularly exciting development is the discovery of the emerging role of GnIH, which will stimulate further research into understanding how gonadotrophin release, and particularly FSH release, is controlled. This may involve further research on the relationship between the common α-subunit expression and gonadotrophe function and lead to a better understanding of how the complex functions of the ovary are controlled using a limited repertoire of primary signalling hormones.

The HPGA functions normally in egg-laying hens but in an abnormal manner in meat-type hens. In meat-type birds the environmental manipulation of the HPGA is currently the only tool available to sustain current production, but new genetic methods are likely to come from the sequencing of the chicken genome. One of the intriguing aspects of the avian HPGA is the apparent absence of components that are present in the mammalian HPGA. Specifically, in mammals, both leptin and kisspeptin regulate HPGA by mediating the effects of food intake on reproductive function (Fernandez-Fernandez *et al.*, 2006). However, genes encoding homologues of kisspeptin and leptin have not been found in the chicken genome. This may be because the genomic regions containing these genes have not yet been sequenced, or because birds and mammals do not use the same neuroendocrine and endocrine mechanisms in the metabolic control of reproductive activity. Even where a peptide involved in the regulation of metabolic responses in mammals, for example ghrelin, occurs in chickens it does not function in the same way as in mammals (Furuse *et al.*, 2001). It is therefore concluded that it cannot be assumed that mechanisms controlling the interactions between energy balance and the mammalian HPGA operate in poultry. In future it should be possible not only to better understand which genes are critical to control the poultry HPGA but to identify at the molecular level genetic variation responsible for breed differences in activity of the axis.

ACKNOWLEDGEMENT

ICD is supported by DEFRA and NTJ by a BBSRC studentship. NAC was supported by a BBSRC/COBB poultry studentship.

REFERENCES

Advis, J.P. and Contijoch, A.M. (1993) The median-eminence as a site for neuroendocrine control of reproduction in hens. *Poultry Science* 72, 932–939.

Ahn, J., You, S., Kim, H., Chaiseha, Y. and El Halawani, M. (2001) Effects of active immunization with inhibin alpha subunit on reproductive characteristics of turkey hens. *Biology of Reproduction* 65, 1594–1600.

Ando, H. and Ishii, S. (1994) Molecular-cloning of complementary deoxyribonucleic acids for the pituitary glycoprotein hormone alpha-subunit and luteinizing-hormone beta-subunit precursor molecules of Japanese quail (*Coturnix coturnix japonica*). *General and Comparative Endocrinology* 93, 357–368.

Arora, K.K., Cheng, Z.Y. and Catt, K.J. (1996) Dependence of agonist activation on an aromatic moiety in the DPLIY motif of the gonadotropin-releasing hormone receptor. *Molecular Endocrinology* 10, 979–986.

Arora, K.K., Cheng, Z.Y. and Catt, K.J. (1997) Mutations of the conserved DRS motif in the second intracellular loop of the gonadotropin-releasing hormone receptor affect expression, activation, and internalization. *Molecular Endocrinology* 11, 1203–1212.

Baines, E., Boswell, T., Dunn, I.C., Sharp, P.J. and Talbot, R.T. (1999) The effect of photostimulation on the levels of gonadotropin hormone releasing hormone (GnRH) mRNA in the hypothalamus of Japanese quail (*Coturnix coturnix japonica*). *Journal of Reproduction and Fertility* Abstract Series Number 24, 25.

Ball, G.F., Bernard, D.J., Foidart, A., Lakaye, B. and Balthazart, J. (1999) Steroid sensitive sites in the avian brain: does the distribution of the estrogen receptor alpha and beta types provide insight into their function? *Brain Behavior and Evolution* 54, 28–40.

Bedecarrats, G., Shimizu, M. and Guemene, D. (2006) Gonadotropin releasing hormones and their receptors in avian species. *The Journal of Poultry Science* 43, 199–214.

Bhatt, R., Youngren, O., Kang, S. and El Halawani, M. (2003) Dopamine infusion into the third ventricle increases gene expression of hypothalamic vasoactive intestinal peptide and pituitary prolactin and luteinizing hormone beta subunit in the turkey. *General and Comparative Endocrinology* 130, 41–47.

Blomqvist, A.G., Soderberg, C., Lundell, I., Milner, R.J. and Larhammar, D. (1992) Strong evolutionary conservation of neuropeptide-Y – sequences of chicken, goldfish, and Torpedo-Marmorata DNA clones. *Proceedings of the National Academy of Sciences of the USA B* 89, 2350–2354.

Boswell, T., Dunn, I.C. and Corr, S.A. (1999) Hypothalamic neuropeptide Y mRNA is increased after feed restriction in growing broilers. *Poultry Science* 78, 1203–1207.

Bruggeman, V., D'Hondt, E., Berghman, L., Onagbesan, O., Vanmontfort, D., Vandesande, F. and Decuypere, E. (1998a) The effect of food intake from 2 to 24 weeks of age on LHRH-I content in the median eminence and gonadotrophin levels in pituitary and plasma in female broiler breeder chickens. *General and Comparative Endocrinology* 112, 200–209.

Bruggeman, V., Onagbesan, O., Vanmontfort, D., Berghman, L., Verhoeven, G. and Decuypere, E. (1998b) Effect of long-term food restriction on pituitary sensitivity to cLHRH-I in broiler breeder females. *Journal of Reproduction and Fertility* 114, 267–276.

Buchanan, S., Robertson, G.W. and Hocking, P.M. (2002) Comparative changes in plasma concentrations of progesterone, oestradiol and LH during the ovulatory cycle in a multiple ovulating male line and a single ovulating traditional line of turkeys. *Reproduction* 123, 127–133.

Burger, L.L., Dalkin, A.C., Aylor, K.W., Haisenleder, D.J. and Marshall, J.C. (2002) GnRH pulse frequency modulation of gonadotropin subunit gene transcription in normal gonadotropes – assessment by primary transcript assay provides evidence for roles of GnRH and follistatin. *Endocrinology* 143, 3243–3249.

Chen, C.C. and Johnson, P.A. (1996) Molecular cloning of inhibin activin beta(a)-subunit complementary deoxyribonucleic acid and expression of inhibin activin alpha- and beta(a)-subunits in the domestic hen. *Biology of Reproduction* 54, 429–435.

Chowdhury, V.S. and Yoshimura, Y. (2002) Cell proliferation and apoptosis in the anterior pituitary of chicken during inhibition and resumption of laying. *General and Comparative Endocrinology* 125, 132–141.

Ciccone, N.A., Sharp, P.J., Wilson, P.W. and Dunn, I.C. (2003) Expression of neuroendocrine genes within the hypothalamic–pituitary–ovarian axis of end of lay broiler breeder hens. *British Poultry Science* 44, S11–S12.

Ciccone, N.A., Dunn, I.C., Boswell, T., Tsutsui, K., Ubuka, T., Ukena, K. and Sharp, P.J. (2004) Gonadotrophin inhibitory hormone depresses gonadotrophin alpha and follicle-stimulating hormone beta subunit expression in the pituitary of the domestic chicken. *Journal of Neuroendocrinology* 16, 999–1006.

Ciccone, N.A., Sharp, P.J., Wilson, P.W. and Dunn, I.C. (2005) Changes in reproductive neuroendocrine mRNAs with decreasing ovarian function in ageing hens. *General and Comparative Endocrinology* 144, 20–27.

Ciccone, N.A., Dunn, I.C., Boswell, T. and Sharp, P.J. (2007) Increased food intake stimulates GnRH-I, glycoprotein hormone alpha-subunit and follistatin mRNAs, and ovarian follicular numbers in laying hens. *Domestic Animal Endocrinology* 32, 155–166.

Conneely, O.M., Sullivan, W.P., Toft, D.O., Birnbaumer, M., Cook, R.G., Maxwell, B.L., Zaruckischulz, T., Greene, G.L., Schrader, W.T. and Omalley, B.W. (1986) Molecular cloning of the chicken progesterone receptor. *Science* 233, 767–770.

Contijoch, A.M., Gonzalez, C., Singh, H.N., Malamed, S., Troncoso, S. and Advis, J.P. (1992) Dopaminergic regulation of luteinizing-hormone-releasing hormone release at the median eminence level – immunocytochemical and physiological evidence in hens. *Neuroendocrinology* 55, 290–300.

Contijoch, A.M., Malamed, S., McDonald, J.K. and Advis, J.P. (1993a) Neuropeptide-Y regulation of LHRH release in the median eminence – immunocytochemical and physiological evidence in hens. *Neuroendocrinology* 57, 135–145.

Contijoch, A.M., Malamed, S., Sarkar, D.K. and Advis, J.P. (1993b) Beta-endorphin regulation of LHRH release at the median eminence level – immunocytochemical and physiological evidence in hens. *Neuroendocrinology* 57, 365–373.

Davis, A.J., Brooks, C.F. and Johnson, P.A. (2001a) Activin A and gonadotropin regulation of follicle-stimulating hormone and luteinizing hormone receptor messenger RNA in avian granulosa cells. *Biology of Reproduction* 65, 1352–1358.

Davis, A.J., Brooks, C.F. and Johnson, P.A. (2001b) Follicle-stimulating hormone regulation of inhibin alpha- and beta (B)-subunit and follistatin messenger ribonucleic acid in cultured avian granulosa cells. *Biology of Reproduction* 64, 100–106.

Decuypere, E., Bruggeman, V., Onagbesan, O. and Safi, M. (2002) Endocrine physiology of reproduction in the female chicken: old wine in new bottles. *Avian and Poultry Biology Reviews* 13, 145–153.

Dunn, I.C. and Millam, J.R. (1998) Gonadotropin releasing hormone: forms and functions in birds. *Poultry and Avian Biology Reviews* 9, 61–85.

Dunn, I.C. and Sharp, P.J. (1990) Photoperiodic requirements for LH-release in juvenile broiler and egg-laying strains of domestic chickens fed ad-libitum or restricted diets. *Journal of Reproduction and Fertility* 90, 329–335.

Dunn, I.C. and Sharp, P.J. (1999) Photo-induction of hypothalamic gonadotrophin releasing hormone-I mRNA in the domestic chicken: a role for oestrogen? *Journal of Neuroendocrinology* 11, 371–375.

Dunn, I.C., Chen, Y., Hook, C., Sharp, P.J. and Sang, H.M. (1993) Characterization of the chicken preprogonadotrophin-releasing hormone-I gene. *Journal of Molecular Endocrinology* 11, 19–29.

Dunn, I.C., Beattie, K.K., Maney, D., Sang, H.M., Talbot, R.T., Wilson, P.W. and Sharp, P.J. (1996) Regulation of chicken gonadotropin-releasing hormone-I mRNA in incubating, nest-deprived and laying Bantam hens. *Neuroendocrinology* 63, 504–513.

Dunn, I.C., Lewis, P.D., Wilson, P.W. and Sharp, P.J. (2003) Acceleration of maturation of FSH and LH responses to photostimulation in prepubertal domestic hens by oestrogen. *Reproduction* 126, 217–225.

Dunn, I.C., Miao, Y.W., Morris, A., Romanov, M.N., Wilson, P.W. and Waddington, D. (2004) A study of association between genetic markers in candidate genes and reproductive traits in one generation of a commercial broiler breeder hen population. *Heredity* 92, 128–134.

Dunn, I.C., Sharp, P.J. and Ciccone, N.A. (2006) Is gonadotrophin α-subunit expression a limiting step in avian gonadotrophin secretion? *Proceedings of the 10th European Conference of Comparative Endocrinology*. Manchester, 29 August–2 September, p. 96.

Eitan, Y. and Soller, M. (2001) Effect of photoperiod and quantitative feed restriction in a broiler strain on onset of lay in females and onset of semen production in males: a genetic hypothesis. *Poultry Science* 80, 1397–1405.

Elamraoui, A. and Dubois, P.M. (1993) Experimental evidence for an early commitment of gonadotropin-releasing-hormone neurons, with special regard to their origin from the ectoderm of nasal cavity presumptive territory. *Neuroendocrinology* 57, 991–1002.

Evans, N.P., Dahl, G.E., Mauger, D. and Karsch, F.J. (1995) Estradiol induces both qualitative and quantitative changes in the pattern of gonadotropin-releasing-hormone secretion during the presurge period in the ewe. *Endocrinology* 136, 1603–1609.

Fernandez-Fernandez, R., Martini, A.C., Navarro, V.M., Castellano, J.M., Dieguez, C., Aguilar, E., Pinilla, L. and Tena-Sempere, M. (2006) Novel signals for the integration of energy balance and reproduction. *Molecular and Cellular Endocrinology* 254, 127–132.

Fraley, G.S. and Kuenzel, W.J. (1993) Immunocytochemical and histochemical analyses of gonadotropin releasing hormone, tyrosine hydroxylase, and cytochrome oxidase reactivity within the hypothalamus of chicks showing early sexual maturation. *Histochemistry* 99, 221–229.

Fraser, H.M. and Sharp, P.J. (1978) Prevention of positive feedback in hen (*Gallus domesticus*) by antibodies to luteinizing-hormone releasing hormone. *Journal of Endocrinology* 76, 181–182.

Furuse, M., Tachibana, T., Ohgushi, A., Ando, R., Yoshimatsu, T. and Denbow, D.M. (2001) Intracerebroventricular injection of ghrelin and growth hormone releasing factor inhibits food intake in neonatal chicks. *Neuroscience Letters* 301, 123–126.

Garcia-Segura, L.M. and McCarthy, M.M. (2004) Minireview: role of glia in neuroendocrine function. *Endocrinology* 145, 1082–1086.

Gault, P.M., Maudsley, S. and Lincoln, G.A. (2003) Evidence that gonadotropin-releasing hormone II is not a physiological regulator of gonadotropin secretion in mammals. *Journal of Neuroendocrinology* 15, 831–839.

Graham, G., Boswell, T., Talbot, R., Wilson, P., Downing, A., Li, Q. and Sharp, P.J. (2003) Photoperiod and hypothalamic gene expression in the chicken – insights gained from microarrays. *Proceedings of the Annual Meeting of the British Society for Neuroendocrinology*. Manchester, 7–9 September, p. 10.

Guemene, D. and Williams, J.B. (1999) LH responses to chicken luteinizing hormone-releasing hormone I and II in laying, incubating, and out of lay turkey hens. *Domestic Animal Endocrinology* 17, 1–15.

Hayes, F.J., McNicholl, D.J., Schoenfeld, D., Marsh, E.E. and Hall, J.E. (1999) Free alpha-subunit is superior to luteinizing hormone as a marker of gonadotropin-releasing hormone despite desensitization at fast pulse frequencies. *Journal of Clinical Endocrinology and Metabolism* 84, 1028–1036.

Hecht, D.J., Davis, A.J., Brooks, C.F. and Johnson, P.A. (2000) Molecular cloning and expression analysis of the complementary deoxyribonucleic acid for chicken inhibin/activin beta(B) subunit. *Biology of Reproduction* 62, 1128–1134.

Heding, A., Vrecl, M., Bogerd, J., McGregor, A., Sellar, R., Taylor, P.L. and Eidne, K.A. (1998) Gonadotropin-releasing hormone receptors with intracellular carboxyl-terminal tails undergo acute desensitization of total inositol phosphate production and exhibit accelerated internalization kinetics. *Journal of Biological Chemistry* 273, 11472–11477.

Hernandez, A.G. and Bahr, J.M. (2003) Role of FSH and epidermal growth factor (EGF) in the initiation of steroidogenesis in granulosa cells associated with follicular selection in chicken ovaries. *Reproduction* 125, 683–691.

Hocking, P.M. (2004) Roles of body weight and feed intake in ovarian follicular dynamics in broiler breeders at the onset of lay and after a forced molt. *Poultry Science* 83, 2044–2050.

Hocking, P.M. (2005) Review of QTL mapping results in chickens. *World's Poultry Science Journal* 61, 215–226.

Hocking, P.M. and McCormack, H.A. (1995) Differential sensitivity of ovarian follicles to gonadotropin stimulation in broiler and layer lines of domestic fowl. *Journal of Reproduction and Fertility* 105, 49–55.

Hocking, P., Wilson, P., Robertson, G. and Dunn, I. (2007) Heritability of body weight, organ size and ovarian follicle numbers at the onset of lay in a broiler × layer advanced intercross. *British Poultry Science Abstracts* 3, 12.

Ikemoto, T. and Park, M.K. (2005) Chicken RFamide-related peptide (GnIH) and two distinct receptor subtypes: identification, molecular characterization, and evolutionary considerations. *Journal of Reproduction and Development* 51, 359–377.

Ikemoto, T. and Park, M.K. (2006) Molecular and evolutionary characterization of the GnRH-II gene in the chicken: distinctive genomic organization, expression pattern, and precursor sequence. *Gene* 368, 28–36.

Johnson, A.L. and Advis, J.P. (1985) Changes in luteinizing-hormone-releasing hormone content of discrete hypothalamic areas associated with spontaneous and induced preovulatory luteinizing-hormone surges in the domestic hen. *Biology of Reproduction* 32, 813–819.

Johnson, A.L. and Bridgham, J.T. (2001) Regulation of steroidogenic acute regulatory protein and luteinizing hormone receptor messenger ribonucleic acid in hen granulosa cells. *Endocrinology* 142, 3116–3124.

Johnson, A.L. and Wang, S.Y. (1993) The molecular biology and endocrinology of inhibin in the domestic hen. In: Sharp, P.J. (ed.) *Avian Endocrinology. Journal of Endocrinology*, Bristol, pp. 297–308.

Johnson, A.L., Bridgham, J.T. and Wagner, B. (1996) Characterization of a chicken luteinizing hormone receptor (cLH-R) complementary deoxyribonucleic acid, and expression of cLH-R messenger ribonucleic acid in the ovary. *Biology of Reproduction* 55, 304–309.

Johnson, A.L., Bridgham, J.T. and Woods, D.C. (2004) Cellular mechanisms and modulation of activin A and transforming growth factor beta-mediated differentiation in cultured hen granulosa cells. *Biology of Reproduction* 71, 1844–1851.

Johnson, P.A., Brooks, C., Wang, S.Y. and Chen, C.C. (1993) Plasma concentrations of immunoreactive inhibin and gonadotropins following removal of ovarian follicles in the domestic hen. *Biology of Reproduction* 49, 1026–1031.

Jozsa, R., Mess, A., Gladwell, R.T., Cunningham, F.J. and Sharp, P.J. (1997) The stimulatory action of glutamate agonist, N-methyl aspartate, on luteinizing hormone release in the cockerel with immunocytochemical observations on its mode of action. In: Korf, H.W. and Usadel, K.H. (eds) *Neuroendocrinology Retrospectives and Perspectives*. Springer Verlag, Berlin, pp. 151–162.

Kang, S.W., Thayananuphat, A., Rozenbolm, I., Millam, J.R., Proudman, J.A. and El Halawani, M.E. (2006) Expression of hypothalamic GnRH-I mRNA in the female turkey at different reproductive states and following photostimulation. *General and Comparative Endocrinology* 146, 91–99.

Kauffman, A.S. (2004) Emerging functions of gonadotropin-releasing hormone II in mammalian physiology and behaviour. *Journal of Neuroendocrinology* 16, 794–806.

Knight, P.G. (1983) Variations in hypothalamic luteinizing-hormone releasing hormone content and release in vitro and plasma-concentrations of luteinizing-hormone and testosterone in developing cockerels. *Journal of Endocrinology* 99, 311–319.

Knight, P.G., Cunningham, F.J. and Gladwell, R.T. (1983) Concentrations of immunoreactive luteinizing hormone releasing hormone in discrete brain regions of the cockerel – effects of castration and testosterone replacement therapy. *Journal of Endocrinology* 96, 471–480.

Knight, P.G., Gladwell, R.T. and Cunningham, F.J. (1985) Release of LHRH *in vitro* and anterior pituitary responsiveness to LHRH *in vivo* during sexual maturation in pullets (*Gallus domesticus*). *Journal of Reproduction and Fertility* 74, 145–151.

Kobayashi, M. and Ishii, S. (2002) Effects of starvation on gonadotropin and thyrotropin subunit mRNA levels and plasma hormone levels in the male Japanese quail (*Coturnix coturnix japonica*). *Zoological Science* 19, 331–342.

Kobayashi, M., Cockrem, J.F. and Ishii, S. (2002) Effects of starvation and refeeding on gonadotropin and thyrotropin subunit mRNAs in male Japanese quail. *Zoological Science* 19, 449–461.

Kobayashi, M., Ito, T., Ishii, S. and Wada, M. (2004) Seasonal change in luteinizing hormone subunit mRNA in Japanese quail and effects of short daylength and low temperature. *General and Comparative Endocrinology* 139, 38–47.

Krishnan, K.A., Proudman, J.A., Bolt, D.J. and Bahr, J.M. (1993) Development of an homologous radioimmunoassay for chicken follicle stimulating hormone and measurement of plasma FSH during the ovulatory cycle. *Comparative Biochemistry and Physiology A Physiology* 105, 729–734.

Krust, A., Green, S., Argos, P., Kumar, V., Walter, P., Bornert, J.M. and Chambon, P. (1986) The chicken estrogen receptor sequence – homology with V-Erba and the human estrogen and glucocorticoid receptors. *EMBO Journal* 5, 891–897.

Kuenzel, W.J. and Golden, C.D. (2006) Distribution and change in number of gonadotropin-releasing hormone-1 neurons following activation of the photoneuroendocrine system in the chick, *Gallus gallus*. *Cell and Tissue Research* 325, 501–512.

Kuhnlein, U., Ni, L., Weigend, S., Gavora, J.S., Fairfull, W. and Zadworny, D. (1997) DNA polymorphisms in the chicken growth hormone gene: response to selection for disease resistance and association with egg production. *Animal Genetics* 28, 116–123.

Lakaye, B., Foidart, A., Grisar, T. and Balthazart, J. (1998) Partial cloning and distribution of estrogen receptor beta in the avian brain. *Neuroreport* 9, 2743–2748.

Lal, P., Sharp, P.J., Dunn, I.C. and Talbot, R.T. (1990) Absence of an effect of naloxone, an opioid antagonist, on luteinizing hormone release *in vivo* and luteinizing hormone releasing hormone-I release *in vitro* in intact, castrated, and food restricted cockerels. *General and Comparative Endocrinology* 77, 239–245.

Lewis, P.D., Ciacciariello, M., Ciccone, N.A., Sharp, P.J. and Gous, R.M. (2005) Lighting regimens and plasma LH and FSH in broiler breeders. *British Poultry Science* 46, 349–353.

Liu, H.K., Lilburn, M.S., Koyyeri, B., Anderson, J.W. and Bacon, W.L. (2004) Preovulatory surge patterns of luteinizing hormone, progesterone, and estradiol-17 beta in broiler breeder hens fed *ad libitum* or restricted fed. *Poultry Science* 83, 823–829.

Liu, J.L. and Cui, S. (2005) Ontogeny of estrogen receptor (ER) alpha and its co-localization with pituitary hormones in the pituitary gland of chick embryos. *Cell and Tissue Research* 320, 235–242.

Lovell, T.M., Knight, P.G., Groome, N.P. and Gladwell, R.T. (2000a) Measurement of dimeric inhibins and effects of active immunization against inhibin alpha-subunit on plasma hormones and testis morphology in the developing cockerel. *Biology of Reproduction* 63, 213–221.

Lovell, T.M., Vanmontfort, D., Bruggeman, V., Decuypere, E., Groome, N.P., Knight, P.G. and

Gladwell, R.T. (2000b) Circulating concentrations of inhibin-related proteins during the ovulatory cycle of the domestic fowl (*Gallus domesticus*) and after induced cessation of egg laying. *Journal of Reproduction and Fertility* 119, 323–328.

Lovell, T.M., Knight, P.G., Groome, N.P. and Gladwell, R.T. (2001) Changes in plasma inhibin A levels during sexual maturation in the female chicken and the effects of active immunization against inhibin alpha-subunit on reproductive hormone profiles and ovarian function. *Biology of Reproduction* 64, 188–196.

Lovell, T.M., Knight, P.G. and Gladwell, R.T. (2005) Variation in pituitary expression of mRNAs encoding the putative inhibin co-receptor (betaglycan) and type-I and type-II activin receptors during the chicken ovulatory cycle. *Journal of Endocrinology* 186, 447–455.

Maseki, Y., Nakamura, K., Iwasawa, A., Zheng, J., Inoue, K. and Sakai, T. (2004) Development of gonadotropes in the chicken embryonic pituitary gland. *Zoological Science* 21, 435–444.

McArdle, C.A., Davidson, J.S. and Willars, G.B. (1999) The tail of the gonadotrophin-releasing hormone receptor: desensitization at, and distal to, G protein-coupled receptors. *Molecular and Cellular Endocrinology* 151, 129–136.

McArdle, C.A., Franklin, J., Green, L. and Hislop, J.N. (2002) Signalling, cycling and desensitisation of gonadotrophin-releasing hormone receptors. *Journal of Endocrinology* 173, 1–11.

Meddle, S.L. and Follett, B.K. (1997) Photoperiodically driven changes in Fos expression within the basal tuberal hypothalamus and median eminence of Japanese quail. *Journal of Neuroscience* 17, 8909–8918.

Millam, J.R., Faris, P.L., Youngren, O.M., Elhalawani, M.E. and Hartman, B.K. (1993) Immunohistochemical localization of chicken gonadotropin releasing hormone-I and hormone-II (cGnRH-I and cGnRH-II) in turkey hen brain. *Journal of Comparative Neurology* 333, 68–82.

Millam, J.R., Wang, R., Craig-Veit, C.B. and Siopes, T.D. (2002) Apposition of enkephalinergic axons with cGnRH I-containing perikarya in turkey hen brain. *General and Comparative Endocrinology* 127, 128–135.

Millar, R.P., Milton, R.C.D., Follett, B.K. and King, J.A. (1986) Receptor binding and gonadotropin releasing activity of a novel chicken gonadotropin releasing hormone ([His5, Trp7, Tyr8]GNRH) and a D-ARG6 analog. *Endocrinology* 119, 224–231.

Millar, R.P., Lu, Z.L., Pawson, A.J., Flanagan, C.A., Morgan, K. and Maudsley, S.R. (2004) Gonadotropin-releasing hormone receptors. *Endocrine Reviews* 25, 235–275.

Miyamoto, K., Hasegawa, Y., Nomura, M., Igarashi, M., Kangawa, K. and Matsuo, H. (1984) Identification of the 2nd gonadotropin releasing hormone in chicken hypothalamus – evidence that gonadotropin secretion is probably controlled by 2 distinct gonadotropin releasing hormones in avian species. *Proceedings of the National Academy of Sciences of the USA B* 81, 3874–3878.

Morgan, K. and Millar, R.P. (2004) Evolution of GnRH ligand precursors and GnRH receptors in protochordate and vertebrate species. *General and Comparative Endocrinology* 139, 191–197.

Morris, T.R., Sharp, P.J. and Butler, E.A. (1995) A test for photorefractoriness in high-producing stocks of laying pullets. *British Poultry Science* 36, 763–769.

Norgren, R.B. (1996) Development of LHRH neurons in the avian embryo. *Poultry and Avian Biology Reviews* 7, 163–182.

Norgren, R.B. and Lehman, M.N. (1991) Neurons that migrate from the olfactory epithelium in the chick express luteinizing hormone releasing hormone. *Endocrinology* 128, 1676–1678.

Onagbesan, O.M., Mast, J., Goddeeris, B. and Decuypere, E. (2000) Effect of TNF-alpha on LH and IGF-I modulated chicken granulosa cell proliferation and progesterone production during follicular development. *Journal of Reproduction and Fertility* 120, 433–442.

Onagbesan, O.M., Metayer, S., Tona, K., Williams, J., Decuypere, E. and Bruggeman, V. (2006) Effects of genotype and feed allowance on plasma luteinizing hormones, follicle-stimulating hormones, progesterone, estradiol levels, follicle differentiation, and egg production rates of broiler breeder hens. *Poultry Science* 85, 1245–1258.

Padmanabhan, V., Karsch, F.J. and Lee, J.S. (2002) Hypothalamic, pituitary and gonadal regulation of FSH. *Reproduction* Suppl. 59, 67–82.

Palmer, S.S. and Bahr, J.M. (1992) Follicle-stimulating-hormone increases serum oestradial-17-beta concentrations, number of growing follicles and yolk deposition in aging hens (*Gallus gallus domesticus*) with decreased egg production. *British Poultry Science* 33, 403–414.

Papavasiliou, S.S., Zmeili, S., Khoury, S., Landefeld, T.D., Chin, W.W. and Marshall, J.C. (1986) Gonadotropin releasing hormone differentially regulates expression of the genes for luteinizing hormone alpha and beta subunits in male rats. *Proceedings of the National Academy of Sciences of the USA B* 83, 4026–4029.

Pawson, A.J. and McNeilly, A.S. (2005) The pituitary effects of GnRH. *Animal Reproduction Science* 88, 75–94.

Pawson, A.J., Katz, A., Sun, Y.M., Lopes, J., Illing, N., Millar, R.P. and Davidson, J.S. (1998) Contrasting internalization kinetics of human and chicken gonadotropin-releasing hormone receptors mediated by C-terminal tail. *Journal of Endocrinology* 156, R9–R12.

Pawson, A.J., Maudsley, S.R., Lopes, J., Katz, A.A., Sun, Y.M., Davidson, J.S. and Millar, R.P. (2003) Multiple determinants for rapid agonist-induced internalization of a nonmammalian gonadotropin-releasing hormone receptor: a putative palmitoylation site and threonine doublet within the carboxyl-terminal tail are critical. *Endocrinology* 144, 3860–3871.

Peralta, I., Romano, M.C. and Velazquez, P.N. (2004) Proliferative and steroidogenic effects of follicle-stimulating hormone on cultured chick embryo testis cells. *Poultry Science* 83, 1193–1198.

Perera, A.D. and Follett, B.K. (1992) Photoperiodic induction *in vitro* – the dynamics of gonadotropin releasing hormone release from hypothalamic explants of the Japanese Quail. *Endocrinology* 131, 2898–2908.

Preisinger, R. and Flock, D.K. (2000) Genetic changes in layer breeding: historical trends and future prospects. *Occasional Publications of the British Society of Animal Science* 27, 20–28.

Proudman, J.A., Vandesande, F. and Berghman, L.R. (1999) Immunohistochemical evidence that follicle-stimulating hormone and luteinizing hormone reside in separate cells in the chicken pituitary. *Biology of Reproduction* 60, 1324–1328.

Proudman, J.A., Scanes, C.G., Johannsen, S.A., Berghman, L.R. and Camp, M.J. (2006) Comparison of the ability of the three endogenous GnRHs to stimulate release of follicle-stimulating hormone and luteinizing hormone in chickens. *Domestic Animal Endocrinology* 31, 141–153.

Puebla-Osorio, N., Proudman, J.A., Compton, A.E., Clements, K.E., Decuypere, E., Vandesande, F. and Berghman, L.R. (2002) FSH- and LH-cells originate as separate cell populations and at different embryonic stages in the chicken embryo. *General and Comparative Endocrinology* 127, 242–248.

Rangel, P.L., Sharp, P.J. and Gutierrez, C.G. (2006) Testosterone antagonist (flutamide) blocks ovulation and preovulatory surges of progesterone, luteinizing hormone and oestradiol in laying hens. *Reproduction* 131, 1109–1114.

Rombauts, L., Berghman, L.R., Vanmontfort, D., Decuypere, E. and Verhoeven, G. (1993) Changes in immunoreactive FSH and inhibin in developing chicken embryos and the effects of estradiol and the aromatase inhibitor R76713. *Biology of Reproduction* 49, 549–554.

Ruf, F., Fink, M.Y. and Sealfon, S.C. (2003) Structure of the GnRH receptor-stimulated signaling network: insights from genomics. *Frontiers in Neuroendocrinology* 24, 181–199.

Sakai, H. and Ishii, S. (1983) Radioimmunoassay of avian FSH. In: Mikami, S., Homma, K. and Wada, M. (eds) *Avian Endocrinology*. Japan Scientific Societies Press, Tokyo, pp. 125–134.

Sanchez-Bringas, G., Salazar, O., Pedernera, E. and Mendez, C. (2006) Follicle-stimulating hormone treatment reverses the effect of hypophysectomy on cell proliferation in the chicken embryo ovary. *General and Comparative Endocrinology* 149, 134–140.

Satterlee, D.G., Cadd, G.G. and Fioretti, W.C. (2002) Active immunization of broiler breeder hens with a recombinant chicken inhibin fusion protein enhances egg lay. *Poultry Science* 81, 519–528.

Satterlee, D.G., Castille, S.A. and Fioretti, W.C. (2006) Active immunization of broiler breeder cockerels against chicken inhibin accelerates puberty and prevents age-induced testicular involution. *Poultry Science* 85, 1087–1094.

Scanes, C.G. (1984) Hypothalamic, pituitary and gonadal hormones. In: Cunningham, F., Lake, P. and Hewitt, D. (eds) *Reproductive Biology of Poultry; 17th Poultry Science Symposium*. British Poultry Science Ltd, Abingdon, UK, pp. 1–14.

Senthilkumaran, C., Peterson, S., Taylor, M. and Bedecarrats, G. (2006) Use of a vascular access port for the measurement of pulsatile luteinizing hormone in old broiler breeders. *Poultry Science* 85, 1632–1640.

Sharp, P.J., Dunn, I.C. and Talbot, R.T. (1987a) Sex differences in the LH responses to chicken LHRH-I and LHRH- II in the domestic fowl. *Journal of Endocrinology* 115, 323–331.

Sharp, P.J., Dunn, I.C. and Talbot, R.T. (1987b) Sex differences in the LH responses to chicken LHRH-I and LHRH-II in the domestic fowl. *Journal of Endocrinology* 115, 323–331.

Sharp, P.J., Talbot, R.T., Main, G.M., Dunn, I.C., Fraser, H.M. and Huskisson, N.S. (1990) Physiological roles of chicken LHRH-I and LHRH-II in the control of gonadotropin release in the domestic chicken. *Journal of Endocrinology* 124, 291–299.

Sharp, P.J., Dunn, I.C. and Cerolini, S. (1992) Neuroendocrine control of reduced persistence of egg laying in domestic hens – evidence for the development of photorefractoriness. *Journal of Reproduction and Fertility* 94, 221–235.

Sharp, P., Dunn, I., Sang, H., Mess, A. and Gladwell, R. (1994) Inhibitory effects of steroids on gonadotrophin-releasing hormone-I mRNA and neurone numbers. *Neuroendocrinology* 60 (Suppl.1), 26.

Shen, S.T. and Yu, J.Y.L. (2002) Cloning and gene expression of a cDNA for the chicken follicle-stimulating hormone (FSH)-beta-subunit. *General and Comparative Endocrinology* 125, 375–386.

Sherwood, N.M., Tello, J.A. and Roch, G.J. (2006) Neuroendocrinology of protochordates: insights from Ciona genomics. *Comparative Biochemistry and Physiology A-Molecular and Integrative Physiology* 144, 254–271.

Shimizu, M. and Bedecarrats, G.Y. (2005) Identification and expression patterns of a novel gonadotropin releasing hormone receptor and its splice variants in the chicken. *Biology of Reproduction* 74, 106–106.

Shimizu, M. and Bedecarrats, G.Y. (2006) Identification of a novel pituitary-specific chicken gonadotropin-releasing hormone receptor and its splice variants. *Biology of Reproduction* 75, 800–808.

Siegel, P.B. (2002) Body weight, obesity, and reproduction in meat-type chickens. *Avian and Poultry Biology Reviews* 13, 209–213.

Stern, C.D., Yu, R.T., Kakizuka, A., Kintner, C.R., Mathews, L.S., Vale, W.W., Evans, R.M. and Umesono, K. (1995) Activin and its receptors during gastrulation and the later phases of mesoderm development in the chick embryo. *Developmental Biology* 172, 192–205.

Sun, Y.M., Dunn, I.C., Baines, E., Talbot, R.T., Illing, N., Millar, R.P. and Sharp, P.J. (2001a) Distribution and regulation by oestrogen of fully processed and variant transcripts of gonadotropin releasing hormone I and gonadotropin releasing hormone receptor mRNAs in the male chicken. *Journal of Neuroendocrinology* 13, 37–49.

Sun, Y.M., Flanagan, C.A., Illing, N., Ott, T.R., Sellar, R., Fromme, B.J., Hapgood, J., Sharp, P., Sealfon, S.C. and Millar, R.P. (2001b) A chicken gonadotropin-releasing hormone receptor that confers agonist activity to mammalian antagonists – identification of D-Lys(6) in the ligand and extracellular loop two of the receptor as determinants. *Journal of Biological Chemistry* 276, 7754–7761.

Sweeney, S.A. and Johnson, P.A. (2005) Messenger RNA and protein expression analysis of betaglycan in the pituitary and ovary of the domestic hen. *Biology of Reproduction* 72, 172–178.

Talbot, R.T., Dunn, I.C., Wilson, P.W., Sang, H.M. and Sharp, P.J. (1995) Evidence for alternative splicing of the chicken vasoactive intestinal polypeptide gene transcript. *Journal of Molecular Endocrinology* 15, 81–91.

Teruyama, R. and Beck, M.M. (2001) Double immunocytochemistry of vasoactive intestinal peptide and cGnRH-I in male quail: photoperiodic effects. *Cell and Tissue Research* 303, 403–414.

Thayananuphat, A., Kang, S.W., Bakken, T., Millam, J.R. and El Halawani, M. (2007) Rhythmic dependent light induction of gonadotrophin-releasing hormone-I expression and activation of dopaminergic neurones within the premammillary nucleus of the turkey hypothalamus. *Journal of Neuroendocrinology* 19, 399–406.

Tsutsui, K., Saigoh, E., Ukena, K., Teranishi, H., Fujisawa, Y., Kikuchi, M., Ishii, S. and Sharp, J.P. (2000) A novel avian hypothalamic peptide inhibiting gonadotropin release. *Biochemical and Biophysical Research Communications* 275, 661–667.

Tsutsui, K., Ubuka, T., Yin, H., Osugi, T., Ukena, K., Bentley, G.E., Ciccone, N., Inoue, K., Chowdhury, V.S., Sharp, P.J. and Wingfield, J.C. (2006) Mode of action and functional significance of avian gonadotropin-inhibitory hormone (GnIH): a review. *Journal of Experimental Zoology A – Comparative Experimental Biology* 305A, 801–806.

Ubuka, T., Bentley, G.E., Ukena, K., Wingfield, J.C. and Tsutsui, K. (2005) Melatonin induces the expression of gonadotropin-inhibitory hormone in the avian brain. *Proceedings of the National Academy of Sciences of the USA B* 102, 3052–3057.

Ubuka, T., Ukena, K., Sharp, P.J., Bentley, G.E. and Tsutsui, K. (2006) Gonadotropin-inhibitory hormone inhibits gonadal development and maintenance by decreasing gonadotropin synthesis and release in male quail. *Endocrinology* 147, 1187–1194.

Ukena, K., Ubuka, T. and Tsutsui, K. (2003) Distribution of a novel avian gonadotropin-inhibitory hormone in the quail brain. *Cell and Tissue Research* 312, 73–79.

Vanmontfort, D., Berghman, L.R., Rombauts, L., Verhoeven, G. and Decuypere, E. (1994) Changes of immunoreactive inhibin, follicle-stimulating-hormone, luteinizing-hormone, and progesterone in plasma after short-term food deprivation and during the ovulatory cycle of the domestic hen. *General and Comparative Endocrinology* 95, 117–124.

Vanmontfort, D., Berghman, L.R., Rombauts, L., Verhoeven, G. and Decuypere, E. (1995) Developmental changes in immunoreactive inhibin and FSH in plasma of chickens from hatch to sexual maturity. *British Poultry Science* 36, 779–790.

Vizcarra, J.A., Kreider, D.L. and Kirby, J.D. (2004) Episodic gonadotropin secretion in the mature fowl: serial blood sampling from unrestrained male broiler breeders (*Gallus domesticus*). *Biology of Reproduction* 70, 1798–1805.

Wang, S.Y. and Johnson, P.A. (1993) Complementary deoxyribonucleic acid cloning and sequence analysis of the alpha subunit of inhibin from chicken ovarian granulosa cells. *Biology of Reproduction* 49, 453–458.

Wilson, S. and Cunningham, F.J. (1984) Endocrine control of the ovulation cycle. In: Cunningham, F., Lake, P. and Hewitt, D. (eds) *Reproductive Biology of Poultry; 17th Poultry Science Symposium*. British Poultry Science Ltd, Abingdon, UK, pp. 29–49.

Wilson, S.C. and Sharp, P.J. (1973) Variations in plasma LH levels during the ovulatory cycle of the hen, *Gallus domesticus*. *Journal of Reproduction and Fertility* 35, 561–564.

Wilson, S.C. and Sharp, P.J. (1975) Changes in plasma concentrations of luteinizing-hormone after injection of progesterone at various times during the ovulatory cycle of domestic hen (*Gallus domesticus*). *Journal of Endocrinology* 67, 59–70.

Wilson, S.C., Chairil, R.A., Cunningham, F.J. and Gladwell, R.T. (1990) Changes in the hypothalamic contents of LHRH-I and LHRH-II and in pituitary responsiveness to synthetic chicken LHRH-I and LHRH-II during the progesterone-induced surge of LH in the laying hen. *Journal of Endocrinology* 127, 487–496.

Witkin, J.W., Dao, D., Livne, I., Dunn, I.C., Zhou, X.L., Pula, K. and Silverman, A.J. (2003) Early expression of chicken gonadotropin-releasing hormone-1 in the developing chick. *Journal of Neuroendocrinology* 15, 865–870.

Yamamura, T., Yasuo, S., Hirunagi, K., Ebihara, S. and Yoshimura, T. (2006) T-3 implantation mimics photoperiodically reduced encasement of nerve terminals by glial processes in the median eminence of Japanese quail. *Cell and Tissue Research* 324, 175–179.

Yang, J.Y., Long, D.W. and Bacon, W.L. (1997) Changes in plasma concentrations of luteinizing hormone, progesterone, and testosterone in turkey hens during the ovulatory cycle. *General and Comparative Endocrinology* 106, 281–292.

Yin, H., Ukena, K., Ubuka, T. and Tsutsui, K. (2005) A novel G protein-coupled receptor for gonadotropin-inhibitory hormone in the Japanese quail (*Coturnix japonica*): identification, expression and binding activity. *Journal of Endocrinology* 184, 257–266.

You, S., Bridgham, J.T., Foster, D.N. and Johnson, A.L. (1996) Characterization of the chicken follicle-stimulating hormone receptor (cFSH-R) complementary deoxyribonucleic acid, and expression of cFSH-R messenger ribonucleic acid in the ovary. *Biology of Reproduction* 55, 1055–1062.

Chapter 7

Control of Follicular Development: Intra-ovarian Actions of Transforming Growth Factor-β (TGF-β) Superfamily Members

P.G. Knight,[1] S.L. Al-Musawi,[2] T.M. Lovell[1] and R.T. Gladwell[1]

[1]*School of Biological Sciences, The University of Reading, UK;* [2]*Department of Veterinary Basic Sciences, The Royal Veterinary College, London*

ABSTRACT

Elucidation of the mechanisms that regulate the orderly maturation and ovulation of follicles in the hen ovary remains a major challenge for avian biologists. Whilst pituitary gonadotrophins have long been recognized as primary regulators of ovarian follicle function, an increasing number of intra-ovarian growth factors have been implicated as autocrine and paracrine modulators of follicle cell proliferation and differentiation. Amongst these are members of the transforming growth factor-β (TGF-β) superfamily, including activins, inhibins and bone morphogenetic proteins (BMPs). Recent studies have shown that mRNA transcripts encoding various TGF-β-related ligands, including inhibin-A, inhibin-B, activin-A, activin-B and BMP-2, -3, -4, -5, -6, -7, are expressed to different extents by granulosa (GC) and theca cells (TC) of prehierarchical and pre-ovulatory hen follicles. Similarly, there are differences in expression of various receptors and binding proteins for these ligands during follicle development. Evidence supporting a functional involvement of these signalling systems in follicle development has mainly come from *in vitro* studies on isolated chicken GC and TC. Within the hen follicle, most activin-A is synthesized by the TC layer; treatment of cultured GC with activin-A can promote 'basal' secretion of inhibin-A, upregulate expression of gonadotrophin receptors and potentiate gonadotrophin-induced secretion of progesterone and inhibin-A. This indicates a paracrine role for TC-derived activin-A in modulating GC function. Conversely, gonadotrophin treatment has been shown to modulate expression of inhibin and activin subunits, activin receptors and betaglycan (inhibin co-receptor) by GC and TC, suggesting a possible mechanism for fine-tuning follicle responsiveness to intra-ovarian activins and inhibins. BMP-4,

 Biology of Breeding Poultry (ed. P.M. Hocking)

-6 and -7 have also been shown to enhance gonadotrophin-induced progesterone secretion by hen GC, while BMP-6 has additionally been shown to upregulate inhibin-A secretion and expression of mRNAs for inhibin and activin subunits, gonadotrophin receptors and cytochrome P450 side-chain cleavage (P450scc). Despite these recent observations, when considered together with earlier reports that other locally produced growth factors (e.g. insulin-like growth factors, epidermal growth factor, transforming growth factor-alpha, basic fibroblast growth factor) can modulate GC proliferation and steroidogenesis, our understanding of the integrated endocrine, paracrine and autocrine control systems that coordinate avian follicle development is far from complete. The challenge remains to define the key intra-ovarian signalling networks that are obligatory for successful follicle progression and distinguish these from those that are dispensable.

INTRODUCTION

Ovarian follicles

The functional left ovary of a laying domestic hen is endowed with several thousand follicles at different developmental stages, ranging from 'small white' follicles (SWF; less than 1 mm), 'large white' follicles (LWF; 2–4 mm), 'small yellow' follicles (SYF; 4–8 mm) and 'large yellow' follicles (LYF; 8–40 mm). As follicles advance along this developmental continuum, taking some 2–3 weeks to grow from 1.5 to 40 mm, there is a progressive decrease in the number of follicles belonging to a particular size class (Perry *et al.*, 1983). Typically, the largest four to eight rapidly growing LYFs are arranged in a distinct size hierarchy that corresponds to the order in which they will ovulate; ovulation of the largest of these (termed F1) occurs at intervals of 24–27 h (Bahr and Johnson, 1984; Cunningham, 1987; Etches, 1996). Succeeding follicles (F2, F3, F4, etc.) then advance one place in the pre-ovulatory hierarchy and an additional follicle (6–8 mm diameter) is recruited from the SYF pool (Gilbert *et al.*, 1983; Etches and Schoch, 1984). The remaining follicle wall after ovulation is referred to as the post-ovulatory follicle.

The yolk-filled oocyte of a mature follicle is enclosed by a concentric arrangement of tissues constituting the follicle wall. Moving outwards, this comprises the vitelline membrane, perivitelline layer, granulosa layer, basement membrane, theca interna, theca externa and vascularized connective tissue representing an extension of the follicle stalk (Wyburn *et al.*, 1965). The germinal disc (GD) contains the female pronucleus and the majority of the oocyte's cytoplasmic organelles required for metabolic activity. A very thin layer of cytoplasm and vitelline membrane stretches around the yolk. The GD is visible as a white plaque of approximately 2–3 mm lying on the surface of a mature ovum; there is intimate contact between the GD and adjacent granulosa cells, indicative of functional interaction (Romanoff, 1960; Perry *et al.*, 1978).

The rapid growth of hierarchical follicles is associated with the extensive uptake of lipid-rich yellow yolk precursors and requires major remodelling of the tissues comprising the follicle wall (Gilbert, 1971). The delivery of blood-borne yolk precursors is dependent on a well-vascularized theca interna. The rate of granulosa cell proliferation in rapidly growing follicles of the pre-ovulatory hierarchy is much less than that in slow-growing 1–8 mm follicles (Tilly *et al.*, 1992). The majority of unselected prehierarchical follicles (less than 6 mm in diameter) eventually undergo atresia, attributed to apoptosis and cellular deletion, whereas follicles recruited to the pre-ovulatory hierarchy rarely undergo atresia (Gilbert *et al.*, 1983; Tilly *et al.*, 1992).

The laying hen as a model

Typically, oviposition precedes ovulation of the subsequent egg in the laying sequence by 15–30 min. Eggs belonging to one sequence ('clutch') are laid on consecutive days at progressively later times. An exception arises on a 'pause day', when following the terminal oviposition in a sequence there is no accompanying ovulation due to the absence of an LH surge (Cunningham, 1987; Etches, 1996). Thus, by monitoring ovipositions, follicles at known developmental stages relative to the time of ovulation can be readily identified. The anatomy of the hen follicle permits granulosa and theca cell layers to be easily separated from one another. Moreover, both cell types are amenable to cell culture, providing investigators with valuable model systems for functional *in vitro* studies to explore the actions and interactions of various signalling molecules on aspects of follicle cell function such as cell proliferation, steroidogenesis and apoptosis.

Regulation of follicle development and ovulation

Follicle development culminating in ovulation is primarily regulated by the temporally coordinated actions of hypothalamic gonadotrophin-releasing hormone-I (GnRH-I), pituitary gonadotrophins (LH, FSH) and ovarian steroids (androgens, oestrogens, progesterone), the secretion of which varies in a cyclic manner during the ovulatory cycle (Johnson and van Tienhoven, 1980; Bahr *et al.*, 1983; Cunningham *et al.*, 1984; Etches and Duke, 1984; Cunningham, 1987). Gonadotrophins bind to specific LH and FSH receptors on both TC and GC (Johnson *et al.*, 1996; You *et al.*, 1996). Gonadotrophin-dependent actions include the promotion of cell proliferation, steroidogenesis, peptide hormone production (e.g. inhibin), plasminogen activator production and ovulation (Bahr and Johnson, 1984; Cunningham, 1987; Tilly *et al.*, 1991a,b; Lovell *et al.*, 2002a).

The expression of gonadotrophin receptors (LH-R, FSH-R; Johnson *et al.*, 1996; You *et al.*, 1996) and steroidogenic capacity (Bahr *et al.*, 1983; Etches and Duke, 1984) of TC and GC varies during follicle development. There is a shift from FSH responsiveness to LH responsiveness as follicles are selected into the pre-ovulatory hierarchy (Tilly *et al.*, 1991a,b; Johnson *et al.*, 1996; You *et al.*, 1996), accompanied by a decline in expression of

P450 aromatase and oestrogen production by the theca externa. Expression of P450c17 and production of androgen by the theca interna fall sharply during the F2 to F1 transition and are accompanied by a rise in progesterone production by the GC. Indeed, progesterone secreted by the GC of the F1 follicle has a critical positive feedback action in promoting the pre-ovulatory LH surge responsible for inducing its own ovulation some 4 h later (Etches, 1996). However, the processes that coordinate and 'fine-tune' follicle progression, permitting only some follicles to attain ovulatory status, are poorly understood; they are thought to involve locally produced signals emanating from the ovary itself, including steroids and proteins.

Intra-ovarian growth factors and follicle development

Evidence has accrued from studies on both mammals and birds to implicate a range of local intra-ovarian growth factor signalling systems in the regulation of follicle development and ovulation. Amongst these are the insulin-like growth factor (IGF) system (Adashi, 1994; Onagbesan and Peddie, 1995; Fortune, 2003), the epidermal growth factor/transforming growth factor-α system (Li and Johnson, 1993; Onagbesan *et al.*, 1994; Peddie *et al.*, 1994; Woods *et al.*, 2005), basic fibroblast growth factor (Li and Johnson, 1993) and several members of the transforming growth factor-β (TGF-β) superfamily (reviews: Decuypere *et al.*, 1997; Shimasaki *et al.*, 2004; Juengel and McNatty, 2005; Knight *et al.*, 2005; Knight and Glister, 2006); the latter superfamily has been under the spotlight in recent years and provides the principal focus for the present chapter.

THE TGF-β SUPERFAMILY

Ligands

The TGF-β superfamily includes over 35 structurally related proteins that function as extracellular ligands in a diverse range of tissues and organ systems to modulate cellular processes such as proliferation, differentiation, survival and apoptosis (Massague and Wotton, 2000; Knight and Glister, 2006). Members of the superfamily include TGF-β isoforms, activins, inhibins, bone morphogenetic proteins (BMPs), growth differentiation factors (GDFs) and anti-Mullerian hormone (AMH, also known as Mullerian-inhibiting substance, MIS). These molecules are widely represented throughout the animal kingdom and the amino acid sequences of homologous TGF-β proteins are highly conserved amongst different species. A common feature shared by TGF-β superfamily members is that the mature forms are homo- or heterodimers of the carboxy-terminal regions of larger pre-proproteins. Typically, these dimers are covalently linked by an interchain disulfide bond between conserved cysteine residues; in some cases (e.g. GDF3, GDF9) constituent chains are non-covalently linked.

Receptors, signal transduction and binding proteins

Most TGF-β superfamily ligands exert their effects on target cells by cooperative binding to two types of signalling receptor (designated type-I and -II), each of which possesses an extracellular ligand-binding domain, a single transmembrane domain and an intracellular serine/threonine kinase domain. Typically, ligand binding to type-II receptor induces recruitment and transphosphorylation of type-I receptor to generate a heterotetrameric signalling complex (Massague and Wotton, 2000; Chang *et al.*, 2002; Miyazawa *et al.*, 2002). In mammals there are seven known type-I and five type-II receptors associated with TGF-β signal transduction; in birds six type-I and five type-II receptors are known. Different TGF-β superfamily ligands form active signalling complexes by binding to one or more combinations of type-I and type-II receptors, suggesting an element of redundancy.

Receptor activation leads to the phosphorylation of signalling intermediaries called receptor-regulated Smads (R-Smads). These associate with a common partner Smad (co-Smad) and translocate to the nucleus to alter gene expression through interaction with various transcription factors and co-activators or co-repressors of transcription (Massague and Wotton, 2000). Inhibitory Smads (I-Smads) can attenuate TGF-β ligand-induced signal transduction pathways at the intracellular level while several secreted binding proteins (including follistatin, noggin, chordin and gremlin) bind to some TGF-β members extracellularly, thus disrupting their interaction with cell surface receptors (Gumienny and Padgett, 2002). In addition, other cell surface molecules, including betaglycan (also known as TGF-β type-III receptor), function as co-receptors for some ligands including TGF-β and inhibin. Betaglycan binds TGF-β and inhibin with high affinity and enhances the presentation of these ligands to type-II receptors on the cell surface, thus potentiating their action (Lewis *et al.*, 2000; Phillips and Woodruff, 2004).

TGF-β SUPERFAMILY MEMBERS IN THE AVIAN OVARY

Expression of TGF-β superfamily members in developing follicles

The expression of mRNA transcripts encoding TGF-β-related ligands has been reported in hen ovarian follicles, including inhibin α subunit, inhibin and activin β_A and β_B subunits (Chen and Johnson, 1996a,b; Davis and Johnson, 1998; Davis *et al.*, 2001; Knight *et al.*, 2005), TGF-β1, -β2, -β3 isoforms (Law *et al.*, 1995), various BMPs (Onagbesan *et al.*, 2003; Al-Musawi *et al.*, 2007) and GDF9 (Johnson *et al.*, 2005). Moreover, the above studies have revealed a number of interesting differences in expression pattern according to follicle cell type and developmental stage. Similarly, follicular GC and TC contents of several of the translated and post-translationally-processed protein products, including inhibin-A, inhibin-B, activin-A and activin-binding protein (follistatin), have been shown to vary in a coordinated fashion (Lovell *et al.*, 1998, 2003).

The progression of follicles into the pre-ovulatory hierarchy is accompanied by a dramatic shift from inhibin-B to inhibin-A production (Lovell *et al.*, 1998, 2003), associated with concomitant changes in GC expression of mRNAs for the corresponding β_B and β_A subunits (Davis and Johnson, 1998; Hecht *et al.*, 2000; Slappey and Davis, 2003; Knight *et al.*, 2005). In prehierarchical (8–10 mm) follicles, GC selectively produce inhibin-B, while in pre-ovulatory follicles, especially the F1, GC selectively produce high levels of inhibin-A (Lovell *et al.*, 2003). The functional significance of this striking shift from inhibin-B to inhibin-A production remains obscure. As follicles are recruited into the pre-ovulatory hierarchy they appear to acquire a high 'activin tone', reflected by an increase in TC activin-A production, decreased GC inhibin-B production and a decrease in follistatin production. During the final stages of the pre-ovulatory hierarchy, particularly at the F2 and F1 transition, 'activin tone' is markedly reduced, accompanied by a significant rise in inhibin-A and progesterone production with little change in follistatin level (Lovell *et al.*, 2003). These sequential shifts in the balance between inhibin and activin during follicular development imply functional roles for the inhibin–activin system necessary for the progression of follicles into the pre-ovulatory hierarchy and acquisition of ovulable status. As yet there have been no comparable studies to quantify ovarian expression of BMPs at the protein level and, to our knowledge, immunoassays for avian BMPs have not been developed.

Further reports have documented follicle cell- and stage-dependent changes in expression of mRNAs for activin receptors (Slappey and Davis, 2003; Lovell *et al.*, 2006), BMP receptors (Al-Musawi *et al.*, 2007), inhibin co-receptor, betaglycan (Knight *et al.*, 2005; Sweeney and Johnson, 2005; Lovell *et al.*, 2006) and follistatin (Davis and Johnson, 1998). Collectively, the above evidence reinforces the concept that multiple TGF-β-related signalling systems (ligands, receptors, binding proteins) are operational at the intra-ovarian level and are likely to contribute to the regulation of follicle progression. What evidence is there from functional studies to support such an involvement?

Evidence for functional roles of activins and inhibins in hen follicles

Inhibins and activins have been implicated as direct (paracrine, autocrine) and indirect (endocrine) regulators of ovarian follicle development in birds (Vanmontfort *et al.*, 1994; Davis *et al.*, 2001; Lovell *et al.*, 2002a,b; Johnson *et al.*, 2004; reviews: Decuypere *et al.*, 1997; Knight *et al.*, 2005). Both LH and FSH promote inhibin-A secretion by GC from hen hierarchical follicles *in vitro* (Lovell *et al.*, 2002a,b), and the fall in plasma inhibin-A concentration that accompanies follicle regression induced by food restriction is associated with a reciprocal rise in plasma FSH (Lovell *et al.*, 2000). Active immunization against inhibin α subunit has been shown to advance the onset of lay and increase laying performance in Japanese quail (Moreau *et al.*, 1998) and broiler breeder pullets (Satterlee *et al.*, 2002). Inhibin-α immunization of pullets has also been shown to increase the number of 8–10 mm follicles at the bottom of the pre-ovulatory hierarchy, the number of recent post-ovulatory follicles and the incidence of double ovipositions (Lovell *et al.*, 2001). Despite the above evidence that active immunization against inhibin can perturb follicle development, Lovell *et al.* (2001) were unable to detect any effect on plasma gonadotrophin profiles, and the mechanism underlying the ovarian response has not been resolved. As discussed below, inhibin-α antibodies may be acting at an intrafollicular level.

Are activin and inhibin involved in selection of prehierarchical follicles?

It is intriguing that GC expression of inhibin β_B subunit mRNA (Hecht *et al.*, 2000; Knight *et al.*, 2005) and inhibin-B protein (Lovell *et al.*, 2003) is maximal in 7–9 mm follicles at the point of entry into the pre-ovulatory hierarchy and that the number of follicles in this same size class was twofold higher in inhibin-α immunized birds (Lovell *et al.*, 2001). This raises the possibility that partial immunoneutralization of inhibin-B in these follicles tips the prevailing intrafollicular inhibin:activin balance in favour of activin. In this regard, activin-A has been shown to augment FSH and LH receptor mRNA expression in undifferentiated GCs from prehierarchical hen follicles (Johnson *et al.*, 2004; Woods and Johnson, 2005; Johnson *et al.*, 2006), potentially sensitizing them to gonadotrophic stimulation, facilitating their acquisition of steroidogenic competence and subsequent recruitment to the pre-ovulatory hierarchy.

Lovell *et al.* (2007) examined whether gonadotrophins, in a reciprocal manner, influence expression of activin receptors and inhibin co-receptor (betaglycan) in GC from prehierarchical 6–8 mm follicles. Both gonadotrophins enhanced activin receptor-IIA (ActRIIA) mRNA expression; LH also increased activin receptor-I and -IIB mRNA levels, suggesting increased GC responsiveness to activin, which requires both type-I and -II activin receptors to form an active signalling complex. However, FSH decreased ActR-I and -IIB mRNA levels, and both gonadotrophins increased expression of betaglycan (inhibin co-receptor), possibly indicating a lower GC responsiveness to activin (and increased responsiveness to inhibin) in prehierarchical follicles. Both LH and FSH also raised betaglycan mRNA levels in TC from prehierarchical follicles without affecting expression of activin receptors, suggesting an increased TC sensitivity to inhibin. As mentioned above, GC inhibin-B production is maximal in prehierarchical follicles (Lovell *et al.*, 2003). Whether this inhibin-B has a role in promoting thecal androgen production by hen prehierarchical follicles has not been established, although a pilot study in the authors' laboratory has provided some evidence to support this. An antibody against inhibin α subunit was shown to inhibit LH-induced androstenedione secretion by 6–8 mm follicle explants (i.e. TC plus GC) without affecting LH-induced progesterone secretion (Fig. 7.1). Studies in several mammalian species indicate that inhibin can upregulate thecal androgen production while activin has the opposite effect (see Knight and Glister, 2003).

Collectively, the above observations support the notion that activin-A, predominantly secreted by the TC of avian prehierarchical follicles, exerts a local paracrine action on GC to sensitize them to gonadotrophins and facilitate

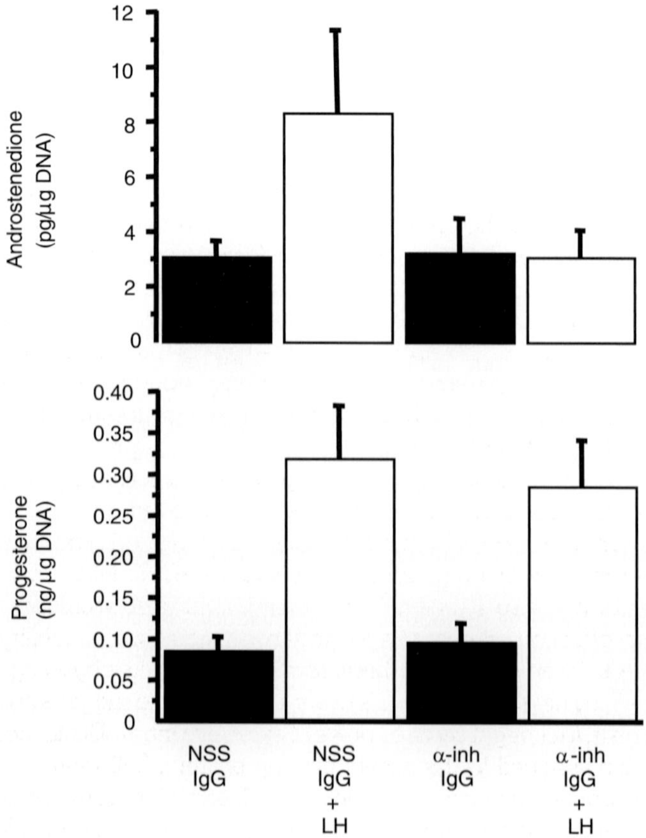

Fig. 7.1. Preliminary evidence that endogenous inhibin promotes LH-induced androgen production by theca cells of prehierarchical follicles. Follicle wall explants were cultured overnight with or without ovine LH (10 ng/ml) in medium supplemented with an IgG fraction of normal sheep serum (NSS) or sheep inhibin α subunit (α-inh) antiserum. The inhibin antibody abolished LH-induced androstenedione secretion but did not affect basal or LH-induced progesterone secretion. Values are means ± SEM (n = 5).

follicle advancement to the pre-ovulatory hierarchy. Mutually, gonadotrophins (particularly LH) may reinforce this signal by upregulating GC responsiveness to TC-derived activin. A pronounced though transient rise in inhibin-B expression by growing prehierarchical follicles may serve as a 'brake' on further development and restrict entry into the pre-ovulatory hierarchy. On the reasonable assumption that, *in vivo*, each member of the cohort of 6–9 mm prehierarchical follicles is exposed to the same level of gonadotrophin supplied by the thecal vasculature, we hypothesize that the follicle that is promoted to the hierarchy has an optimal intrafollicular balance of stimulatory (activin-A?) versus inhibitory (inhibin-B? follistatin?) factors that maximizes gonadotrophin sensitivity and is most conducive to rapid cell proliferation. One can envisage that an individual 6–9 mm follicle with one or more of the following characteristics – high intrafollicular activin-A content, high activin receptor expression, low follistatin content, low inhibin-B content, low inhibin co-receptor expression – would have a selective advantage and a high chance of joining the rapidly growing pre-ovulatory hierarchy.

Activin and inhibin in pre-ovulatory follicles

In contrast to mammalian follicles, in which GCs are the main source of activins, the TC layer of chicken pre-ovulatory (F1 to F4) follicles has an approximately 40-fold higher concentration of activin-A protein than the corresponding GC layer (Lovell *et al.*, 1998, 2003). Moreover, activin type-I and -II receptors are expressed by pre-ovulatory GC at both the mRNA (Lovell *et al.*, 2007) and protein (Lovell *et al.*, 2002b) level. This suggests that TC-derived activin-A may function as a local paracrine regulator of neighbouring GCs. In support of this hypothesis Lovell *et al.* (2002b) found that *in vitro* treatment of GCs from hen pre-ovulatory follicles (F1 to F3) with activin-A alone greatly enhanced secretion of inhibin-A in F1 follicles, with a progressively smaller response in F2 and F3 follicles; progesterone secretion was not affected. Furthermore, activin-A greatly enhanced FSH-induced secretion of inhibin-A and progesterone, the relative response being greatest in F1 follicles and smallest in F3 follicles (Fig. 7.2). Activin-A also enhanced LH-induced inhibin-A secretion but had only a marginal effect on LH-induced progesterone secretion. The ability of TC-derived activin-A to promote the synthesis and secretion of its physiological antagonist, inhibin-A, by neighbouring GC is of some interest and suggests an autoregulatory mechanism that might serve to prevent excessive intrafollicular activin signalling, a role also assigned to the activin binding protein, follistatin.

Additional evidence for a functional interaction between gonadotrophin-dependent and activin-dependent signalling pathways was provided by Johnson

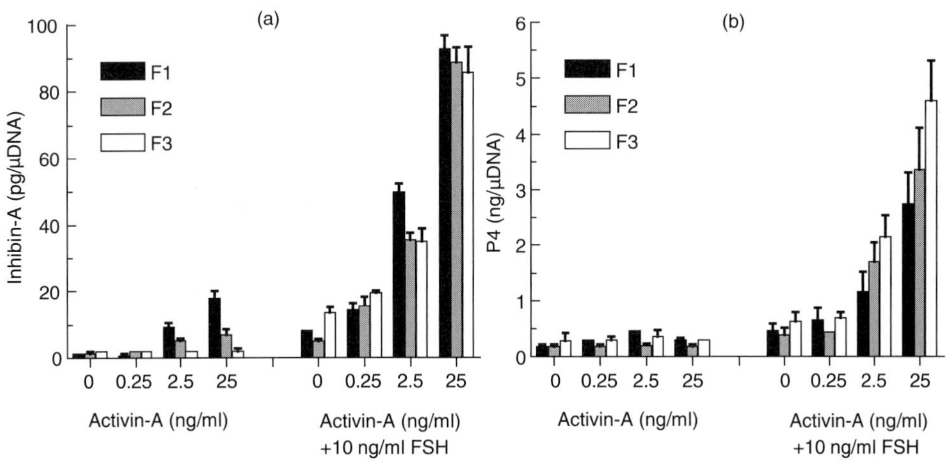

Fig. 7.2. Stimulatory effect of activin-A on 'basal' and FSH-induced secretion of (a) inhibin-A and (b) progesterone by granulosa cells from F1, F2 and F3 follicles (data replotted from Lovell *et al.*, 2002b).

et al. (2006), who showed that activin-A enhanced both FSH- and LH-receptor mRNA expression in GC from pre-ovulatory follicles. In a reciprocal manner, both LH and FSH augment expression of ActRI and ActRIIB mRNA by GC from pre-ovulatory F1 follicles (Lovell *et al.*, 2007).

A dramatic increase in GC progesterone output, inhibin-A content (Lovell *et al.*, 1998, 2003) and inhibin β_A subunit mRNA expression (Chen and Johnson, 1996a; Knight *et al.*, 2005) occurs as hierarchical follicles advance to the F1 position. Both GC inhibin-A protein content and β_A subunit mRNA expression in the F1 follicle fall sharply after the LH surge. It remains to be established whether the sharp increase in inhibin-A contributes to the rise in progesterone output by the F1 follicle that plays such a vital role in promoting the pre-ovulatory LH surge that triggers its ovulation. In view of recent evidence that TC-derived testosterone facilitates LH-induced progesterone secretion by GC from pre-ovulatory hen follicles (Rangel *et al.*, 2007) and that administration of an anti-androgen, flutamide, abolishes the pre-ovulatory surge of progesterone and blocks ovulation (Rangel *et al.*, 2006), we speculate that recurrent increases in inhibin-A production by GCs of successive F1 follicles generate cyclic increases in TC androgen output from other follicles (F2 to Fn hierarchical follicles and perhaps prehierarchical follicles) and thus facilitate the pre-ovulatory progesterone surge. Further research is clearly needed to evaluate this hypothesis but previous *in vitro* studies have shown that inhibin can enhance TC androgen production in mammals (Hillier, 1991; Wrathall and Knight, 1995) and embryonic chicken gonad (Rombauts *et al.*, 1996).

Evidence for functional roles of BMPs and GDF9 in hen follicles

Studies over the last decade have revealed the existence of a functional BMP system in the mammalian ovary that appears to have a number of critical roles in the regulation of folliculogenesis (reviewed by Shimasaki *et al.*, 2004; Knight and Glister, 2006). Various BMP ligands, receptors, binding proteins and signalling intermediaries are expressed in a cell and follicle stage-specific manner, and *in vitro* studies of both GC and TC have revealed BMP-induced alterations in cell proliferation, steroidogenesis and gonadotrophin responsiveness (rodent: Shimasaki *et al.*, 1999; Otsuka *et al.*, 2001; Erickson and Shimasaki, 2003; porcine: Brankin *et al.*, 2005; ovine: Bodensteiner *et al.*, 1999; McNatty *et al.*, 2005; bovine: Bodensteiner *et al.*, 1999; Glister *et al.*, 2004, 2005).

While there is a general paucity of information on BMPs in the avian ovary, as discussed below, a few recent studies have examined the expression of BMP ligands and receptors in hen follicles (Onagbesan *et al.*, 2003; Johnson *et al.*, 2005; Al-Musawi *et al.*, 2007). In addition, Onagbesan *et al.* (2003) demonstrated stimulatory effects of BMP-4 and -7 on GC cell proliferation and gonadotrophin-induced progesterone production, while Al-Musawi *et al.* (2007) reported stimulatory effects of BMP-6 on basal and gonadotrophin-induced secretion of inhibin-A and progesterone by GC, and on expression of mRNAs for gonadotrophin receptors, inhibin/activin subunits and P450scc. Johnson *et al.* (2005) showed that GDF9 is expressed in the avian oocyte and provided evidence supporting a stimulatory role for oocyte-derived GDF9 in hen GC proliferation.

BMPs

In a recent study undertaken in the authors' laboratory (Al-Musawi *et al.*, 2007) RT-PCR was used to compare the pattern of expression of BMP ligands and receptors in cultured GC and TC from prehierarchical and pre-ovulatory (F1) hen follicles. BMP-2 transcript was detected in GC from prehierarchical follicles, while BMP-2 and -4 (plus very weak BMP-3 and -5) transcripts were detected in GC from F1 follicles. Similarly, expression of BMP-2, -4, -6 and -7 mRNA was reported in chicken GCs from F1 to F3 pre-ovulatory follicles (Onagbesan *et al.*, 2003). Thus an autocrine and paracrine action of GC-derived BMPs in pre-ovulatory follicle function is a distinct possibility. In comparison with GC, however, Al-Musawi *et al.* (2007) found that cultured chicken TC showed abundant expression of a more extensive range of BMP ligands at both prehierarchical (BMP-3, -4, -5, -6, -7 and -15) and pre-ovulatory (BMP-2, -3, -4, -5, -6 and -7) follicle stages. At both stages TC also expressed the three BMP receptor forms, compatible with an autocrine and paracrine role of BMPs in modulating theca cell function from the prehierarchical stage through to ovulation. This finding is in agreement with Onagbesan *et al.* (2003), who reported that TC from F1 to F3 pre-ovulatory follicles express several BMP ligands (BMP-2, -4, -6 and -7) as well as BMPR-IA, -IB, -II. To our knowledge the potential action of BMPs on TC has yet to be explored in the chicken, but recent studies in cattle (Glister *et al.*, 2005) and sheep (Campbell *et al.*, 2006) have demonstrated BMP-induced suppression of both basal and LH-induced androgen secretion by TC.

The above findings prompted Al-Musawi *et al.* (2007) to undertake functional studies on GCs from hierarchical follicles to test the hypothesis that TC-derived BMPs modulate GC function in a paracrine manner. Secretory responses of GCs to BMP-6, in terms of inhibin-A, inhibin-B and progesterone secretion, were evaluated. Quantitative RT-PCR was used to examine the effect of BMP-6 on basal, gonadotrophin- and cyclic adenosine monophosphate analogue 8-Br-cAMP-induced expression of mRNA transcripts for gonadotrophin receptors, cytochrome P450scc and the inhibin and activin subunits (α, β_A, β_B). The study showed that BMP-6 increased 'basal' and gonadotrophin-induced inhibin-A and progesterone secretion but did not enhance the effect of 8-Br-cAMP (Fig. 7.3). This suggested that the observed synergism between BMP-6 and gonadotrophins might involve BMP-induced upregulation of gonadotrophin receptors. In support of this theory, treatment with BMP-6 alone increased LHR mRNA in F1 cells and FSHR mRNA in F1, F2 and F3/4 cells. BMP-6 also enhanced LH- and FSH-induced LHR expression in each cell type but did not raise FSHR expression above that induced by BMP-6 alone.

To further explore the action of BMP-6 on inhibin-A secretion, Al-Musawi *et al.* (2007) examined the effect of BMP-6 on expression of mRNAs encoding inhibin and activin subunits (α, β_A, β_B). Consistent with its effect on inhibin-A secretion, BMP-6 enhanced 'basal' expression of α and β_A subunit mRNA in

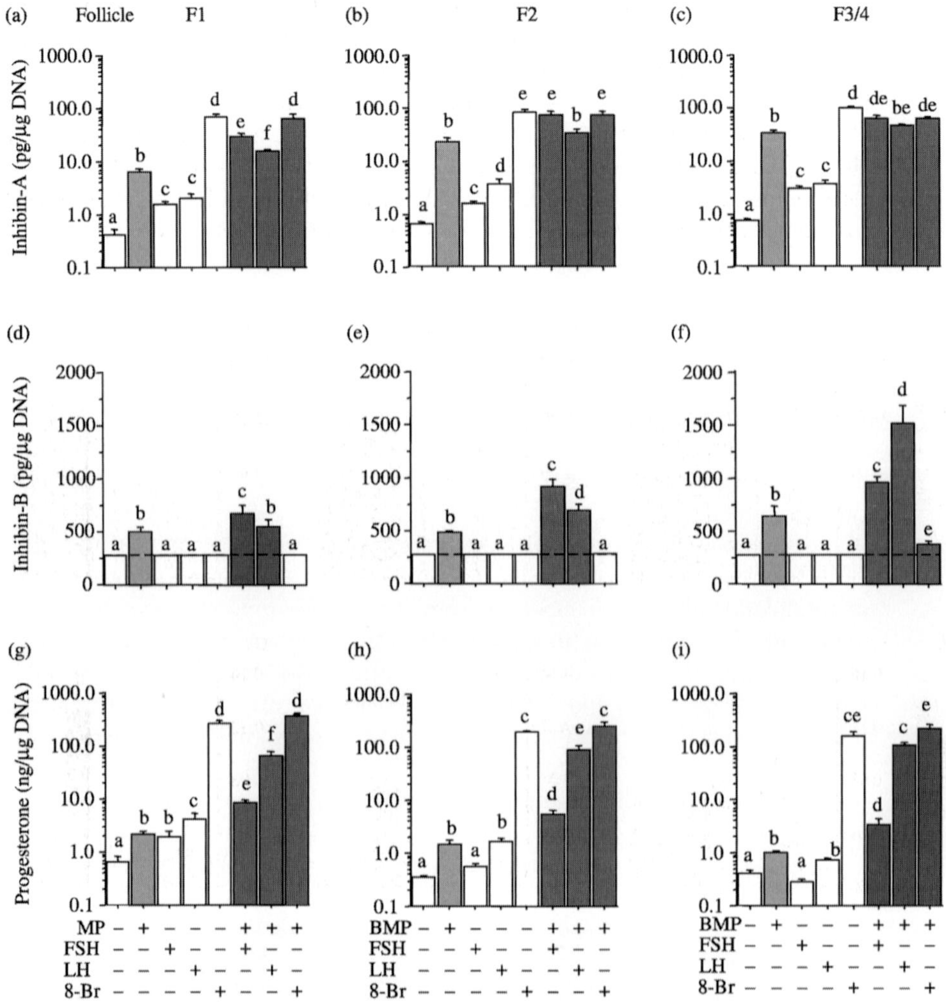

Fig. 7.3. Effect of BMP-6 (50 ng/ml) with or without FSH/LH (100 ng/ml) or 8-Br-cAMP (0.5 mM) on inhibin-A (a–c; log scale), inhibin-B (d–f) and progesterone (g–i; log scale) secretion by cultured granulosa cells derived from F1, F2 and F3/4 follicles. Values are means ± SEM (n = 6 independent cultures); within each panel, means without a common letter are significantly different ($P < 0.05$). The hashed line in d–f indicates the effective detection limit of the inhibin-B assay. (From Al-Musawi *et al.*, 2007, *Reproduction* 134, 293–306; © Society for Reproduction and Fertility. Reproduced by permission.)

F1, F2 and F3/4 GCs. BMP-6 markedly enhanced FSH- and LH-induced expression of the α subunit in GCs from all follicles and FSH-induced β_A subunit in GCs from F2 and F3/4 follicles but not from F1 follicles. Treatment of GCs with either BMP-6 alone or FSH and LH had no effect on 'basal' βB mRNA abundance. However, co-treatment of GCs with gonadotrophin and BMP-6 greatly increased β_B subunit expression, the response being greatest in F3/4 follicles and lowest in F1 follicles (Fig. 7.4). Collectively, these findings, and those of Onagbesan *et al.* (2003), constitute direct evidence to support the existence of a functional BMP system in the avian ovary in which TC-derived BMP(s) exert a major influence on GC function. Studies to investigate potential effects of intrafollicular BMPs on TC function in birds have not been reported, although, as mentioned above, hen TCs express both type-I and -II receptors for BMPs (Onagbesan *et al.*, 2003; Al-Musawi *et al.*, 2007) consistent with such an action. BMP-4, -6 and -7 have been shown to suppress 'basal' and

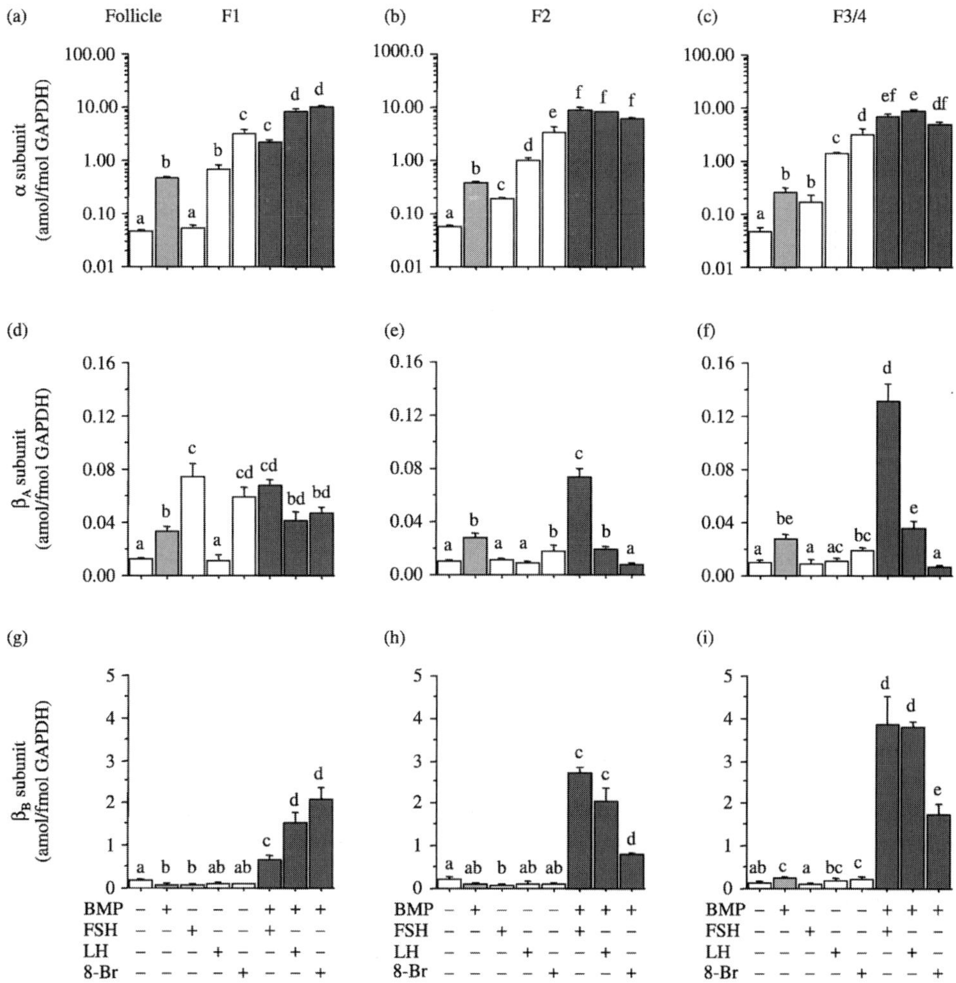

Fig. 7.4. Effect of BMP-6 (50 ng/ml) with or without FSH/LH (100 ng/ml) or 8-Br-cAMP (0.5 mM) on relative amounts of mRNA transcript encoding the inhibin/activin α (a–c; log scale), β_A (d–f) and β_B (g–i) subunits in the F1, F2 and F3/4 granulosa cells. Values are means ± SEM (n = 3 independent cultures); within each panel, means without a common letter are significantly different ($P < 0.05$). (From Al-Musawi *et al.*, 2007, *Reproduction* 134, 293–306; © Society for Reproduction and Fertility. Reproduced by permission.)

LH-induced androgen secretion by bovine TC in primary culture (Glister *et al.*, 2005) and so a comparable action in the avian ovary is a distinct possibility.

GDF9

Ablation of the GD region (GD plus adjacent GCs) of a hen follicle 1–2 days before ovulation induces follicle atresia, associated with a blockage of the pre-ovulatory LH and progesterone surges (Yoshimura *et al.*, 1994; Yao *et al.*, 1998). Signals emanating from the GD have a role in promoting GC proliferation and follicle development (Yao *et al.*, 1998); one such signal appears to be EGF (Yao and Bahr, 2001). Another signal may be GDF9, an oocyte-specific member of the TGF-β superfamily first identified in the mouse as playing an obligatory role in the growth of early-stage (primary) follicles (Dong *et al.*, 1996; Shimasaki *et al.*, 2004). As reported by Johnson *et al.* (2005) chicken GDF9 mRNA and protein are expressed in hen follicles, with the greatest concentration in oocytes from SWF (<1 mm). In larger, yellow yolk-filled oocytes, GDF9 appeared to be predominantly localized at the periphery of the oocyte, separated from the granulosa cell layer by the vitelline membrane. The observation that SWF-conditioned culture medium promotes proliferation of hen GC, and that this action was blocked by a GDF9 antibody (Johnson *et al.*, 2005), supports a functional role for chicken GDF9 in promoting GC proliferation, akin to its role in mammalian follicles (see Shimasaki *et al.*, 2004).

CONCLUSIONS

Ovarian follicle development in the hen is closely regulated by a variety of hormonal and intra-ovarian signals. Emerging evidence indicates that multiple TGF-β superfamily ligands, receptors and extracellular binding proteins are expressed by different cell types in avian ovarian follicles. This evidence has prompted functional *in vitro* studies that have begun to unravel the physiological roles of these intra-ovarian factors in follicle growth, selection, steroidogenesis and ovulation. However, current information arising from studies on isolated cells is far from complete and there is a need for further investigation to delineate the key players involved and to resolve the issue of how such a diverse array of signalling systems (including gonadotrophins, metabolic hormones, ovarian and adrenal steroids, ovarian growth factors) is integrated at the 'whole animal' level to bring about orderly follicle progression culminating in ovulation of fertile oocytes.

ACKNOWLEDGEMENTS

The authors are indebted to Professor Frank Cunningham, Dr Gerald Pope, Professor Nigel Groome, Dr John Proudman and Dr Al Parlow for providing valuable research reagents. Much of the authors' work cited here was supported by grants from the Biotechnology and Biological Sciences Research Council.

REFERENCES

Adashi, E.Y. (1994) Growth factors and ovarian function: the IGF-I paradigm. *Hormone Research* 42, 44–48.

Al-Musawi, S.L., Gladwell, R.T. and Knight, P.G. (2007) Bone morphogenetic protein-6 (BMP-6) enhances gonadotrophin-dependent progesterone and inhibin secretion and expression of mRNA transcripts encoding gonadotrophin receptors and inhibin/activin subunits in chicken granulosa cells. *Reproduction* 134, 293–306.

Bahr, J.M. and Johnson, A.L. (1984) Regulation of the follicular hierarchy and ovulation. *Journal of Experimental Zoology* 232, 495–500.

Bahr, J.M., Wang, S.C., Huang, M.Y. and Calvo, F.O. (1983) Steroid concentrations in isolated theca and granulosa layers of preovulatory follicles during the ovulatory cycle of the domestic hen. *Biology of Reproduction* 29, 326–334.

Bodensteiner, K.J., Clay, C.M., Moeller, C.L. and Sawyer, H.R. (1999) Molecular cloning of the ovine growth/differentiation factor-9 gene and expression of growth/differentiation factor-9 in ovine and bovine ovaries. *Biology of Reproduction* 60, 381–386.

Brankin, V., Quinn, R.L., Webb, R. and Hunter, M.G. (2005) BMP-2 and -6 modulate porcine theca cell function alone and co-cultured with granulosa cells. *Domestic Animal Endocrinology* 29, 593–604.

Campbell, B.K., Souza, C.J., Skinner, A.J., Webb, R. and Baird, D.T. (2006) Enhanced response of granulosa and theca cells from sheep carriers of the FecB mutation *in vitro* to gonadotropins and bone morphogenic protein-2, -4, and -6. *Endocrinology* 147, 1608–1620.

Chang, H., Brown, C.W. and Matzuk, M.M. (2002) Genetic analysis of the mammalian transforming growth factor-beta superfamily. *Endocrine Reviews* 23, 787–823.

Chen, C.C. and Johnson, P.A. (1996a) Expression of inhibin a and inhibin/activin βA subunits in the granulosa layer of the large preovulatory follicles of the hen. *Biology of Reproduction* 55, 450–454.

Chen, C.C. and Johnson, P.A. (1996b) Molecular cloning of inhibin/activin beta A-subunit complementary deoxyribonucleic acid and expression of inhibin/activin alpha- and beta A-subunits in the domestic hen. *Biology of Reproduction* 54, 429–435.

Cunningham, F.J. (1987) Ovulation in the hen: neuroendocrine control. *Oxford Reviews of Reproductive Biology* 9, 96–136.

Cunningham, F.J., Wilson, S.C., Knight, P.G. and Gladwell, R.T. (1984) Chicken ovulation cycle. *Journal of Experimental Zoology* 232, 485–494.

Davis, A.J. and Johnson, P.A. (1998) Expression pattern of messenger ribonucleic acid for follistatin and the inhibin/activin subunits during follicular and testicular development in *Gallus domesticus*. *Biology of Reproduction* 59, 271–277.

Davis, A.J., Brooks, C.F. and Johnson, P.A. (2001) Follicle-stimulating hormone regulation of inhibin α- and β_B-subunit and follistatin messenger ribonucleic acid in cultured avian granulosa cells. *Biology of Reproduction* 64, 100–106.

Decuypere, E., Romauts, D., Vanmontfort, D. and Verhoeven, G. (1997) Inhibin from embryo to adult hen. In: Harvey, S. and Etches, R.J. (eds) *Perspectives in Avian Endocrinology*. Journal of Endocrinology Ltd, Bristol, UK, pp. 71–80.

Dong, J., Albertini, D.F., Nishimori, K., Kumar, T.R., Lu, N. and Matzuk, M.M. (1996) Growth differentiation factor-9 is required during early ovarian folliculogenesis. *Nature* 383, 531–535.

Erickson, G.F. and Shimasaki, S. (2003) The spatiotemporal expression pattern of the bone morphogenetic protein family in rat ovary cell types during the estrous cycle. *Reproductive Biology and Endocrinology* 1, 9 (http://www.RBEj.com/content/1/1/9).

Etches, R. (1996) The ovary. In: Etches, R. (ed.) *Reproduction in Poultry*. CAB International, Wallingford, UK, pp. 125–166.

Etches, R.J. and Duke, C.E. (1984) Progesterone, androstenedione and estradiol content of theca and granulosa tissues of the 4 largest ovarian follicles during the ovulatory cycle of the hen (*Gallus domesticus*). *Journal of Endocrinology* 103, 71–76.

Etches, R.J. and Schoch, J.P. (1984) A mathematical representation of the ovulatory cycle of the domestic hen. *British Poultry Science* 25, 65–76.

Fortune, J.E. (2003) The early stages of follicular development: activation of primordial follicles and growth of preantral follicles. *Animal Reproduction Science* 78, 135–163.

Gilbert, A. (1971) The ovary. In: Bell, D.J. and Freeman, B.M. (eds) *Physiology and Biochemistry of the Domestic Fowl*, Vol. 3. Academic Press, London, pp. 1163–1202.

Gilbert, A.B., Perry, M.M., Waddington, D. and Hardie, M.A. (1983) Role of atresia in establishing the follicular hierarchy in the ovary of the domestic hen (*Gallus domesticus*). *Journal of Reproduction and Fertility* 69, 221–227.

Glister, C., Kemp, C.F. and Knight, P.G. (2004) Bone morphogenetic protein (BMP) ligands and receptors in bovine ovarian follicle cells: actions of BMP-4, -6 and -7 on granulosa cells and differential modulation of Smad-1 phosphorylation by follistatin. *Reproduction* 127, 239–254.

Glister, C., Richards, S.L. and Knight, P.G. (2005) Bone morphogenetic proteins (BMP)-4, -6, and -7 potently suppress basal and luteinizing hormone-induced androgen production by bovine theca interna cells in primary culture: could ovarian hyperandrogenic dysfunction be caused by a defect in thecal BMP signaling? *Endocrinology* 146, 1883–1892.

Gumienny, T.L. and Padgett, R.W. (2002) The other side of TGF-beta superfamily signal regulation: thinking outside the cell. *Trends in Endocrinology and Metabolism* 13, 295–299.

Hecht, D.J., Davis, A.J., Brooks, C.F. and Johnson, P.A. (2000) Molecular cloning and expression analysis of the complementary deoxyribonucleic acid for chicken inhibin/activin β_B subunit. *Biology of Reproduction* 62, 1128–1134.

Hillier, S.G. (1991) Regulatory functions for inhibin and activin in human ovaries. *Journal of Endocrinology* 131, 171–175.

Johnson, A.L. and van Tienhoven, A. (1980) Plasma concentrations of six steroids and LH during the ovulatory cycle of the hen, *Gallus domesticus*. *Biology of Reproduction* 23, 386–393.

Johnson, A.L., Bridgham, J.T. and Wagner, B. (1996) Characterization of a chicken luteinizing hormone receptor (cLH-R) complementary deoxyribonucleic acid, and expression of cLH-R messenger ribonucleic acid in the ovary. *Biology of Reproduction* 55, 304–309.

Johnson, A.L., Bridgham, J.T. and Woods, D.C. (2004) Cellular mechanisms and modulation of activin A- and transforming growth factor beta-mediated differentiation in cultured hen granulosa cells. *Biology of Reproduction* 71, 1844–1851.

Johnson, P.A., Dickens, M.J., Kent, T.R. and Giles, J.R. (2005) Expression and function of growth differentiation factor-9 in an oviparous species, *Gallus domesticus*. *Biology of Reproduction* 72, 1095–1100.

Johnson, P.A., Woodcock, J.R. and Kent, T.R. (2006) Effect of activin A and inhibin A on expression of the inhibin/activin beta-B-subunit and gonadotropin receptors in granulosa cells of the hen. *General and Comparative Endocrinology* 147, 102–107.

Juengel, J.L. and McNatty, K.P. (2005) The role of proteins of the transforming growth factor-beta superfamily in the intraovarian regulation of follicular development. *Human Reproduction Update* 11, 143–160.

Knight, P.G. and Glister, C. (2003) Local roles of TGFbeta superfamily members in the control of ovarian follicle development. *Animal Reproduction Science* 78, 165–183.

Knight, P.G. and Glister, C. (2006) TGFbeta superfamily members and ovarian follicle development. *Reproduction* 132, 191–206.

Knight, P.G., Gladwell, R.T. and Lovell, T.M. (2005) The inhibin–activin system and ovarian folliculogenesis in the chicken. In: Dawson, A. and Sharp, P.J. (eds) *Functional Avian Endocrinology*. Narosa Publishing House, New Delhi, pp. 323–337.

Law, A.S., Burt, D.W. and Armstrong, D.G. (1995) Expression of transforming growth factor-beta mRNA in chicken ovarian follicular tissue. *General and Comparative Endocrinology* 98, 227–233.

Lewis, K.A., Gray, P.C., Blount, A.L., MacConell, L.A., Wiater, E., Bilezikjian, L.M. and Vale, W. (2000) Betaglycan binds inhibin and can mediate functional antagonism of activin signalling. *Nature* 404, 411–414.

Li, Z. and Johnson, A.L. (1993) Expression and regulation of cytochrome P450 17-alpha-hydroxylase messenger ribonucleic acid levels and androstenedione production in hen granulosa cells. *Biology of Reproduction* 49, 1293–1302.

Lovell, T.M., Gladwell, R.T., Cunningham, F.J., Groome, N.P. and Knight, P.G. (1998) Differential changes in inhibin A, activin A, and total α-subunit levels in granulosa and thecal layers of developing preovulatory follicles in the chicken. *Endocrinology* 139, 1164–1171.

Lovell, T.M., Vanmontfort, D., Bruggeman, V., Decuypere, E., Groome, N.P., Knight, P.G. and Gladwell, R.T. (2000) Circulating concentrations of inhibin-related proteins during the ovulatory cycle of the domestic fowl (*Gallus domesticus*) and after induced cessation of egg laying. *Journal of Reproduction and Fertility* 119, 232–238.

Lovell, T.M., Knight, P.G., Groome, N.P. and Gladwell, R.T. (2001) Changes in plasma inhibin A levels during sexual maturation in the female chicken and the effects of active immunization against inhibin α-subunit on reproductive hormone profiles and ovarian function. *Biology of Reproduction* 64, 188–196.

Lovell, T.M., Gladwell, R.T., Groome, N.P. and Knight, P.G. (2002a) Modulatory effects of gonadotrophins and insulin-like growth factor on the secretion of inhibin A and progesterone by granulosa cells from chicken preovulatory (F1–F3) follicles. *Reproduction* 123, 291–300.

Lovell, T.M., Gladwell, R.T., Groome, N.P. and Knight, P.G. (2002b) Activin exerts differential effects on basal and gonadotrophin-induced secretion of inhibin A and progesterone by granulosa cells from chicken preovulatory (F1–F3) follicles. *Reproduction* 124, 649–657.

Lovell, T.M., Gladwell, R.T., Groome, N.P. and Knight, P.G. (2003) Ovarian follicle development in the laying hen is accompanied by divergent changes in inhibin A, inhibin B, activin A and follistatin production in granulosa and theca layers. *Journal of Endocrinology* 177, 45–55.

Lovell, T.M., Knight, P.G. and Gladwell, R.T. (2006) Differential expression of mRNAs encoding the putative inhibin co-receptor betaglycan and activin type-I and type-II receptors in preovulatory and prehierarchical follicles of the laying hen ovary. *Journal of Endocrinology* 188, 241–249.

Lovell, T.M., Al-Musawi, S.L., Gladwell, R.T. and Knight, P.G. (2007) Gonadotrophins modulate hormone secretion and steady-state mRNA levels for activin receptors (type I, IIA, IIB) and inhibin co-receptor (betaglycan) in granulosa and theca cells from chicken prehierarchical and preovulatory follicles. *Reproduction* 133, 1159–1168.

Massague, J. and Wotton, D. (2000) Transcriptional control by the TGF-beta/Smad signaling system. *EMBO Journal* 19, 1745–1754.

McNatty, K.P., Juengel, J.L., Reader, K.L., Lun, S., Myllymaa, S., Lawrence, S.B., Western, A., Meerasahib, M.F., Mottershead, D.G., Groome, N.P., Ritvos, O. and Laitinen, M.P. (2005) Bone morphogenetic protein 15 and growth differentiation factor 9 co-operate to regulate granulosa cell function. *Reproduction* 129, 473–480.

Miyazawa, K., Shinozaki, M., Hara, T., Furuya, T. and Miyazono, K. (2002) Two major Smad pathways in TGF-beta superfamily signaling. *Genes and Cells* 7, 1191–1204.

Moreau, J.D., Satterlee, D.G., Rejman, J.J., Cadd, G.G., Kousoulas, K.G. and Fioretti, W.C. (1998) Active immunization of Japanese quail hens with a recombinant chicken inhibin fusion protein enhances production performance. *Poultry Science* 77, 894–901.

Onagbesan, O.M. and Peddie, M.J. (1995) Effects of insulin-like growth factor I and interactions with transforming growth factor alpha and LH on proliferation of chicken granulosa cells and production of progesterone in culture. *Journal of Reproduction and Fertility* 104, 259–265.

Onagbesan, O.M., Gullick, W., Woolveridge, I. and Peddie, M.J. (1994) Immunohistochemical localization of epidermal growth factor receptors, epidermal growth factor-like and transforming growth factor alpha-like peptides in chicken ovarian follicles. *Journal of Reproduction and Fertility* 102, 147–153.

Onagbesan, O.M., Bruggeman, V., Van As, P., Tona, K., Williams, J., and Decuypere, E. (2003) BMPs and BMPRs in chicken ovary and effects of BMP-4 and -7 on granulosa cell proliferation and progesterone production *in vitro*. *American Journal of Physiology – Endocrinology and Metabolism* 285, 973–983.

Otsuka, F., Moore, R.K. and Shimasaki, S. (2001) Biological function and cellular mechanism of bone morphogenetic protein-6 in the ovary. *The Journal of Biological Chemistry* 276, 32889–32895.

Peddie, M.J., Onagbesan, O.M. and Williams, J. (1994) Chicken granulosa cell proliferation and progesterone production in culture: effects of EGF and theca secretions. *General and Comparative Endocrinology* 94, 341–356.

Perry, M.M., Gilbert, A.B. and Evans, A.J. (1978) Structure of germinal disk region of hen's ovarian follicle during rapid growth phase. *Journal of Anatomy* 127, 379–392.

Perry, M.M., Waddington, D., Gilbert, A.B. and Hardie, M.A. (1983) Growth rates of the small yolky follicles in the ovary of the domestic fowl. *IRCS Medical Science – Biochemistry* 11, 979–980.

Phillips, D.J. and Woodruff, T.K. (2004) Inhibin: actions and signalling. *Growth Factors* 22, 13–18.

Rangel, P.L., Sharp, P.J. and Gutierrez, C.G. (2006) Testosterone antagonist (flutamide) blocks ovulation and preovulatory surges of progesterone, luteinizing hormone and oestradiol in laying hens. *Reproduction* 131, 1109–1114.

Rangel, P.L., Rodriguez, A. and Gutierrez, C.G. (2007) Testosterone directly induces progesterone production and interacts with physiological concentrations of LH to increase granulosa cell progesterone production in laying hens (*Gallus domesticus*). *Animal Reproduction Science* 102, 56–65.

Romanoff, A. (1960) *The Avian Embryo: Structural and Functional Development*. Macmillan, New York.

Rombauts, L., Vanmontfort, D., Decuypere, E. and Verhoeven, G. (1996) Inhibin and activin have antagonistic paracrine effects on gonadal steroidogenesis during the development of the chicken embryo. *Biology of Reproduction* 54, 1229–1237.

Satterlee, D.G., Cadd, G.G. and Fioretti, W.C. (2002) Active immunization of broiler breeder hens with a recombinant chicken inhibin fusion protein enhances egg lay. *Poultry Science* 81, 519–528.

Shimasaki, S., Zachow, R.J., Li, D., Kim, H., Iemura, S., Ueno, N., Sampath, K., Chang, R.J. and Erickson, G.F. (1999) A functional bone morphogenetic protein system in the ovary. *Proceedings of the National Academy of Sciences USA* 96, 7282–7287.

Shimasaki, S., Moore, R.K., Otsuka, F. and Erickson, G.F. (2004) The bone morphogenetic protein system in mammalian reproduction. *Endocrine Reviews* 25, 72–101.

Slappey, S.N. and Davis, A.J. (2003) Expression pattern of messenger ribonucleic acid for the activin type II receptors and the inhibin/activin subunits during follicular development in broiler breeder hens. *Poultry Science* 82, 338–344.

Sweeney, S.A. and Johnson, P.A. (2005) Messenger RNA and protein expression analysis of betaglycan in the pituitary and ovary of the domestic hen. *Biology of Reproduction* 72, 172–178.

Tilly, J.L., Kowalski, K.I. and Johnson, A.L. (1991a) Cytochrome P450 side-chain cleavage (P450scc) in the hen ovary. II. P450scc messenger RNA, immunoreactive protein, and enzyme activity in developing granulosa cells. *Biology of Reproduction* 45, 967–974.

Tilly, J.L., Kowalski, K.I. and Johnson, A.L. (1991b) Stage of ovarian follicular development associated with the initiation of steroidogenic competence in avian granulosa cells. *Biology of Reproduction* 44, 305–314.

Tilly, J.L., Kowalski, K.I., Li, Z., Levorse, J.M. and Johnson, A.L. (1992) Plasminogen activator activity and thymidine incorporation in avian granulosa cells during follicular development and the periovulatory period. *Biology of Reproduction* 46, 195–200.

Vanmontfort, D., Berghman, L.R., Rombauts, L., Verhoeven, G. and Decuypere, E. (1994) Changes of immunoreactive inhibin, follicle-stimulating hormone, luteinizing hormone, and progesterone in plasma after short-term food deprivation and during the ovulatory cycle of the domestic hen. *General and Comparative Endocrinology* 95, 117–124.

Woods, D.C. and Johnson, A.L. (2005) Regulation of follicle stimulating hormone receptor messenger RNA in hen granulosa cells relative to follicle selection. *Biology of Reproduction* 72, 643–650.

Woods, D.C., Haugen, M.J. and Johnson, A.L. (2005) Opposing actions of TGF beta and MAP kinase signaling in undifferentiated hen granulosa cells. *Biochemical and Biophysical Research Communications* 336, 450–457.

Wrathall, J.H.M. and Knight, P.G. (1995) Effects of inhibin-related peptides and oestradiol on androstenedione and progesterone secretion by bovine theca cells in vitro. *Journal of Endocrinology* 145, 491–500.

Wyburn, G., Aitken, R. and Johnston, H. (1965) The ultrastructure of the zona radiata of the ovarian follicle of the domestic fowl. *Journal of Anatomy* 99, 468–484.

Yao, H.H. and Bahr, J.M. (2001) Chicken granulosa cells show differential expression of epidermal growth factor (EGF) and luteinizing hormone (LH) receptor messenger RNA and differential responsiveness to EGF and LH dependent upon location of granulosa cells to the germinal disc. *Biology of Reproduction* 64, 1790–1796.

Yao, H.H., Volentine, K.K. and Bahr, J.M. (1998) Destruction of the germinal disc region of an immature preovulatory chicken follicle induces atresia and apoptosis. *Biology of Reproduction* 59, 516–521.

Yoshimura, Y., Tischkau, S.A. and Bahr, J.M. (1994) Destruction of the germinal disc region of an immature preovulatory follicle suppresses follicular maturation and ovulation. *Biology of Reproduction* 51, 229–233.

You, S., Bridgham, J.T., Foster, D.N. and Johnson, A.L. (1996) Characterization of the chicken follicle stimulating hormone receptor (cFSH-R) complementary deoxyribonucleic acid, and expression of cFSH-R messenger ribonucleic acid in the ovary. *Biology of Reproduction* 55, 1055–1062.

Part IV
Mating Behaviour and Fertility

CHAPTER 8
Mating Behaviour and Fertility

I.J.H. Duncan
Department of Animal and Poultry Science, University of Guelph, Canada

And with that word he flew down from the beam,
For it was day, and down went his hens all;
And with a cluck he them began to call,
For he had found some corn within the yard.
Regal he was, and fears he did discard.
He feathered Pertelote full many a time
And twenty times he trod her ere 'twas prime.
He looked as if he were a grim lion
As on his toes he strutted up and down;
He deigned not set his foot upon the ground.
He clucked when any grain of corn he found,
And all his wives came running at his call.

Geoffrey Chaucer (1340?–1400)
The Canterbury Tales
The Nun's Priest's Tale
(lines 3172–3183).

ABSTRACT

Mating behaviour in domestic fowl can be divided into an appetitive phase, courtship, and a consummatory phase, copulation. Courtship displays serve the important functions of arousing and synchronizing the participants. The main courtship displays are performed by the cock and comprise waltzing, wing-flapping, titbitting, cornering, strutting and rear approach. These have all been observed in the progenitor of the domestic fowl, the red jungle fowl (*Gallus gallus spadiceus*). The development of mating behaviour is described. The presence of males during rearing has an accelerating effect on the sexual development of hens. Mate choice within a breeding flock of domestic fowl is not a random process. Both male and female fowl show mate preferences and these are probably based more on physical attributes than on the behaviour of the opposite-sex individual. Females may also choose males according to certain fitness characteristics, but it is argued that fluctuating asymmetry is probably not a factor under commercial conditions. There is a strong diurnal pattern of mating behaviour, with a peak in the late afternoon. Mating efficiency also increases at this time of day. There is a wide variation in both mating frequency and mating efficiency between cocks and little relationship between either of those traits and fertility. All these findings mean that the mating dynamics in large commercial flocks of breeders is extremely complicated. A ratio of about 1:10 or 1:11 of cocks:hens usually gives the best fertility, but the effective ratio is probably nearer 1:20, with many cocks not inseminating hens. The evidence suggests that the decline in fertility in broiler breeders after 50 weeks of age is probably due to the conformation of the cocks preventing cloacal contact rather than a reduction in libido. The recent problem of hyperaggressiveness in broiler breeder males is discussed and a possible cause suggested. Differences in the mating systems of turkeys and ducks compared with that of the domestic fowl are described.

INTRODUCTION

In comparison with other poultry science topics, mating behaviour has received little attention in the past 40 years. Most of the detailed descriptions of mating behaviour were published in the 1950s and 1960s (e.g. Wood-Gush, 1954, 1956; Lake and Wood-Gush, 1956; Lill and Wood-Gush, 1965) and, while things were running smoothly within the poultry industry, there seemed no need to investigate this area in any more depth. However, in the early 1990s, problems associated with mating were reported (Mench, 1993) and these have stimulated a renewed interest in this area.

It is often helpful to think of behaviour as being made up of two phases, the appetitive phase and the consummatory phase. In the poultry industry, producers are generally concerned about what takes place as a result of consummatory behaviour, e.g. the amount of food that the birds consume or the number of eggs that are collected or the percentage fertility reached by a flock. However, the correct expression of consummatory behaviour largely depends on it being preceded by appropriate appetitive behaviour. It is often aberrant appetitive behaviour that can account for poor consummatory behaviour. When considering mating behaviour, it is therefore important to pay particular attention to the appetitive elements, which are the various courtship displays, and not just to the consummatory act of treading, ejaculation and transmission of semen.

COURTSHIP AND MATING

Domestic fowl have a fairly complex courtship pattern (the appetitive phase), which appears to serve several functions: it attracts females to the male, assures the dominance of the male over the female and thereby facilitates mating, arouses both participants and synchronizes the act of copulation in both participants.

These functions are extremely important. Compared with most mammalian species, birds have to overcome several handicaps in order to transfer semen from male to female. When a male mounts a female, their cloacas point in the same direction and generally there is no penile organ to lock them together. Therefore, both the male and the female have to be well organized for semen to be transferred. They must be highly aroused and very well synchronized for a successful mating to take place. And that is what courtship behaviour is all about. Some avian species have evolved extremely elaborate courtship displays, such as the bowerbirds (e.g. *Sericulus* spp.) and the lyrebirds (*Menura* spp.). But even the humble domestic fowl (*Gallus gallus domesticus*) has a complex courtship composed of several elements.

Many courtship displays are thought to be derived from displacement activities, i.e. behaviour patterns which occur out of context when the animal is in a conflict situation. These displacement activities gradually become emancipated from their original motivational systems and incorporated into the mating system (Bastock, 1967). The following description of mating behaviour in the fowl is based on the classic research of Wood-Gush (1954, 1956).

Waltzing is generally regarded as a potent courtship display, during which the cock moves round the hen with short shuffling steps and drops the wing furthest from her. It is thought that waltzing occurs when there is a strong conflict in the cock between the tendencies to approach the hen sexually, to attack her aggressively and to flee.

During wing-flapping both wings are raised above the level of the back and flapped. Wing-flapping can be performed at various intensities. In its most vigorous form the cock raises himself to his full height and beats his wings so that they clap. There is therefore probably an acoustic as well as a visual component to this display. In its more subdued form, the wings may be raised just above the level of the back and then dropped. It is thought that wing-flapping also denotes conflict between sex, aggression and fear but with a smaller sexual component.

During titbitting the cock pecks and scratches the ground, often in a crouched posture, and gives food calls. The cock often moves away from the hen he appears to be interested in and titbits in an exaggerated way and so it is easy to spot. It is probably shown when the cock is highly aroused sexually and his other displays have been unsuccessful in stimulating a female. It nearly always results in hens moving towards the cock.

Cornering is a variation of titbitting in which the cock runs to a corner, stamps his feet, lowers his body to the ground and pecks at the substrate. Occasionally titbitting and cornering can switch to dust-bathing behaviour. This presumably happens through ‘postural facilitation’; what starts out as a courtship display can switch to dust-bathing behaviour when sexual motivation is relatively low compared with dust-bathing motivation and when the cock adopts a posture that is common to all three behaviour patterns.

During strutting the cock runs about with his legs slightly bent, his wings drooped and his neck slightly retracted. This display is often carried out in a jerky manner and the cock often chases the hens while strutting.

The rear approach or high-step advance is self-explanatory. The cock approaches the hen from the rear with slow, high steps.

If the hen is receptive, these courtship displays lead to her crouching with her wings slightly spread.

During mounting the cock steps on to the hen's back (often, but not always, after a rear approach) and grips her wing bases with his feet and her comb or neck feathers with his beak. This is usually followed by treading, during which the cock makes small foot movements, depresses and spreads his tail to one side while the hen moves her tail to the other side. The male then flexes his pelvis and slides over the rear of the hen. This is usually followed by cloacal contact and semen is ejaculated through the engorged phallic folds into the everted cloaca of the hen (Etches, 1996). Although it is not always possible to see the actual cloacal contact there are some good indicators that this has taken place: the cock always stops treading and makes a definite backwards and downwards thrust with his pelvis and the hen nearly always gives a very characteristic high-intensity feather-ruffle immediately after the cock dismounts.

In addition to these obvious courtship displays, Wood-Gush (1954) observed that cocks engaged in courtship behaviour performed occasional feather-ruffles, tail-wags, head-shakes and preens. In feather-ruffling, the neck is stretched, the ruff and other body feathers are raised and the whole body is shaken. Tail-wagging consists of the tail being moved rapidly from side to side in a horizontal plane. In head-shaking the head is tilted to one side and shaken with rotating movements. Preening is the manipulation of feathers using the beak. All these behaviour patterns have been observed as displacement activities when a domestic fowl is slightly frustrated or in conflict (e.g. Duncan and Wood-Gush, 1972). It is likely that their occurrence during courtship is as displacement activities and not as true courtship displays.

It is obvious from the foregoing description that the cock plays the more active role in mating behaviour. However, in order for mating to be successful, it is necessary for the hen to behave appropriately. Three characteristics are used to describe female sexual behaviour: (i) attractivity or the stimulus value of a female for a given male; (ii) proceptivity or the extent to which a female initiates copulation, including appetitive activities aimed at establishing contact with males; and (iii) receptivity or the willingness of a female to accept courtship and copulatory behaviour of a male. Attractivity will be discussed in a later section of this chapter. There is no evidence of female domestic fowl showing proceptivity; however, there has been interest over the years in receptivity and particularly its relationship to fecundity or rate of lay (see also Birkhead amd Pizzari, Chapter 9, this volume).

Using earlier unpublished data (Wilkins, 1915; Heuser, 1916) together with results from a small experiment, Lamoreux (1940) suggested that sexual receptivity and fecundity are positively correlated. However, research by others on this topic could find no such relationship (Upp, 1928; Guhl and Warren, 1946; Wood-Gush, 1958a).

The relationship between a hen's position in the social hierarchy and her receptivity has also been investigated. There is some indication that hens high in the hierarchy have an above-average rate of lay but are less likely to crouch to cocks than birds lower in the hierarchy (Guhl *et al.*, 1945; Guhl, 1950).

It should be noted that all the courtship displays described above have also been observed in the red jungle fowl (*Gallus gallus spadiceus*) with any differences being in relative frequencies and intensities of the displays (Kruijt, 1964; Lill, 1966; Collias and Collias, 1967).

DEVELOPMENT OF MATING BEHAVIOUR

It has been known for a long time that the process of imprinting in the first few days of a chick's life can have a profound effect on sexual responses later in life (e.g. Guiton, 1962, 1966). The imprinting process occurs when a very young chick, capable of following its mother, follows the first large moving object that it encounters and forms a strong attachment to it (Lorenz, 1937; Sluckin, 1964; Bateson, 1966). One of the unique characteristics of the imprinting process is that, in addition to learning to recognize its mother individually, the chick learns the broad features of the species and later in life directs its sexual behaviour towards an animal similar to its parent. It should be noted that the actual sexual behaviour displayed in adulthood does not seem to be affected by imprinting; it is the bird's perception of what is a suitable target for its sexual advances that is changed (Guiton and Wood-Gush, 1962).

Although the imprinting process is of great theoretical interest, it is of little importance practically. Breeding stock are generally reared socially from day-old but in the absence of a mother hen or surrogate, and this does not lead to problems later in the birds' breeding lives. However, Guiton and Wood-Gush (1962) suggest that cockerels that are destined to be semen donors in artificial insemination programmes might become better donors if imprinted to human beings while chicks. Guiton and Wood-Gush (1962) also suggest that, where breeding practice will involve the mating of lines differing markedly in plumage and morphology, it might be advisable to rear the chicks in mixed broods to avoid infertility at the start of breeding. While there may be value in this suggestion, the process involved probably has nothing to do with imprinting (since chicks of a few days old bear no resemblance to their adult appearance). It is more likely to work by getting birds accustomed to different plumage and morphological types so that they are not frightened by the novelty of 'strange birds' when the breeding flock is assembled (Bateson, 1964).

The normal development of sexual behaviour is, of course, under hormonal control, and this is usually managed through manipulations of the photoperiod. This is such a large topic, which has received so much attention recently (e.g. Sharp, 1993; Etches, 1996; Lewis and Morris, 2006; Sharp, Chapter 11 and Lewis, Chapter 14, in this volume), that it will not be discussed in this chapter. However, it should be pointed out that sexual behaviour, at least in the male, is not entirely dependent on post-pubertal hormone levels. Andrew (1966) was able to elicit copulatory behaviour in male chicks when only 48 h old by thrusting his hand with fingers extended towards the chick. The chick would climb on to his hand, make grasping movements with the beak, raise the wings slightly, lower the pelvis and show treading movements. This sequence usually ended with quivering movements of the body. Andrew (1966) also observed titbitting in these very young males. This suggests that the neural mechanisms necessary for the performance of some appetitive and consummatory elements of mating are available from a very early age.

The development of normal sexual behaviour has not received much attention. Kruijt (1964), in his study of Burmese red jungle fowl, does give some detailed descriptions of the development of sexual displays. Kruijt (1964) reports that, between 30 and 80 days of age, elements of sexual behaviour, namely rear approach and treading by males and copulatory sitting by males and females, did occur. Occasionally, these elements occurred quite separately from agonistic behaviour patterns, which were the focus of Kruijt's study. At other times, a mixture of sexual and agonistic behaviour was observed. Interestingly, the sexual behaviour performed by males seemed to be directed at both males and females. Complete copulatory responses were not seen at this age, even when crouching models of birds of the same age were provided. Kruijt (1964) goes on to report that this early copulatory behaviour does not change much in form between 80 and 120 days, although more of it progresses to the treading stage. During this time, the birds' juvenile vocalizations are developing into the adult type (suggesting hormonal changes) but complete copulations are still not performed. From about 120 days, males start to perform complete copulations, and titbitting and cornering appear in the jungle fowl's repertoire. During this stage, sexual behaviour is often mixed with aggressive attacking behaviour. Kruijt (1964) speculates that this is a result of sexual, aggressive and avoidance behaviour all being activated simultaneously during male–female encounters.

Does the presence of the opposite sex during rearing have any effect on the development of sexual behaviour? It is known that in many species of non-gallinaceous birds, females require the presence of a male to stimulate ovarian growth and egg laying (Lehrman, 1959). It is known that domestic fowl and turkey hens will lay eggs in the complete absence of males. However, at least in turkeys, male presence has been shown to have some effect, as turkey hens will lay at a higher rate when housed in the presence of toms (Jones and Leighton, 1987; Felts *et al.*, 1992). Early experiments with domestic fowl suggested that the presence of males had no effect on the onset of puberty or rate of lay in hens (Venkataramaiah and Choudary, 1973; Bhagwhat and Craig, 1979). However, in both these studies hens in the all-female flocks were housed in the same room as the mixed-sex flocks. The hens were therefore exposed to visual and auditory stimuli from the cocks. Widowski *et al.* (1998) carried out a more controlled investigation into the presence of males on sexual maturation and sexual responses towards humans in domestic fowl laying-strain females. In this study hens in mixed-sex groups matured earlier. They had larger combs at 15 and 20 weeks and came into lay about 1 week sooner than hens in single-sex groups. They also laid more eggs up to 22 weeks. However, by 30 weeks, hens in single-sex groups had overtaken those in mixed-sex groups. Rearing adjacent to mixed-sex groups was just as effective in accelerating sexual maturity as being in mixed-sex groups. This suggests that it is visual or auditory cues from the males that are important in triggering the acceleration effect and not actual physical contact with them. However, Widowski *et al.* (1998) point out that physical contact with the males *did* have some effect. Over 80% of hens in separate-sex pens and pens adjacent to mixed-sex showed crouching to an approaching human being whereas less than 20% of hens in mixed-sex groups showed this response.

In a study of White Leghorns, Leonard *et al.* (1993) found that, for both males and females, being given early experience of the opposite sex results in more successful mating in adulthood. In this study, a successful mating was defined as the male mounting, treading and making cloacal contact. An unsuccessful mating was defined as the male mounting, or mounting and treading, but making no cloacal contact. The success rate was the number of successful copulations expressed as a percentage of mating attempts. When the males and females had both had early experience of the opposite sex, the success rate was 88% compared with 68% when neither had early experience of the opposite sex. The rates were even lower when only one sex had early experience of the opposite sex.

CHOICE OF MATE

Mate choice within *Gallus* species is not a random process. The profound effects of imprinting on mate choice later in life have already been mentioned. However, there are other processes operating as well. In order to understand some of these, it is necessary to consider the natural social system in which mating takes place. Domestic fowl are polygamous, as are their progenitors the Burmese red jungle fowl (*Gallus gallus spadiceus*). The natural social system, which has been observed in jungle fowl (Collias *et al.*, 1966; Collias and Collias, 1967) and in feral domestic fowl (McBride *et al.*, 1969), is for dominant cocks to establish territories and maintain a small harem of hens and their followers within those territories. Within one of these groups, the dominant cock performs all, or very nearly all, the matings. Small mixed-sex flocks of domestic fowl kept under semi-natural conditions, say in a barnyard, probably operate similarly. In small groups there is no doubt that the dominant male mates much more frequently than do the subordinates. However, as the group size increases this pattern becomes less distinct. It used to be thought that, even in large commercial flocks, males would set up territories or home ranges, bird movements in general would be fairly restricted, and dominant males would do most of the mating within their territories (McBride and Foenander, 1962). However, careful observation has shown that domestic fowl in large flocks show much less territoriality and home ranging than had previously been thought. For example, in one study of a flock of 3600 female and 349 male broiler breeders in a barn measuring 46 m × 15 m, the males ranged over 16–72% and the females ranged over 11–61% of the total area (Appleby *et al.*, 1985). Within commercial breeder flocks, therefore, it is likely that there will be matings amongst many different pair combinations.

However, this does not mean that mate choice is a random process. Upp (1928) was one of the first to show that domestic fowl have preferences for members of the opposite sex. In a series of experiments, Lill investigated non-random mating in detail and showed that both male and female domestic fowl have mate preferences (Lill and Wood-Gush, 1965; Lill, 1968a,b). However, it should be pointed out that Lill did not favour using the word 'preference' to describe this phenomenon; he suggested that it would be better categorized as 'a response to signals irrespective of the identity of the "signaller"' (Lill, 1966). An interesting finding was that males discriminated amongst females on visual cues and in particular on plumage colour (Lill and Wood-Gush, 1965; Lill, 1968a). Females discriminated amongst males on physical characteristics rather than on quantitative differences in male courtship (Lill and Wood-Gush, 1965; Lill, 1968b). Both males and females of pure breeds showed preferences for birds of the plumage type they had been reared with. Females of a broiler strain, on the other hand, showed preferences for Brown Leghorn males, which Lill and Wood-Gush (1965) thought might be a preference for 'primitive-type plumage'. Lill (1966) also investigated this phenomenon in Burmese red jungle fowl (*Gallus gallus spadiceus*) and showed that both males and females exhibited non-random mating. For both males and females, the behaviour of the opposite-sex individual was of only minor importance whereas aspects of physical appearance played a more crucial role. However, other studies have shown that male behaviour does have a role, probably combined with morphological characteristics, in determining female mate choice (Graves *et al.*, 1985; Leonard and Zanette, 1998).

The finding, in some of these studies, that physical appearance was more important than quality or quantity of courtship is surprising. In recent years, there have been further investigations into mate selection in both jungle fowl and domestic fowl, particularly with regard to fitness characteristics. Zuk *et al.* (1990a) and Ligon and Zwartjes (1995) showed that red jungle fowl females chose males with larger combs but paid little attention to plumage colour. Since males infected with nematode parasites have smaller combs (Zuk *et al.*, 1990b) the implication is that the females were using comb size as an indicator of fitness. Sexual ornaments are thought to be reliable indicators of male quality because the costliness of carrying these traits prevents cheating. The immunocompetence handicap hypothesis states that males carry ornaments at the expense of their resistance to disease and parasites and there is evidence that this holds true in the domestic fowl (Verhulst *et al.*, 1999). There is also evidence to support this hypothesis in wild turkeys (*Meleagris gallopavo*), where coccidial infections reduce plumage iridescence and, it is assumed, the attractiveness of turkey stags (Hill *et al.*, 2005).

A related aspect of physical appearance that has received some attention recently is fluctuating asymmetry. Fluctuating asymmetry is the deviation from perfect symmetry in bilaterally symmetrical traits (Møller and Höglund, 1991; Møller and Swaddle, 1997). Deviations from perfect symmetry are thought to reflect inability in an individual to compensate for stress during development. Therefore, those individuals who are better able to cope with stress during development should show less asymmetry (Parsons, 1992). Moreover, ornamental traits should be more affected than non-ornamental traits.

Several studies have claimed to show sexual selection for individuals showing greater symmetry (e.g. Møller, 1990, 1992; Møller and Pomiankowski, 1993). However, some caution is needed when interpreting these results. For example, in a detailed investigation of fluctuating asymmetry in red jungle fowl, measures of asymmetry in both ornamental and non-ornamental traits did not correlate well with other indicators of male quality such as comb size, body size and body condition (Kimball *et al.*, 1997). Therefore, a relationship between fluctuating asymmetry and quality of a potential sexual partner cannot be assumed. In addition, it has been shown that starlings (*Sturnus vulgaris*) in the laboratory have difficulty in detecting small asymmetrical differences, a process that would be necessary for such visual cues to be used in sexual selection (Swaddle, 1999; Swaddle and Ruff, 2004). Finally, there must be some doubt about whether a species such as the domestic fowl, which has been domesticated for 7000–8000 years, under some degree of artificial selection during that time (West and Zhou, 1989) and under intense artificial selection in the past 60 years, would have retained a mechanism that probably requires continual fine-tuning.

MATING PATTERNS

There is a strong diurnal pattern of mating behaviour, with a peak during the late afternoon (e.g. Upp, 1928; Parker *et al.*, 1940; Lake and Wood-Gush, 1956; Guhl and Fischer, 1969). There is also an increase in the percentage of completed matings at this time. This is termed the mating efficiency and is the ratio of attempted matings to completed matings (Wood-Gush and Osborne, 1956). This pattern ties in quite nicely with artificial insemination (AI) studies which have found a greater semen yield and greater absolute spermatozoa numbers in the late afternoon (Penquite *et al.*, 1929; Lake and Wood-Gush, 1956). AI studies have also shown that insemination success rate is lower if there is a hard-shelled egg in the shell gland (Giesen *et al.*, 1980). Since there is less chance of there being an egg in the shell gland in the late afternoon, this diurnal rhythm of mating makes sense in evolutionary terms.

MATING FREQUENCY AND SEX DRIVE

There can be wide variation in mating frequency between cocks (Penquite *et al.*, 1929; Skard, 1937; Guhl *et al.*, 1945; Wood-Gush and Osborne, 1956; Siegel, 1965), with Guhl (1951) reporting a range of 15 to 41 or more matings per day. There is also wide variation in mating efficiency between cocks (Wood-Gush and Osborne, 1956). And, just to complicate things even further, there is little relationship between either of those traits and fertility. Even 'completed' matings (i.e. matings in which there appears to be cloacal contact) do not always result in the transfer of spermatozoa. Studies which have taken smears from hens after apparently complete matings have found that 14% of matings result in no sperm transfer (Parker *et al.*, 1940; see also Birkhead and Pizzari, Chapter 9, this volume).

As well as individual differences in mating frequencies, Wood-Gush and Osborne (1956) found differences in mating frequencies between six sire families of cockerels. They also found a negative relationship between comb size and mating frequency amongst the families, suggesting that differences in mating frequencies were not due to testosterone differences. In another study,

Wood-Gush (1958b) confirmed that differences in libido were under genetic control and that social experiences during rearing had only a slight effect.

In order to investigate the genetic basis of rate of copulation, Wood-Gush (1960) carried out an experiment in which he selected two strains from these families for a high and low rate of mating. By the second generation, the high strain was mating twice as frequently as the low strain. The two strains did not differ in aggression towards other males, crowing rate, acceptance of unusual feeds or frequency of alarm calling. Neither was there any indication of differences in testosterone levels between the strains. It thus appeared that the differences in mating rate were due to central nervous factors and not to hormonal or experiential factors. There was a significant difference in semen yield between the strains, with the low strain producing almost three times the volume of ejaculate from manual massage compared with the high strain. Siegel (1965) carried out a similar selection experiment on rate of mating, but with many more birds and over six generations. He corroborated the findings of Wood-Gush (1960) and was able to estimate a heritability of 0.18 ± 0.05 for increased mating rate and 0.13 ± 0.11 for decreased mating rate. Siegel (1965) also found a negative genetic correlation between mating rate and semen volume and sperm concentration.

Before discussing the practical implications of all these findings, one other phenomenon should be described. Guhl *et al.* (1945) described subordinate males being attacked frequently by dominants while attempting to mate as young cocks. Fear became so conditioned to sexual attempts that, later in life, they did not even try to approach females sexually although dominant males were absent. Guhl *et al.* (1945) described this as 'psychological castration'.

PRACTICAL IMPLICATIONS

All these findings mean that the mating dynamics in a large commercial flock of breeders are extremely complicated. By trial and error, producers have found that a ratio of about 1:10 or 1:11 of cocks:hens usually gives the best fertility. Results from a Dutch study of White Leghorn mating behaviour suggested an upper limit of 15 for the number of females which one male could mate and obtain optimum fertility (Brantas *et al.*, 1972). However, this estimate was made for traditional breeder flocks of several thousand birds. In an investigation with much smaller groups of one male and 15 broiler breeder hens, Duncan *et al.* (1990) recorded the number of matings and fertility rates. From their results they predicted that fertility could be maintained at a high level with a ratio of 1:40 or 1:50 cocks:hens, but this prediction would probably only work in groups with one male present. The problem then is that if something untoward happens to that male (he becomes lame or injured or ill) there would be a catastrophic decline in fertility in 40–50 hens. The industry has therefore spread the risk and opted for a system with many hens and cocks and a ratio, reached by trial and error, of around 1:10. Of course, the *effective* ratio under commercial conditions is probably nearer to 1:20 or perhaps even higher. It is likely that many cocks in breeder flocks will not be inseminating hens. There will be males present who are prevented from mating by dominant flock-mates. There will be males who are psychological castrates. There will be males who are rejected by the hens. There will be males who are infertile for a variety of reasons.

It has long been known that meat strains of domestic fowl are less fertile than egg-laying strains (Soller *et al.*, 1965; Rappaport and Soller, 1966). By the 1970s and 1980s increased selection for fast growth and breast width had led to severe fertility problems in broiler breeders after 50 weeks of age (Kirk *et al.*, 1980). There could be several reasons for this. The bulk or conformation of males might make it difficult for them to achieve cloacal contact. Male turkeys have already gone down that route and can now only achieve very low levels of fertility by natural mating. A common thought in the industry is that a drop in libido associated with the males gaining too much weight could be responsible for the reduction in fertility. Musculo-skeletal disease, which is common in male broiler breeders, could cause pain or mechanical weakness and interfere with copulation (Duff and Hocking, 1986; Hocking and Duff, 1989). Duncan *et al.* (1990) investigated the decrease in fertility in broiler breeders in the later part of the laying year by monitoring sexual behaviour and fertility at 28, 38 and 58 weeks of age. The males in this study were also fed different energy levels in order to reach different target body weights. The frequency of all sexual behaviour, including courtship and copulations, showed a large decrease with age, which suggested a decline in libido. However, rather unexpectedly, the heavier males showed less of a decrease in sexual behaviour than lighter males. Moreover, the reduction in sexual behaviour between 28 and 38 weeks of age had little effect on fertility, which was comparable with commercial fertility rates. However, there was a big drop in fertility at 58 weeks, particularly in groups with very heavy or very light males. The birds were inspected regularly and musculo-skeletal disease was ruled out as a cause. The very light-weight cocks also showed a bigger decline in courtship at this time and it was concluded that their energy intake was too low to maintain full sexual behaviour. The heaviest birds also showed a decline in courtship, but, rather unexpectedly, this was less of a decline than in the other groups. Being ‘overweight’ by industry standards did *not* result in a drastic reduction of libido. In addition to showing more courtship than the other groups, they were also seen to mount and tread the females more. The reduction in their fertility was therefore due to their bulk or their conformation precluding cloacal contact. An earlier study by Soller *et al.* (1965) reported that males of the Cornish breed had a lower fertility than those of the White Rock breed in natural matings, and the authors attributed this to the special body conformation of the Cornish breed and not to body weight *per se*. In a subsequent study, Wilson *et al.* (1979) examined the relationship between body weight and several other physical characteristics in broiler breeders for both natural matings and artificial insemination. For natural matings, they found a tendency for a negative relationship between fertility and body weight, breast angle and hip width. Hip width at 8 weeks of age was the only body trait consistently negatively correlated with fertility. For artificial insemination, there was little relationship between fertility and any of these traits, suggesting that the effect was not due to some reduction in semen quality but to difficulty in completing a successful mating. The authors speculated that the combination of broad breast and wide hips might lead to difficulty in mounting or difficulty with balance, or lead to a higher incidence of leg weakness, all of which might interfere with successful mating. The more likely reason would seem to be that the combination of broad breast and wide hips will reduce flexibility in the pelvic girdle and prevent cloacal contact when the male thrusts downwards. The fact that fertility in broiler breeders commonly drops off later in the breeding year, as they become broader breasted and wider hipped, would support this contention. Further corroboration of this hypothesis, that the problem stems from broad breastedness and reduced flexibility of the pelvic girdle, was produced by Hocking and Bernard (1997) in an experiment involving manipulation of body weight in two strains of broiler breeder males. One strain had lower fertility than the other and this was associated with a greater weight of breast muscle. The strain with the lower fertility also showed a higher number of incomplete matings. In this experiment there was a fairly low level of musculo-skeletal disease and this could not explain differences in fertility. In a more recent series of experiments, Hocking and Bernard (2000) investigated the separate effects of age of male and female broiler breeders on fertility, hatchability and sexual behaviour when male body weight was carefully controlled. They compared three ages of birds, namely 27–29, 35–37 and 55–57 weeks of age. Although the younger cocks were observed to mate more frequently than the two older age groups, this was not reflected in fertility, which was not affected by age. Also, there were no differences in the frequencies of male courtship behaviour at any age. These results suggest that ageing may have little effect on the fertility of male and female broiler breeders when male body weight is adequately controlled to 60 weeks of age.

A common practice to improve fertility of broiler breeders used to be 'spiking' the flock or replacing some (25–30%) of the older males with younger ones (North and Bell, 1990; Leeson and Summers, 2000). However, with the advent of separate-sex feeding this practice is much less common. The beneficial results which are reported to follow spiking are generally attributed to the younger age of the introduced cocks, with the industry assuming that the mechanism is through increased libido, whereas the real reason is probably due to increased flexibility of the younger cocks. However, there might be another mechanism adding to the improved fertility, and that is that placing the new cocks in an environment with many unknown hens may trigger the Coolidge effect, i.e. the phenomenon whereby males of nearly every species show continuously high sexual performance given the introduction of new receptive females (Brown, 1974).

HYPERAGGRESSIVE BEHAVIOUR IN BROILER BREEDER MALES

During the 1990s, an increasing number of reports described broiler breeder males showing high levels of aggression towards females (Mench, 1993). Typically, in an affected flock, the males would patrol the litter area and the females would restrict themselves to the raised slatted area in a house. Any female that descended to the litter, say to feed, would be chased and forcibly mounted. When this syndrome occurred, fertility levels would plummet, and females would be injured, often with severe lacerations on the backs of their heads where cocks had pecked and grabbed them with their beaks, and on their torsos under their wings where the cocks' inside claws had ripped the skin during forced mounts. Some females would even be killed by this behaviour. Aggressive behaviour by males towards females is very unusual; male domestic fowl dominate females passively and seldom show any overt aggression towards them (Wood-Gush, 1956, 1958c; Craig and Bhagwat, 1974; Rushen 1983/84). Hyperaggressive behaviour is both a production problem, because of much lower fertility rates, and a welfare problem, because females are being harassed, badly injured and even killed by males. At first, only one strain of broiler breeder males seemed to be affected, but within a year or two most strains were showing similar symptoms (Millman *et al.*, 2000). In a series of experiments, Millman and Duncan (2000a,b,c) showed that the hyperaggressiveness towards females could not be explained in terms of a general increase in aggression. Game fowl males which have been bred for fighting, and are much more aggressive towards other males than are broiler breeder males (Millman and Duncan, 2000a), court females very nicely and show little, if any, aggression towards them (Millman and Duncan, 2000b). Broiler breeder males are very severely food restricted in order to keep them reproductively fit. Perhaps this results in high levels of frustration. It is known that frustrated feeding behaviour can lead to increased aggression in domestic fowl (Duncan and Wood-Gush, 1971). However, when this possibility was tested in male broiler breeders, food restriction did not seem to be a cause of the increased aggression, either during the rearing phase (Millman and Duncan, 2000b) or during the adult phase (Millman *et al.*, 2000). In fact, fully fed males were more aggressive than restricted-fed males. It should also be stated that the problem definitely resides with the males; broiler breeder females respond normally to the courtship of laying-strain males (Millman and Duncan, 2000c).

It has been shown that broiler breeder males are deficient in certain elements of courtship behaviour. Compared with laying-strain males, broiler breeders show a similar frequency of waltzing but a much lower frequency of titbitting and high-step advancing (Millman *et al.*, 2000). The result is that the females do not always react appropriately when males approach but often move away and avoid them (Millman and Duncan, 2000c), and this probably exacerbates the situation by frustrating the males and causing them to be more aggressive.

More recently, Doherty (2006) has carried out further studies on this topic and has generally confirmed Millman's results. She has also shown that male broiler breeders can learn to perform more effective courtship, but this is not sufficient to eliminate their hyperaggression.

At the moment, it is not at all clear where the hyperaggressive behaviour in broiler breeder males has come from. It almost certainly has a genetic basis. It is also unclear whether the courtship deficiency and the hyperaggressiveness are separate or linked problems. There is a possibility that these traits are genetically linked to a production trait, such as broad-breastedness, for which the breeding companies have been selecting. On the other hand, it may be the result of a misguided attempt by geneticists to improve fertility, which is poor in broiler breeders, particularly towards the end of the breeding year. As has already been stated, there is a misconception in the industry that this drop in fertility is due to decreased libido. If the breeding companies have been selecting males that approach females very quickly in the mistaken belief that this is a sign of sexiness, they may have been selecting males that are, in fact, very aggressive. The fact that male broiler breeders are deficient in titbitting, a courtship behaviour pattern that involves moving *away* from the females, is evidence that points in the same direction.

There are two pieces of information that should be added to this section. Until recently, all reports of hyperaggressiveness in broiler breeders came from North America and none came from Europe. In 2006, a survey of broiler breeder welfare was carried out in Denmark, and as part of this survey a few breeder farms were visited, farm managers interviewed, and some casual observations of the birds were made (Duncan and Forkman, 2006). Farm managers reported that they had never seen male hyperaggressive behaviour in their flocks and this was confirmed by the casual observations. Typical domestic fowl courtship and mating were seen. No chasing or forced copulations were observed nor were hens with tell-tale lacerations. The interesting thing is that all broiler breeders in Denmark are the Ross 308 strain, which was the first strain to show problems in North America. It is known that American and European stock are derived from the same foundation stock. This happened about 22 years ago and selection procedures have been the same on both continents (J.C. McKay, Alabama, 2006, personal communication). This large difference in behaviour between North American and Danish Ross broiler breeders requires some explanation. It is difficult to imagine that such a big difference could be due to environmental effects alone. Environmental differences between the continents tend to be slight. For example, North American diets are based on maize/soybean formulations whereas European poultry rations have a wheat base. Probably more significant is the fact that genetic selection in North America has probably taken place under 'skip-a-day' feeding regimens whereas this has not been the case in Europe. It is possible that selection under a skip-a-day programme might produce a more aggressive bird and lead to genetic differences between American and European stock. It certainly appears that the two stocks have drifted apart genetically.

More difficult to explain, and more worrying, is the recent report of sexual aggression in broiler breeders in the Netherlands (De Jong *et al.*, 2006). The Dutch birds involved are Ross 308, exactly the same strain as used in Denmark. The worrying feature is that the Dutch and Danish stocks should be almost identical genetically. Either this is not the case, and there is considerable variation within the European Ross 308 strain, or environmental factors play a bigger role in causing hyperaggressiveness than previously thought. The whole topic of genotype–environment interaction with regard to hyperaggressiveness in broiler breeders obviously requires further investigation.

The second piece of information is that investigations into hyperaggressiveness in broiler breeder males have concentrated on measuring the aggressive tendency and the sexual tendency. However, it should be remembered that there is another motivational system involved and that is fear or avoidance. It was proposed more than 50 years ago that conflicts between motivational systems could lead to 'displacement activities' and that these could become emancipated during evolution and thereafter serve as social signals (Tinbergen, 1952). Wood-Gush (1956) thought that at least some of the courtship displays he observed in domestic cocks could originally have been displacement activities resulting from a conflict between sex, aggression and avoidance. Bastock (1967) gave this idea more formal standing. As pointed out by Doherty (2006), it is possible that the tendencies for sex and aggression are still the same in broiler breeder strains but they have lost some of their fearfulness and tendency to avoid the females. Avoidance is an extremely important part of the three-way conflict normally present during sexual encounters (see Fig. 8.1). It is the tendency to avoid that keeps the male from immediately approaching the

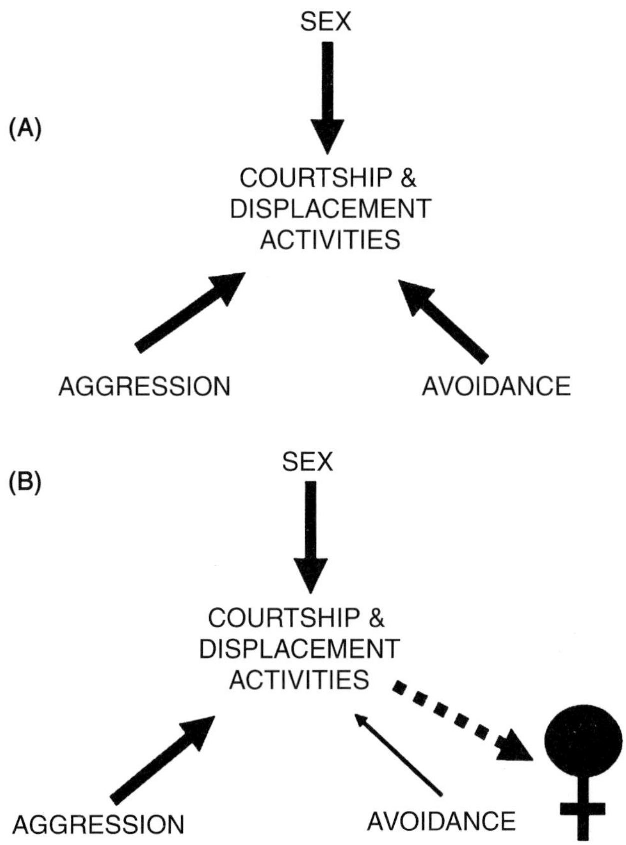

Fig. 8.1. A possible explanation of hyperaggressiveness in male broiler breeders. (A) is adapted from Bastock (1967) and represents the motivational conflict normally present in male domestic fowl during sexual encounters. Sex, aggression and avoidance are in balance and lead to displacement activities and courtship displays. (B) is adapted from Doherty (2006) and represents the motivational conflict present in male broiler breeders. Sex and aggression are unchanged but avoidance is reduced. This leads to an unbalanced conflict, a reduction in displacement activities and courtship displays, and with both aggression and sex stimulating the male to approach the female immediately.

female and it is during this conflict that the courtship displays are shown. If avoidance is reduced, then the tendencies for both sex and aggression drive the male towards the female and possibly do that without certain courtship elements being performed. It would be worthwhile to investigate this possibility in the future.

MATING BEHAVIOUR IN OTHER POULTRY SPECIES

There has been little recent investigation into the mating behaviour of the other main poultry species, namely turkeys and ducks. In the case of turkeys, the reason for the lack of studies is simple: because of the conformation of commercial turkey strains, the industry has been obliged to use artificial insemination since the 1960s, and so there has been little demand for information on natural mating. However, there is currently a renewed interest in heritage (relatively unselected) breeds of turkey to meet certain organic and animal welfare standards.

There are full descriptions of mating behaviour in the domesticated turkey stemming from observations taken in the 1950s (e.g. Hale and Schein, 1962). However, even at that time there was marked variation in the ability of male turkeys to complete attempted matings (Smyth and Leighton, 1953; Hale, 1955), presumably because selection for breast width had made cloacal contact difficult.

Compared with domestic fowl, turkey courtship is characterized by slow and restricted movements of the male combined with elaborate feather display. The snood becomes elongated and turgid and the naked skin areas of the head and neck change from red to bright blue. Females often crouch after moving towards the male, and males never attempt to mount a non-crouching female (Hale and Schein, 1962). The female head is an extremely important visual stimulus for arousing and orientating male turkeys during mating (Schein and Hale, 1957; Schoettle and Schein, 1959), whereas in domestic fowl the female body is as important a stimulus as the head (Carbaugh *et al.*, 1962). Another interesting difference from domestic fowl is that female turkeys become non-receptive to males for several days after complete eversion of the oviduct during copulation (Hale, 1955). Complete eversion of the oviduct occurs in response to tactile stimulation of the base of the tail or area to the side of the cloaca (Hale, 1959) and so occurs whether or not the mating is complete. This can obviously exacerbate infertility in flocks with a high incidence of incomplete matings (Hale, 1955).

Mating behaviour of the wild turkey (*Meleagris gallopavo*) is of particular interest because the mating system has a superficial resemblance to lekking behaviour. A lek is a group of males of a species such as black grouse (*Tetrao tetrix*) which gathers at a traditional location for the purposes of performing a competitive mating display. Each day during the breeding season, the same group of males take up the same individual positions on an arena, each occupying and defending a small territory. The females choose males according to their displays (Kruijt and Hogan, 1967). In the case of wild turkeys, however, males are typically seen courting in pairs. Genetic analysis of pairs of males courting together has shown that they are close relatives with half of their genetic material being identical. The more dominant male sires almost all of the progeny of this cooperative effort. The theory is that the less dominant male has a greater chance of passing along genetic material that is identical to his than he would if he was courting alone (Krakauer, 2005). If the turkey breeding industry decides to revert to natural mating, it will be an interesting challenge to make use of this tendency to mate cooperatively.

In the case of ducks, the challenge is somewhat different. All domestic breeds of duck (apart from Muscovy ducks, *Cairina moschata*) are derived from the mallard (*Anas platyrhychos*). Mallard typically show pair formation, which is not a characteristic favouring domestication; there are obvious advantages to human beings in having promiscuous domestic species (Hale, 1962). Desforges and Wood-Gush (1976) compared the sexual behaviour of a particular breed of domestic duck, the Aylesbury, with the mallard when both were kept under captive conditions. The social displays of the Aylesbury ducks were less intense than those of the mallards, and the down–up display was performed less frequently. Desforges and Wood-Gush (1976) suggested that in the wild mallard these displays probably act as interspecific ethological isolating mechanisms, which become unimportant under domestic conditions. Another notable difference in sexual behaviour was observed in that Aylesbury females incited several drakes whereas mallard females incited only one. Also, the preen-behind-wing display was performed to several females by Aylesbury drakes but only to one female by mallard males. This strongly suggested that the mallards were forming pairs whereas the Aylesbury ducks were not. Further evidence for pair formation was the fact that mallard pairs rested in close proximity and fed together without showing agonistic behaviour whereas this was not seen with Aylesbury ducks (Desforges and Wood-Gush, 1975). However, although the Aylesbury ducks tended towards promiscuity, both males and females directed their sexual behaviour to only two or three individuals. Desforges and Wood-Gush (1976) suggest that, since complete promiscuity would be an advantage in commercial practice, this phenomenon of limited promiscuity requires further study.

REFERENCES

Andrew, R.J. (1966) Precocious adult behaviour in the young chick. *Animal Behaviour* 14, 485–500.

Appleby, M.C., Maguire, S.N. and McRae, H.E. (1985) Movement by domestic fowl in commercial flocks. *Poultry Science* 64, 48–50.

Bastock, M. (1967) *Courtship, a Zoological Study*. Heinemann Educational Books Ltd, London.

Bateson, P.P.G. (1964) Effect of similarity between rearing and testing conditions in chicks' following and avoidance responses. *Journal of Comparative and Physiological Psychology* 57, 100–103.

Bateson, P.P.G. (1966) The characteristics and context of imprinting. *Biological Reviews* 41, 177–220.

Bhagwhat, A.L. and Craig, J.V. (1979) Effects of male presence on agonistic behaviour and productivity of White Leghorn hens. *Applied Animal Behaviour Science* 5, 267–282.

Brantas, G.C., Dennert, H.G. and Dennert-Distelbrink, A.L. (1972) The influence of the number of cocks on the conception rate among White Leghorns. *Archiv für Geflügelkunde* 36, 16–21.

Brown, R.E. (1974). Sexual arousal, the Coolidge effect and dominance in the rat (*Rattus norvegicus*). *Animal Behaviour* 22, 634–637.

Carbaugh, B.T., Schein, M.W. and Hale, E.B. (1962) Effects of morphological variations of chicken models on sexual responses of cocks. *Animal Behaviour* 10, 235–238.

Collias, N.E. and Collias, E.C. (1967) A field study of the red jungle fowl in north-central India. *The Condor* 69, 360–386.

Collias, N.E., Collias, E.C., Hunsaker, D. and Minning, L. (1966) Locality fixation, mobility and social organization within an unconfined population of red junglefowl. *Animal Behaviour* 14, 550–559.

Craig, J.V. and Bhagwat, A.L. (1974) Agonistic and mating behaviour of adult chickens modified by social and physical environments. *Applied Animal Ethology* 1, 57–65.

De Jong, I.C., Wolthuis, M. and Van Emous, R.A. (2006) Development of sexual behaviour in commercially housed broiler breeders. *Proceedings of the 40th International Congress of the International Society for Applied Ethology*. Bristol, UK, p. 143.

Desforges, M.F. and Wood-Gush, D.G.M. (1975) A behavioural comparison of domestic and Mallard ducks: spatial relationships in small flocks. *Animal Behaviour* 23, 698–705.

Desforges, M.F. and Wood-Gush, D.G.M. (1976) Behavioural comparison of Aylesbury and Mallard ducks: sexual behaviour. *Animal Behaviour* 24, 391–397.

Doherty, C.M. (2006) The effect of learning, morphology and behaviour on female-directed aggression in male broiler breeders. MSc thesis, University of Guelph, Canada.

Duff, S.R.I. and Hocking, P.M. (1986) Chronic orthopaedic disease in adult male broiler breeding fowls. *Research in Veterinary Science* 41, 340–348.

Duncan, I.J.H. and Forkman, B. (2006) Report on Broiler Breeder Welfare in Denmark. Unpublished report for Foreningen til Dyrenes Beskyttelse (Danish Animal Protection Society), Frederiksberg, Denmark.

Duncan, I.J.H. and Wood-Gush, D.G.M. (1971) Frustration and aggression in the domestic fowl. *Animal Behaviour* 19, 500–504.

Duncan, I.J.H. and Wood-Gush, D.G.M. (1972) An analysis of displacement preening in the domestic fowl. *Animal Behaviour* 20, 68–71.

Duncan, I.J.H., Hocking, P.M. and Seawright, E. (1990) Sexual behaviour and fertility in broiler breeder domestic fowl. *Applied Animal Behaviour Science* 26, 201–213.

Etches, E.J. (1996) *Reproduction in Poultry*. CAB International, Wallingford, Oxon, U.K.

Felts, J.V., Leighton, A.T., Denbow, D.M. and Hulet, R.M. (1992) Effects of light sources and the presence or absence of males on reproduction of female breeder turkeys. *Poultry Science* 71, 1817–1822.

Giesen, A.F., McDaniel, G.R. and Sexton, T.J. (1980) Effect of time of day of artificial insemination and oviposition–insemination interval on the fertility of broiler breeder hens. *Poultry Science* 59, 2544–2549.

Graves, H.B., Hable, C.P. and Jenkins, T.H. (1985) Sexual selection in *Gallus*: effects of morphology and dominance on female spatial behaviour. *Behavioural Processes* 11, 189–197.

Guhl, A.M. (1950) Social dominance and receptivity in the domestic fowl. *Physiological Zoology* 23, 361–366.

Guhl, A.M. (1951) Measurable differences in mating behaviour of cocks. *Poultry Science* 30, 687–693.

Guhl, A.M. and Fischer, G.J. (1969) The behaviour of chickens. In: Hafez, E.S.E. (ed.) *The*

Behaviour of Domestic Animals (2nd edn). Baillière, Tindall and Cassell, London, pp. 515–553.

Guhl, A.M. and Warren, D.C. (1946) Number of offspring sired by cockerels related to social dominance in chickens. *Poultry Science* 25, 460–472.

Guhl, A.M., Collias, N.E. and Allee, W.C. (1945) Mating behaviour and the social hierarchy in small flocks of White Leghorns. *Physiological Zoology* 18, 365–390.

Guiton, P. (1962) The development of sexual responses in the domestic fowl in relation to the concept of imprinting. *Symposia of the Zoological Society of London* 8, 227–234.

Guiton, P. (1966) Early experience and sexual object-choice in the Brown Leghorn. *Animal Behaviour* 14, 534–538.

Guiton, P. and Wood-Gush, D.G.M. (1962) The sexual behaviour of domestic birds in relation to early experience. *Proceedings of the 12th World's Poultry Congress*, pp. 516–518.

Hale, E.B. (1955) Defects in sexual behaviour as factors affecting fertility in turkeys. *Poultry Science* 34, 1059–1067.

Hale, E.B. (1959) Stimuli terminating sexual receptivity in female turkeys. *Anatomical Record* 134, 577 (Abstract).

Hale, E.B. (1962) Domestication and the evolution of behaviour. In: Hafez, E.S.E. (ed.) *The Behaviour of Domestic Animals*. Baillière, Tindall and Cox, London, pp. 21–53.

Hale, E.B. and Schein, M.W. (1962) The behaviour of turkeys. In: Hafez, E.S.E. (ed.) *The Behaviour of Domestic Animals*. Baillière, Tindall and Cox, London, pp. 531–564.

Heuser, G.F. (1916) A study of the mating behaviour of the domestic fowl. Unpublished PhD thesis, Cornell University.

Hill, G., Doucet, S.M. and Buchholz, R. (2005) The effect of coccidial infection on iridescent plumage coloration in wild turkeys. *Animal Behaviour* 69, 387–394.

Hocking, P.M. and Bernard, R. (1997) Effects of male body weight, strain and dietary protein content on fertility and musculo-skeletal disease in naturally mated broiler breeder males. *British Poultry Science* 38, 29–37.

Hocking, P.M. and Bernard, R. (2000) The effects of the age of male and female broiler breeders on fertility and hatchability of eggs. *British Poultry Science* 41, 370–377.

Hocking, P.M. and Duff, S.R.I. 1989. Musculo-skeletal lesions in adult male broiler breeder fowls and their relationships with body weight and fertility at 60 weeks of age. *British Poultry Science* 30, 779–786.

Jones, M.C. and Leighton, A.T. (1987) Research note: effect of presence or absence of the opposite sex on egg production and semen quality of breeder turkeys. *Poultry Science* 66, 2056–2059.

Kimball, R.T., Ligon, J.D. and Merola-Zwartjes, M. (1997) Fluctuating asymmetry in red junglefowl. *Journal of Evolutionary Biology* 10, 441–457.

Kirk, S., Emmans, G.C., McDonald, R. and Arnist, D. (1980) Factors affecting the hatchability of eggs from broiler breeders. *British Poultry Science* 21, 37–43.

Krakauer, A.H. (2005) Kin selection and cooperative courtship in wild turkeys. *Nature* 434, 69–72.

Kruijt, J.P. (1964) Ontogeny of social behaviour in Burmese Red Jungle Fowl. *Behaviour* 12 (Supplement), 1–201.

Kruijt, J.P. and Hogan, J.A. (1967) Social behaviour on the lek in the black grouse, *Lyrurus tetrix tetrix* (L.). *Ardea* 55, 203–240.

Lake, P.E. and Wood-Gush, D.G.M. (1956) Diurnal rhythms in semen yields and mating behaviour in the domestic cock. *Nature* 178, 853.

Lamoreux, W.F. (1940) The influence of intensity of egg production upon infertility in the domestic fowl. *Journal of Agricultural Research* 61, 191–205.

Leeson, S. and Summers, J.D. (2000) *Broiler Breeder Production*. University Books, Guelph, Ontario, Canada.

Lehrman, D.S. (1959) Hormonal responses to external stimuli in birds. *Ibis*, 101, 478–496.

Leonard, M.L. and Zanette, L. (1998) Female mate choice and male behaviour in domestic fowl. *Animal Behaviour* 56, 1099–1105.

Leonard, M.L., Zanette, L., Thompson, B.K. and Fairfull, R.W. (1993) Early exposure to the opposite sex affects mating behaviour in White Leghorn chickens. *Applied Animal Behaviour Science* 37, 57–67.

Lewis, P. and Morris, T. (2006) *Poultry Lighting: the Theory and Practice*. Northcot, Andover, UK.

Ligon, J.D. and Zwartjes, P.W. (1995) Ornate plumage of male red junglefowl does not influence mate choice by females. *Animal Behaviour* 49, 117–125.

Lill, A. (1966) Some observations on social organisation and non-random mating in captive Burmese red junglefowl (*Gallus gallus spadiceus*). *Behaviour* 26, 228–242.

Lill, A. (1968a) An analysis of sexual isolation in the domestic fowl: 1. The basis of homogamy in males. *Behaviour* 30, 107–126.

Lill, A. (1968b) An analysis of sexual isolation in the domestic fowl: 2. The basis of homogamy in females. *Behaviour* 30, 127–145.

Lill, A. and Wood-Gush, D.G.M. (1965) Potential ethological isolating mechanisms and assortative mating in the domestic fowl. *Behaviour* 25, 16–44.

Lorenz, K. (1937) The companion in the birds' world. *Auk* 54, 245–273.

McBride, G. and Foenander, F. (1962) Territorial behaviour in flocks of domestic fowls. *Nature* 194, 102.

McBride, G., Parer, L.P. and Foenander, F. (1969) The social organization and behaviour of the feral domestic fowl. *Animal Behaviour Monographs* 2, 127–181.

Mench, J.A. (1993) Problems associated with broiler breeder management. In: Savory, C.J. and Hughes, B.O. (eds) *Proceedings of the 4th European Symposium on Poultry Welfare*. Universities Federation for Animal Welfare, Potters Bar, UK, pp. 195–207.

Millman, S.T. and Duncan, I.J.H. (2000a) Strain differences in aggressiveness of male domestic fowl in response to a male model. *Applied Animal Behaviour Science* 66, 217–233.

Millman, S.T. and Duncan, I.J.H. (2000b) Effect of male-to-male aggressiveness and feed-restriction during rearing on sexual behaviour and aggressiveness towards females by male domestic fowl. *Applied Animal Behaviour Science* 70, 63–82.

Millman, S.T. and Duncan, I.J.H. (2000c) Do female broiler breeder fowl display a preference for broiler breeder or laying strain males in a Y-maze test? *Applied Animal Behaviour Science* 69, 275–290.

Millman, S.T., Duncan, I.J.H. and Widowski, T.M. (2000) Male broiler breeder fowl display high levels of aggression towards females. *Poultry Science* 79, 1233–1241.

Møller, A.P. (1990) Fluctuating asymmetry in male sexual ornaments may reliably reveal male quality. *Animal Behaviour* 40, 1185–1187.

Møller, A.P. (1992) Female swallow preference for symmetrical male sexual ornaments. *Nature* 357, 238–240.

Møller, A.P. and Höglund, J. (1991) Patterns of fluctuating asymmetry in avian feather ornaments: implications for models of sexual selection. *Proceedings of the Royal Society of London B, Biological Sciences* 245, 1–5.

Møller, A.P. and Pomiankowski, A. (1993) Fluctuating asymmetry and sexual selection. *Genetica* 89, 267–279.

Møller, A.P. and Swaddle, J.P. (1997) *Asymmetry, Developmental Stability and Evolution*. Oxford University Press, Oxford.

North, M.O. and Bell, D.D. (1990) *Commercial Chicken Production Manual*. Van Nostrand Reinhold, New York.

Parker, J.E., McKenzie, F.F. and Kempster, H.L. (1940) Observations on the sexual behaviour of New Hampshire males. *Poultry Science* 19, 191–197.

Parsons, P.A. (1992) Fluctuating asymmetry: a biological monitor of environmental and genomic stress. *Heredity* 68, 361–364.
Penquite, R., Craft, W.A. and Thomson, R.B. (1929) Variations in activity and production of spermatozoa by White Leghorn males. *Poultry Science* 9, 247–256.
Rappaport, S. and Soller, M. (1966) Mating behaviour, fertility and rate-of-gain in Cornish males. *Poultry Science* 45, 997–1003.
Rushen, J. (1983/84) The development of sexual relationships in the domestic chicken. *Applied Animal Behaviour Science* 11, 55–66.
Schein, M.W. and Hale, E.B. (1957) The head as a stimulus for orientation and arousal of sexual behaviour in male turkeys. *Anatomical Record* 128, 617.
Schoettle, H.E.T. and Schein, M.W. (1959) Sexual reactions of male turkeys to deviations from a normal female head model. *Anatomical Record* 134, 635.
Sharp, P.J. (1993) Photoperiodic control of reproduction in the domestic hen. *Poultry Science* 72, 897–905.
Siegel, P.B. (1965) Genetics of behaviour: selection for mating ability in chickens. *Genetics* 52, 1269–1277.
Skard, A.G. (1937) Studies in the psychology of needs: observations and experiments of the sexual needs of hens. *Acta Psychologica* 2, 175–232.
Sluckin, W. (1964) *Imprinting and Early Learning*. Methuen, London.
Smyth, J.R. and Leighton, A.T. (1953) A study of certain factors affecting fertility in the turkey. *Poultry Science* 32, 1004–1013.
Soller, M., Schindler, H. and Bornstein, S. (1965) Semen characteristics, failure of insemination and fertility in Cornish and White Rock males. *Poultry Science* 44, 424–432.
Swaddle, J.P. (1999) Limits to length asymmetry detection in starlings: implications for biological signalling. *Proceedings of the Royal Society of London B, Biological Sciences* 266, 1299–1303.
Swaddle, J.P. and Ruff, D.A. (2004) Starlings have difficulty in detecting dot symmetry: implications for studying fluctuating asymmetry. *Behaviour* 141, 29–40.
Tinbergen, N. (1952) Derived activities: their causation, biological significance, origin and emancipation during evolution. *Quarterly Review of Biology* 27, 1–32.
Upp, C.W. (1928) Preferential mating of fowls. *Poultry Science* 7, 225–232.
Venkataramaiah, A. and Choudary, M.R. (1973) Effects of strains and presence of male on agonistic behaviour and part-year egg production in White Leghorns. *Indian Journal of Poultry Science* 8, 258–263.
Verhulst, S., Dieleman, S.J. and Parmentier, H.K. (1999) A trade-off between immunocompetence and sexual ornamentation in domestic fowl. *Proceedings of the National Academy of Sciences of the USA* 96, 4478–4481.
West, B. and Zhou, B.Z. (1989) Did chickens go north? New evidence for domestication. *World's Poultry Science Journal* 45, 205–218.
Widowski, T.M., Lo Fo Wong, D.M.A. and Duncan, I.J.H. (1998). Rearing with males accelerates onset of sexual maturity on female domestic fowl. *Poultry Science* 77, 150–155.
Wilkins, R.H. (1915) Some factors influencing the fertility and hatching power of eggs of the domestic fowl. Unpublished PhD thesis, Cornell University.
Wilson, H.R., Piesco, N.P., Miller, E.R. and Nesbeth, W.G. (1979) Prediction of the fertility potential of broiler breeder males. *World's Poultry Science Journal* 35, 95–118.
Wood-Gush, D.G.M. (1954) The courtship of the Brown Leghorn cock. *British Journal of Animal Behaviour* 2, 95–102.
Wood-Gush, D.G.M. (1956) The agonistic and courtship behaviour of the Brown Leghorn cock. *British Journal of Animal Behaviour* 4, 133–142.
Wood-Gush, D.G.M. (1958a) Fecundity and sexual receptivity in the Brown Leghorn female. *Poultry Science* 37, 30–33.

Wood-Gush, D.G.M. (1958b) Genetic and experiential factors affecting the libido of cockerels. *Proceedings of the Royal Society of Edinburgh* 27, 6–8.

Wood-Gush, D.G.M. (1958c) The effect of experience on the mating behaviour of the domestic cock. *Animal Behaviour* 6, 68–71.

Wood-Gush, D.G.M. (1960) A study of sex drive of two strains of cockerels through three generations. *Animal Behaviour* 8, 43–53.

Wood-Gush, D.G.M. and Osborne, R. (1956) A study in differences in the sex drive of cockerels. *Animal Behaviour* 4, 102–110.

Zuk, M., Thornhill, R., Ligon, J.D., Johnson, K., Austad, S., Ligon S.H., Thirnhill, N. and Costin, C. (1990a) The role of male ornaments and courtship behaviour in female choice of red junglefowl. *American Naturalist* 136, 459–473.

Zuk, M., Thornhill, R. and Ligon, J.D. (1990b) Parasites and mate choice in red junglefowl. *American Zoologist* 30, 235–244.

CHAPTER 9
Sperm Competition and Fertilization Success

T.R. Birkhead[1] and T. Pizzari[2]

[1]*Department of Animal and Plant Sciences, University of Sheffield and* [2]*Edward Grey Institute of Ornithology, Department of Zoology, University of Oxford, UK*

ABSTRACT

Much of our understanding of the reproductive physiology of birds is derived from research by poultry biologists. Starting in the 1970s behavioural ecologists began to explore the evolutionary consequences of female promiscuity across a range of taxa, but mainly insects and birds, focusing on the fact that, if females copulate with two or more males in a single reproductive cycle, selection will favour the male that fertilizes the female's ova. Since male–male competition and female choice can continue after copulation, this area of research is known as post-copulatory sexual selection and comprises both sperm competition and cryptic female choice. Behavioural ecologists concentrated initially on functional (ultimate) issues (the adaptive significance of particular traits) but later also started to address mechanistic (proximate) questions. Information on the reproductive biology of domesticated birds from poultry research and collaboration between poultry biologists and behavioural ecologists provided an unusually productive opportunity for understanding some of the mechanisms of post-copulatory sexual selection in birds within an evolutionary context. We review those studies to identify the factors that determine male and female reproductive success.

INTRODUCTION

Until recently the majority of birds were considered to be monogamous, with a male and female pairing to rear a brood of young together. In his book *The Descent of Man and Selection in Relation to Sex*, Darwin reaffirmed the view that most birds were monogamous, although acknowledging that in pigeons the male might occasionally break his 'marriage vows' (Darwin, 1871). Even the domestic fowl was assumed to be monogamous; William Smellie (1790), paraphrasing Buffon, whose works he had translated, said that 'the dunghill cock and hen, in a natural state, pair [i.e. are monogamous]. In a domestic state, however, the cock is a jealous tyrant, and the hen, a prostitute.' He was wrong to assume that the fowl's sexual behaviour was entirely a consequence of domestication, captivity or artificial selection because it later became clear that the mating system of the red jungle fowl, the ancestor of the domestic fowl, was anything but monogamous, and best described as 'harem polygyny'.

While domestication and artificial selection have moulded the mating system of the fowl, and influenced the evolution of reproductive behaviour, studies of feral populations of domestic fowl and captive populations of red jungle fowl have confirmed that a high degree of sexual promiscuity occurs in 'natural' populations (McBride *et al.*, 1969; Collias and Collias, 1996; Pizzari *et al.*, 2002). These studies confirm that sperm competition is not a by-product of recent domestication but has been an ancestral and important selective episode underlying the evolution of reproductive traits in this species.

The entrenched idea that the majority of birds were sexually monogamous did not begin to change until the late 1960s with the development of 'behavioural ecology' – the evolutionary analysis of animal behaviour (see Krebs and Davies 1978, and subsequent editions; Alcock, 2001 for review). The essence of this new approach was that natural selection operated mostly on individuals rather than for the benefit of the group (Dawkins, 1976). One consequence was that behavioural ecologists started to think about particular behaviours or other traits as adaptations and to consider how they influenced variation in individual fitness or the ability of individuals to pass their genes on to future generations. Until the advent of behavioural ecology it had generally been assumed that the reproductive interests of males and females were identical and that reproduction was a cooperative venture between the sexes.

A major change of this paradigm occurred when Parker (1970) published a paper on insects that set out many of the theoretical foundations for the behavioural ecology of sperm competition in general. Parker defined sperm competition as the competition between the ejaculates of males for the fertilization of a female's eggs and recognized that sperm competition had important evolutionary implications. From an individual selection point of view it was crucial for males to ensure that their sperm and no one else's fertilized a female's eggs. It would pay males to defend their own paternity from competitors and, at the same time, 'steal' paternity from other males. Parker showed that, by copulating with already-mated females, males of some insect species could displace previously stored sperm and replace it with much of their own, fertilizing the majority of a female's eggs. The fact that both securing their own paternity and stealing it from other males was adaptive meant that reproductive traits – adaptations to sperm competition – were likely to evolve rapidly.

Since Parker's pioneering paper, it has become clear that sperm competition is almost ubiquitous across the animal kingdom. The development of molecular methods to assign paternity in the mid-1980s (Burke and Bruford, 1987) confirmed that, even in bird species traditionally thought to have a sexually monogamous mating system, extra-pair copulations and extra-pair fertilization were widespread. Researchers are now careful to distinguish between social monogamy (the observed mating system) and sexual (i.e. genetic) monogamy (Birkhead and Møller, 1992, 1998; Griffith *et al.*, 2002).

The initial focus of the behavioural ecology approach was on the evolutionary consequences of behaviour. These included the idea that frequent copulation within a breeding pair was a male paternity-assurance behaviour and that the close proximity between a male and female during the days the female was egg laying was another kind of paternity guard. As they pursued these questions researchers became interested in more mechanistic issues, such as whether copulation always resulted in ejaculation, how many sperm males transfer during copulation, when exactly a female could be fertilized by a male's sperm, and so on. Once it became clear that in many species females obtained sperm from multiple males and that this often resulted in broods of mixed paternity, researchers began to ask how the sperm from different males compete.

To help answer these questions researchers turned to the poultry biology literature. Until this point there had been very little overlap between the interests of field ornithologists and poultry researchers, but behavioural ecologists soon discovered that not only had poultry researchers already answered some of the key questions (albeit for different reasons), the domestic fowl and other poultry proved to be model organisms for answering additional questions (e.g. Pizzari, 2007).

Those of us working on sperm competition in birds and using the fowl as a study organism have gained enormously from the knowledge and cooperation of poultry researchers. Evolutionary biologists are interested in mechanisms underlying variation in individual fitness because these mechanisms generate Darwinian selection on traits associated with such variation. Poultry researchers are interested in variation in individual reproductive success to maximize the efficiency of breeding programmes and the design of artificial selection regimes. By using an evolutionary framework in which to explore the reproductive biology of birds, by comparing poultry with other birds (notably passerines, in which sperm competition is particularly common), and by considering both ultimate (evolutionary) as well as proximate (mechanistic) aspects of reproductive biology, we have gained a better understanding of the reproductive biology of birds. This interdisciplinary approach has proved particularly successful when investigating the mechanisms through which sperm competition impacts on variation in male and female fertility. Our aim in this chapter is to review our key findings relating to sperm competition and male and female fertility.

SPERM COMPETITION AND MALE FERTILITY

Mechanisms of sperm competition

It was noted by poultry breeders that when one cockerel was replaced by another, the second male fathered the majority of eggs (Crew, 1926; Warren and Kilpatrick, 1929). Similarly, when hens were artificially inseminated sequentially with the semen of two males, the longer the interval between the inseminations the greater proportion of offspring that were fathered by the second insemination (Warren and Gish, 1943; Compton *et al.*, 1978). This phenomenon was referred to as 'last male sperm precedence' and several possible mechanisms were thought to account for it (Lessells and Birkhead, 1990), including: (i) the stratification of sperm (last in, first out) in the female's sperm storage tubules (SSTs); (ii) displacement (second insemination displaces or devalues the previous one); or (iii) passive sperm loss. It was subsequently shown that last male sperm precedence observed in artificial insemination experiments was most consistent with the passive sperm loss model (Birkhead and Biggins, 1998). This did not mean, however, that under natural circumstances the last male always fertilized the majority of eggs. The natural situation was inevitably more complex. Rather, it meant that, all else being equal (i.e. sperm numbers, sperm quality), the greater the interval between two successive inseminations, the greater share of paternity went to the second male. This last male effect occurs because sperm are lost from the sperm storage tubules at a constant rate, and the longer the interval between two inseminations the greater proportion of sperm from the first male that have been lost or utilized before the second insemination. In other words, the passive sperm loss model of last male sperm precedence (all else being equal) hinges on sperm numbers (see also Colegrave *et al.*, 1995). This was consistent with an earlier study in which sperm from two males inseminated in different ratios (9:1, 8:2, 7:3...3:7, 2:8, 1:9) resulted in virtually identical ratios of paternity (Martin *et al.*, 1974). The effect of sperm numbers means that, if the second male inseminated many fewer sperm than the previous male, the first male would father a greater proportion of offspring (Birkhead, 1998). While passive sperm loss appears the most parsimonious and general explanation for variation in fertilization success under sperm competition, other mechanisms may still operate. A recent experimental study in domestic turkeys, *Meleagris gallopavo*, based on the artificial insemination of fluorescently stained sperm indicates that, when multiple successive ejaculates are stored within the same sperm storage tubule, sperm stratification may occur (King *et al.*, 2002). However, the proportion of sperm storage tubules of a female containing sperm from both males was typically low, indicating that, when sperm from different ejaculates are stratified within SSTs but are typically segregated in different SSTs, the ensuing pattern of variation in paternity may match closely that predicted by passive sperm loss alone (Pizzari, 2007). Clearly, more experimental work is needed to fully understand the mechanisms of sperm competition in poultry.

It was well known to early poultry researchers that semen quality varied markedly across cockerels (e.g. Curtis and Lambert, 1929; Romanoff, 1960). In the 1990s Froman and Feltmann (1998) devised a simple assay – referred to as sperm mobility – that in a non-competitive situation predicted fertilization success. In a competitive situation in which equal numbers of sperm from high- and low-mobility males were mixed, the high-mobility male fertilized the majority of eggs. In addition, the greater the disparity in mobility scores between the two males, the greater proportion of eggs the high-mobility male fathered. The differences in sperm mobility between males could be considerable, and in some cases the sperm of a high-mobility male could have a ten times greater probability of fertilizing a female's eggs than a low-mobility male (Birkhead *et al.*, 1999).

Together, these results indicated that sperm numbers and sperm quality could have independent important influences on the outcome of sperm competition. Through artificial insemination experiments, Froman *et al.* (2002) quantified the number of sperm stored in the SSTs and the rate of sperm loss from the SSTs over successive days to disentangle the relative effect of sperm number and sperm mobility on the fertilizing performance of an ejaculate. Viable sperm reaching a freshly ovulated egg in the infundibulum undergo the acrosome reaction and hydrolyse the perivitelline layer (PVL) of the egg, perforating through to the inner PVL (galliformes: Wishart, 1987; Steele *et al.*, 1994; Howarth and Donoghue, 1997; Froman *et al.*, 2006; see Birkhead *et al.*, 1993b for non-galliformes). In the fowl, the probability that an ovum is fertilized is a function of the number of sperm trapped in the PVL (Wishart, 1987, 1997). Therefore, the number of sperm on the PVL of eggs laid by a female on successive days following an insemination typically display a characteristic logarithmic decline, which provides an accurate measure of the number of sperm of an ejaculate that were originally stored in the SSTs (Brillard, 1993) and the rate at which they were depleted over time (Wishart, 1987).

By artificially inseminating single ejaculates of different mobility and sperm numbers, Froman *et al.* (2002) demonstrated that the number of sperm inseminated mostly influenced the number of sperm initially stored in the SSTs, while sperm mobility influenced the rate at which sperm were lost from the SSTs. Recently, Pizzari *et al.* (2008) explored the way in which relative number of sperm inseminated and sperm mobility interact to determine the outcome of sperm competition. Domestic hens were artificially inseminated with heterospermic inseminations containing either four or two times more sperm from a low-mobility male than from a high-mobility male (4:1 and 2:1 treatments, respectively), and paternity was monitored in eggs produced by individual hens over successive days following an insemination. Variation in paternity was determined by a clear mobility and relative number interaction. Generally, low-mobility ejaculates were more likely to fertilize the first eggs produced by a hen following the insemination than the high-mobility ejaculates. In the 4:1 treatment, low-mobility ejaculates retained a fertilizing advantage over the eggs produced in the first 5 days. However, when the numerical advantage to the low-mobility ejaculate was halved in the 2:1 treatment, the fertilizing advantage to the low-mobility ejaculate was restricted to the eggs produced over the first 2 days following an insemination. These results indicate that sperm from low-mobility ejaculates tend to exit the SSTs before sperm from high-mobility ejaculates, resulting in a numerical advantage of the low-mobility ejaculate around the first oocytes ovulated following an insemination. This interpretation is entirely consistent with what we know about sperm mobility and sperm storage within the SST (Bakst *et al.*, 1994; Froman, 2003).

Sperm mobility measures the overall swimming performance of a population of sperm. High-mobility ejaculates tend to contain a greater proportion of sperm cells with straight line velocity (VSL) higher than 30 μm/s than ejaculates with lower mobility (Froman and Feltmann, 2000). The mechanisms through which sperm mobility influences patterns of sperm storage within the SST are unresolved.

The mechanisms underlying sperm storage and utilization in the SSTs of female birds have attracted much attention since the discovery of the SSTs in the early 1960s but remain elusive and debated (Zavaleta and Ogasawara, 1987; Bakst *et al.*, 1994). Van Krey *et al.* (1981) assumed that sperm motility could be inhibited in the SST, leading sperm to agglutinate with each other in a 'quiescent' state within the SST. According to this theory, sperm leave the SST when the agglutinated state can no longer be maintained, and when this happens sperm motility is restored. However, this theory is based on assumptions of the physiology of avian sperm and SST which presently lack empirical support (Froman, 2003). Recently, Froman (2003) proposed a more conservative model and argued that sperm may be maintained in the SST by swimming against a flow generated in the lumen of the tubule. As sperm velocity declines below a threshold, sperm are flushed out of the SST and are able to reach the infundibulum through peristaltic movements of the oviduct for potential fertilization. Because low-mobility ejaculates contain slower-swimming sperm, this model predicts that a greater proportion of a low-mobility ejaculate may reach the infundibulum shortly following insemination than a high-mobility ejaculate, as indeed appears to be the case (Pizzari *et al.*, 2008).

Repeatable variation in sperm mobility across males (Froman *et al.*, 2002; Pizzari *et al.*, 2007) indicates potential for the evolution of alternative male reproductive strategies (see below).

Other factors may also be important in determining the competitive ability of sperm, including the timing of insemination relative to the time of egg laying (oviposition latency). For example, female birds store viable sperm in the SSTs over a prolonged period of time. In the domestic fowl, the median duration of fertility for an inseminated ejaculate is 14 days, for the domestic turkey, 21 days. Obviously, the effect of oviposition latency is in turn influenced by the number of sperm initially inseminated and on sperm mobility (see above).

Male adaptations to sperm competition

The mating system of the red jungle fowl is harem polygyny, and under natural circumstances social groups comprise a dominant male, several females and a number of subordinate males (McBride *et al.*, 1969; Collias and Collias, 1996; Pizzari *et al.*, 2002). Observations of feral fowl show that, while females typically prefer to associate with and copulate with the dominant male, subordinate males make frequent attempts to copulate with females (Pizzari, 2001). As a consequence of the promiscuity in this mating system and the prolonged sperm storage by females, ejaculates typically face a high risk of sperm competition, and as a result male fowl have evolved a number of behavioural, anatomical and physiological adaptations to sperm competition. Most notably, male fowl have relatively large testes, as in other species in which sperm competition is intense (Møller, 1991; Calhim and Birkhead, 2007). Relatively large testes allow males to sustain high sperm output, in part in response to copulating with several females that might be fertile simultaneously but primarily because of sperm competition.

In addition to promoting the evolution of increased testicular investment, sperm competition has also led to the evolution of strategies through which individual males can allocate their limited sperm reserves strategically according to the reproductive value attached to different copulation opportunities (Parker, 1998; Wedell *et al.*, 2002).

The ability of male fowl to ejaculate repeatedly is well known. We developed a simple method for collecting ejaculates (Pizzari and Birkhead, 2000; Pizzari, 2007), based on established techniques (Nishiyama, 1961; Nishiyama and Fujishima, 1961), and were able to demonstrate not only a remarkable capacity for successive ejaculation, albeit with a diminishing number of sperm, but a strong Coolidge effect. That is, after a succession of ejaculates with the same female, on being presented with a new female, the number of sperm in a male's ejaculate increased, suggesting that male fowl allocate sperm strategically to maximize their reproductive success (Pizzari *et al.*, 2003).

Of particular relevance to fertility is the fact that males often tend to mount a female that they previously inseminated without delivering additional sperm (Pizzari *et al.*, 2003). A recent series of experiments demonstrated the functional significance of this non-random occurrence of 'aspermic' copulations. Løvlie *et al.* (2005) showed that females exposed to 'aspermic' copulations, or, in other words, females exposed exclusively to the stimulus of male mounting, were unlikely to copulate again with a new male, while females that were not mounted were more likely to copulate with a new male. These results indicate that mounting *per se* may have strong, albeit transient, inhibitory effects on female promiscuity. Importantly, we were able to demonstrate that female inhibition caused by mounting *per se* was sufficient to reduce the number of sperm from a new male being stored in the female's SSTs (Løvlie *et al.*, 2005). These effects strongly suggest that, by mounting females they previously inseminated, males may be able to defend their paternity without additional ejaculate investment in a female. While this strategy may be beneficial for the male because it increases his share of paternity, females that are prevented from re-mating may pay a fertility cost as the proportion of eggs fertilized may be reduced.

The development of techniques to collect natural ejaculates from male fowl enabled us to conduct a series of experiments to establish how much control males had over the number of sperm they transferred and whether this differed with male social status and female quality. These studies showed that, depending on their social status, male fowl had unprecedented capacity to adjust the numbers of sperm that they transferred to females, according to the social status of a male, the reproductive quality of the female (Pizzari *et al.*, 2003; Cornwallis and Birkhead, 2006, 2007) and her genetic relatedness to a male (Pizzari *et al.*, 2004a; see below).

For example, socially dominant fowl preferentially allocated more sperm to attractive females (i.e. females with larger sexual ornaments such as the comb); subordinate males do not show any preferential allocation, possibly because, by always being in a disfavoured role in terms of sperm competition (that is, almost always facing sperm competition), it pays subordinate males to always transfer as many sperm as possible whenever they obtain a copulation opportunity. This study also showed that there were consistent differences between (socially dominant) individuals in their ability to adjust their ejaculate size (Cornwallis and Birkhead, 2006, 2007). In a further study, it was shown that both the number and, more surprisingly, the quality of sperm (measured as velocity, which influences fertilization success) depend both on male social status and on female attractiveness, suggesting that trade-offs exist between pre- and post-copulatory traits that influence male reproductive success (Cornwallis and Birkhead, 2008). While differential allocation of sperm numbers to individual copulations has been observed across a number of species (e.g. Wedell *et al.*, 2002), rapid change in the sperm mobility of individual males was unexpected. However, Froman *et al.* (2002) demonstrated that, in a random-bred population of domestic fowl, males producing low-mobility ejaculates were significantly more likely to be socially dominant over males producing high sperm mobility, indicating that there may be a trade-off between male investment in sperm mobility and male status. More recently, Pizzari *et al.* (2007) further explored the causal relationship between sperm mobility and male social status, and showed that this trade-off may be phenotypicaly plastic. The study experimentally changed the social status of males and analysed relative changes in sperm mobility before and after status manipulations. The results suggest that, while sperm mobility remains highly repeatable within males, males that become dominant following the experimental status change suffer a decline in sperm mobility, whilst sperm mobility remains constant in males that are socially subdominant.

Fowl spermatids take on average 14 days to fully develop into spermatozoa and reach the distal end of the ductus deferens adjacent to the cloaca (Lake, 1984; Kirby and Froman, 2000), and, since some of the fluctuations in relative sperm mobility observed by Cornwallis and Birkhead (2007) and Pizzari *et al.* (2007) occurred within just 3 days of the status change, these are too rapid to be explained by changes in sperm production occurring during the change in status. Instead, these results indicate that changes in sperm mobility may be determined by rapid changes in the biochemical milieu of the seminal plasma, where sperm mature in the extragonadal sperm stores of a male. Consistent with this idea, Froman *et al.* (2006) provided convincing *in vitro* evidence that seminal plasma may have critical effects on sperm mobility. Seminal plasma is glutamate rich, and Froman *et al.* (2006) indicate that glutamate may enable sperm uptake of Ca^{2+} during maturation in the ductus deferens. Variation in mobility appears to be determined by the fact that high sperm mitochondrial Ca^{2+} levels trigger the formation of mitochondrial permeability transition pores, which accelerate sperm senescence and reduce the metabolic performance of individual sperm, and thus the mobility of an ejaculate.

Because ejaculates of low mobility are competitive only for a brief period of time, a male producing ejaculates of low mobility may need to copulate frequently with a female to 'top up' his sperm reserves in the female's SSTs. Consequently, from an evolutionary point of view, a trade-off between male investment in sperm mobility and in social status may occur, in which dominant males invest preferentially in female monopolization and reduced risk of sperm competition.

FEMALE REPRODUCTIVE SUCCESS

Female reproductive success in birds depends upon fecundity, fertility, hatching success and the subsequent survival of hatched young. Fertility is the proportion of eggs that are fertilized and hatching success (sometimes called hatchability in poultry research) is the proportion of eggs that hatch. Female reproductive success depends on both sperm quantity and quality. The poultry industry has placed great emphasis on female fertility since it often seeks to maximize the number of fertile eggs per female and as a result has selected for female fertility and uses management practices that have little relation to the natural reproductive biology of the species. For example, a female jungle fowl may typically lay a clutch of six to eight eggs two or three times a year whereas commercial fowl typically lay one egg each day throughout the year. Like all birds, female fowl store sperm and can produce fertile eggs for a maximum of 3 or 4 weeks after the last insemination (Romanoff, 1960). However, the duration of fertility (the 'fertile period'; Lake, 1975) is highly dependent upon the numbers of sperm inseminated, with fewer sperm resulting in a shorter fertile period (Taneja and Gowe, 1961). Typically, poultry researchers inseminate about 200×10^6 sperm, resulting in a median fertile period of about 14 days. In natural copulations, sexually rested red jungle fowl usually inseminate fewer sperm than this (usually $<50 \times 10^6$; Pizzari and Birkhead, 2000; Pizzari *et al.*, 2004a). However, under natural conditions, female fowl copulate repeatedly several times each day, both before and during the egg-laying period, so the total sperm load in the sperm storage tubules may not be as dissimilar as the difference in ejaculate size suggests, although, judging from counts of sperm on the perivitelline layers of eggs, they are typically lower than those from artificially inseminated commercial hens. Since the mating system of red jungle fowl is characterized by a mixture of male polygyny and female polyandry and males have an enormous capacity for copulation and insemination, under natural conditions females must rarely suffer infertility as a result of insufficient sperm.

Pre-copulatory mate choice

Under natural conditions female birds may control the quality of the sperm they use to fertilize their eggs through processes that operate before and after copulation. These processes represent a central part of Darwin's concept of sexual selection and are usually referred to as pre-copulatory and post-copulatory sexual selection. In terms of pre-copulatory sexual selection, attempts to test the idea that female red jungle fowl prefer males of different phenotypes consistently showed a female preference for males with a large and brightly coloured comb while female response to other male phenotypic traits proved less consistent (Zuk *et al.*, 1990; Parker and Ligon, 2003). Pizzari *et al.*

(2004b) tested the idea that this female preference may be explained by male comb size being associated with the production of better ejaculates. While males with a larger comb had higher relative testicular mass, there was no evidence that sperm swimming velocity was predicted by comb size. Therefore, males with larger combs may have higher sperm reserves but not necessarily better quality sperm. Inconsistent relationships between male comb size and sperm swimming performance have also been reported by Parker *et al.* (2006) in a population of red jungle fowl, and by McGary *et al.* (2003) in domestic broiler breeder strains. In addition, because male sperm allocation is highly plastic in this species (see above), it would be difficult for females to predict the number of sperm inseminated by a male if his quality was based simply on the expression of a sexual ornament. On the other hand, it appears increasingly likely that male comb expression may reflect important and heritable aspects of the male immune system (Zuk *et al.*, 1995). Because comb phenotypes are heritable, it is possible that female preference for male comb characteristics may be maintained by offspring viability benefits (Parker, 2003; Parker and Garant, 2004).

It is important to bear in mind that these mate choice experiments were designed to detect female preference for certain male physical characteristics while minimizing the effect of social interactions between males. Subsequent studies of free-ranging birds showed that females exhibited a clear preference for socially dominant males (e.g. Wood-Gush, 1971; Leonard and Zanette, 1998; Pizzari and Birkhead, 2000; Pizzari, 2001). The extent to which male comb expression and social status influence female preference in natural groups, independently and through an interaction, remains unclear. In free-ranging red jungle fowl groups an interaction between male comb and status was observed: most females preferred dominant partners but bottom-ranking females were more likely to copulate with subdominant males when these had larger combs (Johnsen *et al.*, 2001). In general, males with larger combs have a competitive advantage (Graves *et al.*, 1985; Ligon *et al.*, 1990; Zuk and Johnsen, 2000), and comb size co-varies positively with male status, although this relationship is often weak (Johnsen *et al.*, 2001; Parker *et al.*, 2002). Experimental manipulations of male status are followed by changes in comb expression (Zuk and Johnsen, 2000; Parker *et al.*, 2002), suggesting a sensitive feedback between perceived social status and ornament expression. Large combs elicit confrontation and comb reduction may avoid costly fights (Parker and Ligon, 2002).

Later studies showed that male status affects both the quantity and quality of sperm that they transfer to females, thereby masking the difference between pre- and post-copulatory processes (see below and Pizzari *et al.*, 2007; Cornwallis and Birkhead, 2007, 2008). Although female fowl show a clear preference for socially dominant males, we do not yet know whether dominance *per se* (that is, after controlling for sperm numbers and quality) influences female reproductive success.

Post-copulatory processes of female choice centre on the storage and utilization of sperm by females. Female birds store only 1 or 2% of inseminated sperm (Bakst and Brillard, 1990; Brillard and Antoine, 1990); the figures for a small passerine bird (mass: 12–17 g), the zebra finch *Taeniopygia guttata*, are similar (Birkhead *et al.*, 1993b).

The larger the insemination the smaller the proportion stored (Brillard and Antoine, 1990), suggesting that the sperm that are stored may be a non-random subset of those inseminated (see also Birkhead and Brillard, 2007). Similar massive reductions in the number of sperm retained by females are known to occur in many other animal taxa, and a long-standing question has been whether the retained portion represents a 'selected' cohort of sperm (Birkhead *et al.*, 1993a). In experiments involving rabbits, Cohen and Tyler (1980) reported pronounced non-random selection of sperm, a result, however, that was not repeatable (Overstreet, 1983). In birds there is circumstantial evidence for non-random selection of sperm. Eslick and McDaniel (1992) looked at fertility and embryo mortality in fowl inseminated with different numbers of sperm and found that both fertility (not surprisingly) and embryo mortality increased the smaller the number of sperm inseminated. One explanation for the increase in embryo mortality suggests that with only a small number of sperm females are forced to utilize suboptimal sperm. However, this idea is currently being tested (Hemmings, Brillard and Birkhead, unpublished).

In an attempt to maximize the duration of fertility (and minimize the number of times hens need to be artificially inseminated), poultry researchers selected for the number of chicks hatched following a single insemination (Brillard *et al.*, 1998). This selection resulted in a marked increase in the duration of fertility, an effect achieved, in part at least, through a 25% increase in the number of SSTs in the selected line compared with the control line, which in turn resulted in a significantly greater number of sperm stored from a standard insemination. It is not known whether the selected lines had larger SSTs (although this is currently being investigated; Hemmings, Brillard and Birkhead, unpublished), which may also increase the capacity for sperm storage. Certainly, in other birds, SSTs differ markedly in length between species (Briskie and Montgomerie, 1992). Nor is it known whether the rate of release of sperm from the SSTs differed between selected and control lines (Brillard *et al.*, 1998).

Post-copulatory phenomena: cryptic female choice

Almost by definition, sperm competition focuses on the male, and certainly for many years it was assumed that females were passive participants in what was seen as an exclusively male exercise. In the early 1900s there was speculation that females may play a more important role in controlling fertilization, and in the mid-1980s the issue was raised again by Thornhill (1983) in his study of the scorpionfly, *Harpobittacus nigriceps*, and referred to as 'cryptic female choice', cryptic because it occurred out of sight inside the female's body, but again there was little attempt to test the idea. In an encyclopedic review of reproductive processes Eberhard (1996) made a strong case for cryptic female choice in animals as a whole, mainly on the grounds of plausibility rather than unequivocal empirical evidence. Indeed, testing cryptic female choice proved to be far from easy because it requires controlling for all possible male effects to be confident that the observed effect is female-mediated (Birkhead, 1998, 2000).

The domestic fowl proved to be an ideal species in which to test cryptic female choice. In multi-male groups, female fowl are routinely subjected to forced copulation attempts by subordinate males. Usually females simply run away from such advances. On other occasions, if the male grasps the female (usually by the nape or comb), she utters a specific distress call that alerts the dominant male, who then approaches rapidly and intervenes (Pizzari, 2001). Experimental playback of these calls confirmed that they serve to attract the dominant male during forced copulation attempts by subordinates. The fact that female feral fowl so obviously attempt to avoid being inseminated by subordinates and preferentially solicit copulations from the dominant males is clear evidence for a preference for the dominant male. The fact that females are susceptible to forced inseminations is precisely the kind of situation in which post-insemination mechanisms of male choice might be expected to evolve (Eberhard, 1996). Thornhill (1988) reported female fowl ejecting semen immediately after insemination, and using habituated feral fowl we confirmed that sperm ejection occurred and did so predominantly after females had been inseminated by subordinate males. We also experimentally manipulated male dominance and found the same effect (Pizzari and Birkhead, 2000). Our results indicated that females based their decision on whether to eject semen or not on male status.

We also tested the hypothesis that female fowl might exert cryptic female choice in the absence of any male phenotypic cues. Using nine pairs of males (taken at random and with no knowledge of their sperm mobility or status), we mixed equal numbers of viable sperm from each cockerel and inseminated them into each of ten females (per male pair) and recorded the paternity of the offspring from each female. We repeated this procedure over several successive months with the same individuals, in order to check for consistency in patterns of paternity. Our prediction was that, if cryptic female choice does not exist, the pattern of paternity should be similar across all females. This was not the case: certain individual females preferentially and consistently fertilized their eggs using sperm of one male while others used the sperm of the other male. This provides strong support for the idea of cryptic female choice (Birkhead *et al.*, 2004). The mechanism by which females use the sperm of one male in preference to another is unknown but may involve an interaction between sperm surface proteins and the female reproductive tract (Birkhead and Brillard, 2007; see below). One of the adaptive explanations for cryptic female choice is that it is a mechanism for avoiding genetic incompatibility (Tregenza and Wedell, 2000).

Clear experimental evidence for cryptic female choice against genetically incompatible partners was provided by a recent study of red jungle fowl. A classic case of genetic incompatibility between reproductive partners is inbreeding, which results in inbreeding depression through loss of overdominance effects and the expression of deleterious recessives (Charlesworth and Charlesworth, 1987). Natural populations of red jungle fowl are often small, isolated and prone to inbreeding. Indeed, as many as 4% of the copulations observed in free-range captive populations of red jungle fowl occurred between first-order relatives (Collias and Collias, 1996). Inbreeding depression has been documented in domestic fowl (Craig and Baruth, 1965; Cheng *et al.*, 1985; Abplanalp *et al.*, 1992). Pizzari *et al.* (2004a) demonstrated that, when exposed to a single prospective partner, either unrelated or full sib, male and female red jungle fowl 'disagree' over inbreeding. When forced to copulate with a brother, females select against sperm of their brother after insemination. They do so in a way that significantly reduces the number of sperm reaching the eggs produced by a female following insemination by a related male relative to the eggs produced by the same female following insemination by an unrelated male. Males, on the other hand, copulate just as readily with sisters as with unrelated females, and in fact appear to inseminate more sperm into their sisters. The functional significance of this male response may be explained by the strong selection against genetically related sperm operated by females. Recent follow-ups of this study have confirmed these antagonistic male and female responses, and shown that, when they are simultaneously exposed to both related and unrelated females, males too display a preference for unrelated females (Løvlie, unpublished data). Together, these results reveal that, because of inbreeding depression, both sexes should favour unrelated partners when given the opportunity to choose. However, because males can produce gametes at a faster rate than females, under some conditions they may also benefit by investing in inbreeding, while females typically avoid inbreeding. These evolutionary dynamics are likely to have profound implications for patterns of variation in fertility and paternity in domestic fowl populations.

CONCLUSION

Prior to the 1970s much of our understanding of the reproductive biology of birds was derived from studies of the domestic fowl and turkeys by poultry biologists. A paradigm shift in evolutionary thinking in the 1970s, and the development of molecular paternity assays in the 1980s that demonstrated that promiscuity and mixed paternity were widespread in birds (despite most species exhibiting a socially monogamous mating system), triggered a study of both the functional (ultimate) and mechanistic (proximate) aspects of sperm competition in birds by evolutionary biologists. Acknowledging the huge body of information on the reproductive biology of domesticated birds derived from the studies of poultry biologists, evolutionary biologists recognized the enormous potential of the domestic fowl as a model organism for the study of post-copulatory sexual selection (i.e. sperm competition and cryptic female choice). These studies, which built on earlier knowledge and have involved collaborations with poultry biologists, and studies of feral fowl in situations that encouraged their natural behaviour have resulted in a considerable advance in our understanding of the reproductive biology of birds, especially the factors that determine both male and female reproductive success.

ACKNOWLEDGEMENTS

Our research on sperm competition in birds has been supported by NERC, BBSRC, the Royal Society and the Leverhulme Trust. We thank our colleagues J.P. Brillard, C.K. Cornwallis, D.P. Froman, N. Hemmings, H. Løvlie and G.J. Wishart for their help.

REFERENCES

Abplanalp, H., Sato, K., Napolitano, D. and Reid, J. (1992) Reproductive performance of inbred Leghorns carrying different haplotypes for the major histocompatibility complex. *Poultry Science* 71, 9–17.

Alcock, J. (2001) *Animal Behaviour: an Evolutionary Approach*. Sinauer, Sunderland, UK.

Bakst, M.R. and Brillard, J.P. (1990) Sperm transport and selection in the turkey oviduct. In: Brillard, J.P. (ed.) *Control of Fertility in Domestic Birds*. INRA, Paris, pp. 113–120.

Bakst, M.R., Wishart, G.J. and Brillard, J.P. (1994) Oviducal sperm selection, transport and storage in poultry. *Poultry Science Reviews* 5, 117–143.

Birkhead, T.R. (1998) Cryptic female choice: criteria for establishing female sperm choice. *Evolution* 52, 1212–1218.

Birkhead, T.R. (2000) Defining and demonstrating post-copulatory female choice – again. *Evolution* 54, 1057–1060.

Birkhead, T.R. and Biggins, J.D. (1998) Sperm competition mechanisms in birds: models and data. *Behavioural Ecology* 9, 253–260.

Birkhead, T.R. and Brillard, J.P. (2007) Postcopulatory prezygotic reproductive isolation in birds. *Trends in Ecology and Evolution* 22, 266–272.

Birkhead, T.R. and Møller, A.P. (1992) *Sperm Competition in Birds: Evolutionary Causes and Consequences*. Academic Press, London.

Birkhead, T.R. and Møller, A.P. (1998) Sperm competition, sexual selection and different routes to fitness. In: Birkhead, T.R. and Møller, A.P. (eds) *Sperm Competition and Sexual Selection*. Academic Press, London, pp. 757–781.

Birkhead, T.R., Møller, A.P. and Sutherland, W.J. (1993a) Why do females make it so difficult for males to fertilize their eggs? *Journal of Theoretical Biology* 161, 51–60.

Birkhead, T.R., Pellatt, E.J. and Fletcher, F. (1993b) Selection and utilization of spermatozoa in the reproductive tract of the female zebra finch *Taeniopygia guttata*. *Journal of Reproduction and Fertility* 99, 593–600.

Birkhead, T.R., Martinez, J.G., Burke, T. and Froman, D.P. (1999) Sperm mobility determines the outcome of sperm competition in the domestic fowl. *Proceedings of the Royal Society of London Series B* 266, 1759–1764.

Birkhead, T.R., Chaline, N., Biggins, J.D., Burke, T.A. and Pizzari, T. (2004) Nontransitivity of paternity in a bird. *Evolution* 58, 416–420.

Brillard, J.P. (1993) Sperm storage and transport following natural mating and artificial insemination. *Poultry Science* 72, 923–928.

Brillard, J.P. and Antoine, H. (1990) Storage of sperm in the uterovaginal junction and its incidence on the numbers of spermatozoa present in the perivitelline layer of hen's eggs. *British Poultry Science* 31, 635–644.

Brillard, J.P., Beaumont, C. and Scheller, M.F. (1998) Physiological responses of hens divergently selected on the basis of the number of chicks obtained from a single insemination. *Journal of Reproduction and Fertility* 114, 111–117.

Briskie, J.V. and Montgomerie, R. (1992) Sperm size and sperm competition in birds. *Proceedings of the Royal Society of London Series B* 247, 89–95.

Burke, T. and Bruford, M.W. (1987) DNA fingerprinting in birds. *Nature* 327, 149–152.
Calhim, S. and Birkhead, T.R. (2007) Testes size in birds – quality vs quantity: assumptions, errors and estimates. *Behavioural Ecology* 18, 271–275.
Charlesworth, D. and Charlesworth, B. (1987) Inbreeding depression and its evolutionary consequences. *Annual Review of Ecology and Systematics* 18, 237–268.
Cheng, K.M., Burns, J.T. and Schoffner, R.N. (1985) Mating behaviour and fertility in domestic chickens. 1. Inbreeding. *Applied Animal Behaviour Science* 13, 371–381.
Cohen, J. and Tyler, K.R. (1980) Sperm populations in the female genital tract of the rabbit. *Journal of Reproduction and Fertility* 60, 213–218.
Colegrave, N., Birkhead, T.R. and Lessells, C.M. (1995) Sperm precedence in zebra finches does not require special mechanisms of sperm competition. *Proceedings of the Royal Society of London Series B* 259, 223–228.
Collias, N.E. and Collias, E.C. (1996) Social organization of a red junglefowl, *Gallus gallus*, population related to evolutionary theory. *Animal Behaviour* 51, 1337–1354.
Compton, M.M., Van Krey, H.P. and Siegel, P.B. (1978) The filling and emptying of the uterovaginal sperm-host glands in the domestic hen. *Poultry Science* 57, 1696–1700.
Cornwallis, C.K. and Birkhead, T.R. (2006) Social status and availability of females determine patterns of sperm allocation in the fowl. *Evolution* 60, 1486–1493.
Cornwallis, C.K. and Birkhead, T.R. (2007) Experimental evidence that female ornamentation increases the acquisition of sperm and signals fecundity. *Proceedings of the Royal Society of London Series B* 274, 583–590.
Cornwallis, C.K. and Birkhead, T.R. (2008) Changes in sperm quality and numbers in response to experimental manipulation of male social status and female attractiveness. *American Naturalist* 170, 758–770.
Craig, J.V. and Baruth, R.A. (1965) Inbreeding and social dominance ability in chickens. *Animal Behaviour* 13, 109–113.
Crew, F.A.E. (1926) On fertility in the domestic fowl. *Proceedings of the Royal Society of Edinburgh Series B* 46, 230–238.
Curtis, V. and Lambert, W.V. (1929) A study of fertility in poultry. *Poultry Science* 8, 142–150.
Darwin, C. (1871) *The Descent of Man and Selection in Relation to Sex*. John Murray, London.
Dawkins, R. (1976) *The Selfish Gene*. Oxford University Press, Oxford.
Eberhard, W.G. (1996) *Female Control: Sexual Selection by Cryptic Female Choice*. Princeton University Press, Princeton.
Eslick, M.L. and McDaniel, G.R. (1992) Interrelationships between fertility and hatchability of eggs from broiler breeder hens. *Applied Poultry Research* 1, 156–159.
Froman, D.P. (2003) Deduction of a model for sperm storage in the oviduct of the domestic fowl (*Gallus domesticus*). *Biology of Reproduction* 69, 248–253.
Froman, D.P. and Feltmann, A. (1998) Sperm mobility: a quantitative trait in domestic fowl *Gallus domesticus*. *Biology of Reproduction* 58, 379–384.
Froman, D.P. and Feltmann, A.J. (2000) Sperm mobility: phenotype in roosters (*Gallus domesticus*) determined by concentration of motile sperm and straight line velocity. *Biology of Reproduction* 63, 303–309.
Froman, D.P., Pizzari, T., Feltmann, A.J., Castillo-Juarez, H. and Birkhead, T.R. (2002) Sperm mobility: mechanisms of fertilizing efficiency, genetic variation and phenotypic relationship with male status in the fowl, *Gallus g. domesticus*. *Proceedings of the Royal Society of London Series B* 269, 607–612.
Froman, D.P., Wardell, J.C. and Feltmann, A.J. (2006) Sperm mobility: deduction of a model explaining phenotypic variation in roosters. *Biology of Reproduction* 74, 487–491.

Graves, H.B., Hable, C.P. and Jenkins, T.H. (1985) Sexual selection in *Gallus*: effects of morphology and dominance on female spatial behaviour. *Behavioural Processes* 11, 189–197.

Griffith, S.C., Owens, I.P.F. and Thurman, K.A. (2002) Extra pair paternity in birds: a review of interspecific variation and adaptive function. *Molecular Ecology* 11, 2195–2212.

Howarth, B. and Donoghue, A.M. (1997) Determination of holes made by sperm in the perivitelline layer of laid eggs. In: Bakst, M.R. and Cecil, H.C. (eds) *Techniques for Semen Evaluation, Semen Storage and Fertility Determination*. Poultry Science Association, Savoy, pp. 93–97.

Johnsen, T.S., Zuk, M. and Fessler, E.A. (2001) Social dominance, male behaviour and mating in mixed-sex flocks of red jungle fowl. *Behaviour* 138, 1–18.

King, L.M., Brillard, J.P., Garrett, W.M., Bakst, M.R. and Donoghue, A.M. (2002) Segregation of spermatozoa within sperm storage tubules of fowl and turkey hens. *Reproduction* 123, 79–86.

Kirby, J.D. and Froman, D.P. (2000) Reproduction in the male. In: Wittow, G.C. (ed.) *Sturkie's Avian Physiology*, 5th edn. Academic Press, San Diego, pp. 597–615.

Krebs, J.R. and Davies, N.B. (1978) *Behavioural Ecology: an Evolutionary Approach*, 1st edn. Blackwell Scientific Publications, Oxford.

Lake, P.E. (1975) Gamete production and the fertile period with particular reference to domesticated birds. *Symposia of the Zoological Society of London* 35, 225–244.

Lake, P.E. (1984) The male in reproduction. In: Freeman, B.M. (ed.) *Physiology and Biochemistry of the Domestic Fowl*, Vol. 5. Academic Press, London, pp. 381–405.

Leonard, M.L. and Zanette, L. (1998) Female mate choice and male behaviour in domestic fowl. *Animal Behaviour* 56, 1099–1105.

Lessells, C.M. and Birkhead, T.R. (1990) Mechanisms of sperm competition in birds: mathematical models. *Behavioural Ecology and Sociobiology* 27, 325–337.

Ligon, J.D., Thornhill, R., Zuk, M. and Johnsen, K. (1990) Male–male competition in red jungle fowl and the multiple roles of testosterone in sexual selection. *Animal Behaviour* 40, 367–373.

Løvlie, H., Cornwallis, C.K. and Pizzari, T. (2005) Male mounting alone reduces female promiscuity in the fowl. *Current Biology* 15, 1222–1227.

Martin, P.A., Reimers, T.J., Lodge, J.R. and Dzuik, P.J. (1974) The effects of ratios and numbers of spermatozoa mixed from two males on the proportion of offspring. *Journal of Reproduction and Fertility* 39, 251–258.

McBride, G., Parer, I.P. and Foenader, F. (1969) The social organisation and behaviour of the feral domestic fowl. *Animal Behaviour Monographs* 2, 125–181.

McGary, S., Estevez, I. and Bakst, M.R. (2003) Potential relationships between physical traits and male broiler breeder fertility. *Poultry Science* 82, 328–337.

Møller, A.P. (1991) Sperm competition, sperm depletion, parental care and relative testis size in birds. *American Naturalist* 137, 882–906.

Nishiyama, H. (1961) On the quality and quantity of the cock semen obtained by different collection methods. *Memoirs of the Faculty of Agriculture, Kagoshima University* 4, 27–42.

Nishiyama, H. and Fujishima, T. (1961) On the ejection of the accessory reproductive fluid of the cock during natural copulation. *Memoirs of the Faculty of Agriculture, Kagoshima University* 4, 43–50.

Overstreet, J.W. (1983) Transport of gametes in the reproductive tract of the female mammal. In: Hartmann, J.F. (ed.) *Mechanism and Control of Animal Fertilization*. Academic Press, San Diego.

Parker, G.A. (1970) Sperm competition and its evolutionary consequences in the insects. *Biological Reviews* 45, 525–567.

Parker, G.A. (1998) Sperm competition and the evolution of ejaculates: towards a theory base. In: Birkhead, T.R. and Møller, A.P. (eds) *Sperm Competition and Sexual Selection*. Academic Press, London.

Parker, T.H. (2003) Genetic benefits of mate choice separated from differential maternal investment in red junglefowl (*Gallus gallus*). *Evolution* 57, 2157–2165.

Parker, T.H. and Garant, D. (2004) Quantitative genetics of sexually dimorphic traits and capture of genetic variance by a sexually-selected condition-dependent ornament in red jungle fowl (*Gallus gallus*). *Journal of Evolutionary Biology* 17, 1277–1285.

Parker, T.H. and Ligon, J.D. (2002) Dominant male red jungle fowl (*Gallus gallus*) test the dominance status of other males. *Behavioral Ecology and Sociobiology* 53, 20–24.

Parker, T.H. and Ligon, J.D. (2003) Female mating preferences in red junglefowl: a meta-analysis. *Ethology, Ecology, and Evolution* 15, 63–72.

Parker, T.H., Knapp, R. and Rosenfield, J.A. (2002) Social mediation of sexually selected ornamentation and steroid hormone levels in male junglefowl. *Animal Behaviour* 64, 291–298.

Parker, T.H., Thompson, D., Ligon, J.D., Schneider, B. and Byrn, F. (2006) Does red junglefowl comb size predict sperm swimming speed and motility? *Ethology, Ecology, and Evolution* 18, 53–60.

Pizzari, T. (2001) Indirect partner choice through manipulation of male behaviour by female fowl, *Gallus g. domesticus*. *Proceedings of the Royal Society of London Series B* 268, 181–186.

Pizzari, T. (2007) Post-insemination sexual selection in birds. In: Roldan, E. and Gomendio, M (eds) *Spermatology*. University Press, Nottingham, pp. 137–154.

Pizzari, T. and Birkhead, T.R. (2000) Female fowl eject sperm of subdominant males. *Nature* 405, 787–789.

Pizzari, T., Froman, D.P. and Birkhead, T.R. (2002) Pre- and post-insemination episodes of sexual selection in the fowl. *Heredity* 89, 112–116.

Pizzari, T., Cornwallis, C.K., Løvlie, H., Jakobsson, S. and Birkhead, T.R. (2003) Sophisticated sperm allocation in male fowl. *Nature* 426, 70–74.

Pizzari, T., Løvlie, H. and Cornwallis, C.K. (2004a) Sex-specific counteracting responses to inbreeding in a bird. *Proceedings of the Royal Society of London Series B* 271, 2115–2121.

Pizzari, T., Jensen, P. and Cornwallis, C.K. (2004b) A novel test of the phenotype-linked fertility hypothesis reveals independent components of fertility. *Proceedings of the Royal Society of London Series B* 271, 51–58.

Pizzari, T., Cornwallis, C.K. and Froman, D.P. (2007) Social competitiveness associated with rapid fluctuations in sperm quality in male fowl. *Proceedings of the Royal Society of London Series B* 274, 853–860.

Pizzari, T., Worley, K., Burke, T.A. and Froman, D.P. (2008) Sperm competition dynamics: ejaculate fertilising efficiency changes differentially with time. *BMC Evolutionary Biology* 8: 332, doi: 10.1186/1471-2148-8-332.

Romanoff, A.L. (1960) *The Avian Embryo*. Macmillan, New York.

Smellie, W. (1790) *The Philosophy of Natural History*. Elliot, Edinburgh, p. 278.

Steele, M.G., Meldrum, W., Brillard, J.P. and Wishart, G.J. (1994) The interaction of avian spermatozoa with the perivitelline layer in vitro and in vivo. *Journal of Reproduction and Fertility* 101, 599–603.

Taneja, G.C. and Gowe, R.S. (1961) The effect of dosage of undiluted semen on fertility in two breeds of fowl. *British Poultry Science* 2, 81–89.

Thornhill, R. (1983) Cryptic female choice and its implications in the scorpionfly *Harpobittacus nigriceps*. *American Naturalist* 122, 765–788.

Thornhill, R. (1988) The jungle fowl hen's cackle incites male competition. *Verhandlungen der Deutschen Zoologischen Gesellschaft* 81, 145–154.

Tregenza, T. and Wedell, N. (2000) Genetic compatibility, mate choice and patterns of parentage. *Molecular Ecology* 9, 1013–1027.

Van Krey, H.P., Balander, R.J. and Compton, M.M. (1981) Storage and evacuation of spermatozoa from the uterovaginal sperm-host glands in domestic fowl. *Poultry Science* 60, 871–877.

Warren, D.C. and Gish, C.L. (1943) Value of artificial insemination in poultry breeding work. *Poultry Science* 22, 108–117.

Warren, D.C. and Kilpatrick, L. (1929) Fertilization in the domestic fowl. *Poultry Science* 8, 237–256.

Wedell, N., Gage, M.J.G. and Parker, G.A. (2002) Sperm competition, male prudence and sperm-limited females. *Trends in Ecology and Evolution* 17, 313–320.

Wishart, G.J. (1987) Regulation of the length of the fertile period in the domestic fowl by numbers of oviducal spermatozoa, as reflected by those trapped in laid eggs. *Journal of Reproduction and Fertility* 80, 493–498.

Wishart, G.J. (1997) Quantitative aspects of sperm:egg interaction in chickens and turkeys. *Animal Reproductive Science* 48, 81–92.

Wood-Gush, D.G.M. (1971) *Behaviour of the Domestic Fowl*. Heinemann, London.

Zavaleta, D. and Ogasawara, F. (1987) A review of the mechanism of the release of spermatozoa from storage tubules in the fowl and turkey oviduct. *World's Poultry Science Journal* 43, 132–139.

Zuk, M. and Johnsen, T.S. (2000) Social environment and immunity in male red jungle fowl. *Behavioural Ecology* 11, 146–153.

Zuk, M., Thornhill, R., Ligon, J.D., Johnson, K., Austad, S., Ligon, S.H., Thornhill, N.W. and Costin, C. (1990) The role of male ornaments and courtship behaviour in female mate choice of red jungle fowl. *American Naturalist* 136, 459–473.

Zuk, M., Johnsen, T.S. and Maclarty, T. (1995) Endocrine-immune interactions, ornaments and mate choice in red jungle fowl. *Proceedings of the Royal Society of London Series B* 260, 205–210.

CHAPTER 10
Semen Quality and Semen Storage

G.J. Wishart
University of Abertay, Dundee, UK

ABSTRACT

Poultry semen quality and semen storage respectively can be considered in terms of the intrinsic components of spermatozoa and seminal plasma and their properties, and of the oviducal sperm-storage systems and their regulation. However, the focus of this chapter is on the technologies for assessing semen quality and for storing semen *in vitro*.

Semen quality is considered as methods for sperm quality evaluation to identify males of different fertilizing abilities. It is shown that sperm quality assays can be used to predict fertilizing ability and to select the 'best breeders' from a flock of chickens or turkeys when artificial insemination is employed. It may also be possible to select for sperm quality traits in breeding programmes. Sperm quality assays have, however, been shown to be poor monitors of the loss of fertilizing ability that occurs in semen during storage.

Semen storage is reviewed in terms of cryopreservation, which enables long-term storage of spermatozoa at −196°C, the temperature of liquid nitrogen, and liquid semen storage, usually at lower than ambient temperatures above 0°C for up to 24 or 48 h.

The most significant advance in sperm cryopreservation technology is the recent establishment of poultry sperm banks in several countries – the result of an international realization of the rate of loss of poultry breeds and lines. The background to the cryopreservation technology used in these cryobanks is considered with a view to providing an appreciation of the opportunities and limitations of poultry sperm cryopreservation.

Liquid semen storage of chicken spermatozoa at temperatures of around 5°C is a well-established, although little-used, technology. Turkey sperm storage is successfully utilized in commercial breeding operations, but the efficiency of such systems in maintaining sperm fertilizing ability *in vitro* for 24 h pale into insignificance in comparison to the *in vivo* oviducal storage system, which can maintain fertilizing ability for many weeks.

INTRODUCTION

The development of systems for semen collection and artificial insemination of poultry from the 1930s onwards provided the opportunity for manipulation of semen for storage and analysis before insemination (see Lake, 1995). The practical outcome of this has included long-term (frozen) and short-term (liquid) storage of semen and the grading of semen quality, especially for the identification of individual male fertility.

Progress in all of these applications has been slow, due to the comparatively limited understanding of the basic biology of poultry sperm and fertility, which has undermined confidence in, for example, sperm cryopreservation and sperm quality testing technologies that have been available since the late 1950s but are only recently finding significant application. There remains plenty of room for improvement, especially in semen storage technology, but, with limited investment and expertise in basic research in this area, it may be more circumspect to characterize and define the possibilities and limitations of the technologies that are currently available.

SEMEN QUALITY

Purity, source and composition of sample

The first and most basic aspect of semen quality is that the sample, on collection, should be free from faecal and urate contamination. The basic cloacal anatomy of birds, whether comprising the phallus of galliformes and most genera or the extended intromissary organ of the anseriformes and ratites, has the external opening of the ductuli deferentia close to the ureter and coprodeum outlets, so that contamination, especially with an inexperienced handler or stressed bird, is not unlikely (see Lake and Stewart, 1978a).

There is a lymph-like exudate that emanates from the folds of the phallus of chickens and turkeys, which has been termed 'transparent fluid', and there are equivalent fluids in other species, such as the particularly exaggerated 'foam' produced by quail. Whether or not the inclusion of this fluid in manually collected ejaculates in chickens is an artefact of the manual collection technique and whether it should be considered as an 'accessory fluid' equivalent to those of the mammalian reproductive tract have been the subject of some debate (see Lake, 1971). However, experiments in which semen was collected during natural mating from the vagina or on to a barrier fixed to the male (see Fujihara *et al.*, 1986) or female (T. Pizzari, Oxford, personal communication, 2004) cloaca demonstrated that transparent fluid as well as opaque ductus deferens semen were transferred during natural copulation. Some evidence suggests that transparent fluid may be deleterious to sperm quality during storage (see Lake, 1971).

Semen volume and concentration

The most straightforward parameter of semen quality is that of sperm concentration – usually routinely measured by light scattering in a 'photometer', colorimeter or spectrophotometer – with the optical density correlated with the sperm concentration of samples estimated using a haemocytometer (see Bakst and Cecil, 1997a). Normal sperm concentration varies among species and is typically 2–4 × 10^9 sperm/ml in chickens and about double that in turkeys (see Lake and Stewart, 1978a). Lower than expected sperm concentration may be the result of inadequate testicular sperm production or concentration of spermatozoa in the ductus deferens and is indicative of an inadequate reproductive tract. If low sperm concentration is found with increased semen volume, then this might indicate that the sample has been excessively diluted with transparent fluid. Naturally, more sperm, whether the result of greater semen volume or sperm concentration, relates to better semen quality, but since the fact that fertilizing ability correlated more significantly with sperm quality than quantity (Cooper and Rowell, 1958; Wishart and Palmer, 1986a; Froman *et al.*, 1997) the rest of this review will consider parameters of sperm quality.

Methods of assessment of sperm quality

The evaluation of sperm quality is performed in an attempt to predict the fertilizing ability of individual birds, or samples of semen from different groups of birds, or samples of semen that have been processed (or stored) differently. The requirements of such tests differ with respect to whether they are to be used in a laboratory situation, for example in developing methods for bird husbandry or semen storage, or whether the tests are to be used routinely 'in the field'. In the former case they should be objective, repeatable and provide as comprehensively as possible a measure of sperm function, and for field use they must also be quick, simple, robust and relatively inexpensive.

Tests of sperm quality have been limited to the assessment of: (i) sperm morphology; (ii) integrity of organelles – plasma membrane, acrosome, mitochondria, nucleus; (iii) normal energy metabolism; (iv) the ability to display appropriate motility; and (v) the ability to bind to and penetrate the egg investments. Brief descriptions of the principal methods that have been employed as *in vitro* tests of semen quality will be given below and are summarized in Table 10.1. A convenient practical outline for most of these methods can be found in Bakst and Cecil (1997b). Unless otherwise stated, the methods refer to the chicken, in which species most work has been done. The few studies that have been performed with turkeys are also mentioned. With other species, such studies generally involve very few individuals, so that these are, in preference, omitted from this review, which will concentrate on comparative and integrative aspects of sperm quality. The review will also focus on studies involving objective methods of sperm quality determination, which have been mostly published from the 1980s onwards. A good review of earlier work and subjective methods is that of Wilson *et al.* (1979). Changes in sperm

Table 10.1. Sperm quality tests.

	Method		Principle
1.	Morphology		Individual sperm morphology assessed by light microscopy in dried semen smears
2.	Dye exclusion/ inclusion	(a)	Assessed microscopically in individual sperm on dried semen smears concurrently with (1) above, or in wet preparations, usually with a fluorescent dye
		(b)	Sperm stained with fluorescent dye and assessed as bulk fluorescence
		(c)	Sperm stained with fluorescent dye and assessed individually using flow cytometry
3.	Motility	(a)	Assessed microscopically as collective movement expressed as a score or individual sperm movement expressed as % motile
		(b)	Assessed as the disturbance of a fine light beam transduced and integrated into an electronic signal
		(c)	Assessed by light scattering of sperm rearranging within, or moving out of, the light path of a spectrophotometer
		(d)	Assessed by computer-assisted sperm analysis (CASA)
		(e)	Assessed (by light scattering) as the movement of sperm populations between solutions of different viscosities
4.	Metabolism	(a)	Dye reduction – generally from colourless oxidized form to coloured reduced form
		(b)	ATP content – measured by luminometry
5.	Sperm:egg interaction	(a)	Measured as the ability of spermatozoa to produce points of hydrolysis in the inner perivitelline layer
		(b)	Measured as the ability of spermatozoa to bind to perivitelline proteins

quality that arise during *in vitro* storage will be dealt with in the relevant sections on semen storage.

Morphology

Morphological assessment (Table 10.1, 1) is usually performed on spermatozoa air-dried on to a background stain such as nigrosin (Lake and Stewart, 1978a), showing up the features of unstained (or more usually eosin-stained – see next section) spermatozoa. Usually morphology of individual spermatozoa is categorized as 'normal' or as one of a range of abnormal types such as 'bent-neck' or 'macrocephalic', and the proportions of each type calculated after assessing at least 100 individual spermatozoa. This is, of course, time-consuming and subjective. It is also possible, with experience, to see some features (such as the presence of an acrosome) of immobilized spermatozoa in wet preparations under phase optics, but this is not useful as a routine sperm quality assay and neither is the use of scanning or transmission electron microscopy, although these techniques can be used to provide a more detailed assessment of surface and internal morphology of individual spermatozoa (e.g. Froman and Kirby, 2005).

Dye exclusion and/or inclusion

The method for assessing sperm morphology against a background stain in dried semen smears is usually used in conjunction with a dye such as eosin, which will only penetrate (and thus stain) sperm organelles that have damaged membranes, to enable a concurrent assessment of sperm morphology and membrane integrity (e.g. Lake and Stewart, 1978a).

As well as the visible 'exclusion' dyes, such as eosin, fluorescent dyes such as propidium iodide have been used, which will penetrate damaged plasma and nuclear membranes to stain nuclei red (Table 10.1, 2a). Propidium iodide is conveniently used in combination with a counterstaining 'inclusion' dye such as carboxyfluorescein diacetate (Bayyari *et al.*, 1990) or SYBR-14 (Chalah and Brillard, 1998), which penetrates the intact plasma membrane to be de-esterified within the sperm cytosol and thus releases, the fluorescent moiety, which, therefore, stains only metabolically intact, membrane-intact cells. Although in one study (Chalah and Brillard, 1998) the fluorescent dye system was considered to be more efficient than that using visible smears, both systems are time-consuming and subjective.

The exclusion dye system has been made objective by using ethidium bromide in a 'tube' test in which the bulk fluorescence emitted by nuclear-stained membrane-permeable sperm is measured with a fluorimeter and expressed as a proportion of the total fluorescence produced by all sperm nuclei in the sample, measured after deliberately permeabilizing all the spermatozoa with detergent (Table 10.1, 2b; Bilgilli and Renden, 1984). A more sophisticated alternative to manual assessment of sperm stained with SYBR-14 and propidium iodide is to subject the stained spermatozoa to flow cytometry (Table 10.1, 2c), as has been performed with turkey spermatozoa (Donoghue *et al.*, 1996) with the advantage that the status of thousands of spermatozoa can be assessed objectively within minutes.

Motility

Motility is the most obvious function of spermatozoa and may be viewed microscopically and, under some conditions, macroscopically. It has been assessed using a variety of subjective and objective qualitative and quantitative assessments. The most fundamental semi-quantitative measure is the 'score system', in which the 'swirling' motion of a sample of diluted or undiluted semen provides a score of between 1 (no movement) and 5 (all of the sperm in vigorous motion) (Table 10.1, 3a; Wheeler and Andrews, 1943). This highly subjective system has been developed into an objective test using the 'sperm quality analyser' (SQA), which measures the degree of interference of a fine light beam as it passes through a 'swirling' sample of spermatozoa, defining this as the 'sperm motility index' (or 'sperm quality index; Neuman *et al.*, 2002a), which is high for vigorous movement and zero for a static sample (Table 10.1, 3b; McDaniel *et al.*, 1998).

The 'proportion of motile spermatozoa' is sometimes estimated by viewing individual sperm motility in highly dilute samples, but unless this is done semi-manually using freeze-frame videomicroscopy (e.g. Ashizawa *et al.*, 1989) it remains highly subjective. The first objective assay of chicken sperm motility, introduced by Wall and Boone (1973) and characterized by Wishart and Ross (1985), was based on the reduction in optical density that occurs when the movement of a sperm suspension in a flow cell is arrested – a phenomenon that may depend on the rheotactic properties of spermatozoa or, perhaps, the attraction of spermatozoa to the sides of the flow cell (G.J. Wishart, personal observation). The method may be used to derive parameters that relate both to the speed of movement of spermatozoa and to the proportion of moving spermatozoa in the sample (Table 10.1, 3c; Wishart and Ross, 1985). Computer-assisted sperm analysis of motility defines a range of more subtle parameters of sperm movement, including the proportion of motile spermatozoa and the straight line and curvilinear velocities (Table 10.1, 3d). It has been used for the assessment of both turkey (Bakst and Cecil, 1992) and chicken (Froman and Feltmann, 2000) sperm. More recently an extensive investigation of sperm motility has been made using a method (Table 10.1, 3e) that measures, as light scattering, the movement of a population of spermatozoa from a low-viscosity medium into a medium of high viscosity containing the polymer Accudenz (Froman and McLean, 1996), defined in later papers as sperm mobility (e.g. Froman *et al.*, 1999).

Metabolism

Oxidoreduction dyes are coloured in one or other of their oxidized or reduced forms and so may enter spermatozoa to be reduced by sperm metabolism to produce a coloured-from-uncoloured or uncoloured-from-coloured product. Although the first dyes used were in the latter category and sperm were assessed subjectively as the time taken to 'bleach' the colour (Cooper and Rowell, 1958), the reduction of tetrazolium dyes, which are colourless in their oxidized and highly coloured in their reduced form, can be assessed objectively (Table 10.1, 4a) as optical density produced following incubation with spermatozoa (Chaudhuri and Wishart, 1988). Alternative methods that do not require the use of metabolic poisons to encourage dye reduction and clarification of sperm components to form an optically clear product, without the need for centrifugation, have also been introduced (Hazary and Wishart, 2000).

Endogenous indicators of sperm metabolic efficiency may be considered to be more 'natural', such as sperm ATP (Table 10.1, 4b), which can be measured by luminometry using firefly luciferase (Wishart and Palmer, 1986a). This basic assay has also been adapted for multiple-sample throughput using a scintillation counter (Long and Guthrie, 2006).

Sperm:egg interaction

Two unrelated methods of assessing sperm quality have been developed based on sperm interaction with the egg investments, the outer and inner perivitelline layers. One method (Table 10.1, 5b) involves assessing the ability of sperm to bind to solubilized (outer and inner) perivitelline proteins in a multi-well plate and assessing the number of sperm bound by reading the fluorescence after staining with a DNA-specific fluorochrome (Barbato *et al.*, 1998).

The second method (Table 10.1, 5a) is based on the finding that sperm can produce, *in vitro*, points of hydrolysis in the inner perivitelline layer of ova removed from the body cavity or infundibulum (Bramwell and Howarth, 1992). This was later characterized as an assay measuring sperm interaction with pieces of inner perivitelline layer separated from laid eggs, and quantified as the number of points of hydrolysis per unit area (Robertson *et al.*, 1998).

Individual bird sperm quality

Sperm quality and fertilizing ability

Whatever the application or parameter, sperm quality tests should ideally be able to predict fertilizing ability, or be shown to be correlated with fertilizing ability, or at least should be able to rank birds according to the measured parameters on the assumption that these are likely to reflect fertilizing ability. In both mammals and birds, tests of semen quality have often been considered to be poor predictors of fertilizing ability. However, the research documented below clearly demonstrates that, in poultry, sperm quality tests are highly significantly correlated with fertilizing ability.

The general principle of sperm quality tests is that they should, as much as possible, reflect the various functions that spermatozoa must display during their interaction with the female reproductive tract and the ovum. In mammals, the reason for poor correlations between sperm quality tests and fertilizing ability is considered to be because these interactions are multifarious and the tests only measure one small aspect of this (Amann and Hammerstedt, 1993). In birds, this argument would also seem to be valid and these functions would include (see Bakst *et al.*, 1994): (i) transport through the vagina and entry into the sperm storage tubules (SSTs), requiring motility and perhaps appropriate surface proteins and glycoproteins; (ii) maintenance in the SSTs, which may depend on internal substrates or a response to regulatory signals; (iii) exit from the tubules, which may require increased or decreased motility (Froman, 2003); (iv) transport from the uterovaginal junction to the infundibulum, which may be passive, dependent on uterine contractions; (v) inner perivitelline-layer binding, acrosome reaction and hypermotility for penetration of the inner perivitelline layer; (vi) binding to proteins for the oolemma receptors for engulfment; and (vii) the ability to form a male pronucleus with an intact genome.

In birds, poor correlations between a sperm quality assessment and fertilizing ability has also been the result of the way in which 'fertility' is measured, in that the relationship between the numbers of inseminated spermatozoa and the proportion of fertile eggs laid by inseminated hens is logarithmic and saturable (e.g. Wishart, 1985). Thus, when high single or multiple doses of spermatozoa are inseminated, it is difficult to distinguish the 'fertility' of samples even when there is a significant innate difference in the numbers or quality of inseminated spermatozoa. Additionally, there is considerable variation in the hens' response to inseminated spermatozoa, in terms of the proportion of fertile eggs laid by hens inseminated with aliquots of the same sample of semen (see Wishart, 1995).

Relationships between sperm quality tests and fertilizing ability can be clarified when fertility trials are conducted using large numbers of birds and a range of insemination doses (Wishart, 1985) or samples with a variable range of quality parameters (Froman *et al.*, 1999) and by taking the non-linearity of the relationship into consideration. A linear relationship may be derived when using suboptimal insemination doses (Wishart and Palmer, 1986a) or when replacing the fertility assessment with the measurement of the number of points of hydrolysis made by spermatozoa in the inner perivitelline layer (Robertson *et al.*, 1998). When this has been taken into consideration, there is a significantly high correlation or regression relationship between the fertilizing ability of samples from individual chicken males and sperm motility (Table 10.1, 3c) and ATP (Wishart and Palmer, 1986a), tetrazolium dye reduction (Chaudhuri *et al.*, 1988; Hazary *et al.*, 2001), and sperm mobility (Froman *et al.*, 1999). Thus, in chickens, single-parameter sperm quality tests can be shown to correlate with sperm fertilizing ability, so either the avian system is less complex than that of mammals or the avian system for measuring 'fertility' is more quantitative.

Interrelationship between sperm quality parameters

The finding that several parameters of sperm quality are correlated with fertilizing ability suggests that these parameters (Table 10.1) may also be correlated with each other, and the available evidence, summarized in Box 10.1, supports this conclusion. Indeed, the results of all sperm quality tests seem to be interrelated, with the exception of one study, which showed the sperm motility index (3b) was poorly correlated with sperm mobility ($r = 0.01$) and motile concentration (3d; $r = 0.14$) (Froman *et al.*, 2003). But, since the sperm motility index (3b) has been shown to be correlated with the proportion of motile spermatozoa deliberately arranged in samples (McDaniel *et al.*, 1998) and has been correlated with motility (3e; $r = 0.78$; Wishart and Wilson, 1997; Y. Wilson and G. Wishart, unpublished results), it seems reasonable to assume that all of these sperm quality parameters are interrelated. This raises the idea that there may be a common factor underpinning sperm characteristics and function in each of the tests. Froman and Feltmann (2005) proposed that between-male differences in sperm mobility were the result of differential mitochondrial ATP synthesis, since the two parameters were correlated and sperm mobility could be increased by protein phosphatase inhibition and was therefore not limiting. However, since substrate-level (glycolytic) ATP production was also related to low and high sperm mobilities (Froman and Feltmann, 1998), the difference is not confined to oxidative phosphorylation. This is also confirmed by the finding that poor sperm quality is measured as low sperm tetrazolium dye reduction in the presence of cyanide (Chaudhuri and Wishart, 1988), which inhibits cytochrome oxidase and thus oxidative phosphorylation. However, the hypothesis that sperm quality is fundamentally dependent on mitochondrial activity remains valid and was given further credence by the finding that, in semen samples from 'low-sperm-mobility' males, 47% of spermatozoa had swollen mitochondria with disorganized cristae, compared with 4% for the high-mobility group (Froman and Kirby, 2005).

Box 10.1. Correlations between indirect (laboratory) measures of fertility.

- Chicken sperm tetrazolium dye reduction with motility (Table 10.1, 3c; r = 0.88), ATP (r = 0.93) and morphology (0.72) (Chaudhuri *et al.*, 1988)
- Chicken sperm perivitelline penetration with ATP (*r* = 0.85) and motility (Table 10.1, 3c; r 0.76) (Robertson *et al.*, 1998)
- Chicken sperm mobility and ATP (r = 0.8) (Froman and Feltmann, 1998)
- Chicken sperm mobility and motile concentration (Table 10.1, 3d; *r* = 0.71) (Froman and Feltmann, 2000)
- Chicken sperm tetrazolium dye reduction and ATP (r = 0.92), mobility (r = 0.79) and perivitelline penetration assay (r = 0.79) (Hazary *et al.*, 2001)
- Chicken sperm mobility with the rate of oxygen consumption (r = 0.83; Froman and Kirby, 2005)

These results suggest that there is a primary lesion in the midpiece, which would affect sperm energy metabolism, although not necessarily restricted to, or focused on, oxidative metabolism. This would lead to a reduction in ion flux, membrane potential and general sperm integrity, as evinced from the finding that general sperm integrity is correlated with motility and metabolic features in samples of spermatozoa (Chaudhuri *et al.*, 1988). It has been proposed that the cause of this mitochondrial abnormality is linked to excessive glutamate found in the excurrent ducts of the male reproductive tract from low-sperm-mobility males, which causes an increase in sperm mitochondrial calcium uptake, leading to mitochondrial toxicity (Froman *et al.*, 2006).

Whilst there is clearly an interrelationship between the above tests of the quality of spermatozoa as a population in a semen sample, we do not know the interrelationship between sperm quality parameters at the level of individual spermatozoa, even when individual sperm parameters are assessed. Nor do we know, when ATP levels of one sample of semen on a per sperm basis are, for example, half that of another sample, whether all spermatozoa in the second sample have half the ATP of the first or whether half of the spermatozoa have no ATP at all – or something in between. It is a reasonable guess, however, that a sperm that is eosin-permeable is likely to be unable to produce ATP and display motility. Most of these tests, therefore, identify a subpopulation of spermatozoa that are incapable of fertilization, the corollary of which is that there is a subpopulation more capable of fertilization, such as those with a linear motile velocity of more than 30 μm/s (Froman *et al.*, 2003). The establishment of subpopulations of spermatozoa with different 'quality' parameters means that the 'primary lesion' is not absolute but initiates a sequence of events and that some spermatozoa may be affected by these adverse effects more than others, perhaps based on their 'age' or environment within the seminiferous tubules or male reproductive tract.

Sperm quality as a trait and heritability

The various sperm quality parameters, sperm mobility (Froman and McLean, 1996), sperm ATP, motility (3e) and perivitelline layer hydrolysis (Robertson *et al.*, 1998), are traits that are repeatable within ejaculations from individual birds, and sperm mobility differences are maintained throughout 20 weeks of the breeding season (Froman *et al.*, 1997; Froman, 2006). Furthermore sperm mobility among males is normally distributed, with an approximately tenfold range (Froman and McLean, 1996). Extrapolating to all assays, sperm quality appears to be an individual male trait, which raises the question regarding its heritability.

The results of genetic selection for sperm quality have been disappointing. For example, selection based on 'good' sperm morphology of White Plymouth Rock chickens yielded variable results over seven generations, while selection for poor sperm quality showed no increase in this parameter over the same period (Haije *et al.*, 1981). Known heritable reduction in sperm quality does exist in, for example, the subfertile Delaware strain, in which the lesion is associated with aberrant morphology of the male excurrent duct (Kirby *et al.*, 1990).

The hypothesis that the primary lesion for poor sperm quality might derive from sperm oxidative metabolism (Froman and Feltmann, 1998), combined with the knowledge that mitochondrial genes, which code for components of the electron transport chain and have been implicated in mammalian infertility (Gemmell *et al.*, 2004), led Froman *et al.* (2002) to perform experiments that suggested that the heritability of sperm mobility might have a significant maternal influence. Furthermore, a single nucleotide polymorphism (SNP) identified by restriction fragment length polymorphism (RFLP) analysis was later shown to be associated with low-, but not high-, sperm-mobility males (Froman and Kirby, 2005). Further research will determine the contributions of the sperm mitochondrial genome and of factors such as the environment of the male reproductive tract to chicken sperm quality.

Turkey sperm quality

Although comparative work on sperm quality assays from individual turkeys has been more limited, sperm motility parameters have also been shown to have apparently normal distributions with a tenfold (3c; Wishart, 1993) or sixfold (3b; Neuman *et al.*, 2002b) range. Sperm mobility also shows a normal distribution among individual turkeys with a fourfold range (Holsberger *et al.*, 1998), and the differences between high- and low-mobility males (Holsberger *et al.*, 1998) and males categorized by the SQA (Neuman *et al.*, 2002b) are maintained throughout several months of the reproductive cycle. Groups of turkeys with high and low sperm mobility have, respectively, higher and lower fertilizing ability (King *et al.*, 2000), and the same has been found for samples of turkey semen with high and low levels of binding to perivitelline proteins (Gill *et al.*, 1999). When semen was pooled from males with different sperm mobilities, it was found that the males with the higher sperm mobility sired the majority of the offspring (Donoghue *et al.*, 1999).Thus the same features of sperm quality found in chicken samples may also be applicable to turkeys.

Conclusion and implications of sperm quality assessment

Despite earlier scepticism, research during the last 20 years has shown that there are objective (and 50 years for subjective) *in vitro* tests of chicken sperm quality that have been shown to reflect the fertilizing ability of semen samples from individual males. Birds from flocks have an approximately tenfold range of these traits, which follows a normal distribution, and individual bird sperm quality is a trait that is maintained throughout the breeding season. On this basis, these tests can be used for selecting males for breeding programmes involving artificial insemination and are therefore relevant to turkey breeding, but in only a few instances to chicken breeding, since fertility in birds breeding by natural mating is compounded by other factors (Bilcik and Russek-Cohen, 2005; see Duncan, Chapter 8, this volume). Poor sperm quality may have its origins in defects in mitochondrial integrity stemming from, or as a result of, limitations in sperm energy metabolism, which has a genetic component that is inherited through the female lineage. If this is, as has been proposed, linked to a mitochondrial gene, then the value of selecting for such a trait may be that it only affects sperm function (Gemmell *et al.*, 2004) and not other parameters of reproduction or growth. The question of which sperm quality test to recommend for genetic selection of males to improve fertility may be judged in terms of price, simplicity, multiple-sample throughput, expense and robustness. In this respect, the sperm mobility and sperm tetrazolium reduction assays score well on these features.

SPERM CRYOPRESERVATION

Background

The study of cryobiology started in earnest when Polge (1951) reported the first live progeny – chickens – from cryopreserved spermatozoa that had, by a serendipitous error, been exposed to the (thus identified) cryopreservative action of glycerol. Since then, there have been many reviews on avian sperm cryopreservation (Lake, 1986; Bakst, 1990; Hammerstedt and Graham, 1992; Bellagamba *et al.*, 1993; Hübner *et al.*, 1994; Hammerstedt, 1995; Surai and Wishart, 1996; Wishart and Hartley, 1998; Donoghue and Wishart, 2000; Wishart, 2003), which document the various different sperm diluents, cryopreservatives, storage vessels, freezing systems, thawing systems and insemination regimens that have been applied mostly to chickens but also some other avian species.

Production of live progeny from many domestic and non-domestic birds has been possible, but routine application of bird-sperm cryopreservation, even to chickens, for which the procedure has been best characterized and may be most efficient, has been slow to be adopted. The reason for this is that the process, in terms of recovery of spermatozoa capable of fertilization, is largely inefficient and quite variable between different lines or breeds. However, recently it has been recognized that different breeds and strains of chickens that may be valuable to agriculture or medical research or to the poultry industry are being lost at an alarming rate, due to lack of funds or infrastructure to maintain them. This has been outlined in publications from both North America (Pisenti *et al.*, 1999) and Europe (ERFP, 2003). Additionally, the increased incidence of avian flu infections, with the policy of mass culling, threatens rare lines of birds that are maintained in small localized groups. These factors led to a recent symposium on the conservation of avian genetic resources, which is summarized in Siegel and Qureshi (2006) and others (*Poultry Science*, 2006, vol. 85, pp. 198–257). Thus, there has been renewed interest in poultry conservation in general and sperm cryopreservation in particular, with the setting up of poultry 'sperm banks' in the USA (Blackburn, 2006; US National Animal Germplasm Program, 2007), France (Blesbois *et al.*, 2007; French National Cryobank, 2007) and The Netherlands (Woelders *et al.*, 2006; Netherlands Centre for Genetic Resources, 2007). The establishment of these gene banks does not reflect any major developments in the fundamental science of avian sperm cryopreservation but the characterization and standardization of protocols and systems to ensure sample identification and biosafety. This part of the review will therefore consider the current status of chicken sperm cryopreservation technology and that of spermatozoa from other poultry species.

Chicken sperm cryopreservation

Cryopreservation systems

The above-mentioned reviews may be consulted for a more detailed and comprehensively referenced description of earlier work. The rest of this section will be restricted to providing a background to aid an appreciation of the methods used in the 'sperm banks' listed above. The major basic systems of all methods involve the steps listed in Box 10.2.

Comparison of the efficacy of the different optional systems in maintaining sperm fertilizing ability (usually expressed as the proportion of eggs laid by inseminated hens) should be confined to those examples where experimental regimes are tested concurrently. Thus, freezing samples as pellets made by dropping aliquots of sperm suspensions into liquid nitrogen, with dimethylsulfoxide as cryoprotectant, has been found to be more efficient than freezing with glycerol as cryoprotectant in straws in a programmable freezer (Chalah *et al.*, 1999; Tselutin *et al.*, 1999). However, pellets do not provide sufficient biosecurity levels necessary for sperm banking, being stored in liquid nitrogen in open containers, whilst straws can be sealed, printed with identification information and take up little space and so are favoured for sperm banking (Blesbois *et al.*, 2007).

A comparison of cryoprotectants used for semen freezing in straws or ampoules, where this has been performed directly, demonstrates that fertilizing ability is better preserved by glycerol rather than dimethylacetamide (Tajima *et al.*, 1990; Tselutin *et al.*, 1999; Blesbois *et al.*, 2007) or dimethylsulfoxide (Haije, 1990; Tajima *et al.*, 1990), and that dimethylformamide is as efficient

Box 10.2. Basic steps in cryopreservation of chicken spermatozoa.

1. Cooling of semen to around 5°C
2. Addition of cryoprotectant
3. Freezing of samples:
   (a) as pellets from around 0.2 ml of sperm suspension dropped into
      (i) solid CO_2
      (ii) a plastic plate held at –70°C in nitrogen vapour
      (iii) liquid nitrogen
   (b) in glass/plastic vials or 'straws' frozen
      (i) in liquid nitrogen vapour
      (ii) in a programmable freezer
4. Storage of samples – submerged in liquid nitrogen at –196°C
5. Thawing of samples at 5°C, 37°C or 60°C
6. Removal of glycerol (when glycerol is used as a cryoprotectant) by:
   (a) centrifugation
   (b) dialysis
7. Insemination

as glycerol (Chalah *et al.*, 1999) but better than dimethylacetamide or ethylene glycol (Schramm, 1991). Thus, even with the inconvenience of having to remove glycerol from samples after thawing, which is necessary since glycerol has a contraceptive action on sperm samples inseminated intravaginally (see Hammerstedt and Graham, 1992), the most efficient straw- or vial-based systems for maintenance of chicken spermatozoa during cryopreservation employ glycerol as a cryoprotectant.

Such a system is used in the French Avian Cryobank (Blesbois *et al.*, 2007), based on the method of Seigneurin and Blesbois (1995), which, in turn, is based on the method of Lake and colleagues (Lake and Stewart, 1978b) but using a faster cooling rate (–7 as opposed to –1°C per min, which improved fertility around twofold), and straws instead of glass ampoules, which does not seem to affect resultant fertility (Ravie and Lake, 1984). The system used in the US National Animal Germplasm Program is based on the same technology, except the cooling rate is defined as the time and depth that samples are held in liquid nitrogen vapour before plunging into liquid nitrogen. Adaptations of this method resulted in fertilization of >90% of eggs from hens inseminated four times with approximately 400 million spermatozoa (Lake *et al.*, 1981), 76% fertile eggs following five inseminations of 300 million spermatozoa (Seigneurin and Blesbois, 1995) and 59% fertile eggs following three inseminations of 600 million spermatozoa from line 'Y33' frozen as part of the French Avian Cryobank programme (Blesbois *et al.*, 2007).

Quantitating fertilizing ability

This system clearly works well but some caution needs to be elicited: one study, employing the method of Lake *et al.* (1981), achieved fertilization of up to

90% of eggs with a single insemination of around 600 million frozen and thawed spermatozoa but also showed that the attainment of 50% fertilized eggs required approximately 60 times more frozen, compared with fresh, spermatozoa. In other words, over 98% of spermatozoa had lost their fertilizing ability as a result of cryopreservation (Wishart, 1985). This level of loss was confirmed by comparing the numbers of spermatozoa associated with the perivitelline layer of eggs from hens inseminated with fresh and cryopreserved spermatozoa (Alexander *et al.*, 1993). The reason that fertility can be achieved with samples that contain such a high proportion of damaged spermatozoa is that fertile eggs typically contain many hundred times more spermatozoa than are necessary for fertilization (Wishart, 1997). Thus, if six sperm points of hydrolysis in the inner perivitelline layer over the germinal disc are typically required for an egg to be fertile (Wishart, 1997), then there should be at least 360 sperm associated with this region following insemination with fresh spermatozoa to ensure that one fertile egg is achievable with spermatozoa from the same sample following freezing. Also, considering that there is a daily loss of 30% of spermatozoa from the oviduct each day after an insemination (Wishart, 1987), then there would need to be 514 sperm associated with the inner perivitelline layer over the germinal disc of the first egg laid after insemination for two fertile eggs to be laid on consecutive days and 735 for three fertile eggs. The appropriate level of fertility should be checked in fresh spermatozoa before attempting to produce progeny from cryopreserved spermatozoa.

Strain differences in response to sperm cryopreservation

The problem for breed or line sperm cryobanking is that most of the cryopreservation protocols have been developed with highly fertile males and females, whereas the lines selected for cryobanking are likely to be of variable innate fertility, which may be compromised in birds from small inbred populations. Not only are there likely to be differences in individual 'fresh' sperm quality (see previous section) but individual birds may produce semen with different responses to cryopreservation – 'freezability'. Ansah *et al.* (1985) assumed that their selection of a line of broiler breeders, based on fertility following insemination with frozen/thawed semen, reflected such 'freezability' but this remains to be demonstrated conclusively, since the line selected for this trait also showed innately higher 'fertility' with unfrozen spermatozoa (see Alexander *et al.*, 1993). Both the above forms of variability have been shown in studies using males and females from the same mutant chicken lines (Tajima *et al.*, 1990). Inseminating spermatozoa from different lines into females from the same lines has been shown to affect (usually reducing) the fertilizing ability of cryopreserved semen compared with the fertility obtained when inseminating into females from a line of known high fertility (Blesbois *et al.*, 2007).

Use of tests of sperm function *in vitro*, with some exceptions (Wishart and Palmer, 1986b; Long, 2006), greatly overestimate the fertilizing ability of cryopreserved spermatozoa (see Donoghue and Wishart, 2000) or are rather crude predictors of the fertilizing ability of cryopreserved spermatozoa (Chalah *et al.*, 1999). Thus, the best strategy for screening males for freezing is to select on the basis of fresh sperm quality, and to choose male–female combinations on the basis of sperm numbers associated with the perivitelline layers of eggs from hens inseminated with fresh spermatozoa (Alexander *et al.*, 1993). However, if these numbers are below the threshold outlined above, that does not mean that the line in question cannot be successfully cryobanked as spermatozoa, as a more invasive insemination protocol may be successful. Bacon *et al.* (1986) demonstrated that cryopreserved spermatozoa from White Leghorn lines that produced no fertile eggs following intravaginal insemination could produce some fertile eggs following transvaginal intrauterine insemination and that all seven lines studied produced eggs following surgical intramagnal insemination.

Cryopreservation of spermatozoa from other poultry species

Turkeys

There have been a few reports of the production of poults from turkey spermatozoa cryopreserved using dimethylsulfoxide (Sexton, 1981; Graham *et al.*, 1982; Zavos and Graham, 1983), ethylene glycol and dimethylacetamide (Schramm and Hubner, 1988; see Surai and Wishart, 1996) or combinations of glycerol and ethylene glycol or dimethylacetamide (Macpherson *et al.*, 1969). In contrast to chicken sperm, there has been little, or no, success in turkeys using glycerol alone, except when intramagnal insemination was employed (Park *et al.*, 1978). In general, reports on turkey sperm cryopreservation have been sporadic and have involved different systems, so an integrated consideration of their comparative efficiency is difficult. The author has also received personal communications from members of several research groups that have had regular success in producing progeny from cryopreserved chicken spermatozoa but have had negative results from turkey trials. So either turkey sperm suffer more damage during cryopreservation or there is a more stringent selectivity of turkey spermatozoa in the female reproductive tract.

With respect to the latter parameter, the rate of release of turkey spermatozoa from uterovaginal sperm storage tubules is slower that that of chickens, so that, for a given insemination dose, approximately three times fewer turkey than chicken spermatozoa are found associated with the egg (Wishart, 1995). However, although turkey and chicken spermatozoa have a similar morphology (Thurston and Hess, 1987) and might therefore be expected to be subjected to the same biophysical stresses during freezing, they have been shown to be less able to maintain viability after a hyperosmotic shock (mimicking the cryopreservation cycle) and are generally more susceptible than chicken spermatozoa to damage caused by cryopreservation in dimethylsulfoxide. Interestingly, chicken spermatozoa are more susceptible to damage caused by slow cooling, whilst turkey spermatozoa are more susceptible to damage caused by fast cooling (Blanco *et al.*, 2000). Blesbois *et al.* (2005) showed that chicken spermatozoa have greater membrane fluidity than turkey spermatozoa, noting that this should promote resistance to cryo-damage. Although the cryopreserved chicken semen samples in their study contained more viable spermatozoa, which relates to the difference in quoted fertility following freezing by pelleting in dimethylacetamide (chicken: 80%; turkeys: 40%), the proportional loss of sperm viability was not much different (36.8% versus 36.2%) in chicken and turkey spermatozoa following cryopreservation (Blesbois *et al.*, 2005). The greater loss of sperm ATP following incubation of frozen and thawed spermatozoa *in vitro* (Wishart and Palmer, 1986b) is perhaps a reflection of their known metabolic differences (Wishart, 1989).

Geese

Geese have been produced following insemination of sperm cryopreserved with dimethylacetamide using pellets (Tselutin *et al.*, 1995), and the process was more efficient than when using ethylene glycol or dimethylsulfoxide as cryopreservatives (Kurbatov and colleagues, see Surai and Wishart, 1996). Gander semen has also shown fertilizing ability following cryopreservation in straws using dimethylformamide, which was considered to be more efficient than when using dimethylsulfoxide (Sakhatsky and colleagues, see Surai and Wishart, 1996) and dimethylacetamide, or dimethylsulfoxide (Tai *et al.*, 2001). Lukaszewicz and colleagues have cryopreserved semen with dimethylformamide in straws, achieving 88% fertility in White Italian geese (Lukaszewicz, 2002) and 37.5% and 25% in Greylag and White Koluda geese, respectively (Lukaszewicz *et al.*, 2004).

Ducks

Ducks have also been produced from semen frozen in vials using dimethylacetamide (Tselutin *et al.*, 1995; Sakhatsky and colleagues, see Surai and Wishart, 1996), as pellets in the same cryoprotectant (Kurbatov and colleagues, see Surai and Wishart, 1996) and in straws with dimethylsulfoxide (Han *et al.*, 2005).

Guinea fowl

Recently guinea fowl progeny have been produced, for the first time, from sperm cryopreserved in straws using dimethylformamide – chosen in preference to glycerol and dimethylacetamide (Seigneurin and Blesbois, 2006). The relatively low fertility achieved may be a reflection of the relatively poor survival rate of guinea fowl spermatozoa compared with chicken spermatozoa, which is possibly a reflection of their lower membrane fluidity (Blesbois *et al.*, 2005).

Conclusions – sperm cryopreservation

The fertility of ducks and geese inseminated with cryopreserved semen has been variable. It is noticeable that, as for turkey spermatozoa, glycerol has not been reported as a successful cryoprotectant as it has for chicken spermatozoa, adding weight to the idea that spermatozoa from different species will react differently to the freezing conditions (Blanco *et al.*, 2000). However, such direct comparisons between spermatozoa from the various species have yet to be made and may be hampered by the fact that guinea fowl (Barna and Wishart, 2003) and ganders (Lukaszewicz, 2006) contain many more pleiomorphic forms of spermatozoa compared with those of chicken, turkey and duck. While the production of progeny from turkeys, geese and ducks is certainly proven, the technology needs to be made more efficient and systems standardized for these to be sufficiently established to warrant 'sperm banking'.

The establishment of the avian 'sperm banks' is a very welcome advance that uses well-established poultry sperm cryopreservation technology. The important technical aspect of these is that the methods should be very well characterized and standardized – and fertility proven for each batch of stored sperm. Such a comprehensive approach has certainly been undertaken by Blesbois *et al.* (2007) for the French National Cryobank, but, at the time of writing, there is no information on fertilizing ability of samples from the chicken sperm samples stored in the US National Animal Germplasm Program. Additionally, although the sperm cryopreservation methodology (freezing in straws with dimethylacetamide) used in the Dutch cryobank (see Netherlands Centre for Genetic Resources) has been reported to maintain 'good' fertilizing ability (Woelders *et al.*, 2006), the system used is very similar to one which did not compare favourably with methods using glycerol (Tselutin *et al.*, 1999; Blesbois *et al.*, 2007). The problem remains that simply reporting 'fertility' produced by one system is unhelpful for quantitative comparison with other systems, for which the solution would be to report egg perivitelline sperm numbers found using fresh and frozen semen. Selecting males on the basis of their sperm quality may be helpful and also perhaps using recipient hens from strains selected on the basis of best 'fecundity' (as response to inseminated spermatozoa). Maybe such strains could be further selected on their genetic proximity to the sperm-banked lines and breeds, since maintaining females of the same line or breed for insemination with banked spermatozoa is contrary to the principle of 'gene-banking'. Of course, strains or breeds cannot be 'reconstituted' by sperm cryopreservation except, as a near proximation, by extensive backcrossing. Such reconstitution would require cryopreservation of embryonic cells (see Petitte, 2006), a technology which is in its infancy. It cannot be said that chicken sperm cryopreservation is in its infancy, but the major problem remains – not so much why so many spermatozoa are damaged by cryopreservation but why a small proportion survive such an unnatural process. The answer lies either, as for sperm quality assessment, in sperm heterogeneity or in the microenvironment during the freezing process.

LIQUID SEMEN STORAGE

Principles and practice of liquid semen storage

The principles and practice of liquid semen storage for poultry have been reviewed in greater detail elsewhere (Lake and Wishart, 1984; Wishart, 1989, 2004; Bakst, 1990; Bootwalla and Miles, 1992; Christensen, 1995; Lake,

1995; Thurston, 1995; Donoghue and Wishart, 2000). The fundamental principle is that chicken or turkey semen is diluted in a medium based on the ionic environment of the male reproductive tract (notably with the main anion as glutamate rather than chloride), although hypertonic compared with seminal plasma, with an added glycolytic substrate and a buffer maintaining a pH of around 7.1. The suspension of spermatozoa in this medium is then held at low temperature – usually around 5°C – to minimize sperm metabolic rate and any deleterious catabolic processes, as well as proliferation of microorganisms. However, chicken spermatozoa may be stored under anaerobic conditions, in which they can maintain ATP levels and fertilizing ability, but turkey spermatozoa, which have poor glycolytic capacity and require oxidative phosphorylation to maintain energy levels, must be stored under aerobic conditions.

In practice, chicken semen, diluted two- or threefold and held in tubes at 5°C, was found to maintain fertility levels of more than 90% of fertile eggs laid by inseminated hens after storage for 24 h (Van Wambeke, 1967; Lake and Ravie, 1979). Turkey semen, held at a similar temperature but under oxygenated conditions provided by 'bubbling' of gas through samples held in tubes or agitation in flasks, with or without the presence of oxygen-carrying perfluorochemicals (see Thurston, 1995), can maintain similar levels of fertility for 24 and even 48 h (Lake *et al.*, 1984; Huyghebaert *et al.*, 1987).

Storage of chicken semen has few practical applications. With efficient storage for 24 h in 'still' tubes held at 5°C, semen could, international biosecurity regulations permitting, be transported globally. Turkey semen has the added disadvantage of having to be agitated, or otherwise aerated, during transit. This is certainly a commercial reality for storage of semen for around 6 h during road transit (within the USA) from 'stud farms' to the insemination site (Christensen, 1995). Despite the successful maintenance of turkey sperm fertilizing ability during storage, there is still a need for improvement to ensure maintenance of acceptable fertilizing ability in stored semen from males with relatively poor initial sperm quality (Thurston, 1995). Towards this end, some analyses of the effects of storage on turkey spermatozoa have been made and conditions of storage manipulated in an attempt to improve the maintenance of fertilizing ability.

Effect of low-temperature liquid storage on turkey sperm fertilizing ability and quality

As for cryopreservation of chicken spermatozoa, the fact that 'good' fertility can be achieved with chicken or turkey semen after low-temperature liquid storage does not indicate that such storage has no deleterious effect on sperm fertilizing ability, which may be masked by insemination of large insemination doses. The effect of 24 h storage at 5°C on the intrinsic function of turkey semen has been assessed as the number of spermatozoa in the outer perivitelline layer of eggs from inseminated hens, which was reduced to 21% under conditions in which the fertility fell from 94 to 88% (Donoghue *et al.*, 1995). In a later experiment, the numbers of holes produced by spermatozoa in the inner perivitelline layer was reduced to 63% by the same storage regime (Donoghue, 1996). Although there is clearly variation in the effect of such storage on different semen samples, the effect is unlikely to be as severe as that of cryopreservation.

As with cryopreservation, the effect of liquid semen storage on turkey *in vitro* sperm quality has been found, as with chicken sperm quality tests (see Wishart, 1989), to be poorly related to its effects on fertility (see Donoghue and Wishart, 2000). It is likely that the standard tests of sperm quality are too insensitive to identify subtle damage in turkey semen samples stored for 24 h at 5°C. Ethidium bromide exclusion by turkey spermatozoa was negligible except when sperm were subjected to an additional hypo-osmotic shock, which reduced this sperm quality parameter to 42% in stored spermatozoa but left fresh spermatozoa unaffected (Bakst *et al.*, 1991). The osmotic-shock technique adapted to the basic test of live versus dead fluorescent stains coupled with flow cytometry resulted in similar observations (Donoghue *et al.*, 1996). The ability of spermatozoa to penetrate the inner perivitelline layer *in vitro* was reduced to 30% under conditions in which sperm eosin permeability and tetrazolium reduction remained unchanged (Wishart *et al.*, 1995). Sperm mobility was reduced to 25% under conditions where fertilizing ability was reduced from 96 to 48% (Parkhurst *et al.*, 2000).

Lipid peroxidation and liquid sperm storage

The requirement for oxygenation of turkey spermatozoa during liquid storage has led to consideration of the possibility that a source of sperm damage might be the toxic effects of oxygen free radicals and resultant lipid peroxidation. Both chicken and turkey spermatozoa have been shown to form lipid peroxides on aerated incubation *in vitro* (see Donoghue and Wishart, 2000), a consequence of their having a high proportion of polyunsaturated fatty acids (see Surai *et al.*, 2001; Brèque *et al.*, 2003). Turkey semen produces the lipid peroxides during storage at low temperatures (Cecil and Bakst, 1993) and the addition of antioxidants has also been shown to greatly improve the maintenance of turkey sperm motility and membrane integrity (Donoghue and Donoghue, 1997). However, in other experiments, vitamin E had no effect on sperm peroxide production, motility or fertilizing ability (Long and Kramer, 2003), or had no effect on sperm peroxide production but aided maintenance of sperm motility and morphology (Douard *et al.*, 2004), or otherwise spermatozoa did not show an increase in lipid peroxide production during 48 h at 4°C (Douard *et al.*, 2005). The lack of consensus in these findings may be the result of using different birds, lipid peroxidation or sperm quality assays and needs to be resolved. Otherwise, for low-temperature semen storage, the link between turkey sperm polyunsaturated fatty acid peroxidation and functional deterioration will remain unproven and the practical application of improved sperm storage systems hindered.

Oxidative substrates and liquid semen storage

Glutamate, acetate and citrate are constituents of the major poultry liquid-semen storage media (see Bakst, 1990; Bootwalla and Miles, 1992). Glutamate is probably not important as an oxidative substrate for chicken spermatozoa (see Lake and Wishart, 1984) and may even have a deleterious effect on stored chicken spermatozoa at high (Froman, 2003; Froman and Kirby, 2005), although not at low (Chaudhuri and Lake, 1988), temperatures. Acetate appears to have a significant impact in improving the maintenance of turkey sperm ATP and motility during storage at 4°C (Douard *et al.*, 2004).

As for chicken semen (Blesbois *et al.*, 1999), turkey spermatozoa, during aerobic storage at 5°C for 48 h, lose phospholipids as the result of their hydrolysis, followed by either metabolism or peroxidation, the former being more likely because the proportion of fatty acids as polyunsaturates remains unchanged during storage (Douard *et al.*, 2000). The presence of seminal plasma (Douard *et al.*, 2005) or added oxidative substrate (Douard *et al.*, 2004) did not appear to 'spare' turkey sperm phospholipids during storage, suggesting to these authors that the seminal plasma does not contribute substrate to incubated spermatozoa and that endogenous phospholipids are not significant substrates for sperm during *in vitro* storage. However, exogenous or endogenous lipid substrates utilized by chicken or turkey spermatozoa during liquid storage remain to be identified.

Conclusions – liquid semen storage

The technology for chicken liquid semen storage is adequate for the limited use to which it might be put and has been for several decades. Turkey semen storage under oxidative conditions can be less than adequate (see Thurston, 1995; Long and Kramer, 2003), especially when samples are from poorly fertile males. Attempts have therefore been made to improve the maintenance of turkey spermatozoa fertilizing ability during storage, but improvements, so far, have been limited.

Whilst cryopreservation is an 'unnatural' process in which sperm damage might be expected, with liquid storage of semen normal sperm metabolic processes are presumed to take place, albeit at a slower rate. The expectation or target for liquid *in vitro* storage might be that it should match *in vivo* storage of spermatozoa, which may be a couple of days in the male excurrent ducts, followed by a couple of weeks in the chicken (or several more weeks in the turkey) oviducal SSTs (see Bakst *et al.*, 1994). The successes and failings of the storage media based on mimicking the fluid environment of the male tract are catalogued above, but, with the technology so well established, a complete change of tack to mimic the environment of the female tract (Bakst and Long, 2004), despite interesting preliminary results (Chaudhuri and Lake, 1988), seems unlikely. Furthermore the microenvironment of the SSTs is not easily accessible for analysis (Holm *et al.*, 2000), and it might not just be the luminal microenvironment of the SSTs that is important for sperm maintenance but the interaction between spermatozoa and the cells of the SSTs. The fertilizing ability of chicken spermatozoa can be maintained for several days *in vitro* at 40°C in the presence of cultured somatic cells (Ashizawa *et al.*, 1976), the cells presumably removing toxic compounds produced by the spermatozoa (such as peroxides or low-molecular-weight seminal plasma factors – see above) or providing factors that maintain sperm viability. Such cell:sperm interaction would be difficult to scale up *in vitro*, although the idea of maintaining spermatozoa in an 'open' system, which would enable dynamic addition and removal of substrates, etc., rather than the 'closed' systems of the current technology, has been considered (Hammerstedt, 1993).

REFERENCES

Alexander, A., Graham, J., Hammerstedt, R.H. and Barbato, G.F. (1993) Effects of genotype and cryopreservation of avian semen on fertility and number of perivitelline spermatozoa. *British Poultry Science* 34, 757–764.

Amann, R.P. and Hammerstedt, R.H. (1993) *In vitro* evaluation of sperm quality: an opinion. *Journal of Andrology* 14, 397–405.

Ansah, G.A., Segura, J.C. and Buckland, R.B. (1985) Semen production, sperm quality, and their heritabilities as influenced by selection of fertility of frozen-thawed semen in the chicken. *Poultry Science* 64, 1801–1803.

Ashizawa, K., Nishiyama, H. and Nagae, T. (1976) Effects of oviducal cells on the survival and fertilising ability of fowl spermatozoa. *Journal of Reproduction and Fertility* 47, 305–311.

Ashizawa, K., Suzuki, K. and Okauchi, K. (1989) Flagellar movement in demembranated preparations of ejaculated fowl spermatozoa. *Journal of Reproduction and Fertility* 86, 263–270.

Bacon, L.D., Salter, D.W., Motta, J.B., Crittenden, L.B. and Ogasawara, F.X. (1986) Cryopreservation of chicken semen of inbred or specialised strains. *Poultry Science* 65, 1965–1971.

Bakst, M.R. (1990) Preservation of avian cells. In: Crawford, R.D. (ed.) *Poultry Breeding and Genetics*. Elsevier, Amsterdam, pp. 91–108.

Bakst, M.R. and Cecil, H.C. (1992). Effect of bovine serum-albumin on motility and fecundity of turkey spermatozoa before and after storage. *Poultry Science* 94, 287–293.

Bakst, M.R. and Cecil, H.C. (1997a) Determination of sperm concentration I, II and III. In: Bakst, M.R. and Cecil, H.C. (eds) *Techniques for Semen Evaluation, Semen Storage and Fertility Determination*. Poultry Science Association, Illinois, pp. 6–22.

Bakst, M.R. and Cecil, H.C. (1997b) *Techniques for Semen Evaluation, Semen Storage and Fertility Determination*. Poultry Science Association, Illinois.

Bakst, M.R. and Long, J. (2004) Oviducal sperm storage: model for *in vitro* sperm storage. *Proceedings of the 22nd World's Poultry Congress*, Istanbul, Turkey, paper 1698.

Bakst, M.R., Cecil, H.C. and Sexton, T.J. (1991) Modification of the ethidium bromide exclusion procedure for evaluation of turkey sperm. *Poultry Science* 70, 336–370.

Bakst, M.R., Wishart, G.J. and Brillard, J.-P. (1994) Oviducal sperm selection, transport and storage in poultry. *Poultry Science Reviews* 5, 117–143.

Barbato, G.F., Cramer, P.G. and Hammerstedt, R.H. (1998) A practical in vitro sperm–egg binding assay that detects subfertile males. *Biology of Reproduction* 58, 686–699.

Barna, J. and Wishart, G.J. (2003) Excess nuclear DNA in spermatozoa of guinea fowl. *Theriogenology* 59, 1685–1691.

Bayyari, G.R., Cook, J.R., Harris, G.C., Macy, L.B., Slavik, M.F. and Skeeles, J.K. (1990) The evaluation of chicken spermatozoa using fluorescent staining in a 96-well format. *Poultry Science* 69, 1602–1605.

Bellagamba, F., Cerolini, S. and Cavalchini, L.G. (1993) Cryopreservation of poultry semen: a review. *World's Poultry Science Journal* 49, 157–166.

Bilcik, B. and Russek-Cohen, E. (2005) Reproductive success of broiler breeders in natural mating systems: the effect of male–male competition, sperm quality and morphological characteristics. *Poultry Science* 84, 1453–1462.

Bilgilli, S.F. and Renden, J.A. (1984) Fluorimetric determination of avian spermatozoa viability and concentration. *Poultry Science* 63, 2275–2277.

Blackburn, H.D. (2006) The Naional Animal Germplasm Program: challenges and opportunities for poultry genetic resources. *Poultry Science* 85, 210–215.

Blanco, J.M., Gee, G.F., Wildt, D.E. and Donoghue, A.M. (2000) Species variation in osmotic, cryoprotectant, and cooling rate tolerance in poultry, eagle, and falcon spermatozoa. *Biology of Reproduction* 63, 1164–1171.

Blesbois, E., Grasseau, I. and Hermier, D. (1999) Changes in lipid content of fowl spermatozoa after liquid storage at 2 to 5 °C. *Theriogenology* 52, 325–334.

Blesbois, E., Grasseau, I. and Seigneurin, F. (2005) Membrane fluidity and the ability of domestic birds spermatozoa to survive cryopreservation. *Reproduction* 129, 371–378.

Blesbois, E., Seigneurin, F., Grasseau, I., Limouzin, C., Besnard, J., Gourichon, D., Coquerelle, G., Rault, P. and Tixier-Boichard, M. (2007) Semen cryopreservation for *ex situ* management of genetic diversity in chicken: creation of the French Avian Cryobank. *Poultry Science* 86, 555–564.

Bootwalla, S.M. and Miles, R.D. (1992) Diluents for domestic fowl semen. *World's Poultry Science Journal* 48, 121–128.

Bramwell, R.K. and Howarth, B. (1992) Preferential attachment of cock spermatozoa to the perivitelline layer directly over the germinal disc of the hen's ovum. *Biology of Reproduction* 47, 1113–1117.

Brèque, C., Surai, P. and Brilllard, J.-P. (2003) Roles of antioxidants on prolonged storage of avian spermatozoa *in vivo* and *in vitro*. *Molecular Reproduction and Development* 66, 314–323.

Cecil, H.C. and Bakst, M.R. (1993) *In vitro* lipid peroxidation of turkey spermatozoa. *Poultry Science* 72, 1370–1378.

Chalah, T. and Brillard J.P. (1998) Comparison of assessment of fowl sperm viability by eosin nigrosin and dual fluorescence (SYBR-14/PI). *Theriogenology* 50, 487–493.

Chalah, T., Seigneurin, F., Blesbois, E. and Brillard, J.P. (1999) *In vitro* comparison of fowl sperm viability in ejaculates frozen by three different techniques and relationship with subsequent fertility *in vivo*. *Cryobiology* 39, 185–191.

Chaudhuri, D. and Lake, P.E. (1988) A new diluent and method of holding fowl semen for up to 17 h at high temperature. *Proceedings of the18th World's Poultry Congress*, Nagoya, Japan, pp. 591–593.

Chaudhuri, D. and Wishart, G.J. (1988) Predicting the fertilizing ability of avian semen: the development of an objective colorimetric method for assessing the metabolic activity of fowl spermatozoa. *British Poultry Science* 29, 837–845.

Chaudhuri, D., Wishart, G.J., Lake, P.E. and Ravie, O. (1988) Predicting the fertilizing ability of avian semen: a comparison of a simple colorimetric test with other methods for predicting the fertilizing ability of fowl semen. *British Poultry Science* 29, 847–851.

Christensen, V.L. (1995) Diluents, dilution and storage of poultry semen for 6 hours. In: Bakst, M.R. and Wishart, G.J. (eds) *Proceedings of the First International Symposium on the Artificial Insemination of Poultry*. Poultry Science Association, Savoy, Illinois, pp. 90–106.

Cooper, D.M. and Rowell, J.G. (1958) Relations between fertility, embryonic survival and some semen characteristics in chicken. *Poultry Science* 37, 699–707.

Donoghue, A.M. (1996) The effect of 24 h *in vitro* storage on sperm hydrolysis through the perivitelline membrane of ovipositioned turkey eggs. *Poultry Science* 75, 1035–1038.

Donoghue, A.M. and Donoghue, D.J. (1997) Effects of water- and lipid-soluble antioxidants on turkey sperm viability, membrane integrity and motility during lipid storage. *Poultry Science* 76, 1440–1445.

Donoghue, A.M and Wishart, G.J. (2000) Preservation of poultry semen. *Animal Reproduction Science*, Special Issue 62, 213–232.

Donoghue, A.M., Bakst, M.R., Holsberger, D.R. and Donoghue, D.J. (1995) Effect of semen storage on sperm numbers in the perivitelline layer of laid turkey eggs. *Journal of Reproduction and Fertility* 105, 221–225.

Donoghue, A.M., Garner, D.L., Donoghue, D.J. and Johnston, L.A. (1996) Flow cytometric assessment of the membrane integrity of fresh and stored turkey sperm using a combination of hypo-osmotic stress and fluorescent staining. *Theriogenology* 46, 153–163.

Donoghue, A.M., Sonstegard, T.D., King, L.M., Smith, E.J. and Burt, D.W. (1999) Turkey sperm mobility influences paternity in the context of competitive fertilization. *Biology of Reproduction* 61, 422–427.

Douard, V., Hermier, D. and Blesbois, E. (2000) Changes in turkey semen lipids during liquid *in-vitro* storage. *Biology of Reproduction* 63, 1450–1456.

Douard, V., Hermier, D., Magistrini, M., Labbé, C. and Blesbois, E. (2004) Impact of changes in composition of storage medium on lipid content and quality of turkey spermatozoa. *Theriogenology* 61, 1–13.

Douard, V., Hermier, D., Labbé, C., Magistrini, M. and Blesbois, E. (2005) Role of seminal plasma in damage to turkey spermatozoa during *in vitro* storage. *Theriogenology* 63, 126–137.

ERFP (2003) Guidelines for the Constitution of National Cryopreservation Programmes for Farm Animals. In: Hiemstra, S.J. (ed.) *Publication 1 of the European Focal Point on Animal Genetic Resources*. http://www.rfp-europe.org/files/file00244.pdf Accessed 30/05/07.

French National Cryobank (2007) http://www.cryobanque.org/index.php?option=com_content&task=view&id=21&Itemid=6&lang=en. Accessed 30/05/07.

Froman, D.P. (2003) Deduction of a model for sperm storage in the oviduct of the domestic fowl (*Gallus domesticus*). *Biology of Reproduction* 69, 248–253.

Froman, D.P. (2006) Application of the sperm mobility assay to primary breeder stock. *Journal of Applied Poultry Research* 15, 280–286.

Froman, D.P. and Feltmann, A.J. (1998) Sperm mobility: a quantitative trait of the domestic fowl (*Gallus domesticus*). *Biology of Reproduction* 58, 379–384.

Froman, D.P. and Feltmann, A.J. (2000) Sperm mobility: phenotype in roosters (*Gallus domesticus*) determined by concentration of motile sperm and straight line velocity. *Biology of Reproduction* 62, 303–309.

Froman, D.P. and Feltmann, A.J. (2005) Fowl (*Gallus domesticus*) sperm motility depends upon mitochondrial calcium cycling driven by extracellular sodium. *Biology of Reproduction* 72, 97–101.

Froman, D.P. and Kirby, J.D. (2005) Sperm mobility: phenotype in roosters (*Gallus domesticus*) determined by mitochondrial function. *Biology of Reproduction* 72, 562–567.

Froman, D.P. and McLean, D.J. (1996) Objective measurement of sperm motility based upon sperm penetration of Accudenz. *Poultry Science* 75, 776–784.

Froman, D.P., Feltmann, A.J. and McLean, D.J. (1997) Increased fecundity resulting from semen donor selection based upon *in vitro* sperm motility. *Poultry Science* 76, 73–77.

Froman, D.P., Feltmann, A.J., Rhoads, M.L. and Kirby, J.D. (1999) Sperm mobility: a primary determinant of fertility in the domestic fowl (*Gallus domesticus*). *Biology of Reproduction* 61, 400–405.

Froman, D.P., Pizzari, T., Feltmann, A.J., Castillo-Juarez, H. and Birkhead, T.R. (2002) Sperm mobility: mechanisms of fertilizing efficiency, genetic variation and phenotypic relationship with male status in the domestic fowl, *Gallus gallus domesticus*. *Proceedings of the Royal Society of London Series B – Biological Sciences* 269, 607–612.

Froman, D.P., Bowling, E.R. and Wilson, J.L. (2003) Sperm mobility phenotype not determined by sperm quality index. *Poultry Science* 82, 496–502.

Froman, D.P., Wardell, J.C., Feltmann, A.J. (2006) Sperm mobility: deduction of a model explaining phenotypic variation in roosters (*Gallus domesticus*). *Biology of Reproduction* 74, 487–491.

Fujihara, N., Nishiyama, H. and Koga, O. (1986) The ejection of transparent fluid into the vagina of the hen. *Japanese Poultry Science* 23, 101–103.

Gemmell, N.J., Metcalf, V.J. and Allendorf, F.W. (2004) Mother's curse: the effect of mtDNA on individual fitness and population viability. *Trends in Ecology and Evolution* 19, 238–244.

Gill, S.P.S., Donoghue, A.M., Holsberger, D.R., Amann, R.P. and Hulet, A.M. (1999) Identifying potentially subfertile toms via a sperm-binding assay. *Poultry Science* 78, 1208–1218.

Graham, E.F., Nelson, D.S. and Schmel, M.K.L. (1982) Development of an extender for frozen turkey semen 2. Fertility trials. *Poultry Science* 61, 558–563.

Haije, U. (1990) Evaluation and cryopreservation of fowl semen (*Gallus domesticus*). PhD thesis, University of Utrecht, the Netherlands.

Haije, U., van Voorst, A. and Pit, R. (1981) Selection for sperm quality in White Plymouth Rocks. *Reproduction Nutrition Dévelopment* 21, 1067–1075.

Hammerstedt, R.H. (1993) Maintenance of bioenergetic balance in sperm and prevention of lipid peroxidation: a review of the effect on design of storage preservation systems. *Reproduction Fertility and Development* 5, 675–690.

Hammerstedt, R.H. (1995) Cryopreservation of poultry semen – current status and economics. In: Bakst, M.R. and Wishart, G.J. (eds) *Proceedings of the First International Symposium on the Artificial Insemination of Poultry*. The Poultry Science Association, Savoy, Illinois, pp. 229–250.

Hammerstedt, R.H. and Graham, J.K. (1992) Cryopreservation of poultry sperm: the enigma of glycerol. *Cryobiology* 29, 26–38.

Han, X.F., Niu, Z.Y., Liu, F.Z. and Yang, C.S. (2005) Effects of diluents, cryoprotectants, equilibration time and thawing temperature on cryopreservation of duck semen. *International Journal of Poultry Science* 4, 197–201.

Hazary, R.C. and Wishart, G.J. (2000) Assay of sperm quality in the domestic fowl by MTT reduction. *British Poultry Science* 42, 111–114.

Hazary, R.C., Chaudhuri, D. and Wishart, G.J. (2001) Application of an MTT reduction assay for assessing sperm quality and predicting fertilising ability of domestic fowl semen. *British Poultry Science* 42, 115–117.

Holm, L., Ekwall, H., Wishart, G.J. and Ridderstråle, Y. (2000) Localisation of calcium and zinc in the sperm storage tubules of chicken, quail and turkey using X-ray microanalysis. *Journal of Reproduction and Fertility* 118, 331–336.

Holsberger, D.R., Donoghue, A.M., Froman, D.P. and Ottinger, M.A. (1998) Assessment of ejaculate quality and sperm characteristics in turkeys: sperm mobility phenotype is independent of time. *Poultry Science* 77, 1711–1717.

Hübner, V.R., Lohle, K. and Schramm, G.-P. (1994) The artificial insemination of poultry in retrospect. Long-term storage of poultry sperm. *Archiv für Tierzucht* 37, 77–87.

Huyghebaert, G.F., Van Wambeke, F., Ketels, E. and De Groote, G. (1987) The effect of storage time, frequency of insemination and numbers of spermatozoa inseminated on reproductive performances of turkeys. *Archiv für Geflügelkunde* 51, 161–168.

King, L.M., Kirby, J.D., Froman, D.P., Sonstegard, T.S., Harry, D.E., Darden, J.R., Marini, P.J., Walker, R.M., Rhoads, M.L. and Donoghue, A.M. (2000) Efficacy of sperm mobility assessment in commercial flocks and the relationship of sperm mobility and insemination dose with fertility in turkeys. *Poultry Science* 79, 1797–1802.

Kirby, J.D., Froman, D.P., Engel, H.N., Bernier, P.E. and Hess, R.A. (1990) Decreased spermatozoal survivability associated with aberrant morphology of the ductuli efferentes proximales of the chicken (*Gallus domesticus*). *Biology of Reproduction* 42, 383–389.

Lake, P.E. (1971) The male in reproduction. In: Bell, D.J. and Freeman, B.M. (eds) *Physiology and Biochemistry of the Domestic Fowl*, Vol. 3. Academic Press, London, pp. 1411–1447.

Lake, P.E. (1986) The history and future of the cryopreservation of avian germplasm. *Poultry Science* 65, 1–15.

Lake, P.E. (1995) Historical perspective of artificial insemination technology. In: Bakst, M.R. and Wishart, G.J. (eds) *Proceedings of the First International Symposium on the Artificial Insemination of Poultry*. Poultry Science Association, Savoy, Illinois, pp. 1–20.

Lake, P.E. and Ravie, O. (1979) Effect on fertility of storing fowl semen for 24 h at 5 °C in fluids of different pH. *Journal of Reproduction and Fertility* 57, 149–155.

Lake, P.E. and Stewart, J.M. (1978a) *Artificial Insemination in Poultry*. Bulletin 213 of the Ministry of Agriculture Fisheries and Food, HMSO, London.

Lake, P.E. and Stewart, J.M. (1978b) Preservation of fowl semen in liquid nitrogen – an improved method. *British Poultry Science* 19, 187–194.

Lake, P.E. and Wishart, G.J. (1984) Comparative physiology of turkey and fowl semen. In: Cunningham, F.J., Lake, P.E. and Hewitt, D. (eds) *Reproductive Biology of Poultry*. Longman, Harlow, UK, pp. 151–160.

Lake, P.E., Ravie, O. and McAdam, J. (1981) Preservation of fowl semen in liquid nitrogen: application to breeding programmes. *British Poultry Science* 22, 71–77.

Lake, P.E., Cherms, F.L. and Wishart, G.J. (1984) Effect of aeration on the fertilizing ability of turkey semen stored for 48 hours at 5 and 15°C: a study from the 33rd to the 47th week of age. *Reproduction Nutrition Développement* 24, 147–153.

Long, J.A. (2006) Avian semen cryopreservation: what are the biological challenges? *Poultry Science* 85, 232–236.

Long, J.A. and Guthrie, H.D. (2006) Validation of a rapid, large-scale assay to quantify ATP concentration in spermatozoa. *Theriogenology* 65, 1620–1630.

Long, J.A. and Kramer, M. (2003) Effect of vitamin E on lipid peroxidation and fertility after artificial insemination with liquid-stored turkey semen. *Poultry Science* 82, 1802–1807.

Lukaszewicz, E. (2002) An effective method for freezing white Italian gander semen. *Theriogenology* 58, 19–27.

Lukaszewicz, E. (2006) Characteristics of fresh gander semen and its susceptibility to cryopreservation in six generations derived from geese inseminated with frozen-thawed semen. *CryoLetters* 27, 51–57.

Lukaszewicz, E., Chrzanowska, M., Jerysza, A. and Chelmonska, B. (2004) Attempts on freezing the Greylag (*Anser anser* L.) gander semen. *Animal Reproduction Science* 80, 163–173.

Macpherson, J.W., Chatterjee, S. and Friars, G.W. (1969) Frozen turkey semen. *Canadian Journal of Comparative Medicine* 33, 37–38.

McDaniel, C.D., Hannah, J.L., Parker, H.M., Smith, T.W., Schultz, C.D. and Zumwalt, C.D. (1998) Use of a sperm analyzer for evaluating broiler breeder males. 1. Effects of altering sperm quality and quantity on the sperm motility index. *Poultry Science* 77, 888–893.

Netherlands Centre for Genetic Resources (2007) Netherlands Centre for Genetic Resources – CGN Animal Resources – Poultry: http://www.cgn.wur.nl/UK/CGN+Animal+Genetic+Resources/Gene+bank+collections/Poultry/. Accessed 30/05/07.

Neuman, S.L., McDaniel, C.D, Frank, L., Radu, J., Einstein, M.E. and Hester, P.Y. (2002a) Utilisation of a sperm quality analyser to evaluate sperm quantity and quality of turkey breeders. *British Poultry Science* 43, 457–464.

Neuman, S.L., McDaniel, C.D., Frank, L., Radu, J. and Hester, P.Y. (2002b) Use of a sperm quality analyser on semen of turkey breeders to monitor storage time effects and age-related changes during a reproductive cycle. *British Poultry Science* 43, 465–471.

Park, R., Ogasawara, F.X. and Marquez, B.J. (1978) Fertility and hatchability of eggs from turkey hens inseminated intravaginally or intramagnally with fresh or frozen-thawed semen. *Proceedings of the XVI World's Poultry Congress*, Vol. 2, Brazil, pp. 197–204.

Parkhurst, A.M., Korn, N. and Thurston, R.J. (2000) The effects of methylxanthines on the mobility of stored turkey sperm. *Poultry Science* 79, 1803–1809.

Petitte, J.N. (2006) Avian germplasm preservation: embryonic stem cells or primordial germ cells. *Poultry Science* 85, 237–242.

Pisenti, J.M., Delany, M.E., Taylor, R.L., Abbott, U.K., Ablplanalp, H., Arthur, J.A., Bakst, M.R., Baxter-Jones, C., Bitgood, J.J., Bradley, F., Cheng, K.M., Dietert, R.R., Dodgson, J.B., Donoghue, A., Emsley, A., Etches, R., Frahm, R.R., Grunder, A.A., Gerrits, R.J., Goetinck, R.F., Harry, D., Lamont, S.J., Martin, G.R., McGuire, P.E., Moberg, G.P., Pieero, L.J., Qualset, C.O., Schultz, F. and Wilson, B.W. (1999) Avian genetic resources at risk. University of California Genetic Resources Conservation Program Report No. 20. Davis, CA.

Polge, C. (1951) Functional survival of fowl spermatozoa after freezing at −70°C. *Nature* 167, 949–950.

Ravie, O. and Lake, P.E. (1984) A comparison of glass ampoules and plastic straws as receptacles for freezing fowl semen, *CryoLetters* 5, 201–208.

Robertson, L., Wilson, Y.I., Lindsay, C. and Wishart, G.J. (1998) Evaluation of semen from individual male domestic fowl by assessment of sperm:perivitelline interaction *in vitro*. *British Poultry Science* 39, 278–291.

Schramm, G.-P. (1991) Suitability of different antifreeze agents for cryoprotection of cock sperm. *Monatshefte für Veterinarmedizin* 46, 438–440.

Schramm, G.-P. and Hubner, R. (1988) Effects of different cryoprotectives and deep freeze techniques on the reproductive potential of turkey sperm after prolonged storage. *Monatshefte für Veterinarmedizin* 43, 426–427.

Seigneurin, F. and Blesbois, E. (1995) Effects of the freezing rate on viability and fertility of frozen-thawed fowl spermatozoa. *Theriogenology* 43, 1351–1358.

Seigneurin, F. and Blesbois, E. (2006) The first method of cryopreservation of guinea fowl semen. *World Poultry Science Association Proceedings of the French Poultry Research Days*. Brittany, March 2005, Article 23. http://www.animalscience.com/uploads/additionalFiles/wpsa.htm. Accessed 30/05/07.

Sexton, T.J. (1981) Development of a commercial method for freezing turkey semen 1. Effect of prefreeze techniques on the fertility of frozen and frozen-thawed semen. *Poultry Science* 60, 1567–1572.

Siegel, P.B. and Qureshi, M.A. (2006) Conservation of avian genetic sources: current opportunities and challenges – a summary. *Poultry Science* 85, 255–257.

Surai, P.F. and Wishart, G.J. (1996) Poultry artificial insemination technology in the countries of the former USSR. *World's Poultry Science Journal* 52, 227–243.

Surai, P.F., Fujihara, N., Speake, B.K., Brillard, J.-P., Wishart, G.J. and Sparks, N.H.C. (2001) Polyunsaturated fatty acids, lipid peroxidation and antioxidant protection in avian semen. *Asian–Australian Journal of Animal Science* 14, 1024–1050.

Tai, J.J.L., Chen, J.C., Wu, K.C., Wang, S.D. and Tai, C. (2001) Cryopreservation of gander semen. *British Poultry Science* 42, 383–388.

Tajima, A., Graham, E.F., Shoffner, R.N., Otis, J.S. and Hawkins, D.M. (1990) Cryopreservation of semen from unique lines of chicken germplasm. *Poultry Science* 69, 999–1002.

Thurston, R.J. (1995) Storage of poultry semen above freezing for twenty-four to forty-eight hours. In: Bakst, M.R. and Wishart, G.J. (eds) *Proceedings of the First International Symposium on the Artificial Insemination of Poultry*. The Poultry Science Association, Savoy, Illinois, pp. 107–122.

Thurston, R.J. and Hess, R.A. (1987) Ultrastructure of spermatozoa from domesticated birds: comparative study of turkey, chicken and guinea fowl. *Scanning Microscopy* 1, 1829–1838.

Tselutin, K., Narubina, L., Mavrodina, T. and Tur, B. (1995) Cryopreservation of poultry semen. *British Poultry Science* 36, 805–811.

Tselutin, K., Seigneurin, F. and Blesbois, E. (1999) Comparison of cryoprotectants and methods of cryopreservation of fowl spermatozoa. *Poultry Science* 78, 586–590.

US National Animal Germplasm Program (2007) US National Animal Germplasm Program – species and breed inventory – breeds/subspecies of chickens: http://www.ars-grin.gov:8080/j2ee/nagppub/jsp/nagp/drilldown2.jsp. Accessed 30/05/07.

Van Wambeke, F. (1967) Storage of fowl sperm. 1. Preliminary results with new diluents. *Journal of Reproduction and Fertility* 13, 571–575.

Wall, K.A. and Boone, M.A. (1973) Objective measurement of sperm motility. *Poultry Science* 53, 657–660.

Wheeler, N.C. and Andrews, F.N. (1943) The influence of season on semen production in the domestic fowl. *Poultry Science* 22, 361–367.

Wilson, H.R., Piesco, N.P., Miller, E.R. and Nesbeth, W.G. (1979) Prediction of the fertility potential of broiler breeder males. *World's Poultry Science Journal* 35, 95–118.

Wishart, G.J. (1985) Quantitation of the fertilizing ability of fresh compared with frozen and thawed fowl spermatozoa. *British Poultry Science* 26, 375–380.

Wishart, G.J. (1987) Regulation of the length of the fertile period in the domestic fowl by numbers of oviducal spermatozoa. *Journal of Reproduction and Fertility* 80, 493–498.

Wishart, G.J. (1989) Physiological changes in fowl and turkey spermatozoa during in vitro storage. *British Poultry Science* 30, 443–454.

Wishart, G.J. (1993) Techniques for semen quality determination. In: *Proceedings of the 3rd International Symposium on Turkey Reproduction*. Raleigh, North Carolina, pp. 83–89.

Wishart, G.J. (1995) New approaches to evaluating male and female fertility. In: Bakst, M.R. and Wishart, G.J. (eds) *Proceedings of the First International Symposium on the Artificial Insemination of Poultry*. The Poultry Science Association, Savoy, Illinois, pp. 207–223.

Wishart, G.J. (1997) Quantitation aspects of sperm:egg interaction in chickens and turkeys. *Animal Reproduction Science* 48, 81–92.

Wishart, G.J. (2003) The cryopreservation of germplasm in domestic and non-domestic birds. In: Watson, P.F. and Holt, W.V. (eds) *Cryobanking the Genetic Resource*. Taylor and Francis, London and New York, pp. 180–200.

Wishart, G.J. (2004) Liquid semen storage: current status and where do we go from here? *Proceedings of the 22nd World's Poultry Congress*, Istanbul, Turkey, paper 1801.

Wishart, G.J. and Hartley, P.S. (1998) Cryoconservation of avian species. *CryoLetters* Supplement 1, 39–46.

Wishart, G.J. and Palmer, F.H. (1986a) Correlation of the fertilizing ability of semen from individual male fowl with sperm motility and ATP content. *British Poultry Science* 27, 97–102.

Wishart, G.J. and Palmer, F.H. (1986b) The effect of cryopreservation at −196°C on the viability of fowl and turkey spermatozoa assessed in vitro. *Animal Reproduction Science* 10, 317–324.

Wishart, G.J. and Ross, F.H. (1985) Characterisation of a spectrophotometric technique for the estimation of fowl and turkey sperm motility. *Gamete Research* 11, 169–178.

Wishart, G.J. and Wilson, Y.I. (1997) Sperm motility analysis using the 'Sperm Quality Analyzer'. In: Bakst, M.R. and Cecil, H.C. (eds) *Techniques for Semen Evaluation, Semen Storage and Fertility Determination*. The Poultry Science Association, Savoy, Illinois, pp. 54–55.

Wishart, G.J., Robertson, L. and Wood, L.K. (1995) Sperm:egg interaction in turkeys – mechanisms and practical applications. *Turkeys* 43, 19–22.

Woelders, H., Zuidberg, C.A. and Hiemstra, S.J. (2006) Animal genetic resources conservation in the Netherlands and Europe: a poultry perspective. *Poultry Science* 85, 216–222.

Zavos, P.M. and Graham, E.F. (1983) Effects of various degrees of supercooling and nucleation temperatures on fertility of frozen turkey spermatozoa. *Cryobiology* 20, 553–559.

PART V
Incubation and Hatching

CHAPTER 11
Broodiness and Broody Control

P.J. Sharp
The Roslin Institute and Royal (Dick) School of Veterinary Sciences, University of Edinburgh, UK

ABSTRACT

Broodiness is a nuisance and sometimes a serious problem in some poultry. The behaviour is recognized as persistent nesting, usually associated with cessation of egg laying. Broodiness has two components: incubation and brooding behaviour; the former can only be induced in laying hens, while the latter can be induced in both laying and non-laying hens and in males, by forced fostering of chicks. The motivation to incubate wanes after 4–5 weeks sitting on infertile eggs, while brooding behaviour gradually declines as the chicks become independent and egg laying resumes. Incubation and brooding behaviour are associated with low concentrations of plasma luteinizing hormone (LH) and ovarian steroids, and with high and low concentrations of plasma prolactin, respectively. The preoptic area of the hypothalamus (POA) forms part of the neural circuitry controlling both incubation and brooding behaviours. High concentrations of prolactin and low concentrations of plasma LH in incubating hens are a consequence, respectively, of increased vasoactive intestinal polypeptide (VIP) and decreased gonadotrophin-releasing hormone (GnRH) release from the hypothalamus. Changes in the release of these two neuropeptides from the hypothalamus are, in part, a consequence of changes in their rates of synthesis. The release of VIP in incubating hens is stimulated by dopaminergic input from the POA, which in turn is stimulated by a serotonergic input. Incubation behaviour is a form of extended nesting behaviour induced by the synergistic action of oestrogen and progesterone. It is transformed into incubation behaviour by increased prolactin secretion stimulated by neural information transmitted through the brood patch from a clutch of eggs to hypothalamic VIP neurones. Broody control is best achieved by good husbandry directed to discouraging the behaviour. Pharmacological and immunological methods can be used to control or prevent broodiness but for most practical purposes are uneconomic and some may delay resumption of egg production. Broodiness is a polygenic trait, and quantitative trait loci analysis can be applied to identify regions of the genome controlling broodiness. Identification of markers in, or linked to, these genes and associated with broody behaviour is likely to be of value in breeding programmes incorporating DNA marker-assisted selection.

INTRODUCTION

The term 'broodiness' is used to describe incubation and brooding behaviours that enable the development of fertilized eggs to hatching and the subsequent care of chicks until they become capable of full independence (Ramsey, 1953). The two behaviours are controlled by different neurocircuitry and associated with different patterns of hormone secretion. Incubation behaviour in poultry is characterized by sitting almost continuously on a clutch of eggs for 3 weeks until they hatch, while brooding behaviour is directed to the care of newly hatched chicks (Richard-Yris *et al.*, 1983; Opel and Proudman, 1988; Ruscio and Adkins-Regan, 2004). In poultry, the expression of incubation behaviour is restricted to the female, while brooding behaviour can be induced in both sexes (Burrows and Byerly, 1938; Nalbandov and Card, 1945; Ruscio and Adkins-Regan, 2004). From the time poultry were first domesticated there was a conflict between selection for appearance, body weight, behaviour and egg production, and the requirement for good broody behaviour to breed the next generation. This conflict arose because broodiness is characterized by ovarian regression and loss of egg production but can be resolved by the use of surrogate mothers with good broody behaviour and, in modern times, by the use of non-broody breeds and artificial incubation. In commercial poultry systems, particularly where birds are not caged and have the opportunity to become attached to nest sites, the broody trait can be a nuisance and sometimes seriously depress flock egg production. This problem can usually be managed by good husbandry, which requires careful attention to be paid to recognizing the first signs of incubation behaviour and acting quickly to discourage it. There is interest in finding less labour-intensive methods of reducing the incidence of broodiness, by pharmacological or genetic means, but a genetic solution is preferable. Research on broodiness in poultry has focused on behavioural analyses (e.g. Richard-Yris *et al.*, 1983; Leboucher *et al.*, 1991) and endocrine control mechanisms (Harvey and Bedrak, 1984; El Halawani *et al.*, 1988, 1997, 2002; Sharp, 1989; El Halawani and Rozenboim, 1993; Sharp *et al.*, 1998). This review will highlight key publications on the behavioural analyses and neuroendocrine control of broodiness, and broody control, with particular emphasis on work published in the past few years. Finally, consideration will be given to the prospects for a genetic solution to broodiness using DNA marker-assisted selection.

BEHAVIOURAL CHANGES ASSOCIATED WITH INCUBATION

The key behavioural change signalling impending broodiness is an increase in nesting behaviour. In bantam hens about to become broody, as each egg is laid on successive days the birds spend more time nesting, starting earlier before, and finishing later after, each oviposition (Lea *et al.*, 1981) (Fig. 11.1). Nesting behaviour increases on successive days as the hens return to their nests and eggs more frequently and for longer periods. A similar increase in nesting frequency is seen in turkeys before the start of incubation (Haller and Cherms, 1961). Finally, nesting behaviour begins at night and progressively extends to occupy most of the day, at which point nesting behaviour has transformed to full incubation behaviour (Lea *et al.*, 1981). The hens sit on their clutches and persistently turn their eggs, rearranging them to ensure they are all well covered. The development of this behaviour is associated with clucking, loss of feathers from the breast to form a brood patch and, usually, cessation of egg laying.

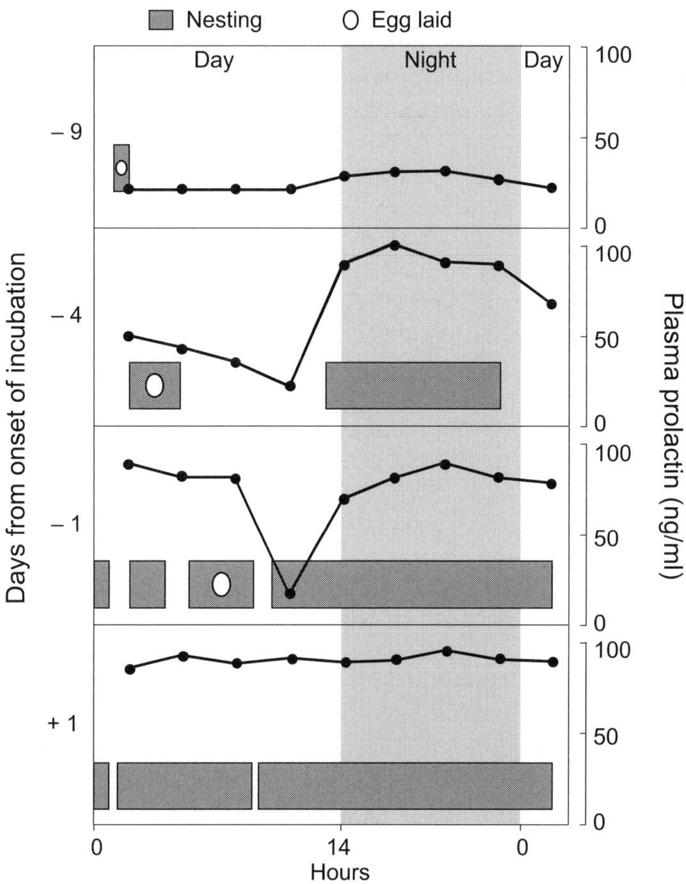

Fig. 11.1. Daily changes in concentrations of plasma prolactin during the 9 days before the onset on incubation in bantam hens in relation to nesting frequency and egg laying. Note that nesting frequency increases in association with the development of nocturnal increases in plasma prolactin until the first day of incubation, when egg laying stops. Thereafter, plasma prolactin is permanently elevated and the hen is almost continuously on the nest (data from Lea *et al.*, 1981).

However, some lines of commercial domestic hens (P.J. Sharp, unpublished observation) and turkeys (Lea and Sharp, 1982) may incubate eggs but not stop laying. In bantam hens, incubation behaviour and cessation of egg laying starts after 10–20 eggs have accumulated in a clutch. It has been suggested that bantam hens, and maybe other birds, stop laying when they have accumulated a full clutch by using an 'egg numerostat'. This allows them to visually compare the size of clutch with an optimal size that is recognized as a result of genetic programming of cognitive function in the higher nervous system (Steen and Parker, 1981). This view is not widely accepted since in some birds, such as the red grouse, in the wild the same number of eggs is laid whether or not eggs are removed from the nests whilst the birds are still laying (Moss and Watson, 1982).

During the normal 3-week incubation period, bantam hens kept in floor pens spend 90–99% of their time on their nests with two breaks (recesses) during the day, one in the morning and one in the afternoon (Lea *et al.*, 1981; Bertrand, 1994). Similar observations have been made on feral chickens (Duncan *et al.*, 1978). The motivation of incubating turkeys or domestic hens to re-nest after nest deprivation is maintained for 3–5 days (El Halawani *et al.*, 1988; Sharp *et al.*, 1988; Richard-Yris *et al.*, 1998). When an incubating domestic hen is off the nest, its behavioural repertoire includes ingestive (feeding, drinking), searching (foraging, walking) and comfort (preening, stretching, defecating, dust bathing, body shaking) activities (McBride *et al.*, 1969; Savory *et al.*, 1978; Bertrand, 1994). During the first recess of the day, incubating bantam hens spend 74% of their time feeding and drinking, or rapidly walking around between the feeders and drinkers, with the remaining time being largely spent on comfort behaviours (Bertrand, 1994). Feeding occurs in bouts of more than a minute's duration. The morning and afternoon recesses are of similar duration (Lea *et al.*, 1981), and, although quantitative observations have not been reported for the afternoon recesses, casual observation suggests that behaviour within them is the same as in the morning (Bertrand, 1994). Incubation is normally terminated when the chicks hatch but may persist if the nest contains unhatched eggs. If a clutch of eggs is removed from an incubating bantam, clucking will stop and egg laying will resume after 8–12 days (Sharp *et al.*, 1979).

If all the eggs are infertile, as in flocks of hens kept for table egg production, incubation may persist for a prolonged period. This is likely to be up to 3–4 months, as was observed at Roslin Institute in bantam hens incubating infertile eggs in floor pens in a temperature-controlled, artificially illuminated poultry hut with minimum disturbance (P.J. Sharp, personal observation). An analysis of many species of wild birds shows that incubation of infertile eggs persists for about 50% longer than is normally required to hatch them (Skutch, 1962). It is possible that prolonged incubation in domestic hens kept in indoor housing and sitting on infertile eggs may be a consequence of environmental conditions being insufficiently variable to distract the hens from abandoning the behaviour, even though motivation to express it is low. This hypothesis was investigated in bantam hens incubating infertile eggs in floor pens (Bertrand, 1994). Behaviours were quantified during the first recess between 2 and 10 weeks of incubation and were seen to begin to change after 4–5 weeks, suggesting decreased motivation to incubate, although there was no change in the duration of the recess (about 24 min). The hens began to start their first recess progressively earlier after the lights came on (4.24 ± 0.3 h at 2 weeks versus 2.78 ± 0.28 h at 10 weeks, $P < 0.001$), while ingestive behaviour progressively became less dominant and searching (Fig. 11.2) and comfort behaviours became more frequent. Ingestive behaviour also changed, with a marked increase in the number of short feeding bouts of less than a minute (0.87 ± 0.21 at 2 weeks versus 11.12 ± 1.70 at 10 weeks, $P < 0.001$) interspersed with foraging behaviours. Casual observation also indicated that after prolonged incubation (i.e. 8–10 weeks) the number of daily recesses increased to about four and that they were all less than an hour (Bertrand, 1994). These observations suggest that after 4–5 weeks' incubation motivation to incubate decreases and are consistent with Skutch's (1962) observation in wild birds that motivation to incubate infertile eggs wanes after the normal incubation time is extended by about 50%. A decline in motivation to incubate after prolonged incubation was confirmed using an operant conditioning method where incubating hens were required to push through a weighted door to regain

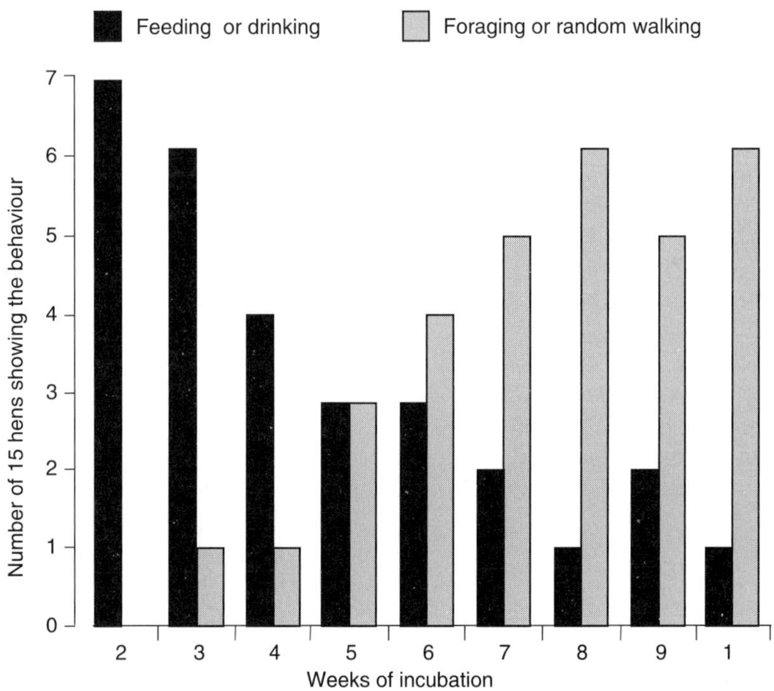

Fig. 11.2. Changes in ingestive (feeding and drinking) or searching behaviours (foraging or random walking) during extended incubation in 15 bantam hens. The behaviours were recorded as the first behaviour shown when the birds leave their nests each morning for a break. Note at about 4–5 weeks, the birds' behaviour becomes less focused on essential ingestive behaviours compared with searching behaviours. This is interpreted as a decrease in motivation to incubate (from Bertrand, 1994).

access to their nest after the first recess. Hens started to push at the door, when given the opportunity to do so, more quickly after 2–4 weeks than after 8–10 weeks of incubation (62.8 ± 20.6 s at 2–4 weeks versus 164.3 ± 13.00 s at 8–10 weeks, $P < 0.005$) (Bertrand, 1994).

BEHAVIOURAL CHANGES ASSOCIATED WITH BROODING

Brooding begins as soon as the chicks hatch. The hen encourages her chicks to nestle underneath her body by crouching in a distinctive manner and slightly raising her wings. She continues to cluck and comes to the assistance of chicks emitting a distress call and gives a special food call (titbitting), picking up and dropping food to encourage them to eat (Ramsey, 1953; Sherry, 1981). The chicks become imprinted on their mother and recognize her cluck vocalization (Kent, 1989). Brooding behaviour can be induced in laying hens and capons by confining them with newly hatched chicks (Burrows and Byerly, 1938) and can be induced in the same way in 'non-broody' strains of domestic hen (Richard-Yris *et al.*, 1983, 1987a,b, 1998). As the chick acquires the ability to thermoregulate, contact with the mother hen progressively decreases (Sherry, 1981) and broody behaviour declines but can be reactivated by introducing newly hatched chicks (Richard-Yris and Leboucher, 1986). Previous experience of brooding behaviour subtly alters the response of a hen to her chicks (Hogan-Warburg *et al.*, 1993). Brooding hens show cyclical sequences of behaviours: brooding, preening, feeding with drinking, exploring, dust bathing and then brooding again. The average cycle length is about 0.5 h, with brooding and feeding taking up about 90% of the time. The cycle length in inexperienced mothers is longer than in experience mothers (32.5 min versus 22.7 min, $P < 0.05$), which can be accounted for by differences in brooding and feeding times.

Brooding behaviour progressively declines over 4–10 weeks, after which laying resumes (Sharp *et al.*, 1979). The time taken to resume egg laying after removal of chicks from brooding bantam hens depends on time elapsed since the chicks hatched. Egg laying resumed in three brooding bantam hens 24, 5 and 7 days, respectively, after removing chicks that had been brooded for 4, 20 and 28 days (Sharp *et al.*, 1979).

NEURAL CIRCUITRY CONTROLLING INCUBATION AND BROODING BEHAVIOURS

Electrolytic lesions at any one of three sites in the hypothalamus of the turkey hen blocks incubation behaviour (Youngren *et al.*, 1989). These sites are in the preoptic area (POA), the basal hypothalamic ventromedial nucleus (VMN) and the lateral hypothalamus. It thus appears that the POA is connected to the VMN by lateral hypothalamic neural pathways to form the core of the neural circuitry that controls incubation behaviour. This view is supported by studies on the ring dove using two techniques: [^{14}C]2-deoxyglucose (2DG) autoradiography and *cfos* immunocytochemistry (Georgiou *et al.*, 1995; Sharp *et al.*, 1996). The 2DG technique involves the systemic injection of 2DG, which is recognized and taken up by cells as glucose to provide energy for increased metabolic activity. 2DG is not metabolized and is retained in the cell, and consequently an increase in cellular 2DG is a marker of increased cellular activity. Using this technique, the uptake of 2DG was seen to increase in the POM and basal hypothalamus at the onset of incubation (Georgiou *et al.*, 1995). The POA identified in this study corresponds to the POA identified in the turkey lesioning study (Belle *et al.*, 2005). In addition, 2DG uptake was also observed to change in two other brain structures in the ring dove at the onset of incubation: the palaeostriatum primitivum and nucleus ovoidalis (Georgiou *et al.*, 1995). The uptake of 2DG was depressed in the palaeostriatum primitivum, a brain area concerned with locomotory activity, and a decrease in activity here may reflect the reduction in movement in incubating birds. In contrast, the uptake of 2DG was increased at the onset of incubation in the nucleus ovoidalis, which is involved in the control of vocalization and aggression, both of which are components of nest defence behaviour. Towards the end of incubation in the ring dove, there is a marked general enhanced uptake of 2DG in the forebrain, preoptic hypothalamus and cerebellum. A similar increase in 2DG uptake is also seen in non-incubating doves after administration of prolactin (Lea *et al.*, 1995). It is suggested that this enhanced neural activity may reflect an effect of increased prolactin secretion at the end of incubation to enhance processing of sensory responsiveness to hatchlings (squabs) and to enhance parental (brooding) behaviours (Lea *et al.*, 1995). It should be noted that, unlike poultry, prolactin secretion does not increase in ring doves until 5 days after the onset of incubation (Goldsmith *et al.*, 1981). It is unknown whether similar changes occur in the brains of incubating poultry.

The second technique used to map neurocircuitry involved in the control of incubation is *cfos* immunocytochemisry. The *cfos* proto-oncogene induces a protein that regulates the transcription of target genes, and its expression is used as a marker of cellular stimulation in the central nervous system. Using this technique, it was found in the ring dove that there was a large increase in *cfos*-labelled cells in the POA and basal hypothalamus at the onset of incubation behaviour in both sexes (Sharp *et al.*, 1996). An increase in *cfos*-labelled cells was also seen in brooding ring doves in the POA and additionally in the lateral hypothalamus when they were reunited with their young after overnight separation (Buntin *et al.*, 2006). The observations on 2GD uptake and *cfos* expression at the onset of incubation and during brooding in the ring dove support the lesioning study in the turkey implicating neural circuitry within, and between, the POA and basal hypothalamus in the control of incubation behaviour (Youngren *et al.*, 1989).

There is less information available on the central nervous system neural circuitry involved in the expression of brooding behaviour. The hypothalamic preoptic area is implicated in the ring dove, where lesions disrupt parental feeding of the young with crop sac 'milk' (Slawski and Buntin, 1995). This view was supported in a study using ZENK (another intermediate early gene) immunocytochemistry in quail brooding foster chicks (Ruscio and Adkins-Regan,

2004). The initiation of brooding behaviour in quail was associated with increased *cfos* expression in the ectostriatum and medial portion of the bed nucleus of the stria terminalis. These two brain areas are thought to be involved, respectively, in focusing visual attention on chicks and in fear responses, and may mediate reduced fearfulness of potential predators (Ruscio and Adkins-Regan, 2004).

MORPHOLOGICAL CHANGES ASSOCIATED WITH BROODINESS

The onset of incubation behaviour is associated with loss of body weight, development of a brood patch, ovarian regression, cessation of laying, narrowing of the gap between the pelvic bones, and loss of redness and shrinkage of the comb (Harvey and Bedrak, 1984). Body weight decreases by 13–15% during normal incubation as a consequence of decreased appetite (Savory, 1979). If the hens are sitting on infertile eggs, this loss of body weight does not continue beyond the time the chicks would normally hatch (Bertrand, 1994). Between 14 and 21 days after the chicks hatch, feathers on the mother's brood patch begin to regrow, and the comb begins to redden as the hens come back into lay (Sharp *et al.*, 1979).

CHANGES IN CIRCULATING HORMONES ASSOCIATED WITH INCUBATION AND BROODING BEHAVIOURS

The onset of incubation in poultry is marked by decreased plasma LH resulting in depressed plasma oestrogen, progesterone and testosterone, by ovarian regression, and an increase in plasma prolactin (Harvey and Bedrak, 1984; Porter *et al.*, 1991a). In the bantam hen, this increase in plasma prolactin is initiated at night, and on subsequent days persists for longer periods into the light period until the first day of full incubation, when plasma prolactin remains constantly elevated (Lea *et al.*, 1983). This pattern of increasing plasma prolactin is correlated with nesting behaviour (Lea *et al.*, 1981). In turkeys, a similar tendency for plasma prolactin to increase at night has been observed immediately before the onset of incubation, although this does not occur prior to an increase in daytime prolactin levels (Proudman, 1998). Plasma prolactin tends to decrease after prolonged incubation in both turkeys (Bédécarrats *et al.*, 1997) and bantam hens (Bertrand, 1994). If the nest and eggs are removed from incubating turkeys and domestic chickens, plasma prolactin immediately decreases but increases after nest return provided incubation resumes (El Halawani *et al.*, 1980; Sharp *et al.*, 1988). If re-nesting does not occur, plasma prolactin remains depressed. In nest-deprived incubating bantam hens, after nest return and resumption of incubation, plasma prolactin increases to a lower level than before nest deprivation (Bertrand, 1994).

Concentrations of plasma prolactin decrease after chicks hatch, but not as rapidly as when incubating hens are deprived of their nests and eggs (Sharp *et al.*, 1988). On the other hand, plasma prolactin does not increase in laying or non-laying domestic hens in which brooding behaviour has been induced by forced adoption of newly hatched chicks (Richard-Yris *et al.*, 1987a). These observations indicate that loss of visual and tactile stimuli from a clutch of unhatched eggs removes a stimulatory control on prolactin secretion that is attenuated by the presence of newly hatched chicks. However, the facilitatory effect of newly hatched chicks on prolactin secretion is conditional on the hen having recently terminated incubation.

The concentrations of plasma prolactin and luteinizing hormone (LH) in incubating hens are inversely related, but this relationship is not precise. In blood samples taken late in the daily photoperiod from bantam hens, plasma LH begins to decrease at the onset of incubation 2 days later than the increase in plasma prolactin (Lea *et al.*, 1981). However, the converse relationship is seen in incubating bantams deprived of their nests, where plasma LH begins to increase about 3 h before the decrease in plasma prolactin (Sharp *et al.*, 1988). In brooding hens, plasma LH remains depressed immediately after the chicks hatch, but, as the chicks become progressively more independent, plasma LH begins to increase in association with increased concentrations of plasma progesterone (Sharp *et al.*, 1979), reflecting the development of pre-ovulatory ovarian follicles (Williams and Sharp, 1977), while plasma prolactin remains low (Sharp *et al.*, 1988). In hens forced to foster young, laying is inhibited and plasma LH and oestrogen are depressed (Richard-Yris *et al.*, 1983, 1987a,b; Leboucher *et al.*, 1990). Conversely plasma LH and oestrogen increase in brooding hens when the chicks are removed (Kuwayama *et al.*, 1992). These observations demonstrate that the mechanisms controlling LH and prolactin release in broody poultry are, in part, independent of each other.

Other hormones measured in incubating domestic hens and turkeys are corticosterone, which is increased, and growth hormone and triiodothyronine, which are both depressed (Harvey and Bedrak, 1984).

NEUROENDOCRINE CONTROL OF PROLACTIN SECRETION IN INCUBATING AND BROODING HENS

In hens coming into lay, increasing oestrogen originating from rapidly growing ovarian follicles (Tanabe *et al.*, 1981) stimulates baseline prolactin secretion by acting directly on the anterior pituitary gland (Knapp *et al.*, 1988; review: El Halawani *et al.*, 2002). The much steeper increase in plasma prolactin secretion associated with the onset of incubation is a consequence of increased pituitary gland prolactin mRNA and prolactin (Talbot *et al.*, 1991; Wong *et al.*, 1991). Using quantitative immunocytochemistry, it has been shown that the numbers of visible cells containing prolactin in the pituitary gland increase at the onset of incubation and that this is partly a consequence of transdifferention of growth hormone cells (somatotrophes) into cells containing both prolactin and growth hormone (mammosomatotrophes) (Ramesh *et al.*, 1998). Decreased prolactin in incubating turkeys deprived of nest and eggs is caused by programmed cell death (apoptosis) of pituitary prolactin cells and the disappearance of mammosomatotrophes (Ramesh *et al.*, 2001). Increased pituitary prolactin and prolactin secretion in incubating poultry is a consequence of increased release of the avian prolactin-releasing factor, vasoactive intestinal polypeptide (VIP), from neurones in the basal hypothalamus (Sharp *et al.*, 1989; reviews: El Halawani and Rozenboim, 1993; El Halawani *et al.*, 1997, 2002; Sharp *et al.*, 1998). Vasoactive intestinal polypeptide directly stimulates the pituitary gland to synthesize and secrete prolactin (Talbot *et al.*, 1991), and the concentration of VIP in the blood draining the median eminence of the hypothalamus into the pituitary gland is threefold higher in incubating than in laying hens (Youngren *et al.*, 1996). As indicated by patterns of prolactin secretion, the release of VIP is pulsatile in laying and incubating turkeys (Proudman and Wentworth, 1996). The stimulatory action of VIP on prolactin release is mediated by the VIP receptors present in the pituitary gland (Chaiseha *et al.*, 2004). The abundance of VIP receptor mRNA in the pituitary gland is higher in incubating and laying turkey hens than in out-of-lay hens, suggesting that changes in prolactin secretion during the reproductive cycles are partly regulated by changes in VIP receptor at the pituitary level.

The homeostatic control of prolactin secretion involves an inhibitory feedback action of prolactin at the level of the hypothalamus to suppress VIP synthesis and release, and possibly at the level of the pituitary to depress VIP receptor abundance (Rozenboim *et al.*, 1993; review: El Halawani *et al.*, 1997). Thus, cerebro-intraventricular administration of ovine prolactin in the laying turkey depresses prolactin secretion, hypothalamic VIP and anterior pituitary VIP receptors (Rozenboim *et al.*, 1993). This inhibitory effect of prolactin is likely to be mediated by prolactin receptor present in both the basal hypothalamus and the anterior pituitary gland (Zhou *et al.*, 1996; Ohkubo *et al.*, 1998). This conclusion is supported by the presence of prolactin receptor in basal hypothalamic VIP neurones (Das, 2008).

Although prolactin secretion is controlled predominantly by the stimulatory action of hypothalamic VIP, there is an inhibitory control system too, mediated by the dopaminergic D_2 receptor in the pituitary gland (Xu *et al.*, 1996; review: El Halawani *et al.*, 2002). In support of this view, a D_2 receptor agonist stimulates prolactin mRNA steady-state levels, and basal and VIP-stimulated prolactin release in cultured laying turkey pituitary cells (Al Kahtane *et al.*, 2003). This inhibitory control system for prolactin release is likely to be suppressed in incubating hens because the expression of pituitary D_2 receptors decreases (MacNamee and Sharp, 1989b).

The activity of basal hypothalamic VIP neurones regulating prolactin secretion in incubating turkeys is stimulated by a dopaminergic input mediated by D_1 receptors located on the VIP neurones (Chaiseha *et al.*, 2001; Youngren *et al.*, 2002; Bhatt *et al.*, 2003; review: El Halawani *et al.*, 2002). Dopaminergic input to the basal hypothalamic VIP neurones probably originates in dopaminergic cells in the basal hypothalamus (Al-Zailaie *et al.*, 2006) controlled by neural pathways from the preoptic hypothalamus (Youngren *et al.*, 2002). Since electrolytic lesions in the preoptic hypothalamus block incubation behaviour (Youngren *et al.*, 1989), it appears that this brain area plays a key role in relaying environmental information required to induce the behaviour to basal hypothalamic VIP neurones. Environmental information required to stimulate incubation behaviour is relayed to dopaminergic neurones controlling VIP release via neural pathways containing serotonin (Youngren *et al.*, 1998) and dynorphin (Youngren *et al.*, 1999). It can therefore be inferred that incubation behaviour is causally related to increased serotonergic and then dopaminergic activity in the preoptic hypothalamus, which, in turn, stimulates the activity of VIP neurones in the basal hypothalamus to increase prolactin secretion. This view is supported by the observation that the turnovers of serotonin (MacNamee and Sharp, 1989a) and dopamine (MacNamee and Sharp, 1989b) in the anterior hypothalamus are increased in incubating bantam hens.

NEUROENDOCRINE CONTROL OF GONADOTROPHIN SECRETION IN INCUBATING AND BROODING HENS

The decrease in plasma LH in incubating poultry is not a consequence of a loss of pituitary responsiveness to gonadotrophin-releasing hormone-I (GnRH-I) or -II (Sharp and Lea, 1981; El Halawani *et al.*, 1987; Guémené and Williams, 1999). However, while GnRH-I is abundant in the median eminence in the turkey and domestic hen, GnRH-II is virtually absent (Sharp *et al.*, 1990; Millam *et al.*, 1995), indicating that GnRH-I rather than GnRH-II directly stimulates pituitary gonadotrophin synthesis (You *et al.*, 1995) and release. The decrease in plasma LH in incubating poultry is, at least in part, a consequence of decreased GnRH-I synthesis, as indicated by the decrease in hypothalamic GnRH-I mRNA in incubating bantam hens and an increase after the hens are deprived of their nests (Dunn *et al.*, 1996). It is likely that GnRH-I release is also directly inhibited, since in the turkey hypothalamus GnRH-I content increases after the onset of incubation, in association with decreased LH secretion (Millam *et al.*, 1995), but subsequently decreases (Rozenboim *et al.*, 1993), presumably as a consequence of decreased GnRH-I synthesis. In the incubating turkey the concentration of GnRH-II increases in the hippocampus but not elsewhere in the brain, implying that this peptide might play a role in the control of gonadotrophin secretion above the level of the pituitary gland or in the expression of incubation behaviour (Millam *et al.*, 1995).

Low concentrations of plasma LH in incubating domesticated hens are partly maintained by an inhibitory feedback action of steroids produced by the regressed ovary, since ovariectomy results in increased LH secretion (Lea *et al.*, 1996), although this does not appear to be the case in incubating turkeys (El Halawani *et al.*, 1993). However, low plasma LH may also be maintained in incubating poultry by increased plasma prolactin acting at the level of the anterior pituitary to inhibit LH synthesis (You *et al.*, 1995), and at the level of the hypothalamus to depress GnRH-I content (Rozenboim *et al.*, 1993). Further evidence for an inhibitory action of prolactin on LH secretion in incubating hens comes from the finding that immunoneutralization of prolactin by an injection of prolactin antibody results in an immediate increase in plasma LH in bantam hens (Lea *et al.*, 1981). Additionally, prolactin administration suppresses ovariectomy-induced increased LH secretion in the turkey (El Halawani *et al.*, 1991) and increased LH secretion in nest-deprived incubating bantams (Sharp *et al.*, 1988). Naturally occurring high concentrations of prolactin appear to attenuate the increase in plasma LH seen after ovariectomy in incubating bantam hens (Lea *et al.*, 1996).

It should be emphasized that LH secretion in broody hens is, at least in part, independent of control by prolactin. This can be deduced by the observations that low concentrations of plasma LH in incubating and brooding hens, respectively, occur in the presence of high and low concentrations of plasma prolactin, while changes in prolactin and LH secretion may not be precisely correlated at the onset of incubation or after nest deprivation and return (see above for references).

A further hypothalamic neuropeptide that may play a role in suppressing LH secretion in broody poultry is gonadotrophin hormone inhibitory hormone (GnIH), discovered in 2001, and this, like GnRH-I and VIP, is abundantly present in the median eminence (Tsutsui *et al.*, 2001). It inhibits LH release by acting directly on the pituitary gland to inhibit gonadotrophin synthesis (Ciccone *et al.*, 2004). The amount of hypothalamic GnIH mRNA is higher in incubating than in laying or non-incubating out-of-lay hens, implying a key role for GnIH in suppressing LH release in incubating hens (Ciccone *et al.*, 2004). This view is reinforced by the finding that the size of visible immunocytochemically identifiable hypothalamic GnIH neurones increases in incubating hens (Das, 2008).

HORMONAL CONTROL OF BROODINESS

The hormonal control of broodiness involves interactions between oestrogen, progesterone and prolactin, and environmental information transmitted through the brood patch (Fig. 11.3). Neural information from the brood patch and concentrations of circulating prolactin and ovarian steroids is suggested to be integrated in the hypothalamus with inputs from the visual and higher central nervous system to induce and maintain broody behaviours, and to suppress ovarian follicular growth by suppressing and increasing, respectively, gonadotrophin and prolactin secretion (Fig. 11.3). Increased plasma prolactin is also likely to contribute to the suppression of ovarian activity by suppressing LH secretion through a direct inhibitory action on the pituitary to inhibit LH synthesis (You *et al.*, 1995). Additionally, prolactin acts directly on the ovary to suppress oestrogen synthesis (Zadworny *et al.*, 1989) and 3β-hydroxysteroid dehydrogenase, a key enzyme required for progesterone synthesis (Taira and Beck, 2006) (Fig. 11.3). However, these direct inhibitory effects of prolactin on the ovary are not required for ovarian regression (Porter *et al.*, 1991a,b), which is likely to be a consequence of decreased follicle-stimulating hormone secretion (N.A. Ciccone, 2006, unpublished observation).

Brood patch formation

The development of a brood patch is one of the earliest morphological signs of the impending development of incubation behaviour. Brood patch development

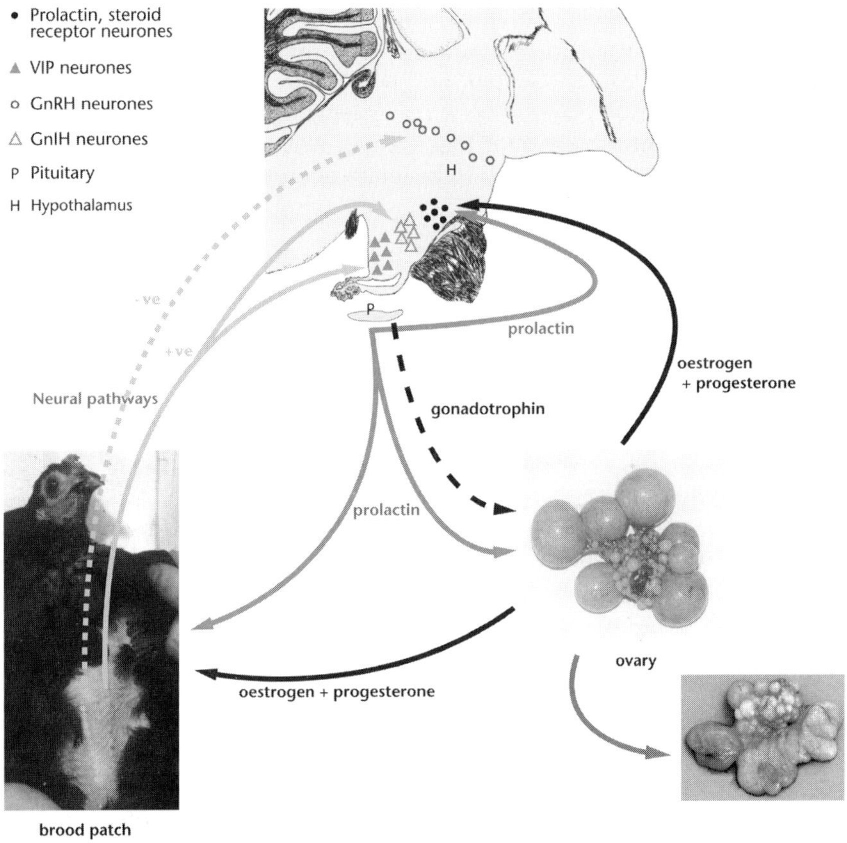

Fig. 11.3. Neuroendocrine interactions in incubating hens. Ovarian steroids stimulate brood patch development in hens about to become broody and prolactin subsequently maintains it. Tactile stimuli from the incubated eggs are transmitted via the brood patch to the hypothalamus to inhibit and stimulate neurones controlling gonadotrophin and prolactin secretion, respectively. Prolactin acts on the hypothalamus to maintain incubation behaviour and, in conjunction with depressed gonadotrophin secretion, on the ovary to suppress ovarian activity. Gonadotrophin secretion is controlled by neurones containing gonadotrophin-releasing hormone (GnRH) and gonadotrophin-inhibiting hormone (GnIH). Prolactin secretion is controlled by vasoactive intestinal polypeptide (VIP) (modified from Sharp, 2004).

is stimulated by increased plasma oestrogen, originating from rapidly growing ovarian follicles acting synergistically with progesterone from pre-ovulatory hierarchical ovarian follicles and prolactin (Drent, 1975; Fig. 11.3). After the onset of incubation, when ovarian steroid secretion decreases, the brood patch is maintained by increased plasma prolactin. Heat transfer from the brood patch to the eggs is achieved by regulating blood flow through vasodilatation (Midtgård *et al.*, 1985) and tachycardia (Gabrielsen and Steen, 1979) in response to egg temperature. The importance of sensory signalling from the brood patch to the central nervous system to initiate the onset of incubation behaviour and associated changes in hormone secretion is illustrated by a study in the turkey (Brook *et al.*, 1991). The eight nerves innervating the brood patch arising from the thoracic and synsacrothoracic vertebrae were severed before the onset of egg production. Although the operated hens laid eggs with associated nesting behaviour, they did not initiate incubation, whereas sham-operated control birds started to incubate after they had laid clutches of eggs. Plasma prolactin increased modestly in control and operated hens after the onset of egg production, reflecting the direct stimulatory effect of increased plasma oestrogen on prolactin secretion from the pituitary (see above), while a further and steeper increase in plasma prolactin was seen in sham-operated hens after the onset of incubation (Brook *et al.*, 1991).

Incubation behaviour

The onset of incubation is conditional on the hens being in laying condition and on the presence of environmental signals encouraging the behaviour. The obvious signals are the availability of a quiet nest site, minimum disturbance and the accumulation of a clutch of eggs. More subtle factors include lighting intensity, temperature and the presence of other incubating hens (review: Sharp, 1989). The influence of housing conditions on the development of incubation behaviour is illustrated by a study in 28-week-old turkeys transferred to floor pens in groups of 12 with nest boxes for each hen; to floor pens singly with a nest box, but no companions; or to individual cages (Bédécarrats *et al.*, 1997). The hens were not allowed to accumulate clutches. No incubation behaviour was seen in the caged hens, but it occurred in the other two groups and at a higher level in communal than in single pens. Further, caging conditions also influenced the increase in plasma prolactin as the hens came into lay, with increases being highest in the communally housed hens and lowest in individually caged hens. The increased plasma prolactin in hens housed individually in floor pens was intermediate between the two other groups. This observation suggests that there is an interaction between increasing plasma oestrogen and environmental signals to control the increase in basal plasma prolactin secretion after the onset of lay.

There is an extensive literature demonstrating a role for prolactin in the control of broodiness (reviews: El Halawani *et al.*, 1988; Buntin, 1996; Sharp, 1997), dating back to the 1930s, when Riddle and colleagues first demonstrated that prolactin induces incubation and cessation of egg laying in broody races of domestic hen, and brooding behaviour in both laying and out-of-lay non-broody races (Riddle *et al.*, 1935). These authors emphasized that the ability of prolactin to induce incubation behaviour is conditional on the hens being in lay and genetically capable of showing the behaviour. With the benefit of hindsight it is now recognized that incubation is an extended form of nesting behaviour. The key prelude to the onset of incubation behaviour is therefore the expression of nesting behaviour induced by the synergistic actions of oestrogen and progesterone (Gilbert and Wood-Gush, 1976; El Halawani *et al.*, 1986; Fig. 11.4). Oestrogen and progesterone receptors located in the hypothalamus are

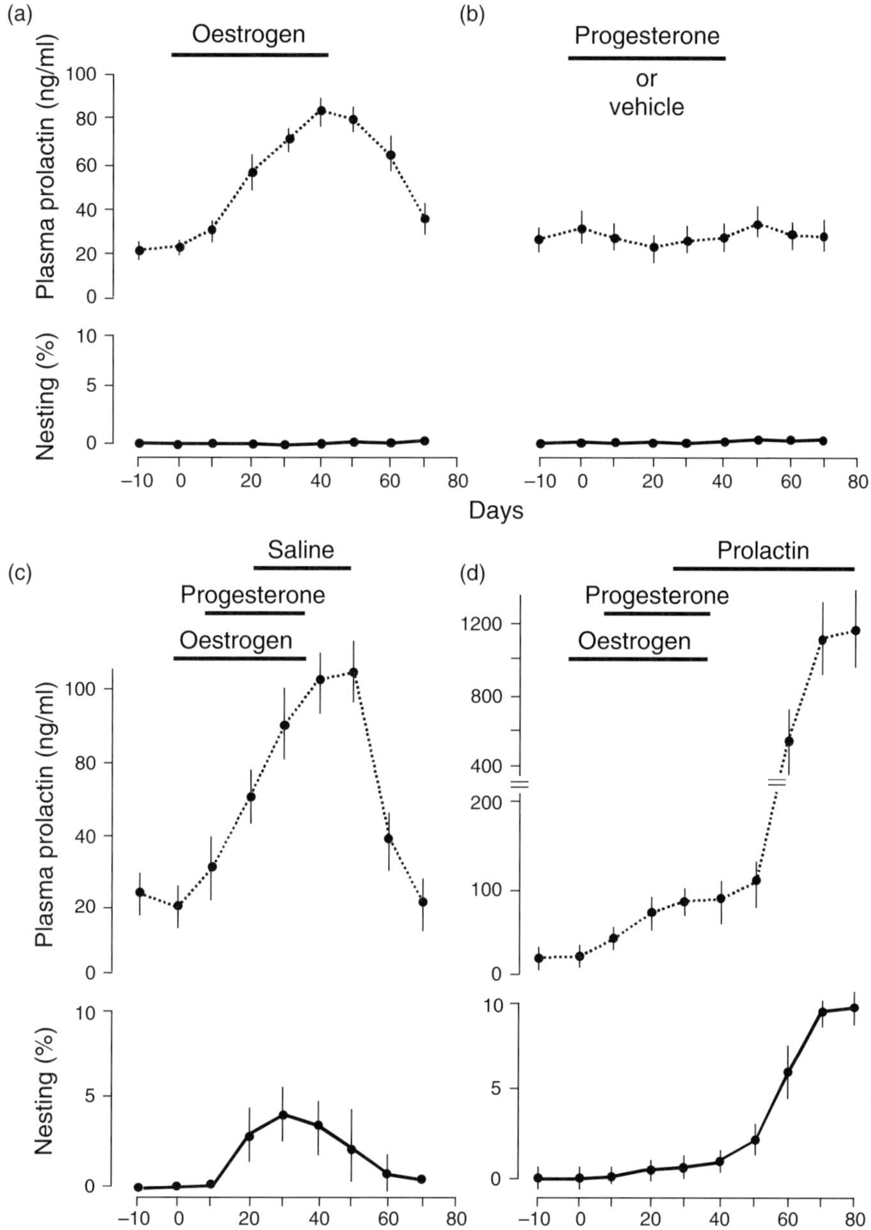

Fig. 11.4. Changes in nesting behaviour and concentrations of plasma prolactin in ovariectomized turkeys treated (horizontal bars) with (a) oestrogen, (b) progesterone or vehicle, (c) oestrogen and progesterone and (d) oestrogen and progesterone followed by prolactin. Note that only a combination of oestrogen and progesterone followed by prolactin is able to increase the time spent nesting into continuous nesting, i.e. incubation behaviour. After incubation behaviour is induced, it is maintained by prolactin alone (redrawn from El Halawani *et al.*, 1986).

likely to mediate the effects of oestrogen and progesterone on nesting behaviour (Sterling *et al.*, 1987; Askew *et al.*, 1997; review: Gahr, 2001). The progressive increase in baseline prolactin secretion induced by oestrogen and environmental signals after the onset of laying is suggested to lower the neural threshold to situational clues such as the presence of an accumulating clutch to encourage prolonged nesting (Sharp, 1989, 1997). The resulting increase in tactile stimulation from the brood patch further increases prolactin secretion, by stimulating VIP release (Fig. 11.3). This increase in plasma prolactin reinforces interest in nesting, progressively transforming it over a few days into full incubation behaviour. Circulating prolactin enters the brain by active transport through the choroid plexus (Buntin *et al.*, 1993), where it probably acts through prolactin receptors in the anterior hypothalamus (Zhou *et al.*, 1996; Ohkubo *et al.*, 1998; review: Buntin, 1996; Fig. 11.3). The observation that intra-cerebroventicular perfusion of prolactin in laying turkeys induces incubation behaviour is consistent with this hypothesis (Youngren *et al.*, 1991).

The interactions between oestrogen, progesterone and prolactin in inducing incubation behaviour have been elegantly demonstrated in a hormone replacement study in ovariectomized turkeys (El Halawani *et al.*, 1986; Fig. 11.4). Treatment with oestrogen alone stimulated increased prolactin, consistent with the facilitatory action of oestrogen on prolactin release from the anterior pituitary gland (see above); but oestrogen, progesterone or prolactin acting alone did not induce nesting behaviour. However, treatment with oestrogen followed by progesterone induced nesting behaviour. When oestrogen–progesterone treatment was followed by prolactin treatment, nesting behaviour was rapidly transformed into incubation behaviour, which was subsequently sustained by prolactin treatment alone (Fig. 11.4).

Tactile information from the brood patch is also proposed to directly inhibit GnRH-I release (Fig. 11.3) and possibly stimulate GnIH release (see above), resulting in depressed gonadotrophin secretion, which results in ovarian regression.

Brooding behaviour

Brooding behaviour can be induced in laying or non-laying poultry by prolactin (Riddle *et al.*, 1935), and plasma prolactin decreases more slowly in nest-deprived incubating hens given chicks than in hens given no chicks (Sharp *et al.*, 1988). However, plasma prolactin does not increase in laying hens forced to adopt chicks (Richard-Yris *et al.*, 1987a). Brooding behaviour is gradually lost as the chicks become independent and circulating ovarian steroids increase (Sharp *et al.*, 1979). These observations suggest that prolactin plays a role in initiating, but not maintaining, brooding behaviour while ovarian steroids depress it.

PHARMACOLOGICAL AND IMMUNOLOGICAL CONTROL OF BROODINESS

Pharmacological methods to disrupt incubation have been extensively investigated, including treatment with gonadotrophins, steroids, anti-oestrogen, GnRH angonist, and inhibitors of serotonergic and dopaminergic function (review: Sharp and Sterling, 1986). Most methods have met with moderate success but are impractical because of the labour and expense involved, or do not return the hens to lay as effectively as simply removing them to a novel environment. Another pharmacological method that might be used to treat broodiness involves the administration of bromocriptine, a dopamine agonist. In mammals, bromocriptine treatment is an effective method to depress prolactin because in mammals, unlike birds, prolactin secretion is under the inhibitory control of dopamine (von Werder *et al.*, 1980). In principle, therefore, it seems unlikely that bromocriptine treatment should inhibit incubation behaviour by depressing plasma prolactin. Furthermore, treatment of nest-deprived incubating domestic hens with bromocriptine delays the resumption of lay (Bedrak *et al.*, 1983). However, the observation in the turkey that dopamine acts directly through a pituitary D_2 receptor to inhibit prolactin secretion (Xu *et al.*, 1996) suggests that in some physiological circumstances bromocriptine treatment might depress plasma prolactin. This is the case in ageing White Leghorn hens in which prolonged treatment with bromocriptine depresses plasma prolactin and increases egg production (Reddy *et al.*, 2007).

Two immunological approaches have been successfully used to inhibit broodiness and involve immunoneutralizing circulating prolactin (March *et al.*, 1994; Crisóstomo *et al.*, 1997) or VIP (El Halawani *et al.*, 1995). However, both methods are labour intensive, particularly since immunizations have to be repeated at frequent intervals, and are probably too expensive for practical use.

Commercially, broodiness can be best prevented by taking precautionary husbandry measures to discourage the behaviour. This involves removing nest sites, which encourage the behaviour, and, at frequent intervals during the day, ensuring that there is no accumulation of clutches. These can accumulate rapidly because some hens lay communally at one site. Clucking and brood patch formation are early signs of impending broodiness in chickens, which are not necessarily associated with decreased egg production, but, if these are associated with a loss of comb redness, it may be too late to prevent ovarian regression and a loss of egg laying. More drastic husbandry methods are then needed to recycle a bird back into lay, which will require a week or more to allow a new crop of follicles to grow in the ovary. This usually involves removing the hen to a separate holding area designed to make the bird's environment mildly uncomfortable.

GENETIC CONTROL OF BROODINESS AND PROSPECTS FOR DNA MARKER-ASSISTED SELECTION AGAINST THE TRAIT

The preferred method for reducing the broody trait is by genetic selection (Hays and Sanborn, 1939; Nestor *et al.*, 1996), or the use of traditional non-broody races such as the White Leghorn (Hutt, 1949). However, the traditional approach to selection could be made more effective by the use of DNA markers of reduced broodiness (see Bijma and Bovenhuis, Chapter 3, this volume). This might be achieved by identifying potential markers in, or associated with, genes encoding hormones such as prolactin, known to control broody behaviour, that occur in broody, but not in non-broody, races (Liang *et al.*, 2006). The problem with this approach is that there can be no guarantee that such markers will be of use for selection purposes in poultry populations, other than those in which they were first discovered. It is also unlikely that one DNA marker will be sufficient for marker-assisted selection because broodiness is a multi-gene trait (Romanov *et al.*, 2002). The preferred approach to marker-assisted selection against broodiness is to identify genes which are functionally important for expression or absence of the trait. One approach is to identify quantitative trait loci (QTL) which associate with the expression or absence of broodiness in the F_2 generation derived from a cross between broody and non-broody races (Sharp, 2004). Identification of functionally significant genes within the QTL may be aided by DNA microarray analyses of changes in gene expression in the hypothalamus and other brain areas associated with the expression of incubation behaviour. If any of the genes identified as being involved in the expression of incubation behaviour by DNA microarray analysis are found to be located at QTL associated with the behaviour, they become candidate genes for marker-assisted selection.

ACKNOWLEDGEMENT

The BBSRC, DEFRA and Roslin Institute are thanked for providing the resources to undertake research on broodiness and to write this review.

REFERENCES

Al Kahtane, A., Chaiseha, Y. and El Halwani, M.E. (2003) Dopaminergic regulation of avian prolactin gene transcription. *Journal of Molecular Endocrinology* 31, 185–196.

Al-Zailaie, K.A., Kang, S.W., Youngren, O.M., Thayananuphat, A., Bakken, T., Chaiseha, Y., Millam, J.R., Proudman, J.A. and El Halawani, M.E. (2006) Identification of dopamine, gonadotrophin-releasing hormone-I and vasoactive intestinal peptide neurones activated by electrical stimulation to the preoptic area of the turkey hypothalamus: a potential reproductive neuroendocrine circuit. *Journal of Neuroendocrinology* 18, 514–525.

Askew, J.A., Georgiou, G.C., Sharp, P.J. and Lea, R.W. (1997) Localization of progesterone receptor in brain and pituitary of the ring dove: influence of breeding cycle and oestrogen. *Hormones and Behavior* 32, 105–113.

Bédécarrats, G., Guémené, D. and Richard-Yris, M.A. (1997) Effects of environment and social factors on incubation behavior, endocrinological parameters, and production traits in turkey hens (*Meleagris gallopavo*). *Poultry Science* 76, 1307–1314.

Bedrak, E., Harvey, S. and Robinzon, B. (1983) Pharmacological disruption of broodiness in White Rock domestic fowl. *British Poultry Science* 24, 573–579.

Belle, M.D.C., Sharp, P.J. and Lea, R.W. (2005) Aromatase inhibition abolishes courtship behaviours in the ring dove (*Streptopelia risoria*) and reduces androgen and progesterone receptors in the hypothalamus and anterior pituitary gland. *Molecular and Cellular Biochemistry* 276, 193–204.

Bertrand, D. (1994) Changes in behaviour and prolactin secretion in bantam hens during extended incubation. MSc thesis, University of Central Lancashire, Preston, UK.

Bhatt, R., Youngren, O.M., Kang, S. and El Halawani, M.E. (2003) Dopamine infusion into the third ventricle increases gene expression of hypothalamic vasoactive intestinal peptide and pituitary prolactin and luteinizing hormone β subunit in the turkey. *General and Comparative Endocrinology* 130, 41–47.

Brook, C.M., Millam, J.R., Guinan, M.J. and Kitchell, R.L. (1991) Brood patch innervation and its role in the onset of incubation in the turkey hen. *Physiology and Behavior* 50, 281–285.

Buntin, J.D. (1996) Neural and hormonal control of parental behaviour in birds. *Advanced Studies in Behaviour* 25, 161–213.

Buntin, J.D., Ruzycki, E. and Witebsky, J. (1993) Prolactin receptors in dove brain: autoradiographic analysis of binding characteristics in discrete brain regions and accessibility to blood-borne prolactin. *Neuroendocrinology* 57, 738–750.

Buntin, L., Berghman, L.R. and Buntin, J.D. (2006) Patterns of *Fos*-like immunoreactivity in the brains of parent doves (*Streptopelia risoria*) given tactile and non-tactile exposure to their young. *Behavioural Neuroscience* 120, 651–664.

Burrows, W.H. and Byerly, T.C. (1938) The effect of certain groups of environmental factors upon the expression of broodiness. *Poultry Science* 17, 324–330.

Chaiseha, Y., Youngren, O.M., Schnell, S.A. and El Halawani, M.E. (2001) Expression of D_1 and D_2 dopamine receptors in the hypothalamus and pituitary during the turkey reproductive cycle. *Poultry Science* 80 (Suppl. 1), 174.

Chaiseha, Y., Youngren, O.M. and El Halawani, M.E. (2004) Expression of vasoactive intestinal peptide messenger RNA in the hypothalamus and pituitary throughout the turkey reproductive cycle. *Biology of Reproduction* 70, 593–599.

Ciccone, N.A., Dunn, I.C., Boswell, T., Tsutsui, K., Ukena, K. and Sharp, P.J. (2004) Gonadotrophin inhibitory hormone depresses gonadotropin α and follicle stimulating hormone β subunit expression in the pituitary of the domestic hen. *Journal of Neuroendocrinology* 16, 999–1006.

Crisóstomo, S., Guémené, D., Garreau-Mills, M. and Zadworny, D. (1997) Prevention of expression of incubation behaviour using passive immunization against prolaction in turkey hens (*Meleagris gallopavo*). *Reproduction, Nutrition and Development* 37, 253–266.

Das, B. (2008) Neuroendocrine regulation of broodiness in the domestic hen. PhD thesis, University of Central Lancashire, Preston, UK.

Drent, R. (1975). Incubation. In: Farner, D.S and King, J.R. (eds) *Avian Biology*, Vol. 5. Academic Press, New York, pp. 333–420.

Duncan, I.J.H., Savory, C.J. and Wood-Gush, D.G.M. (1978) Observations on the reproductive behaviour of domestic fowl in the wild. *Applied Animal Ethology* 4, 29–42.

Dunn, I.C., Beattie, K.K., Maney, D., Sang, H.M., Talbot, R.T., Wilson, P.W. and Sharp, P.J. (1996) Regulation of chicken gonadotrophin releasing hormone-I mRNA in incubating, nest deprived and laying bantam hens. *Neuroendocrinology* 63, 504–513.

El Halawani, M.E. and Rozenboim, I. (1993) Incubation behaviour in the turkey: molecular and endocrinological implications. In: Sharp, P.J. (ed.) *Avian Endocrinology*. Journal of Endocrinology Ltd, Bristol, pp. 99–110.

El Halawani, M.E., Burke, W.H. and Dennison, P.T. (1980) Effect of nest deprivation on serum prolactin in nesting female turkeys. *Biology of Reproduction* 23, 118–123.

El Halawani, M.E., Silsby, J.L., Behnke, E.J. and Fehrer, S.C. (1986) Hormonal induction of incubation behaviour in ovariectomized female turkeys (*Meleagris gallopavo*). *Biology of Reproduction* 35, 59–67.

El Halawani, M.E., Fehrer, S.C. and Silsby, J.L. (1987) Enhanced luteinising hormone release by luteinising hormone releasing hormone in incubating female turkeys. *Biology of Reproduction* 36, 884–889.

El Halawani, M.E., Fehrer, S.C., Hargis, B.M. and Porter, T.E (1988) Incubation behaviour in the domestic turkey: physiological correlates. *CRC Critical Reviews in Poultry Biology* 1, 285–314.

El Halawani, M.E., Silsby, J.L., Youngren, O.M. and Phillips, R.E. (1991) Exogenous prolactin delays photoinduced sexual maturity and suppresses ovariectomy-induced luteinizing hormone secretion in the turkey (*Meleagris gallopavo*). *Biology of Reproduction* 44, 420–424.

El Halawani, M.E., Silsby, J.L., Foster, L.K., Rozenboim, I. and Foster, D.N. (1993) Ovarian involvement in the suppression of luteinizing hormone in the incubating turkey (*Meleagris gallopavo*). *Neuroendocrinology* 58, 35–41.

El Halawani, M.E., Silsby, J.L. and Rozenboim, I. (1995) Increased egg production by active immunization against vasoactive intestinal peptide in the turkey (*Meleagris gallopavo*). *Biology of Reproduction* 52,179–183.

El Halawani, M.E., Youngren, O.M. and Pitts, G.R. (1997) Vasoactive intestinal peptide as the avian prolactin releasing factor. In: Etches, R. and Harvey, S. (eds) *Perspectives in Avian Endocrinology*. Journal of Endocrinology Ltd, Bristol, pp. 403–416.

El Halawani, M.E., Youngren, O.M. and Chaiseha, Y. (2002) Neuroendocrinology of prolactin regulation in the domestic turkey. In: Dawson, A. and Chaturvedi, C.M. (eds) *Avian Endocrinology*. Narosa Publishing House, New Delhi, pp. 233–244.

Gabrielsen, G. and Steen, J.B. (1979) Tachycardia during egg-hypothermia in incubating ptarmigan (*Lagopus lagopus*). *Acta Physiologica Scandinavica* 107, 273–277.

Gahr, M. (2001) Distribution of sex steroid hormone receptors in the avian brain: functional implications for neural sex differences and sexual behaviors. *Microscopy Research and Techniques* 55, 1–11.

Georgiou, G.C., Sharp, P.J. and Lea, R.W. (1995) [14C]-Deoxyglucose uptake in the brain of the ring dove (*Streptopelia risoria*) II. Differential uptake at the onset of incubation. *Brain Research* 700, 137–141.

Gilbert, A.B. and Wood-Gush, D.G.M. (1976) The effect of exogenous oestrogen and progesterone on laying and nesting behaviour in the domestic hen. *British Poultry Science* 17, 13–17.

Goldsmith, A.R., Edwards, C., Koprucu, M. and Silver, R. (1981) Concentrations of prolactin and luteinising hormone in plasma of doves in relation to incubation and development of the crop gland. *Journal of Endocrinology* 90, 437–443.

Guémené, D. and Williams, J.B. (1999) LH responses to chicken luteinizing hormone releasing hormone I and II in laying, incubating, and out of lay turkey hens. *Domestic Animal Endocrinology* 17, 1–15.

Haller, R.W. and Cherms, F.L. (1961) A comparison of several treatments on terminating broodiness in broad breasted bronze turkeys. *Poultry Science* 40, 155–163.

Harvey, S. and Bedrak, E. (1984) Endocrine basis of broodiness in poultry. In: Cunningham, F.J., Lake, P.E. and Hewitt, D. (eds) *Reproductive Biology of Poultry*. British Poultry Science Ltd, Longmans, Harlow, UK, pp. 111–132.

Hays, F.A. and Sanborn, R. (1939) Breeding for egg production. Massachusetts Agricultural Experimental Station Bulletin 307.

Hogan-Warburg, A.J., Panning, L. and Hogan, J.A. (1993) Analysis of the brooding cycle of broody jungle fowl with chicks. *Behaviour* 125, 21–37.

Hutt, F.B. (1949) *Genetics of the Fowl*. McGraw-Hill Book Company, New York.

Kent, J.P (1989) On the acoustic basis of recognition of the mother hen by the chick in the domestic fowl (*Gallus gallus*). *Behaviour* 102, 1–14.

Knapp, T.R., Fehrer, S.C., Silsby, J.L., Porter, T.E., Behnke, E.J. and El Halawani, M.E. (1988) Gonadal steroid modulation of basal and vasoactive intestinal polypeptide-stimulated prolactin release by turkey anterior pituitary cells. *General and Comparative Endocrinology* 72, 226–236.

Kuwayama, T., Shimada, K., Saito, N., Ohkubo, T., Sato, K., Wada, M. and Ichinoe, K. (1992) Effects of removal of chicks from hens on concentrations of prolactin, luteinizing hormone and oestradiol of brooding Gifujidori hens. *Journal of Reproduction and Fertility* 95, 617–622.

Lea, R.W. and Sharp, P.J. (1982) Plasma concentrations of prolactin in broody turkeys: lack of agreement between homologous chicken and turkey prolactin radioimmunoassays. *British Poultry Science* 23, 451–459.

Lea, R.W., Dods, A.S.M., Sharp, P.J. and Chadwick, A. (1981) The possible role of prolactin in the regulation of nesting behaviour and secretion of luteinizing hormone in broody bantams. *Journal of Endocrinology* 91, 89–97.

Lea, R.W., Sharp, P.J. and Chadwick, A. (1983) Daily variations in the concentrations of plasma prolactin in broody bantams. *General and Comparative Endocrinology* 48, 275–284.

Lea, R.W., Georgiou, G.C., Li, Q., Sterling, R.J. and Sharp, P.J. (1995) [14C]-Deoxyglucose uptake in the brain of the ring dove (*Streptopelia risoria*) I. Prolactin-induced uptake. *Brain Research* 700, 129–136.

Lea, R.W., Richard-Yris, M.A. and Sharp, P.J. (1996) The effect of ovariectomy on concentrations of plasma prolactin and LH and parental behaviour in the domestic fowl. *General and Comparative Endocrinology* 101, 115–121.

Leboucher, G., Richard-Yris, M.A, Williams, J. and Chadwick, A. (1990) Incubation and maternal behaviour in domestic hens: influence of the presence of chicks on circulating LH, prolactin, oestradiol and behaviour. *British Poultry Science* 31, 851–862.

Leboucher, G., Richard-Yris, M.A., Williams, J.B. and Chadwick, A. (1991). Maternal behaviour in domestic hens: does previous incubating and maternal experience influence maternal responsiveness? *Ethology, Ecology and Evolution* 3, 101–111.

Liang, Y., Cui, J., Yang, G., Leung, F.C.C. and Zhang, X. (2006) Polymorphisms of 5' flanking region of chicken prolactin gene. *Domestic Animal Endocrinology* 30, 1–16.

MacNamee, M.C. and Sharp, P.J. (1989a) The functional activity of hypothalamic 5-hydroxytryptamine neurones in broody bantam hens. *Journal of Endocrinology* 120, 124–134.

MacNamee, M.C. and Sharp, P.J. (1989b) The functional activity of hypothalamic dopamine in broody bantam hens. *Journal of Endocrinology* 121, 67–74.

March, J.B., Sharp, P.J., Wilson, P.W. and Sang, H.M. (1994) Effect of active immunization against recombinant-derived chicken prolactin fusion protein on the onset of broodiness and photoinduced egg laying in the bantam hens. *Journal of Reproduction and Fertility* 101, 227–233.

McBride, G.J.P., Parer, J.P. and Foenander, F. (1969) The social organisation and behaviour of the feral fowl. *Animal Behaviour Monograph* 2, 125–181.

Midtgård, U., Sejrsen, P. and Johansen, K. (1985) Blood flow in the brood patch of bantam hens: evidence of cold vasodilation. *Journal of Comparative Physiology B* 155, 703–709.

Millam, J.R., Craig-Veit, C.B. and Faris, P.L. (1995) Concentration of chicken gonadotropin-releasing hormones I and II in microdissected areas of turkey hen brain during the reproductive cycle. *Domestic Animal Endocrinology* 12, 1–11.

Moss, R. and Watson, A. (1982) Determination of clutch size in Red Grouse without implying the egg-numerostat. A reply to Steen and Parker. *Ornis Scandinavica* 13, 249–250.

Nalbandov, A.V. and Card, L.E. (1945) Endocrine identification of the broody genotype of cocks. *Journal of Heredity* 36, 35–39.

Nestor, K.E., Noble, D.O., Zhu, J. and Moritsu, Y. (1996) Direct and correlated responses to long-term selection for increased body weight and egg production in turkeys. *Poultry Science* 75, 1180–1191.

Ohkubo, T., Tanaka, M., Nakshima, K. and Sharp, P.J. (1998) Relationship between prolactin receptor mRNA in the anterior pituitary gland and hypothalamus and reproductive state in male and female bantams (*Gallus domesticus*). *General and Comparative Endocrinology* 111, 167–176.

Opel, H. and Proudman, J.A. (1988) Effects of poults on plasma concentrations of prolactin in turkey hens incubating with or without eggs or nest. *British Poultry Science* 31, 791–800.

Porter, T.E., Silsby, J.L., Behnke, E.J., Knapp, T.R. and El Halawani, M.E. (1991a) Ovarian steroid production *in vitro* during gonadal regression in the turkey. I. Changes associated with incubation behaviour. *Biology of Reproduction* 45, 581–586.

Porter, T.E., Silsby, J.L., Hargis, B.M., Fehrer, S.C. and El Halawani, M.E. (1991b) Ovarian steroid production *in vitro* during gonadal regression in the turkey. II. Changes induced by forced molting. *Biology of Reproduction* 45, 587–591.

Proudman, J.A. (1998) Circulating prolactin levels at the end of the photophase and the end of the scotophase throughout the reproductive cycle of the turkey hen. *Poultry Science* 77, 303–308.

Proudman, J.A. and Wentworth, B.C. (1996) Pulsatile secretion of prolactin in laying and incubating turkey hens. *Domestic Animal Endocrinology* 13, 277–282.

Ramesh, R., Solow, R. and Proudman, J.A. (1998) Identification of mammosomatotrophs in the turkey hen pituitary: increased abundance during hyperprolactinemia. *Endocrinology* 139, 781–786.

Ramesh, R., Kuenzel, W.J. and Proudman, J.A. (2001) Increased proliferative activity and programmed cellular death in the turkey hen pituitary following interruption of incubation behaviour. *Biology of Reproduction* 64, 611–618.

Ramsey, A.O. (1953) Variations in the development of broodiness in the fowl. *Behaviour* 5, 51–57.

Reddy, I.J., David, C.G. and Raju, S.S. (2007) Effect of suppression of plasma prolactin on luteinising hormone concentration, intersequence pause days and egg production in domestic hen. *Domestic Animal Endocrinology* 33, 167–175.

Richard-Yris, M.A. and Leboucher, G. (1986) Comportement maternel induit chez la poule domestique: influence du développement du poussin sur la qualité des réponses maternelles. *Biology of Behaviour* 11, 157–166.

Richard-Yris, M.A., Garnier, D.H. and Leboucher, G. (1983) Induction of maternal behaviour and some hormonal and physiological correlates in the domestic hen. *Hormones and Behavior* 17, 345–355.

Richard-Yris, M.A., Leboucher, G., Williams, J. and Garnier, D.H. (1987a) Induction of maternal behaviour in incubating and non-incubating hens: influence of hormones. *Physiology and Behavior* 40, 193–199.

Richard-Yris, M.A., Leboucher, G., Williams, J. and Garnier, D.H. (1987b) Influence of food restriction and the presence of chicks on the reproductive system of the domestic hen. *British Poultry Science* 28, 251–260.

Richard-Yris, M.A., Guémené, D., Lea, R.W., Sharp, P.J. Bédécarrats, G., Forasté, M. and Wauters, A.M. (1998) Behaviour and hormone concentrations in nest-deprived and renesting hens. *British Poultry Science* 39, 309–317.

Riddle, O., Bates, R.W. and Lahr, E.L. (1935) Prolactin induces broodiness in fowl. *American Journal of Physiology* 111, 352–360.

Romanov, M.M., Talbot, R.T., Wilson, P.W. and Sharp, P.J. (2002) Genetic control of incubation behavior in the domestic hen. *Poultry Science* 81, 928–931.

Rozenboim, I., Tabibzadeh, J.L., Silsby, J.L. and El Halawani, M.E. (1993) Effect of ovine prolactin administration on hypothalamic vasoactive intestinal peptide (VIP), gonadotropin releasing hormone I and II content and anterior pituitary VIP receptors in laying turkey hens. *Biology of Reproduction* 48, 1246–1250.

Ruscio, M.G. and Adkins-Regan, E. (2004) Immediate early gene expression associated with induction of brooding behavior in Japanese quail. *Hormones and Behavior* 46, 19–29.

Savory, C.J. (1979) Changes in food intake and body weight of bantam hens during breeding. *Applied Animal Ethology* 5, 283–288.

Savory, C.J., Duncan, I.J. and Wood-Gush, D.G.M. (1978) Feeding behaviour in a population of domestic fowl in the wild. *Applied Animal Behaviour Science* 4, 13–27.

Sharp, P.J. (1989) Physiology of egg production. In: Nixey, C. and Grey, T.C. (eds) *Recent Advances in Turkey Science*. British Poultry Science Symposium 21, Butterworth & Co Ltd, Sevenoaks, UK, pp. 31–54.

Sharp, P.J. (1997) Neurobiology of the onset of incubation behaviour in birds. In: Maitra, S.K. (ed.) *Frontiers in Environmental and Metabolic Endocrinology*. The University of Burdwan, Burdwan, India, pp. 193–202.

Sharp, P.J. (2004) Genes for persistency of egg laying: White Leghorns and broodiness. Roslin Institute Annual Report 2003/2004, pp. 38–43.

Sharp, P.J. and Lea, R.W. (1981) The response of the pituitary gland to luteinizing hormone releasing hormone in broody bantams. *General and Comparative Endocrinology* 45, 131–133.

Sharp, P.J. and Sterling, R.J. (1986) Approaches to the pharmacological control of broodiness. *Proceedings of the 7th European Poultry Conference*. World's Poultry Science Association, French Branch, Paris, pp. 1029–1033.

Sharp, P.J., Scanes, C.G., Williams, J.B., Harvey, S. and Chadwick, A. (1979) Variations in concentrations of prolactin, luteinizing hormone, growth hormone and progesterone in the plasma of broody bantams (*Gallus domesticus*). *Journal of Endocrinology* 80, 51–57.

Sharp, P.J., MacNamee, M.C., Sterling, R.J., Lea, R.J. and Pederson, H.C. (1988) Relationships between prolactin, luteinising hormone and broody behaviour in bantam hens. *Journal of Endocrinology* 118, 279–286.

Sharp, P.J., Sterling, R.J., Talbot, R.T. and Huskisson, N.S. (1989) The role of hypothalamic vasoactive intestinal polypeptide in the maintenance of prolactin secretion in incubating bantam hens: observations using passive immunisation, radioimmunoassay and immunohistochemistry. *Journal of Endocrinology* 122, 5–13.

Sharp, P.J., Talbot, R.T., Main, G.M., Dunn, I.C., Fraser, H.M. and Huskisson, N.S. (1990) Physiological roles of chicken LHRH-I and II in the control of gonadotrophin release in the domestic chicken. *Journal of Endocrinology* 124, 291–299.

Sharp, P.J., Li, Q., Georgiou, G. and Lea, R.W. (1996) Expression of cfos-like immunoreactivity in the hypothalamus of the ring dove (*Streptopelia risoria*) at the onset of incubation. *Journal of Neuroendocrinology* 8, 291–298.

Sharp, P.J., Dawson, A. and Lea, R.W. (1998) Control of luteinising hormone and prolactin in birds. *Comparative Biochemistry and Physiology Part* C 119, 275–282.

Sherry, D.F. (1981) Parental care and the development of thermoregulation in red jungle fowl. *Behaviour* 76, 250–279.

Skutch, A.F. (1962) The constancy of incubation. *Wilson Bulletin* 74, 115–151.

Slawski, B.A. and Buntin, J.D. (1995) Preoptic lesions disrupt prolactin-induced parental feeding behaviour in ring doves. *Hormones and Behavior* 29, 248–266.

Steen, J.B. and Parker, H. (1981) The egg-numerostat. A new concept in the regulation of clutch size. *Ornis Scandinavica* 12, 109–110.

Sterling, R.J., Gasc, J.M., Sharp, P.J., Renoir, J.M., Tuohimaa, P. and Baulieu, E.E. (1987) The distribution of nuclear progesterone receptor in the hypothalamus and forebrain of the domestic hen. *Cell and Tissue Research* 248, 201–205.

Taira, H. and Beck, M.M. (2006) Activity of three-β-hydroxysteroid dehydrogenase in granulosa cells treated in vitro with luteinising hormone, follicle stimulating hormone, prolactin, or a combination. *Poultry Science* 85, 1769–1774.

Talbot, R.T., Hanks, M.C., Sterling, R.J., Sang, H. and Sharp, P.J. (1991) Pituitary prolactin messenger ribonucleiuc acid levels in incubating and laying hens: effects of manipulating plasma levels of vasoactive intestinal polypeptide. *Endocrinology* 129, 496–502.

Tanabe,Y., Nakamura, T., Tanase, H. and Dio, O. (1981) Comparisons of plasma LH, progesterone, testosterone and estradiol concentrations in male and female chickens (*Gallus domesticus*) from 28 to 1141 days of age. *Endocrinologica Japonica* 28, 605–613.

Tsutsui, K., Saigoh, E., Ukena, K., Teranishi, H., Fujisawa, Y., Kikuchi, M., Ishii, S. and Sharp, P.J. (2001) A novel avian hypothalamic peptide inhibiting gonadotropin release. *Biochemical and Biophysical Research Communications* 275, 661–667.

von Werder, K., Eversmann, T., Fahlbusch, R. and Rjosk, H.K. (1980) Treatment of hyperprolactinemia. In: MacLeod, R.M. and Scapugnini, U. (eds) *Central and Peripheral Regulation of Prolactin Function*. Raven Press, New York, pp. 253–369.

Williams, J.B. and Sharp, P.J (1977) A comparison of plasma progesterone and luteinising hormone in growing hens from eight weeks to sexual maturity. *Journal of Endocrinology* 73, 447–448.

Wong, E.A., Ferrin, N.H., Silsby, J.L. and El Halawani, M.E. (1991) Cloning of a turkey prolactin cDNA: expression of prolactin mRNA throughout the reproductive cycle of the domestic turkey (*Meaeagris gallopavo*). *General and Comparative Endocrinology* 83, 18–26.

Xu, M., Proudman, J.A., Pitts, G.R., Wong, E.A., Foster, D.N. and El Halawani, M.E. (1996) Vasoactive intestinal peptide stimulates prolactin mRNA expression in turkey pituitary cells. Effects of dopaminergic drugs. *Proceedings of the Society for Experimental Biology and Medicine* 212, 52–62.

You, S., Foster, L.K., Silsby, J.L., El Halawani, M.E. and Foster, D.N. (1995) Sequence analysis of the turkey LH beta subunit and its regulation by gonadotropin releasing hormone and prolactin in cultured pituitary cells. *Journal of Molecular Endocrinology* 14, 117–129.

Youngren, O.M., El Halawani, M.E., Phillips, R.E. and Silsby, J.L. (1989) Effects of preoptic and hypothalamic lesions in female turkeys during a photoinduced reproductive cycle. *Biology of Reproduction* 41, 610–617.

Youngren, O.M., El Halawani, M.E., Silsby, J.L. and Phillips, R.E. (1991) Intracranial prolactin perfusion induces incubation behaviour in turkey hens. *Biology of Reproduction* 44, 425–431.

Youngren, O.M., Chaiseha, Y., Phillips, R. and El Halawani, M.E. (1996) Vasoactive intestinal peptide concentrations in turkey hypophysial portal blood differ across the reproductive cycle. *General and Comparative Endocrinology* 103, 323–330.

Youngren, O.M., Chaiseha, Y. and El Halawani, M.E. (1998) Serotonergic stimulation of avian prolactin secretion requires an intact dopaminergic system. *General and Comparative Endocrinology* 112, 63–68.

Youngren, O.M., Pitts, G.R., Chaiseha, Y. and El Halawani, M.E. (1999) An opioid pathway in the hypothalamus of the turkey that stimulates prolactin secretion. *Neurendocrinology* 70, 317–323.

Youngren, O., Chaiseha,Y., Al-Zailaie, K., Whiting S., Kang, S.W. and El Halawani, M.E. (2002) Regulation of prolactin secretion by dopamine at the level of the hypothalamus in the turkey. *Neuroendocrinology* 75, 185–192.

Zadworny, D., Shimada, K., Ishida, H. and Sato, K. (1989) Gonadotropin-stimulated estradiol production in small ovarian follicles of the hen is suppressed by physiological concentrations of prolactin in vitro. *General and Comparative Endocrinology* 74, 468–473.

Zhou, J.F., Zadworny, D., Guémené, D. and Kuhnlein, U. (1996) Molecular cloning, tissue distribution, and expression of the prolactin receptor during various reproductive states in *Melagris gallopavo. Biology of Reproduction* 55, 1081–1090.

Chapter 12
Incubation and Hatching

N.A. French
Aviagen Turkeys Ltd, Tattenhall, Cheshire, UK

ABSTRACT

The emphasis of poultry incubation research since the late 1980s has focused on four different themes, namely: (i) the relationship between incubation temperature and embryo temperature; (ii) how to engineer commercial incubators to meet the requirements of the embryo; (iii) the effect of genetic changes to poultry breeds on their incubation requirements; and (iv) the effect of incubation environment on the post-hatch performance or epigenetic adaptation. It is the intention of this short review to focus on these four areas of research and to discuss how they are affecting commercial practice.

A simple systems model is used to describe how embryo growth and metabolism, the properties of the egg in which the embryo develops and the environment of the incubator interact to determine the embryo temperature. The model shows that the relationship between the egg and its incubation environment is a dynamic process, with the embryo responding to its incubation environment but also affecting the environment. Our understanding of these interrelationships has resulted in better design of artificial incubators and the development of methods to directly measure the response of the egg to its environment.

The third section of the review discusses whether the genetic improvements to poultry in both egg and meat production have altered the incubation environment of the embryo. It is concluded that there are differences in embryo metabolic rate between genetic lines, which may impact on the optimum incubation environment. It was also noted that the mechanism by which genetic selection had affected embryo metabolism was not yet understood. The last section of the chapter discusses the possibility that post-hatch performance can be affected by the incubation environment and presents evidence to show that this may indeed be the case.

INTRODUCTION

In 1989, two symposia that extensively reviewed the then current research on egg incubation were held in the UK, the first covering the artificial incubation of poultry species (Tullett, 1991) and the second covering both artificial and natural incubation of all avian and reptilian species (Deeming and Ferguson, 1991). These symposia reflected some of the major themes of incubation research in the 1970s and 1980s, stimulated by the pioneering work of Professor Rahn and co-workers into the relationships between eggshell conductance, egg water balance and respiration.

Since these symposia, the emphasis of poultry incubation research has shifted and four different themes have come to the fore, namely: (i) the relationship between incubation temperature and embryo temperature; (ii) how to engineer commercial incubators to meet the requirements of the embryo; (iii) the effect of genetic changes to poultry breeds on their incubation requirements; and (iv) the effect of incubation environment on post-hatch performance or epigenetic adaptation. It is the intention of this short review to focus on these four areas of research and to discuss how they are affecting commercial practice. The subject of incubation by wild birds was extensively reviewed in Deeming (2002).

DETERMINATES OF EMBRYO TEMPERATURE

Of all the factors that determine incubation success, temperature is the most critical: even small deviations from the optimum can affect hatchability (Wilson, 1991). The temperature experienced by the embryo during incubation is dependent on the temperature of the incubator air, the metabolic heat production of the embryo, the slight cooling effect of water being lost from the egg and the efficiency of heat transfer between the egg and the incubator air (French, 1997). Figure 12.1 shows a simple system model to describe how these and other factors interact to determine the embryo temperature. The inner box within the model shows the factors that are determined by the egg and embryo, and the outer box represents the incubator environment and the parameters that need to be controlled for successful incubation to take place.

A major determinant of embryo temperature is the metabolic heat production of the embryo, and its effect will increase as the embryo grows through the incubation period. Measured values of chick embryo heat production increase from less than 2 mW on day 2 of incubation to approximately 140 mW by day 18 of incubation (Romijn and Lokhorst, 1960). Values for the metabolic heat production of turkey, guinea fowl, quail and duck eggs have also been published (for example, Dietz *et al.*, 1998; Harun *et al.*, 2001) and show similar increases during incubation. At the start of incubation, the negligible level of metabolic heat production results in the embryo temperature being primarily determined by the incubator air temperature. However, as the heat production of the growing embryo increases during incubation, the temperature of the egg increases above the incubator air temperature (Tazawa and Nakagawa, 1985; French, 1997).

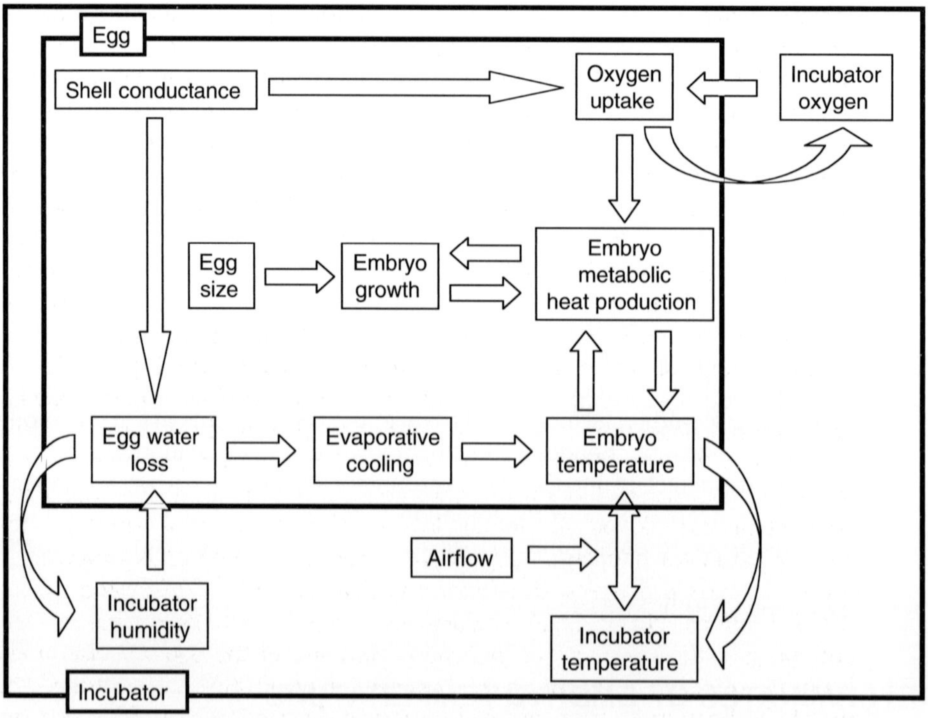

Fig. 12.1. A model to show the interactions between embryo temperature, embryo metabolic heat production, egg characteristics and the incubation environment.

The extent that the egg temperature increases depends on the efficiency of heat transfer between the egg and the surrounding air (French, 1997). Theoretical models of heat transfer between the egg and incubator have been proposed by several authors (Kashkin, 1961; Sotherland *et al.*, 1987; Meijerhof and van Beek, 1993) and these studies concluded that the main determinant of heat transfer in an artificial incubator was the rate of airflow over the eggs. The importance of airflow in determining heat transfer was confirmed in the recent studies by Van Brecht *et al.* (2003, 2005); however, they found that actual heat transfer coefficients were higher than those predicted from the theoretical models that assumed the egg was a sphere.

Depending on the efficiency of the heat transfer process, the heat lost from the eggs in the incubator can also have localized effects on the air temperature within the incubator, resulting in temperature variations within the machine (French, 1997, 2002; Lourens, 2001). Factors such as the spacing between egg trays and speed of the incubator ventilator have been shown to affect the airflow over the eggs and as a consequence the air temperature surrounding the eggs (Kaltofen, 1969; French, 1997; Van Brecht *et al.*, 2003).

The embryo's metabolic heat production will, in turn, be affected by three factors: (i) oxygen uptake into the egg from the incubator air; (ii) embryo growth; and (iii) the feedback effect of embryo temperature. Embryo metabolism requires oxygen, which needs to pass by diffusion from the incubator air into the egg via the respiratory pores in the shell (Rahn *et al.*, 1979). In addition to the uptake of oxygen there will be a concomitant production of carbon dioxide, which must diffuse out of the egg through the same respiratory pores. The rate at which oxygen can diffuse into the egg will depend on the oxygen demand of the embryo, the oxygen partial pressure in the incubator air and the conductance of the eggshell to oxygen (Wangensteen *et al.*, 1970). Therefore, either shell conductance or incubator oxygen concentration can limit the rate of oxygen uptake into the egg and hence limit embryo metabolism. The reducing embryo metabolism will decrease the rate of embryo growth, and as embryo mass determines metabolic rate a feedback loop will be created, by which the metabolic rate will also determine the rate of embryo growth (Fig. 12.1).

It has been shown in several poultry species that shell conductance can limit embryo growth at the end of incubation (Tullett and Deeming, 1982; Burton and Tullett, 1983, 1985), where both wet and dry embryo mass increased with increasing shell conductance to a plateau, above which shell conductance is no longer the limiting factor. Other studies have also shown that altering incubator oxygen levels will also result in changed embryo growth or metabolic heat production (Lokhorst and Romijn, 1967; Tullett and Burton, 1987; Lourens, 2006). Shell conductance is not the only factor that can limit embryo growth at the end of incubation, the size of the egg will clearly also have an effect. In a recent study, Lourens *et al.* (2006) demonstrated that metabolic heat production of chick embryos developing in a 70 g egg was greater than that of a 56 g egg from day 15 of incubation onwards but not at earlier ages.

The third factor that will affect metabolic heat production of the embryo is its body temperature. The precocial avian embryo is unable to thermoregulate until around the time of hatching and effectively behaves as a poikilotherm. Early in incubation the embryo acts as a true poikilotherm and metabolic rate increases with increasing body temperature (Whittow and Tazawa, 1991). At the plateau phase of development, just prior to pipping, a slight homeothermic response is seen, but the metabolic rate still increases with body temperature, following an exponential or parabolic function (Nichelmann *et al.*, 1998, 2001; Janke *et al.*, 2002). It is only after hatching that full homeothermy develops (Dietz and van Kampen, 1994). As a consequence there is a positive feedback from embryo temperature to embryo metabolic rate over the normal range of incubation temperatures (Fig. 12.2). As the embryo's metabolic rate increases, the heat generated will result in a rise in embryo temperature unless the extra heat can be lost from the egg. A rise in embryo body temperature will further increase metabolic heat production, although there does appear to be an upper limit of a body temperature close to 40°C, beyond which metabolic rate starts to decrease (Fig. 12.2; Janke *et al.*, 2002).

The model in Fig. 12.1 also shows how egg water loss is related to embryo temperature. As eggs have porous shells to allow the diffusion of respiratory gases, they will also lose water as vapour that diffuses out through the shell. The rate of water loss will depend on the humidity of the incubator and the conductance of the shell. This loss of water from the egg is necessary in most

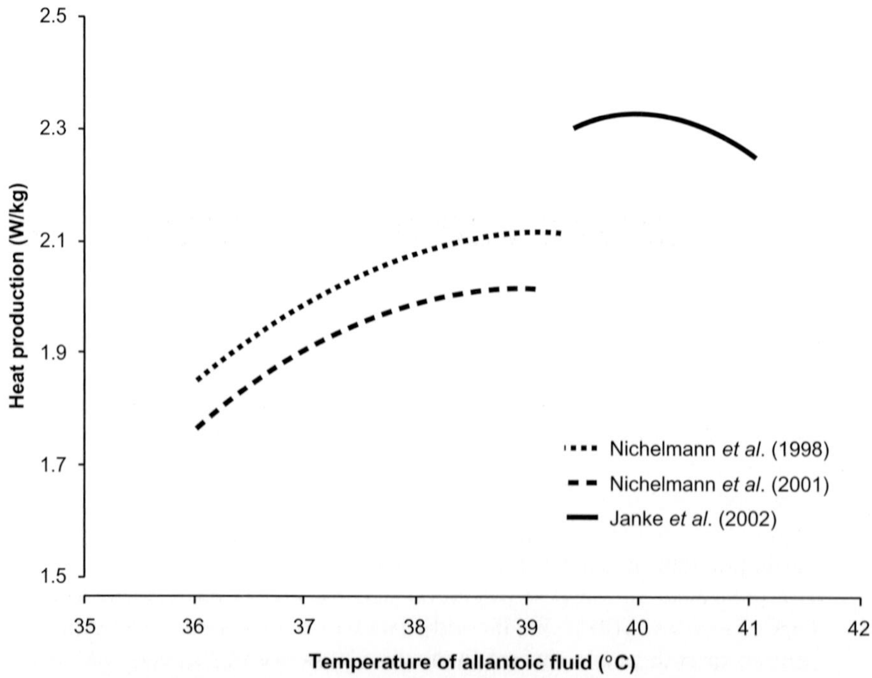

Fig. 12.2. The relationship between internal egg temperature, as measured in the allantoic fluid, and metabolic heat production of 21-day chicken embryos.

avian species to provide an air chamber in the egg to allow lung inflation just prior to hatch and to lose the water generated by lipid metabolism (Ar, 1991). The phase change of liquid water to water vapour requires heat, so as water vapour is lost from the egg some heat is also lost due to evaporative cooling (typically 11–12 mW for a chicken egg). While the heat loss from evaporative cooling is small by comparison to the metabolic heat production at the end of incubation, this is not the case during the first half of incubation, when the effect of evaporative cooling is greater than metabolic heat production (Romijn and Lokhorst, 1960). This explains the observations that internal egg temperature can be slightly cooler than the surrounding air temperature during the first half of incubation (Tazawa and Nakagawa, 1985; French, 1997).

Incubator humidity could, in theory, have a second effect on embryo temperature by affecting the efficiency of heat transfer between the egg and the incubator air. Water vapour has a higher thermal conductivity than air and so the higher the humidity, the higher the thermal conductivity of the air and water vapour mix and the more efficient the heat transfer away from the egg, resulting in a lower egg temperature. However, these effects are small and would be lower than the opposing effect of increased evaporative cooling with lower humidity, as was confirmed in a recent study that was unable to show any relationship between the efficiency of heat transfer and incubator humidity (Van Brecht *et al.*, 2005).

The model shown in Fig. 12.1 is an attempt to show how all the above factors relate to each other. Metabolic heat production is determined by the mass of the embryo and increases with time of incubation. Increasing metabolic heat production results in an increase in embryo body temperature, the extent of which will depend on the efficiency of heat transfer between the egg and the surrounding incubator air. If embryo temperature rises then this will result in a positive feedback on metabolic heat production. Embryo metabolism requires oxygen to diffuse into the egg through the porous shell, and this can be limited by either the shell conductance or the oxygen concentration in the incubator air. If metabolic rate is limited this will reduce embryo growth. Shell conductance, in combination with incubator humidity, also affects the rate of water loss from the egg, which has a small evaporative cooling effect on the temperature of the egg. To affect this process, the poultry incubationist has three areas of control: the incubator air temperature, humidity and oxygen concentration.

The model also illustrates how the egg, or eggs, within the artificial incubator have an effect on the environment within the incubator. As an example, the eggs within a 10,000 capacity, single-stage chicken incubator at the end of the incubation period would add to the incubator environment approximately 1.4 kW/day of heat, 4 kg/day of water and 4 m^3/day of carbon dioxide, and remove 6 m^3/day of oxygen. The relationship between the egg and its incubation environment is a dynamic process, and it is our understanding of this process that has resulted in better artificial incubator design.

DEVELOPMENTS IN ARTIFICIAL INCUBATOR DESIGN

For many years the incubation of poultry eggs was relatively simple and mostly used a constant incubator air temperature, humidity and ventilation rate (to supply oxygen and remove carbon dioxide). A 6-month study of temperatures within an incubator that used a constant machine setting showed variations in air temperatures that depended on the total metabolic heat production of the eggs within the machine (French and Houlbrooke, 2004). More recent incubator designs have attempted to use control systems that monitor the response of the egg to its incubation environment so that it can alter conditions to better match the needs of the embryo.

The first development in incubation control systems was to directly measure egg water loss by weighing eggs to control incubator humidity. It has been understood for a long time that eggs of poultry species need to lose between 12 and 14% of their fresh egg weight through water loss prior to pipping to have the best chance of hatching successfully (Ar, 1991). Several studies have shown that monitoring egg weight loss during incubation so that eggs achieve 12–14% water loss results in improved hatchability (Meir *et al.*, 1984; Hulet *et al.*, 1987; Meir and Ar, 1991; Ar *et al.*, 1996).

In many hatcheries egg weighing was done manually, but now incubator manufacturers have developed automatic weighing systems that can monitor water loss on a daily basis and adjust humidity to ensure the correct level of water loss is maintained. Studies have shown that it is not necessary for the rate of water loss to be constant through the incubation process. Indeed, a low level of water loss during the first half of incubation followed by a higher water loss for the rest of incubation, but still achieving the final target of 12–14% at pipping, has been shown to give the best hatch results (Snyder and Birchard, 1982; Meir *et al.*, 1984). The use of automatic egg-weighing systems makes a non-linear water loss target easier to achieve within large commercial hatcheries.

The use of carbon dioxide sensors to control ventilation levels in artificial incubators has started to become a widespread feature of new machines. Carbon dioxide sensors have been chosen rather than oxygen sensors because they are more economic and reliable. Maintaining carbon dioxide levels will ensure oxygen levels are also maintained (Owen, 1991). What is lacking is research to show the range of carbon dioxide levels that are required in the incubator to achieve maximum hatchability.

Early studies suggest that carbon dioxide levels above 1% are detrimental to hatch, but many of the results were confounded due to differences in other environmental factors between experimental treatments (Lundy, 1969). Similarly, studies looking at the effects of low oxygen levels have tended to produce inconsistent findings and use levels of hypoxia lower (<16%) than are likely to be found in commercial incubators (Chan and Burggren, 2005). However, some studies have suggested that carbon dioxide levels between 0.3 and 1.5% early in incubation improve hatch (Gildersleeve and Boeschen, 1983) and stimulate embryo growth (De Smit *et al.*, 2006). There is growing commercial experience to suggest that high carbon dioxide levels (*circa* 1%) in the first half of incubation are beneficial for hatch (Hogg, 1997), but this still needs to be confirmed experimentally.

Traditionally, incubation temperature has been controlled by a sensor located in free air within the incubator, and this has been shown to differ by up to 3°C from the temperature of the air surrounding the eggs (French, 1997, 2002; French and Houlbrooke, 2004) and up to 5°C variation compared with egg temperature (Lourens, 2001). While it is difficult to directly measure embryo temperature in commercial incubators, shell surface temperatures have been shown to be close to internal temperature, as the main barrier to heat loss from the egg is the surrounding air (Sotherland *et al.*, 1987; French 1997). Van Brecht *et al.* (2002) demonstrated that infrared thermometry could be used to measure shell surface temperature and several studies have used this method to investigate the effect of egg temperature on embryo development (Lourens *et al.*, 2005, 2006). The use of infrared thermometry has now been incorporated into the design of some commercial incubators so that they are able to directly control shell temperature rather than air temperature.

While direct measurement of shell temperature can ensure that the incubator provides the correct temperature for embryo development, it is not possible to measure the temperature of every egg within an incubator and so a sample of eggs need to be chosen for monitoring. This gives rise to two potential problems in large commercial incubators, which can contain between 10,000 and 100,000 eggs. First, how to batch eggs with similar characteristics (e.g. egg mass or shell conductance) to the sampled eggs together within one machine

(French, 2002). Second, to ensure that the incubation environment is uniform throughout the incubator so that the temperature applied to the sampled eggs also applies to eggs at other locations within the machine.

Achieving a uniform thermal environment within an incubator is dependent on achieving a uniform airflow within the machine (Owen, 1991; Meijerhof and van Beek, 1993; French, 1997). New techniques for measuring and modelling the airflow and temperature gradients within commercial incubators have recently been developed (Van Brecht *et al.*, 2003, 2005) and these methodologies are being applied to improving the design of incubators.

These recent developments in incubator technology have provided the hatchery manager with better tools with which to match the incubator environment to the requirements of the developing embryo within the egg. However, the extent that science can inform the hatchery manager regarding the ideal environments for the developing embryo is very variable. The incubation humidity environment requirement and how it controls egg water loss is well understood but the optimum oxygen and carbon dioxide levels within the incubator still need to be defined.

The most important incubation parameter, temperature, is also well understood in terms of control but the majority of the research has studied the effect of air temperature rather than egg temperature on embryo development (French, 1997, 2002; Meijerhof, 2002; Lourens *et al.*, 2006). The optimum egg temperature still needs to be defined and alternative options have been suggested, to a large extent based on practical experience in commercial hatcheries (Boerjan *et al.*, 2006). An example of the incubation environment in a typical broiler incubator is shown in Fig. 12.3, and these programmes have been developed in part through our understanding of incubation science but largely through trial and error in commercial hatcheries.

CHANGES TO INCUBATION REQUIREMENTS DUE TO GENETIC CHANGE IN POULTRY SPECIES

The performance of poultry strains used today differs dramatically from that of the strains used 50 years ago, with a remarkable improvement in the efficiency of egg and meat production (Havenstein *et al.*, 2003; Flock *et al.*, 2005; Hunton, 2006). For example, the 2001 broiler reaches a weight of 1.8 kg in 32 days compared with 101 days for the 1957 broiler, and most of this improvement has been as a result of genetic selection (Havenstein *et al.*, 2003). This raises the question of whether the significant changes in post-hatch performance have had any consequences for the incubation requirements of today's commercial poultry strains. For example, has the selection for a bird that grows faster after hatching resulted in an increase in embryo growth rate and metabolic heat production?

Genetic selection for either egg or meat production could impact the incubation requirements of the embryo, either by a direct impact on the growth and physiology of the embryo itself or by altering the egg in which it develops.

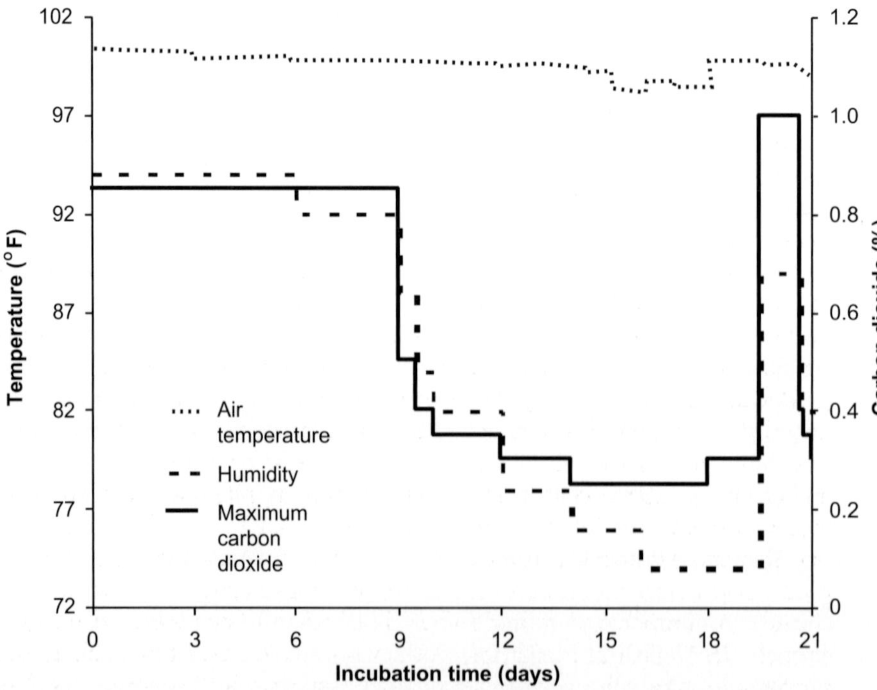

Fig. 12.3. An example of a single-stage incubation programme used to incubate broiler eggs. Humidity is shown as wet bulb temperature and carbon dioxide levels are shown as maximum level before ventilation will increase. Programme supplied courtesy of Petersime.

Studies have shown that genetic selection for either egg production or post-hatch growth can result in correlated changes to egg mass and shell conductance (French and Tullett, 1991; Christensen *et al.*, 1993, 1995; Christensen and Nestor, 1994; Tona *et al.*, 2004). In the model shown in Fig. 12.1, the shell conductance, egg size, embryo growth and embryo metabolic rate could all be changed by genetic selection for growth or egg production. One of the difficulties of attempting to evaluate the impact of genetic selection on the physiology of the embryo is that it is hard to separate these effects from those as a result of changes to the egg in which the embryo develops. A second problem when investigating the effects of genetic selection on the growth or metabolic rate of embryos is that in the majority of studies comparisons are not made at the same egg temperature (Lourens *et al.*, 2006).

The concept that post-hatch growth rate affects embryo growth rate has been investigated in a few studies. Tullett and Burton (1983) found that embryo growth differed between broiler lines with different post-hatch growth rates once egg weight and shell conductance had been accounted for. However, the relationship between pre- and post-hatch growth was not straightforward and one line was more developed at the start of incubation, which resulted in faster embryo growth and an earlier hatch time, even though it did not have the fastest post-hatch growth. A study comparing eggs of the same weight from a broiler and a layer chicken line showed faster embryo growth in the broiler line (Pal *et al.*, 2002) but this study did not control for shell conductance. In addition to differences in overall embryo growth rate, there is evidence that fast-growing chicken lines allocate more resources to muscle than other tissues during incubation compared with slow-growing lines (Clum *et al.*, 1995). Whether these differences in relative tissue composition between fast- and slow-growing lines of chicken are sufficient to alter the total metabolic heat production of the embryo per gram of tissue remains to be shown.

Some experimental work on the effect of long-term selection for either egg production or growth on chick embryo metabolic rate has been reported. One study compared two broiler lines and a White Leghorn layer line and showed no difference between the two broiler lines, whereas the metabolic heat production of both was significantly higher than the layer line (Janke *et al.*, 2004). Interestingly the metabolic heat production of the two broiler lines in this study was very similar to measurements made 20 years previously (Tullett and Deeming, 1982; Burton and Tullett, 1983), despite the significant gains in broiler growth rate in the intervening period (Janke *et al.*, 2004). Similar observations were made in another study that compared the metabolic heat production of broiler embryos from a high breast meat yield line, a modern commercial broiler and a broiler line that had not been selected for growth since 1978 (O'Dea *et al.*, 2004). Two other studies have been able to detect significant differences in metabolic heat production at the end of incubation between different broiler strains (Tona *et al.*, 2004; De Smit *et al.*, 2005), although in both these studies there were significant differences in both egg weight and shell conductance between the lines.

While there is evidence of a relationship between pre- and post-hatch growth rates in chickens, well-constructed studies of turkey lines have not shown this; where embryos of significantly different post-hatch genetic potential due to the sire were incubated in eggs laid by the same female, a similar relationship could not be demonstrated (Bielfelt and Shultz, 1989; Christensen *et al.*, 2004). A study comparing growth and metabolic heat production of embryos from turkey lines selected for 30 years for growth or egg production against their random-bred controls did show differences but there were also significant differences in both egg mass and shell conductance between the lines (Christensen *et al.*, 1993). Similarly, increased embryo growth has been shown in quail lines selected for higher post-hatch growth (Lilja and Olsson, 1987).

The evidence to date would suggest that there can be differences between genetic lines of poultry in both growth and metabolic rate of the embryos, although whether these differences are due to changes in embryo physiology, development rate or the egg in which the embryo is developing has yet to be ascertained. The practical outcome of these studies is that the genetic selection of poultry lines can affect the total metabolic heat production of the egg, which in turn would increase egg temperature if the incubation environment is not adjusted accordingly. However, it should also be noted that greater changes in metabolic heat production of the egg can occur within genetic lines because of differences in egg mass due to the age of the breeder flock (O'Dea *et al.*, 2004; Lourens *et al.*, 2006).

POST-HATCH EFFECTS OF INCUBATION TEMPERATURE – EPIGENETIC ADAPTATION

There have been a series of recent studies that have investigated whether incubation temperature manipulation during development can result in permanent changes to the structure or physiology of the embryo that last into the post-hatching phase. The term epigenetic adaptation has been used to describe the process by which altering temperature at critical embryo stages results in a permanent change in the expression of genes that control the development process (Nichelmann and Tzschentke, 2002; Nichelmann, 2004; Tzschentke and Plagemann, 2006).

One of the main pieces of evidence that epigenetic adaptation can occur has come from studies of the development of the thermoregulatory system of poultry species. Several studies have shown that a persistent change in the thermoregulatory abilities of newly hatched chicks, turkeys and Muscovy ducks can be affected by the temperature experienced during incubation (Decuypere, 1984; Minne and Decuypere, 1984; Nichelmann and Tzschentke, 1999; Yahav *et al.*, 2004a). Exposing embryos to high incubation temperatures during the later stages of incubation resulted in chicks that were more tolerant of high environmental temperatures post-hatch (Moraes *et al.*, 2003; Yahav *et al.*, 2004b; Collin *et al.*, 2005), although this relationship may be further influenced by the age of the breeder parent (Yalçin *et al.* 2005). Turkey poults from eggs incubated at a higher temperature were shown to prefer a higher ambient temperature for at least their first 10 days post-hatch than the normally incubated controls (Nichelmann, 2004). Similarly, low incubation temperatures at the end of incubation have also been shown to affect the development of thermoregulation, through delaying the developmental stage at which thermoregulatory ability develops (Black and Burggren, 2004), decreasing the thermoregulatory set-point (Tzschentke and Nichelmann, 1999), increasing metabolic heat production (Geers *et al.*, 1983; Minne and Decuypere, 1984) and lowering the preferred ambient temperature of the hatchling (Tzschentke and Nichelmann, 1999).

Further studies into the mechanisms by which incubation temperature can modify the thermoregulatory set-point of the hatchling have found differences in the warm- and cold-sensitive neurones in the hypothalamus that are responsible for integrating the afferent temperature signals and controlling the thermoregulatory response (Yahav, 2007). Muscovy ducks incubated at both high and low incubation temperatures from day 17 to 21 of incubation were found, when compared with normal temperature, to have a higher proportion of temperature-sensitive neurones just prior to hatch (Loh *et al.*, 2004) and altered ratios of cold- and warm-sensitive neurones post-hatch (Tzschentke and Basta, 2002; Tzschentke *et al.*, 2004).

The thermoregulatory system is further mediated by the endocrine system, both directly through the hypothalamus–pituitary–thyroid axis and indirectly through the hypothalamus–pituitary–adrenal axis, which affects the stress response of the bird (Yahav, 2007). The rate of development and maturation of both systems and the level of hormone production have been shown to be affected by changes in incubation temperature (Nvota *et al.*, 1980; Minne and

Decuypere, 1984; Iqbal *et al.*, 1990; Moraes *et al.*, 2003, 2004; Yahav *et al.*, 2004a).

The development of other systems has also been shown to be affected by altering incubation temperature. High and low incubation temperatures early in incubation have been shown to alter the number of fibres in the leg muscles of 16-day-old turkey poults (Maltby *et al.*, 2004), and high temperatures during late incubation have been shown to delay muscle satellite cell differentiation of broilers (Halevy *et al.*, 2006) and increase breast meat yield at 42 days (Collin *et al.*, 2007). Short periods of high and low temperature during incubation have been shown to affect broiler lung and skeletal development at hatch (Yalçin and Siegel, 2003). Studies have also shown that high or low egg temperature during incubation can result in lower post-hatch growth in broilers (Lourens *et al.*, 2005; Joseph *et al.*, 2006).

Decuypere and Michels (1992), in a review of the effects of incubation temperature, suggested that in the future the success of the incubation process would not be measured only in terms of improving hatchability but also in terms of how it affects the potential performance of the broiler post-hatch. To date most of the above studies have demonstrated effects of incubation environment on short-term performance post-hatch and further studies are needed to show whether these changes can last to typical slaughter ages (Yahav, 2007). Indeed, a recent study was unable to demonstrate improved thermotolerance in 42-day broilers when they had been subjected to high temperatures during early incubation (Collin *et al.*, 2007). For this reason, this concept has not been implemented in commercial practice and most poultry companies still regard the number of chicks, poults and ducklings hatched as the primary measure of success.

FINAL OBSERVATIONS

Tullett (1990) suggested in his review of incubation research that the basic parameters of successful incubation were well understood and in general the poultry industry was achieving good hatchability. Future work would be the fine-tuning of the incubation process to customize the environment to meet the needs of individual batches of eggs. This observation still holds and the process of fine-tuning is now easier to achieve owing to the developments in technology that allow the incubator to continually monitor the response of the egg to its environment and make the necessary adjustments.

A consequence of the new incubator control mechanisms is a better definition of the questions that need to be answered in order to achieve the best results. For example, what are the optimum egg temperatures rather than the optimum incubation temperatures for best hatchability and what are the correct oxygen and carbon dioxide concentrations at all stages of incubation? It is probable that the answers to these questions will come from the poultry industry rather than the research community because of the large numbers of eggs required to detect the small differences in performance that are likely to be achieved by this fine-tuning process. To detect at the $P = 0.05$ level of significance a 1% improvement on 80% hatchability requires in excess of 12,000 eggs per treatment (Laughlin and Lundy, 1976).

The last part of this chapter briefly discussed the possibility that post-hatch performance can be modified through manipulating the incubation environment. Although the poultry industry has not, as yet, attempted to modify embryo and post-hatch development through altering the incubation environment, there is increasing evidence that the incubation process must be regarded as having a much greater effect than just determining hatching success. Perhaps this should not be a surprise when one considers that today's broiler spends one-third of its total life as an embryo.

REFERENCES

Ar, A. (1991) Egg water movements during incubation. In: Tullett, S.G. (ed.) *Avian Incubation*. Butterworth-Heinemann, London, pp.157–174.

Ar, A., Meir, M., Aizik, N. and Campi, D. (1996) Standard values and proper ranges of ostrich egg parameters as a basis for proper artificial incubation. In: Deeming, D.C. (ed.) *Improving our Understanding of Ratites in a Farming Environment*. Ratite Conference, Banbury, UK, pp. 144–145.

Bielfelt, S.W. and Shultz, F.T. (1989) Genetic effects on embryonic growth in domestic turkeys. *Poultry Science* 68, 315–318.

Black, J.L. and Burggren, W.W. (2004) Acclimation to hypothermic incubation in developing chicken embryos (*Gallus domesticus*) I. Development effects and chronic and acute metabolic adjustments. *The Journal of Experimental Biology* 207, 1543–1552.

Boerjan, M.L., Hogg, A., Lourens, A., Tzschentke, B. and Janke, O. (2006) Roundtable on 'What is incubation temperature?' *Avian and Poultry Biology Reviews* 17, 72–76.

Burton, F.G. and Tullett, S.G. (1983) A comparison of the effects of egg shell porosity on the respiration and growth of domestic fowl, duck and turkey embryos. *Comparative Biochemistry and Physiology* 75A, 167–174.

Burton, F.G. and Tullett, S.G. (1985) Respiration of avian embryos. *Comparative Biochemistry and Physiology* 82A, 735–744.

Chan, T. and Burggren, W. (2005) Hypoxic incubation creates differential morphological effects during specific developmental critical windows in the embryo of the chicken (*Gallus gallus*). *Respiratory Physiology and Neurobiology* 145, 251–263.

Christensen, V.L. and Nestor, K.E. (1994) Changes in functional qualities of turkey eggshells in strains selected for increased egg production or growth. *Poultry Science* 73, 1458–1464.

Christensen, V.L., Donaldson, W.E. and Nestor, K.E. (1993) Embryonic viability and metabolism in turkey lines selected for egg production and growth. *Poultry Science* 72, 829–838.

Christensen, V.L., Havenstein, G.B. and Davis, G.S. (1995) Egg characteristics, carbohydrate metabolism, and thyroid hormones in late chick embryos from different genetic lines. *Poultry Science* 74, 551–562.

Christensen, V.L., Ort, D.T., Wineland, M.J. and Grimes, J.L. (2004) Turkey sire effects on embryonic survival and physiology. *International Journal of Poultry Science* 3, 80–88.

Clum, N.J., McClearn, D.K. and Barbato, G.F. (1995) Comparative embryonic development in chickens with different patterns of post natal growth. *Growth, Development & Aging* 59, 129–138.

Collin, A., Picard, M. and Yahav, S. (2005) The effect of duration of thermal manipulation during broiler chick embryogenesis on body weight and body temperature of post-hatched chicks. *Animal Research* 54, 105–111.

Collin, A., Berri, C., Tesseraud, S., Rodon, F.E., Skiba-Cassy, S., Crochet, S., Duclos, M.J., Rideau, N., Tona, K., Buyse, J., Bruggeman, V., Decuypere, E., Picard, M. and Yahav, S. (2007) Effects of thermal manipulation during early and late embryogenesis on thermotolerance and breast muscle characteristics in broiler chickens. *Poultry Science* 85, 795–800.

Decuypere, E. (1984) Incubation temperature in relation to postnatal performance in chickens. *Archive of Experimental Veterinary Medicine* 38, 439–449.

Decuypere, E. and Michels, H. (1992) Incubation temperature as a management tool: a review. *World's Poultry Science Journal* 48, 127–138.

Deeming, D.C. (2002) *Avian Incubation: Behaviour, Environment and Evolution*. Oxford University Press, Oxford, UK.

Deeming, D.C. and Ferguson, M.W.J. (1991) *Egg Incubation*. Cambridge University Press, Cambridge, UK.

De Smit, L., Tona, K., Bruggeman, V., Onagbesan, O., Hassanzadeh, M., Arckens, L. and Decuypere, E. (2005) Comparison of three lines of broilers differing in ascites susceptibility or growth rate. 2. Egg weight loss, gas pressures, embryonic heat production, and physiological hormone levels. *Poultry Science* 84, 1446–1452.

De Smit, L., Bruggeman, V., Tona, J.K., Debonne, M., Onagbesan, O., Arckens, L., De Baerdemaeker, J. and Decuypere, E. (2006) Embryonic developmental plasticity of the chick: increased CO_2 during early stages of incubation changes developmental trajectories during prenatal and postnatal growth. *Comparative Biochemistry and Physiology A* 145, 166–175.

Dietz, M.W. and van Kampen, M. (1994) The development of thermoregulation in turkey and guinea fowl hatchlings: similarities and differences. *Journal of Comparative Physiology* 164B, 69–75.

Dietz, M.W., van Kampen, M., van Griensven, M.J.M. and van Mourik, S. (1998) Daily energy budgets of avian embryos: the paradox of the plateau phase in egg metabolic rate. *Physiological Zoology* 71, 147–156.

Flock, D.K., Laughlin, K.F. and Bentley, J. (2005) Minimising losses in poultry breeding and production: how breeding companies contribute to poultry welfare. *World's Poultry Science Journal* 61, 227–237.

French, N.A. (1997) Modeling incubation temperature: the effects of incubator design, embryonic development, and egg size. *Poultry Science* 76, 124–133.

French, N.A. (2002) Managing the incubation environment in commercial hatcheries to meet the requirements of the embryo. In: William, J. (ed.) *Avian and Poultry Biology Reviews, Special Issue, International Congress on Bird Reproduction* 13, 179–186.

French, N.A. and Houlbrooke, R.J. (2004) A long-term study of the environment within a tunnel incubator for turkey eggs. *Journal of Applied Poultry Science* 13, 77–84.

French, N.A. and Tullett, S.G. (1991) Variation in the eggs of poultry species. In: Tullett, S.G. (ed.) *Avian Incubation*. Butterworth-Heinemann, London, pp. 59–77.

Geers, R., Michels, H., Nackaerts, G. and Konings, F. (1983) Metabolism and growth of chickens before and after hatch in relation to incubation temperatures. *Poultry Science* 62, 1869–1875.

Gildersleeve, R.P. and Boeschen, D.P. (1983) The effects of incubator carbon dioxide level on turkey hatchability. *Poultry Science* 62, 779–784.

Halevy, O., Lavi, M. and Yahav, S. (2006) Enhancement of meat production by thermal manipulations during embryogenesis of broilers. In: Yahav, S. and Tzschentke, B. (eds) *New Insights into Fundamental Physiology and Peri-natal Adaptation of Domestic Fowl*. Nottingham University Press, Nottingham, UK, pp. 77–88.

Harun, M.A.S., Veeneklaas, R.J., Visser, G.H. and Van Kampen, M. (2001) Artificial incubation of Muscovy duck eggs: why some eggs hatch and others do not. *Poultry Science* 80, 219–224.

Havenstein, G.B., Ferket, P.R. and Qureshi, M.A. (2003) Growth, livability, and feed conversion of 1957 versus 2001 broilers when fed representative 1957 and 2001 broiler diets. *Poultry Science* 82, 1500–1508.

Hogg, A. (1997) Single stage incubation trials. *Avian and Poultry Biology Reviews* 8, 168.

Hulet, R.M., Christensen, V.L. and Bagley, L.G. (1987) Controlled weight loss during incubation of turkey eggs. *Poultry Science* 66, 428–432.

Hunton, P. (2006) 100 years of poultry genetics. *World's Poultry Science Journal* 62, 417–428.

Iqbal, A., Decuypere, E., Abd el Azim, A. and Kühn, E.R. (1990) Pre- and post-hatch high temperature exposure affects the thyroid hormone and corticosterone response to acute heat stress in growing chickens (*Gallus domesticus*). *Journal of Thermal Biology* 15, 149–153.

Janke, O., Tzschentke, B., Höchel, J. and Nichelmann, M. (2002) Metabolic responses of chicken and Muscovy duck embryos to high incubation temperatures. *Comparative Biochemistry and Physiology Part A* 131, 741–750.

Janke, O., Tzschentke, B. and Boerjan, M. (2004) Comparative investigations of heat production and body temperature in embryos of modern breeds. *Avian and Poultry Biology Reviews* 15, 191–196.

Joseph, N.S., Lourens, A. and Moran, E.T. Jr. (2006) The effects of suboptimal eggshell temperature during incubation on broiler chick quality, live performance, and further processing yield. *Poultry Science* 85, 932–938.

Kaltofen, K.S. (1969) The effects of air movements on hatchability and weight loss of chicken eggs during artificial incubation. In: Carter, T.C. and Freeman, B.M. (eds) *The Fertility and Hatchability of the Hen's Egg*. Oliver and Boyd, Edinburgh, UK, pp. 177–190.

Kashkin, V.V. (1961) Heat exchange of bird eggs during incubation. *Biophysica* 6, 57–63.

Laughlin, K. and Lundy, H. (1976) The influence of sample size on the choice of method and interpretation of incubation experiments. *British Poultry Science* 17, 53–57.

Lilja, C. and Olsson, U. (1987) Changes in embryonic development associated with long-term selection for high growth rate in Japanese quail. *Growth* 51, 301–308.

Loh, B., Maier, I., Winar, A., Janke, O. and Tzschentke, B. (2004) Prenatal development of epigenetic adaptation processes in poultry: changes in metabolic rate and neuronal thermoregulatory mechanisms. *Avian and Poultry Biology Reviews* 15, 119–128.

Lokhorst, W. and Romijn, C. (1967) Energy metabolism of eggs incubated in air with a low oxygen content. In: Blaxter, K.L., Kielanowski, J. and Thorbeck, G. (eds) *Energy Metabolism of Farm Animals,* EAAP no. 12. Oriel Press, Newcastle, UK, pp. 331–338.

Lourens, A. (2001) The importance of air velocity in incubation. *World Poultry* 17(3), 29–30.

Lourens, A. (2006) Why eggs from different breeds need different eggshell temperatures for best results. *Avian and Poultry Biology Reviews* 17, 74–75.

Lourens, A., van den Brand, H., Meijerhof, R. and Kemp, B. (2005) Effect of eggshell temperature during incubation on embryo development, hatchability, and posthatch development. *Poultry Science* 84, 914–920.

Lourens, A., Molenaar, R., van den Brand, H., Heetkamp, M.J.W., Meijerhof, R. and Kemp, B. (2006) Effect of egg size on heat production and the transition of energy from egg to hatching. *Poultry Science* 85, 770–776.

Lundy, H. (1969) A review of the effects of temperature, turning, and gaseous environment on the hatchability of the hen's egg. In: Carter, T.C. and Freeman, B.M. (eds) *The Fertility and Hatchability of the Hen's Egg*. Oliver and Boyd, Edinburgh, UK, pp. 143–176.

Maltby, V., Somaiya, A., French, N.A. and Stickland, N.C. (2004) *In ovo* temperature manipulation influences post-hatch muscle growth in the turkey. *British Poultry Science* 45, 491–498.

Meijerhof, R. (2002) Design and operation of commercial incubators. In: Deeming, D.C. (ed.) *Practical Aspects of Commercial Incubation in Poultry*. Ratite Conference Books, Linclon, UK, pp. 41–46.

Meijerhof, R. and van Beek, G. (1993) Mathematical modeling of temperature and moisture loss of hatching eggs. *Journal of Thermal Biology* 165, 27–41.

Meir, M. and Ar, A. (1991) Compensation for seasonal changes in eggshell conductance and hatchability of goose eggs by dynamic control of egg water loss. *British Poultry Science* 28, 723–732.

Meir, M., Nir, A. and Ar, A. (1984) Increasing hatchability of turkey eggs by matching incubator humidity to shell conductance of individual eggs. *Poultry Science* 63, 1489–1496.

Minne, B. and Decuypere, E. (1984) Effects of late prenatal temperatures on some thermoregulatory aspects in young chickens. *Archive of Experimental Veterinary Medicine* 38, 374–383.

Moraes, V.M.B., Malherios, R.D., Bruggeman, V., Collin, A., Tona, K., Van As, P., Onagbesan, O.M., Buyse, J., Decuypere, E. and Macari, M. (2003) Effect of thermal conditioning during embryo development on aspects of physiological responses of broilers to heat stress. *Journal of Thermal Biology* 28, 133–140.

Moraes, V.M.B., Malherios, R.D., Bruggeman, V., Collin, A., Tona, K., Van As, P., Onagbesan, O.M., Buyse, J., Decuypere, E. and Macari, M. (2004) The effect of timing of thermal conditioning during incubation on embryo physiological parameters and its relationship to thermotolerance in adult broiler chickens. *Journal of Thermal Biology* 29, 55–61.

Nichelmann, M. (2004) Perinatal epigenetic temperature adaptation in avian species: comparison of turkey and Muscovy duck. *Journal of Thermal Biology* 29, 613–619.

Nichelmann, M. and Tzschentke, B. (1999) Thermoregulation in precocial embryos. *Ornis Fennica* 76, 177–187.

Nichelmann, M. and Tzschentke, B. (2002) Ontogeny of thermoregulation in precocial birds. *Comparative Biochemistry and Physiology A* 131, 751–763.

Nichelmann, M., Janke, O., Höchel, J. and Tzschentke, B. (1998) Avian embryonic temperature regulation: role of Q_{10} in interpretation of endothermic reactions. *Journal of Thermal Biology* 23, 369–376.

Nichelmann, M., Janke, O. and Tzschentke, B. (2001) Efficiency of thermoregulation in precocial avian species during the prenatal period. *Journal of Thermal Biology* 26, 273–280.

Nvota, J., Vyboh, P., Jurani, M., Lamosova, D., Boda, K. and Baroskova, Z. (1980) The influence of environment in the early ontogenesis on the development of the endocrine functions and body growth in fowls. *Proceedings of the 28th International Congress of Physiological Science*. Pergamon, Budapest, p. 616A.

O'Dea, E.E., Fasenko, G.M., Feddes, J.J.R., Robinson, F.E., Segura, J.C., Ouellette, C.A. and van Middlekoop, J.H. (2004) Investigating the eggshell conductance and embryonic metabolism of modern and unselected domestic avian genetic strains at two flock ages. *Poultry Science* 83, 2059–2070.

Owen, J. (1991) Principles and problems of incubator design. In: Tullett, S.G. (ed.) *Avian Incubation*. Butterworth-Heinemann, London, pp. 205–226.

Pal, A.K., Quadri, M.A., Jadhao, S.B., Kumbhakar, J., Kushwah, H.S. and Datta, I.C. (2002) Genotype influences body composition of developing chicken embryo. *Pakistan Journal of Nutrition* 2, 82–84.

Rahn, H., Ar, A. and Paganelli, C.V. (1979) How bird eggs breathe. *Scientific American* 240, 46–55.

Romijn, C. and Lokhorst, W. (1960) Foetal heat production in the fowl. *Journal of Physiology* 150, 239–249.

Snyder, G.K. and Birchard, G.F. (1982) Water loss and survival in embryos of the domestic chicken. *Journal of Experimental Zoology* 219, 115–117.

Sotherland, P.R., Spotila, J.R. and Paganelli, C.V. (1987) Avian eggs: barriers to the exchange of heat and mass. *Journal of Experimental Zoology* Supplement 1, 81–86.

Tazawa, H. and Nakagawa, S. (1985) Response of egg temperature, heart rate and blood pressure in the chick embryo to hypothermal stress. *Journal of Comparative Physiology* 155B, 195–200.

Tona, K., Onagbesan, O., Jego, Y., Kamers, B., Decuypere, E. and Bruggeman, V. (2004) Comparison of embryo physiological parameters during incubation, chick quality, and growth performance of three lines of broiler breeders differing in genetic composition and growth rate. *Poultry Science* 83, 507–513.

Tullett, S.G. (1990) Science and the art of incubation. *Poultry Science* 69, 1–15.

Tullett, S.G. (1991) *Avian Incubation*. Butterworth-Heinemann, London.

Tullett, S.G. and Burton, F.G. (1983) An investigation within a species (*Gallus domesticus*) of the hypothesis that embryonic and juvenile growth rates in birds are correlated. *Reproduction Nutrition Development* 23, 525–535.

Tullett, S.G. and Burton, F.G. (1987) Effect of two gas mixtures on growth of domestic fowl embryos from 14 through 17 days of incubation. *Journal of Experimental Zoology* Supplement 1, 347–350.

Tullett, S.G. and Deeming, D.C. (1982) The relationship between eggshell porosity and oxygen consumption in the domestic fowl. *Comparative Biochemistry and Physiology* 72A, 529–533.

Tzschentke, B. and Basta, D. (2002) Early development of neuronal hypothalamic thermosensitivity in birds: influence of epigenetic adaptation. *Comparative Biochemistry and Physiology A* 131, 825–832.

Tzschentke, B. and Nichelmann, M. (1999) Development of avian thermoregulatory system during the early postnatal period: development of the thermoregulatory set-point. *Ornis Fennica* 76, 189–198.

Tzschentke, B. and Plagemann, A. (2006) Imprinting and critical periods in early development. *World's Poultry Science Journal* 62, 626–637.

Tzschentke, B., Basta, D., Janke, O. and Maier, I. (2004) Characteristics of early development of body functions and epigenetic adaptation to the environment in poultry: focused on development of central nervous mechanisms. *Avian and Poultry Biology Reviews* 15, 107–118.

Van Brecht, A., Aerts, J.M., Degraeve, P. and Berckmans, D. (2002) Quantification of the spatio-temporal eggshell temperature and convection coefficient by infrared thermography. *Transactions of the American Society of Agricultural Engineers* 45, 1951–1957.

Van Brecht, A., Aerts, J.M., Degraeve, P. and Berckmans, D. (2003) Quantification and control of the spatiotemporal gradients of air speed and air temperature in an incubator. *Poultry Science* 82, 1677–1687.

Van Brecht, A., Hens, H., Lemaire, J.-L., Aerts, J.M., Degraeve, P. and Berckmans, D. (2005) Quantification of the heat exchange of chicken eggs. *Poultry Science* 84, 353–361.

Wangensteen, O.D., Wilson, D. and Rahn, H. (1970) Diffusion of gases across the shell of the hen's egg. *Respiration Physiology* 11, 16–30.

Whittow, G.C. and Tazawa, H. (1991) The early development of thermoregulation in birds. *Physiological Zoology* 64, 1371–1390.

Wilson, H.R. (1991) Physiological requirements of the developing embryo: temperature and turning. In: Tullett, S.G. (ed.) *Avian Incubation*. Butterworth-Heinemann, London, pp. 145–156.

Yahav, S. (2007) Thermal manipulation during the perinatal period – does it improve thermotolerance and performance of broiler chickens? *Australian Poultry Science Symposium* 19, 1–8.

Yahav, S., Sasson-Rath, R. and Shinder, D. (2004a) The effect of thermal manipulations during embryogenesis of broiler chicks (*Gallus domesticus*) on hatchability, body weight and thermoregulation after hatch. *Journal of Thermal Biology* 29, 245–250.

Yahav, S., Collin, A., Shinder, D. and Picard, M. (2004b) Thermal manipulations during broiler chick embryogenesis: effects of timing and temperature. *Poultry Science* 83, 1959–1963.

Yalçin, S. and Siegel, P.B. (2003) Exposure to cold or heat stress during incubation on development of stability of broiler embryos. *Poultry Science* 82, 1388–1392.

Yalçin, S., Özkan, S., Çabuk, M., Buyse, J., Decuypere, E. and Siegel, P.B. (2005) Pre- and postnatal conditioning induced thermotolerance on body weight, physiological responses, and relative asymmetry of broilers originating from young and old breeder flocks. *Poultry Science* 84, 967–976.

CHAPTER 13

Hatching Egg and Chick Quality

V. Bruggeman,[1] K. Tona,[2] O. Onagbesan[1] and E. Decuypere[1]

[1]*K.U. Leuven, Laboratory of Livestock Genetics, Immunology and Physiology, Leuven, Belgium;* [2]*University of Lomé, Faculty of Sciences, Department of Physiology, Lomé, Togo*

ABSTRACT

Increasing evidence shows that factors which can alter in the internal and external environment of the hatching egg have repercussions not only on embryonic development, hatchability and day-old-chick quality but most probably also on post-hatch growth performance. Incubation of fertile eggs, and thus embryonic development, takes place under specific conditions of turning, temperature, humidity and gaseous environment. In addition to these variables, other characteristics of the hatching egg can vary due to differences in maternal feeding, breeder age and storage conditions. All these egg- and incubation-related factors can, potentially, influence the development and metabolism of the embryo, its hatching process and the day-old-chick quality. Moreover, an overview of the current systems of scoring 1-day-old chick quality will be given. It will become clear that adaptations of components in the hatching egg or of the incubation and early post-hatch environment to the requirements of the hatching egg of modern breeds could lead to improved post-hatch performance.

INTRODUCTION

The post-hatch performance of broilers is determined by genotype, environmental and feed-related factors. However, increasing evidence shows that other factors which can alter the internal and external environment of the hatching egg have repercussions not only on embryonic development, hatchability and day-old-chick quality but most probably also on post-hatch growth performance. This suggests that some physiological imprinting mechanisms exist in the embryo, resulting in alterations in post-hatch life.

Embryonic development of birds takes place in a fixed container (the shell), which provides an optimal content of feed components (in albumen and yolk) and adequate water and gas exchange with the environment through the eggshell via specialized systems. Incubation of fertile eggs, and thus embryonic development, takes place under specific conditions of turning, temperature, humidity and gaseous environment (see also French, Chapter 12, this volume). In addition to these variables, other characteristics of the hatching egg can vary due to differences in maternal feeding, breeder age and storage conditions. All these egg- and incubation-related factors can, potentially, influence the development and metabolism of the embryo, its hatching process and the day-old-chick quality (Fig. 13.1). This overview will focus on some of the factors which appear to impact on chick quality and parameters related to post-hatch performance; for additional information, we refer to Tazawa and Whittow (2000), Decuypere *et al.* (2001), Tona *et al.* (2005) and Decuypere and Bruggeman (2007).

DEFINING CHICK QUALITY

Good hatchling quality is very important in commercial poultry production for minimizing mortality and maximizing post-hatch performance, yet it is very difficult to define. The major objective of a hatchery is to obtain a high hatchability of marketable chicks at a low spread of hatching time. For the

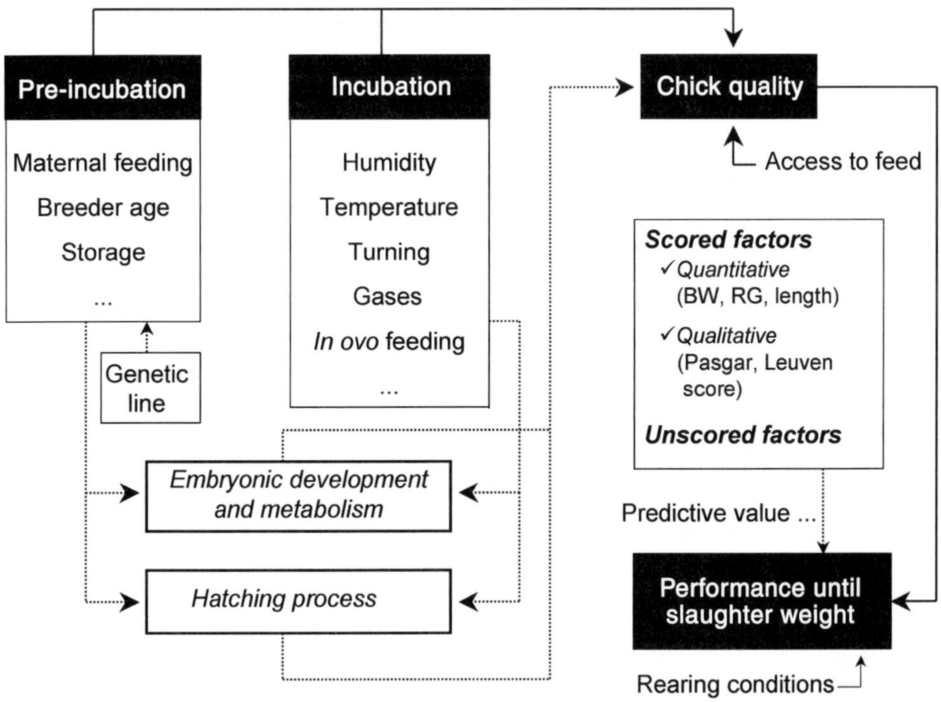

Fig. 13.1. Diagram illustrating the relationships between factors in the breeder hen and the incubation of hatching eggs that affect chick quality and broiler performance. BW = body weight, RG = relative growth.

farmers these chicks have to perform well, as shown by high viability, growth rate and breast meat yield combined with a low feed-conversion rate. A good-quality 1-day-old chick is thus a crucial hinge between the hatchery and the broiler farm. Ultimately day-old-chick quality at take-off is an all-or-none question: marketable or non-marketable chicks. Moreover, it has been reported that good hatchability does not necessarily correlate positively with a high percentage of good-quality chicks and that maximal hatchability is not always linked with the highest post-hatch viability and growth of the chick. The parameters used for quality selection are neither well defined nor standardized and depend on the judgement of individual persons. In the hatchery, chick appearance, vitality and alertness are often used as criteria for assessing chick quality.

Assessment of chick quality can be done by taking into account quantitative, measurable traits such as weight and length or qualitative parameters based on the classification of several observed criteria. Chick hatching weight is often used as a quantitative measurement of chick quality but appears to have limited value as an indicator of overall chick quality or predictor of body weight at slaughter age. The most important factor in determining hatchling mass in birds is the initial egg mass at laying (Deeming and Birchard, 2006). There are also reports stating that 7- or 10-day-old body weight may be more related to slaughter weight than 1-day-old chick weights. The use of growth potential measured as relative growth up to 7 days of age is also used as an *a posteriori* quantitative assessment of chick quality, predicting performance at 6 weeks of age better than 1-day-old chick weights (Tona *et al.*, 2003a). Some emphasis has recently been put on chick length as an indicator of chick quality. Hill (2001) described a relationship between day-old chick length and rate of yolk absorption and the age of breeders or incubation conditions. The studies of Wolanski *et al.* (2003, 2006) showed that chick length (length from the tip of the beak to the middle toe) correlated with yolk-free body mass at hatch in eight different broiler pure lines (r = 0.56). The correlation between chick length at hatch and chick weight has been reported to be, on average, 0.57 at day 7, 0.38 at week 2 and only 0.20 at week 6 (Wolanski *et al.*, 2003, 2006). It has been claimed that one extra cm at hatch would result in 21.7 g extra body weight at day 7 (www.hatchtech.nl) but no results for day 42 were reported. Other authors (Deeming, 2005) doubt the usefulness of chick length as a meaningful indicator of chick quality. Recent results of our group (Willemsen *et al.*, 2008) support this statement. One hundred first-grade broiler chicks were followed in two separate trials from hatch until 6 weeks of age and several body measurements were taken. In both experiments, chick length showed a significant correlation (average r = 0.33) with body weight at day 7 but this correlation disappeared thereafter. The parameter which showed the highest significant correlation with body weight at week 6 was body weight at day 7 (average r = 0.43). Further scientific evidence is needed to support claims of the value of measuring chick length (in relation to egg size) as an indicator of post-natal performance.

In addition to these quantitative measurements of chick quality, two scoring systems have been developed to convert differences in qualitative morphological parameters (such as chick viability, yolk sac uptake, navel area conditions) of the hatchling into a quantitative score. In the Pasgar score, chicks lose points from a maximum score of 10 when abnormalities are observed (Boerjan, 2002). The group at Leuven (Tona *et al.*, 2003a) developed a trapped scoring system (differential importance of different parameters), with a total score between 0 and 100, based on a wide range of parameters each with a hedonic score. Chicks of optimal quality being free from any abnormalities have a maximum score of 100. Those chicks with a score of 100 showed the highest relative growth up to 7 days of age and the highest body weights at 6 weeks of age. These results indicate that scoring of day-old-chick qualitative aspects is relevant for predicting broiler performance.

FACTORS AFFECTING CHICK QUALITY: PRE-INCUBATION FACTORS

Maternal feeding

Good nutrition of the hen is essential for the correct formation of the egg, success of embryonic development and subsequent viability of the hatched chick (Whitehead, 1991). Nutrient deposition in the egg is the result of specific synthesis, metabolization and transport processes, resulting in a relatively constant composition of the egg, except for those nutrients where the body pool of the hen is relatively small or can be influenced by the hen's diet. Protein and amino acid profiles of the egg are rather stable, whereas lipid, vitamin and mineral content can be changed through maternal feeding, which in some cases affects hatchability (reviewed by Wilson, 1997). Recent studies have proved the beneficial effects of supplementation with nutrients (e.g. vitamin E, carotenoids and selenium) that stimulate anti-oxidant systems in the developing embryo (see also Kidd, Chapter 19, this volume). Supplementation results in better hatchability and an improvement in defence mechanisms against oxidative stress (Pappas *et al.*, 2006; Surai *et al.*, 2006).

Maternal age

It is well known that incubating egg weight and therefore day-old-chick weight at hatch depend on the age of the breeder. Hill (2001) reported an increase in chick length with increasing age of the breeder, which suggests that chick length may be related to egg size. The incidence of chicks of subnormal quality is higher in chicks of hatching eggs of older breeders (Tona *et al.*, 2001, 2004a; Boerjan, 2002). Fresh eggs from young breeders had better albumen quality, hatched better and produced higher percentages of high-quality 1-day-old chicks, with lower weights at hatch but higher post-hatch growth rate compared with older breeders (Tona *et al.*, 2004a).

There are several reports indicating differences in plasma hormones and metabolites between chicks originating from a young or old broiler breeder flock, although in these studies no direct relationship with chick quality was observed. Christensen *et al.* (1996) showed that the physiology of turkey embryos originating from breeders of different ages is different in terms of glycogen concentrations in tissues, blood glucose plasma concentrations and thyroid hormone concentrations. Research by Noble *et al.* (1986) and Latour *et al.* (1998) indicates that, besides glucose, the lipids and fatty acid profiles in the developing embryo are affected by broiler breeder age. Weytjens *et al.* (1999) reported that the thermoregulatory ability of broiler chickens originating from a young or old flock was also different.

The age effect of the hen on chick quality could be reversed by moulting the breeders, leading to an improvement in hatchability and chick quality in terms of growth performance (Tona *et al.*, 2002). This suggests that the effect of moulting on the 1-day-old chick could be due to an improvement in egg quality.

Storage time

Hatching eggs can be successfully stored under optimal conditions of humidity and temperature for up to 7 days with little or no effect on hatchability. However, when stored for more than 1 week, embryonic abnormalities and mortality increase and cause a decline in hatchability. Moreover, incubation time is delayed when eggs are stored for a longer time, with each extra day of storage being associated with one additional hour of incubation time.

Long-term egg storage is known to affect general egg quality (yolk membranes, yolk, perivitelline layers and albumen quality). The incidence of more abnormal and dead embryos may be related to a higher number of embryonic cells with necrotic nuclei and an increase in the number of apoptotic cells (cells programmed to die) as a result of storage. The initiation of embryonic development is delayed and could be related to the delay in hatching time. This delay is seen in a later occurrence of the initiation of internal pipping (IP) and a prolonged IP stage (Tona *et al.*, 2003b). Moreover, the rate of embryonic development is slower due to longer storage, but this phenomenon is not observed for all hatching eggs, indicating that not all embryos are affected by storage in the same way (reviewed by Fasenko *et al.*, 2002). Not only growth but also metabolism is found to be influenced by storage time. It was shown that the metabolism of embryos, as measured indirectly by embryonic CO_2 output, proceeds at a slower rate. Interestingly, Fasenko (1996) reported that turkey embryos from eggs stored for 14 days relied more on gluconeogenesis during pipping and hatching than embryos from eggs stored for 4 days. Christensen *et al.* (2001a) showed that embryos from a line that resisted storage mortality maintained higher glycogen concentrations in muscle and heart tissues than those from a line susceptible to storage. The ability of the embryos to metabolize adequate carbohydrate reserves at the time of hatching seems to be an advantage for embryonic growth and survival.

Day-old-chick quality is the result of events during embryonic development, and from the foregoing it is clear that storage affects embryonic development in different ways. Storage for longer periods resulted in a greater occurrence of poor-quality chicks (Boerjan, 2002; Fasenko *et al.*, 2002; Tona *et al.*, 2003b) than from eggs stored for shorter duration, and this was further exacerbated by storage of eggs from older breeders. However, the poor growth of a flock that hatched from eggs stored for a longer period could not be totally ascribed to the greater occurrence of poor-quality chicks (Tona *et al.*, 2004a), suggesting that other intrinsic or extraneous factors, which are as yet unidentified, also influence chick performance.

Embryo physiology and chick quality

The hatched 1-day-old chick is the result of 21 days of development, in which several physiological and endocrine changes take place in the embryo when incubated under controlled conditions. Heat production and metabolism, hormonal balances of thyroid hormones and corticosterone, and gas exchanges (O_2 and CO_2) are of fundamental importance for embryonic development and survival during incubation, and their levels may affect the liveability of the embryo and therefore chick quality. Results from our studies indicate that embryos with higher pCO_2 levels in the air cell, and higher T_3/T_4 ratios at internal pipping or in newly hatched chicks, had better hatchability, chick quality and post-hatch chick growth up to 7 days (Tona *et al.*, 2005). In addition, the work of Christensen *et al.* (2001a) showed that adequate carbohydrate metabolism in the embryo is important for development and survival.

FACTORS AFFECTING CHICK QUALITY: INCUBATION CONDITIONS

Humidity

Incubator humidity is a controlling factor for the evaporation of water vapour from the egg and must be properly controlled. The mass loss by the egg is dependent on the incubator humidity and the eggshell conductance, and it is known that a high natural variation of conductance exists in eggs. This causes a loss of between 5 and 20% (optimally 11–13%) in egg mass under standard conditions. One well-known factor that affects eggshell conductance is the age of the breeder. There is increasing conductance with age, suggesting that eggs from older flocks should need a higher humidity during incubation. Due to the high variation in eggshell conductance between incubating eggs, the optimal humidity level in the incubator for maximal hatchability may also be very variable (a range of 40–60% relative humidity is reported in the literature). To optimize the mass loss of the eggs, it is therefore advisable to match the incubator humidity with the eggshell conductance to decrease embryonic mortality and increase hatchability. The efficacy of this optimization has been reported in the hatchability results from a variety of species (Meir *et al.*, 1984; Meir and Ar, 1987, 1991; Peebles and Brake, 1987).

Reports on the effect of incubator humidity on chick quality are scarce. Only one report of Meir and Ar (1987) showed that adaptation of relative humidity to eggshell conductance resulted in higher hatchability of turkey poults with improved poult quality. In contrast, the work of Bamelis (2003) showed that, when relative humidity was matched to eggshell conductance, no significant improvement of chick quality could be observed.

Temperature

The optimum operating temperature for poultry species during incubation appears to be between 37 and 38°C. Incubation temperature not only is important for normal embryonic development and hatching success but also affects post-hatch performance (Decuypere and Michels, 1992). It is recommended that incubation temperature is regulated according to the 'natural' heat production pattern of the incubating egg to obtain the highest hatchability with good-quality chicks. During early incubation, embryonic temperature (measured by eggshell temperature) is close to incubator temperature, but, from mid-incubation onwards, the embryonic heat production increases and egg temperature rises above incubator temperature. It is therefore important to have a good, standardized measurement of embryonic temperature (eggshell temperature), which can be done by infrared measurements, and accurate regulation of air temperature and flow in the incubator. This issue becomes even more important in modern incubation practice as it is becoming increasingly clear that metabolic heat production differs between different breeds. Changes in embryo metabolism are the result of differential selection for production traits and this is reflected in the higher metabolic heat production during embryonic development in breeds with a higher growth potential (Janke *et al.*, 2004; Tona *et al.*, 2004b). Moreover, egg size also affects embryonic heat dissipation (French, 1997; Lourens *et al.*, 2006). Due to these differences between the metabolism of embryos originating from different strains or differences in egg size, there are also different needs for incubation temperature regulation during incubation.

Hulet (2000) showed that, when conditions of the setter in real time were changed according to metabolic activity of the embryos (in terms of CO_2 production), an improvement in hatchability was recorded. A recent publication of Lourens *et al.* (2005) also shows that the highest hatchability, embryo development (higher embryo length and yolk-free embryo weight) and post-hatch performance were found when eggshell temperature was maintained at 37.8°C constantly throughout incubation. Similar results in large-scale trials are reported by Banwell (2007). Additionally, Joseph *et al.* (2006) stressed the importance of maintaining eggshell temperature of 37.8°C, especially from day 0 to day 10 of incubation, to optimize hatchability, performance to 6 weeks of age, carcass weight and breast fillet yield. Changes in eggshell temperature in the beginning, and also during the later phases, of incubation exert effects on post-hatch growth and performance. Eggs for which eggshell temperature was kept at 38.6°C (regulated with an air temperature of 37.6°C) from day 16 onwards performed better in terms of body weight gain compared with the eggs kept at a higher eggshell temperature (39.7°C) (Hulet *et al.*, 2007). Boerjan (2002), using the Pasgar score, reported an improvement in quality score (9.1 versus 8.6) when eggs from hens of 45 weeks of age were incubated at higher temperature (0.11°C from day 10 to 12 and 0.56°C at day 18 of incubation) compared with standard incubation conditions.

Incubation temperature may also interact with storage conditions. An increase in incubation temperature during the initial stages of development of eggs stored for 15 days improved liveability of turkey embryos by accelerating embryonic growth and development early in the incubation period (37.8°C versus 37.5°C). A possible mechanism for the improved liveability might be due to a better utilization of carbohydrates (Christensen *et al.*, 2003a,b). However, it remains to be investigated if incubation under higher temperature during the first period of incubation also favours quality and post-hatch performance of chicks hatched from eggs stored for a longer period.

Recent findings show that deviations in temperature during incubation that cause differences in the rate of development of the embryo may alter proportional growth and influence functional processes of the embryo and the post-hatch chick in a differential way, depending on the period when this temperature deviation is applied. The thermoregulatory system can be influenced by alterations in temperature during specific periods, resulting in alterations in the thermoregulatory ability and body temperature stabilization of the chicken (Decuypere, 1984; Minne and Decuypere, 1984; Black and Burggren, 2004; Yahav *et al.* 2004a,b). Collin *et al.* (2007), however, reported that thermal manipulations during several periods of incubation did not improve acquisition of thermotolerance at week 6 post-hatch. Heat treatments during incubation also affect some endocrine axes, such as the thyroid and adrenal axes, pre- and post-hatch (Avrutina *et al.*, 1985; Decuypere *et al.*, 1988; Iqbal *et al.*, 1989, 1990). Results of Christensen *et al.* (2001b) showed that growth of turkey embryos can be altered by increasing incubator temperatures during the last 3 days of incubation, which was associated with elevated embryonic plasma glucose concentrations and altered plasma insulin-like growth factor concentrations. Finally, temperature during incubation also has an effect on muscle cell development (Maltby *et al.*, 2004) and bone growth (Brookes and May, 1972; Lourens *et al.*, 2005; Banwell, 2007). Thermal manipulation during late embryogenesis significantly improved breast muscle yield (Collin *et al.*, 2007; French, Chapter 12, this volume).

Turning conditions

Egg turning has been reported to reduce malpositions, prevent abnormal adhesion of the embryo or embryonic membranes to the shell membrane, encourage the complete and timely closure of the chorio-allantois at the small end of the egg and, most importantly, achieve an optimal albumen utilization by the embryo (reviewed by Baggott *et al.*, 2002; Deeming, 2002; Tona *et al.*, 2005). Insufficient turning during incubation leads to a delay of hatching and adversely affects day-old-chick qualitative aspects. Hatchability and percentage of high-quality chicks were lower for eggs turned for 15 days compared with those turned for 12 or 18 days (Tona *et al.*, 2003c) (day 15 is in the period of an increasingly functional hypothalamus–pituitary–adrenal axis). Tona *et al.* (2005) hypothesized that discontinuation of turning at this time may be an additional stressor, which can lead to physiological imprinting and altered responsiveness, leading to lower hatchability and chick quality. Other reports, however, did not show any effect on hatchability of stopping egg turning at day 15 (see the review by Deeming, 2002).

Gaseous environment

According to Brian (2000), a CO_2 concentration of 0.1–0.4% would be optimal in a multi-stage setter, rising to 0.8% in the hatcher. Recently, considerable attention has been given to the regulation of the gaseous environment of the hatching egg in order to control and maintain CO_2 at concentrations that have beneficial effects on post-hatch performance of the broiler. The old literature of Taylor and colleagues (Taylor *et al.*, 1956, 1971; Taylor and Kreutziger, 1965, 1966, 1969) and new results of our own group (de Smit *et al.*, 2006; Tona *et al.*, 2006; Bruggeman *et al.*, 2007; Everaert *et al.*, 2007) show that chicken embryos can tolerate quite high CO_2 concentrations, depending on the developmental stage. Hogg (1997) reported that hatchability and chick quality were improved by closing the vents for the first 10 days. This resulted in CO_2 concentrations of 1.5% and humidity of more than 70% at day 10 of incubation. Recent results of de Smit *et al.* (2006) and Tona *et al.* (2006) show that chicks incubated under increased CO_2 during the first 10 days had a faster embryonic growth, hatched earlier and showed a higher growth rate in the first week post-hatch. The mechanisms behind these beneficial effects of CO_2 are still speculative, but it appears from our results that there would be an important role of the albumen in these first stages of embryonic development, since albumen has been shown to react very quickly to higher external CO_2 concentrations by lowering its pH (Bruggeman *et al.*, 2006, 2007). Moreover, both hypercapnia and hypoxia are known to affect different aspects of early development. Further information on the effects of gas concentrations before and during incubation on embryogenesis, hatchability, chick quality and post-hatch growth can be found in Onagbesan *et al.* (2007).

In ovo feeding

During the perinatal period, there is a high caloric demand (in the form of glycogen reserves) to fuel the hatching process, thermogenesis and basal metabolism until access to first feed. The application of *in ovo* feeding (at 17–18 days in chickens) provides an additional energy source for the embryo and allows the late-term embryo to consume externally administered nutrients through drinking of the amniotic fluid. This resulted in several beneficial effects at the level of the gut (e.g. an increased digestive capacity of the hatchlings and increased brush border enzyme activities), which were associated with higher hatchling weights and continued through the first weeks after hatch. In addition, there were positive effects on liver and muscle glycogen status and a stimulation of breast muscle development (reviewed by Ferket and Uni, 2006). Application of *in ovo* feeding seems promising in terms of several aspects of perinatal development of the gut and muscle, which makes the transition from endogenous lipid-rich yolk to utilization of exogenous carbohydrate- and lipid-rich feed easier and gives the hatchlings a good start for later life. However, work still has to be done in order to apply *in ovo* feeding under practical circumstances.

POST-HATCH ENVIRONMENT

Access to first feeding and ambient temperature

Newly hatched chicks do not have immediate access to food and water; moreover, the spread of the hatching curve, together with the position in the sequence of hatching (early or late), creates an additional variation in time between hatching and first food uptake. Several reports have demonstrated that a delay in feed intake after hatch adversely affects post-hatch performance of chicks, especially growth (Pinchasov and Noy, 1993; Bigot *et al.*, 2003; Gonzales *et al.*, 2003; Careghi *et al.*, 2005). This delay in first food intake has repercussions on yolk uptake and utilization, development of the gastrointestinal tract, metabolic status, immune system development and IgG uptake, and overall growth. In addition, some crucial hormone levels and enzyme activities for growth are known to be strongly influenced by post-hatch food intake, such as insulin and p70S6 kinase activity, a rate-limiting step for protein synthesis.

In this transition period from emergence from the egg to the external environment, adequate development of food intake and immunity-related systems is important, and the maturation of the thermoregulatory system (switch from poikilothermy to homeothermy around hatch) can also affect post-hatch performance. Nichelmann and Tzschentke (2002) point to the importance of the acquisition of the capability of the hatchling to maintain its body temperature and that the preferred ambient temperature of hatchlings is situated near the thermoneutral temperature (temperature at which the animal produces the lowest heat). Tazawa *et al.* (2004) even suggested that a different environmental temperature soon after hatch can influence the development of the thermoregulatory competence and heart rate regulation, at least in broilers. Also changes in incubation temperature at the end of the embryonic period (see earlier) or during the early post-hatch period can induce long-lasting cold or heat adaptation post-hatch (Yahav and McMurtry, 2001; Shinder *et al.*, 2002; Moraes *et al.*, 2003). Besides effects on the thermoregulatory system, post-hatch mild heat exposure at an early age had stimulatory effects on skeletal muscle growth due to an increase in satellite cell proliferation and accelerated differentiation (Halevy *et al.*, 2001). An optimal adaptation of the newly hatched chick to these two major environmental changes (food and temperature) is thus important for a good start for subsequent optimal performance.

CONCLUSION

It is clear that many factors affect the development of the chick and determine its final quality and thus growth potential. The main question remains how to score the 1-day-old chick to get the best measure of its quality and how this score can predict the post-hatch performance of the bird. Devising an all-encompassing method that takes into account all the quantitative and qualitative aspects of the day-old chick for estimating day-old-chick quality may provide a method for predicting performance to slaughter. There is still a note of caution in the application of this method since we know that, between chicks (originating from different breeder ages, storage duration or turning duration) of the same high quality, there are still some differences in growth potential, meaning that other intrinsic factors, which cannot yet be identified or 'escape' the scoring system, are involved in growth performance. The mechanisms by which these factors affect performance are still not clear. One can speculate that these factors may have modified the physiology of the embryo through changes in gene expression or changes to downstream transcription, leading to changes in relevant modulators of genes involved in growth. Moreover, although this also remains to be investigated, it is possible that incubation conditions (differential temperatures, humidity and ventilation conditions) also lead to a 1-day-old chick with a different intrinsic 'history' of factors beyond their genetic potential, which cannot be scored and may be involved in differential post-hatch growth processes between chicks. Nevertheless, it is clear that adaptations of components in the hatching egg or of the incubation (and even early post-hatch) environment to the 'needs' of the hatching egg of modern breeds could potentially lead to improved post-hatch performance.

ACKNOWLEDGEMENTS

Veerle Bruggeman is a post-doctoral fellow with support from the Fund for Scientific Research Flanders (FWO – Vlaanderen).

REFERENCES

Avrutina, A.J., Galpern, I.L. and Kisljuk, S.M. (1985) Stimulation of adrenals during the critical periods of development and production in fowls. *World's Poultry Science Journal* 41, 108–114.

Baggott, G.K., Deeming, D.C. and Latter, G.V. (2002) Electrolyte and water balance of the early avian embryo: effects of egg turning. *Avian and Poultry Biology Reviews* 13, 105–119.

Bamelis, F. (2003) Non-invasive assessment of eggshell conductance and different developmental stages during incubation of eggs. PhD thesis, Katholieke Universiteit Leuven, Belgium.

Banwell, R. (2007) Bio-response incubation for better hatch and post-hatch performance. *World Poultry* 23, 20–21.

Bigot, K., Mignon-Grasteau, S., Picard, M. and Tesseraud, S. (2003) Effect of delayed feed intake on body, intestine and muscle development in neonate broilers. *Poultry Science* 82, 781–788.

Black, J.L. and Burggren, W.W. (2004) Acclimation to hypothermic incubation in developing chicken embryos (*Gallus domesticus*) I. Developmental effects and chronic and acute metabolic adjustments. *Journal of Experimental Biology* 207, 1543–1552.

Boerjan, M. (2002) Programs for single stage incubation and chick quality. *Avian and Poultry Biology Reviews* 13, 237–238.

Brian, H. (2000) Incubation, the physical requirements. *International Hatchery Practice* 14, 25.

Brookes, M. and May, K.U. (1972) The influence of temperature on bone growth in the chick. *Journal of Anatomy* 111, 351–363.

Bruggeman, V., de Smit, L., Tona, K., Everaert, N., Witters, A., Debonne, M., Degraeve, P., Garain, P. and Decuypere, E. (2006) Changes in albumen pH due to higher CO_2 concentrations during the first ten days of incubation. In: Romboli, I., Flock, D. and Franchini, A. (eds) *Proceedings of 12th European Poultry Conference*, Verona. Supplement of the *World's Poultry Science Journal* 62, 160.

Bruggeman, V., Witters, A., de Smit, L., Debonne, M., Everaert, N., Kamers, B., Onagbesan, O.M., Degraeve, P. and Decuypere, E. (2007) Changes in parameters related to acid–base balance in chicken embryos (*Gallus domesticus*) incubated under normal or high CO_2 concentrations during the first ten days of incubation. *Respiratory Physiology and Neurobiology* 159, 147–154.

Careghi, C., Tona, K., Onagbesan, O., Buyse, J., Decuypere, E. and Bruggeman, V. (2005) The effects of spread of hatch and interaction with delayed feed access after hatch on broiler performance until seven days of age. *Poultry Science* 84, 1314–1320.

Christensen, V.L., Donaldson, W.E. and McMurtry, J.P. (1996) Physiological differences in late embryos from turkey breeders at different ages. *Poultry Science* 75, 172–178.

Christensen, V.L., Wineland, M.J., Fasenko, G.M. and Donaldson, W.E. (2001a) Egg storage effects on plasma glucose and supply and demand glycogen concentrations of broiler embryos. *Poultry Science* 80, 1729–1735.

Christensen, V.L., McMurtry, J.P., Donaldson, W.E. and Nestor, K.E. (2001b) Incubation temperature affects plasma insulin-like growth factors in embryos from selected lines of turkeys. *Poultry Science* 80, 949–954.

Christensen, V.L., Grimes, J.L., Wineland, M.J. and Davis, G.S. (2003a) Accelerating embryonic growth during incubation following prolonged egg storage 1. Embryonic livability. *Poultry Science* 82, 1863–1868.

Christensen, V.L., Grimes, J.L., Wineland, M.J. and Davis, G.S. (2003b) Accelerating embryonic growth during incubation following prolonged egg storage 2. Embryonic growth and metabolism. *Poultry Science* 82, 1869–1878.

Collin, A., Berri, C., Tesseraud, S., Requena Rodon, F.E., Skiba-Cassy, S., Crochet, S., Duclos, M.J., Rideau, N., Tona, K., Buyse, J., Bruggeman, V., Decuypere, E., Picard, M. and Yahav, S. (2007) Effects of thermal manipulation during early and late embryogenesis on thermotolerance and breast muscle characteristics in broiler chickens. *Poultry Science* 86, 795–800.

Decuypere, E. (1984) Incubation temperature in relation to postnatal performance in chickens. *Archiv für Experimentalle Veterinärmedizin* 38, 439–449.

Decuypere, E. and Bruggeman, V. (2007) The endocrine interface of environmental and egg factors affecting chick quality. *Poultry Science* 86, 1037–1042.

Decuypere, E. and Michels, H. (1992) Incubation temperature as a management tool: a review. *World's Poultry Science Journal* 48, 28–38.

Decuypere, E., Iqbal, A., Michels, H., Kühn, E.R., Schneider, R. and Abd El Azim, A. (1988) Thyroid hormone response to thyrotropin releasing hormone after cold treatment during pre- and postnatal development in the domestic fowl. *Hormone and Metabolism Research* 20, 484–489.

Decuypere, E., Tona, K., Bruggeman, V. and Bamelis, F. (2001) The day-old chick: a crucial hinge between breeders and broilers. *World's Poultry Science Journal*, 57, 127–137.

Deeming, D.C. (2002) Patterns and significance of egg turning. In: Deeming, D.C. (ed.) *Avian Incubation: Behaviour, Environment and Evolution*. Oxford University Press, Oxford, UK, pp. 161–178.

Deeming, D.C. (2005) Yolk sac, body dimensions and hatchling quality of ducklings, chicks and poults. *British Poultry Science* 46, 560–564.

Deeming, D.C. and Birchard, G.F. (2006) Allometry of egg and hatchling mass in birds and reptiles: roles of developmental maturity, eggshell structure and phylogeny. *Journal of Zoology* 270, 1–10.

de Smit, L., Bruggeman, V., Tona, K., Debonne, M., Onagbesan, O., Arkens, L., de Baerdemaker, J. and Decuypere, E. (2006) Embryonic developmental plasticity: increased CO_2 in the incubator during the early stages of incubation changes the developmental trajectories of the chick during prenatal and postnatal growth. *Comparative Biochemistry and Physiology A Molecular Integrative Physiology* 145, 166–175.

Everaert, N., Kamers, B., Witters, A., de Smit, L., Debonne, M., Decuypere, E. and Bruggeman, V. (2007) Effect of 4% CO_2 during the second half of incubation on embryonic development, hatching parameters and post-hatch growth. *Poultry Science* 86, 1372–1379.

Fasenko, G.M. (1996) Factors influencing embryo and poult viability and growth during long term storage of turkey eggs. PhD thesis, North Carolina State University, Raleigh, North Carolina.

Fasenko, G.M., Robinson, F.E. and Christensen, V.L. (2002) How long-term hatching egg storage affects the egg, the embryo and the chick. In: Deeming, D.C. (ed.) *Practical Aspects of Commercial Incubation in Poultry*. Ratite Conference Books, Oxford, UK, pp. 33–39.

Ferket, P. and Uni, Z. (2006) Early feeding – *in ovo* feeding enhances early gut development and digestive capacity of poultry. In: Romboli, I., Flock, D. and Franchini, A. (eds) *Proceedings of 12th European Poultry Conference*, Verona. Supplement of the *World's Poultry Science Journal* 62, 300–301.

French, N. (1997) Modeling incubation temperature: the effects of incubator design, embryonic development and egg size. *Poultry Science* 76, 124–133.

Gonzales, E., Kondo, N., Saldanka, E.S., Loddy, M.M., Careghi, C. and Decuypere, E. (2003) Performance and physiological parameters of broiler chickens subjected to fasting in the neonatal period. *Poultry Science* 182, 1250–1256.

Halevy, O., Krispin, A., Leshem, Y., McMurtry, J.P. and Yahav, S. (2001) Early-age heat exposure affects skeletal muscle satellite cell proliferation and differentiation in chicks. *American Journal of Physiology – Regulatory Integrative Comparative Physiology* 281, 302–309.

Hill, D. (2001) Chick length uniformity profiles as a field measurement of chick quality. *Avian and Poultry Biology Reviews* 12, 188.

Hogg, A. (1997) Single stage incubation trials. *Poultry and Avian Biology Reviews* 8, 168.

Hulet, R.M. (2000) Chick quality, the result of maximizing embryonic metabolism. *Avian and Poultry Biology Reviews* 12, 189.

Hulet, R., Gladys, G., Hill, D., Meijerhof, R. and El-Shiekh, T. (2007) Influence of egg shell embryonic incubation temperature and broiler breeder flock age on posthatch growth performance and carcass characteristics. *Poultry Science* 86, 408–412.

Iqbal, A., Decuypere, E., Abd El Azim, A. and Kühn, E.R. (1989) Plasma iodohormone concentrations in early and late hatched chicks incubated at different temperatures. *Medical Science Research* 17, 269–270.

Iqbal, A., Decuypere, E., Abd El Azim, A. and Kühn, E.R. (1990) Pre-and posthatch high temperature exposure affects the thyroid hormones and corticosterone response to acute heat stress in growing chicken (*Gallus domesticus*). *Journal of Thermal Biology* 15, 149–153.

Janke, O., Tzschentke, B. and Boerjan, M.L. (2004) Comparative investigation of heat production and body temperature in embryos of modern chicken breeds. *Avian and Poultry Biology Reviews* 15, 191–196.

Joseph, N.S., Lourens, A. and Moran, E.T. (2006) The effects of suboptimal eggshell temperature during incubation on broiler chick quality, live performance, and further processing yield. *Poultry Science* 85, 932–938.

Latour, M.A., Peebles, E.D., Doyle, S.M., Pansky, T., Smith, T.W. and Boyle, C.R. (1998) Broiler breeder age and dietary fat influence the yolk fatty acid profiles of fresh eggs and newly hatched chicks. *Poultry Science* 77, 47–53.

Lourens, A., van den Brand, H., Meijerhof, R. and Kemp, B. (2005) Effect of eggshell temperature during incubation on embryo development, hatchability, and posthatch development. *Poultry Science* 84, 914–920.

Lourens A., Molenaar, R., van den Brand, H., Heetkamp, M.J.W., Meijerhof, R. and Kemp, B. (2006) Effect of egg size on heat production and the transition of energy from egg to hatchling. *Poultry Science* 85, 770–776

Maltby, V., Somaiya, A., French, N.A. and Stickland, N.C. (2004) *In ovo* temperature manipulation influences post-hatch muscle growth in the turkey. *British Poultry Science* 45, 491–498.

Meir, M. and Ar, A. (1987) Improving turkey poult quality by correcting incubator humidity to match eggshell conductance. *British Poultry Science* 28, 337–342.

Meir, M. and Ar, A. (1991) Compensation for seasonal changes in eggshell conductance and hatchability of goose eggs by dynamic control of egg water loss. *British Poultry Science* 32, 723–732.

Meir, M., Nir, A. and Ar, A. (1984) Increasing hatchability of turkey eggs by matching incubator humidity to shell conductance of individual eggs. *Poultry Science* 63, 1489–1496.

Minne, B. and Decuypere, E. (1984) Effects of late prenatal temperatures on some thermoregulatory aspects in young chickens. *Archiv für Experimentalle Veterinärmedizin* 38, 374–383.

Moraes, V.M.B., Malheiros, R.D., Bruggeman, V., Collin, A., Tona, K., Van As, P., Onagbesan, O.M., Buyse, J., Decuypere, E. and Macari, M. (2003) Effect of thermal conditioning during embryonic development on aspects of physiological responses of broilers to heat stress. *Journal of Thermal Biology* 28, 133–140.

Nichelmann, M. and Tzschentke, B. (2002) Ontogeny of thermoregulation in precocial birds. *Comparative Biochemistry and Physiology A*, 751–763.

Noble, R.C., Lonsdale, F., Connor, K. and Brown, D. (1986) Changes in the lipid metabolism of the chick embryo with parental age. *Poultry Science* 65, 409–416.

Onagbesan, O., Bruggeman, V., de Smit, L., Debonne, M., Witters, A., Tona, K., Everaert, N. and Decuypere, E. (2007) Gas exchange during storage and incubation of avian eggs: effects on embryogenesis, hatchability, chick quality and post-hatch growth. *World's Poultry Science Journal* 65, 557–573.

Pappas, A.C., Acamovic, T., Sparks, N.H.C., Surai, P.F. and McDevitt, R.M. (2006) Effects of supplementing broiler breeder diets with organoselenium compounds and polyunsaturated fatty acids on hatchability. *Poultry Science* 85, 1584–1593.

Peebles, E.D. and Brake, J. (1987) Eggshell quality and hatchability in broiler breeder eggs. *Poultry Science* 66, 596–604.

Pinchasov, Y. and Noy, Y. (1993) Comparison of post hatch holding time and subsequent early performance of broiler chicks and turkey poults. *British Poultry Science* 34, 111–120.

Shinder, D., Luger, D., Rusal, M., Rzepakovsky, V., Bresler, V. and Yahav, S. (2002) Early age cold conditioning in broiler chickens (*Gallus domesticus*): thermotolerance and growth responses. *Journal of Thermal Biology* 27, 517–523.

Surai, P.F., Karadas, F., Pappas, A. and Sparks, N.H.C. (2006) Effect of organic selenium in quail diet on its accumulation in tissues and transfer to the progeny. *British Poultry Science* 47, 65–72.

Taylor, L.W. and Kreutziger, O. (1965) The gaseous environment of the chick embryo in relation to its development and hatchability. 2. Effects of carbon dioxide and oxygen levels during the period of the fifth through the eighth days of incubation. *Poultry Science* 44, 98–106.

Taylor, L.W. and Kreutziger, O. (1966) The gaseous environment of the chick embryo in relation to its development and hatchability. 3. Effects of carbon dioxide and oxygen levels during the period of the ninth through the twelfth days of incubation. *Poultry Science* 45, 867–884.

Taylor, L.W. and Kreutziger, O. (1969) The gaseous environment of the chick embryo in relation to its development and hatchability. 4. Effects of carbon dioxide and oxygen levels during the period of the thirteenth through the sixteenth days of incubation. *Poultry Science* 48, 871–877.

Taylor, L.W., Sjodin, R.A. and Gunns, C.A. (1956) The gaseous environment of the chick embryo in relation to its development and hatchability. 1. Effects of carbon dioxide and oxygen levels during the first four days of incubation upon hatchability. *Poultry Science* 35, 1206–1215.

Taylor, L.W., Kreutziger, O. and Abercrombie, G.L. (1971) The gaseous environment of the chick embryo in relation to its development and hatchability. 5. Effects of carbon dioxide and oxygen levels during the terminal days of incubation. *Poultry Science* 50, 66–78.

Tazawa, H. and Whittow, G.C. (2000) Incubation physiology. In: Whittow, G.C. (ed.) *Sturkie's Avian Physiology*, 5th edition. Academic Press, London, pp. 617–634.

Tazawa, H., Chiba, Y., Khandoker, A.H., Dzialowski, E.M. and Burggren, W.W. (2004) Early development of thermoregulatory competence in chickens: responses of heart rate and oxygen uptake to altered ambient temperatures. *Avian and Poultry Biology Reviews* 15, 166–176.

Tona, K., Bamelis, F., Coucke, W., Bruggeman, V. and Decuypere, E. (2001) Relationship between broiler breeder's age and egg weight loss and embryonic mortality during incubation in large-scale conditions. *Journal of Applied Poultry Science* 10, 221–227.

Tona, K., Bamelis, F., de Ketelaere, B., Bruggeman, V. and Decuypere, E. (2002) Effect of inducing molting on albumen quality, hatchability and chick body weight from broiler breeders. *Poultry Science* 81, 327–332.

Tona, K., Bamelis, F., de Ketelaere, B., Bruggeman, V., Moraes, V.M.B., Buyse, J., Onagbesan, O. and Decuypere, E. (2003a) Effects of egg storage time on spread of hatch, chick quality and chick juvenile growth. *Poultry Science* 82, 736–741.

Tona, K., Malheiros, R.D., Bamelis, F., Careghi, C., Bruggeman, V., Moraes, V.M.B., Onagbesan, O. and Decuypere, E. (2003b) Effects of storage time on incubating egg gas pressure, thyroid hormones and corticosterone levels in embryos, and their hatching parameters. *Poultry Science* 82, 840–845.

Tona, K., Onagbesan, O., de Ketelaere, B., Decuypere, E. and Bruggeman, V. (2003c) Effects of turning duration during incubation on corticosterone and thyroid hormone levels, gas pressures in air cell, chick quality and juvenile growth. *Poultry Science* 82, 1974–1979.

Tona, K., Onagbesan, O., de Ketelaere, B., Decuypere, E. and Bruggeman, V. (2004a) Effects of age of broiler breeders and egg storage on egg quality, hatchability, chick quality, chick weight and chick post-hatch growth to 42 days. *Journal of Applied Poultry Research* 13, 10–18.

Tona, K., Onagbesan, O., Jego, Y., Kamers, B., Decuypere, E. and Bruggeman, V. (2004b) Comparison of embryo physiological parameters during incubation, chick quality and growth performance of three lines of broiler breeders differing in genetic composition and growth rate. *Poultry Science* 83, 507–513.

Tona, K., Onagbesan, O., Bruggeman, V., Mertens, K. and Decuypere, E. (2005) Effects of turning duration during incubation on embryo growth, utilization of albumen, and stress regulation. *Poultry Science* 84, 315–320.

Tona, K., Onagbesan, O., Bruggeman, V., de Smit, L., Figueiredo, D. and Decuypere, E. (2006) Non-ventilation during early incubation in combination with dexamethasone administration during late incubation 1. Effects on physiological hormone levels, incubation duration and hatching events. *Domestic Animal Endocrinology* 33, 32–46.

Weytjens, S., Meijerhof, R., Buyse, J. and Decuypere, E. (1999) Thermoregulation in chicks originating from breeder flocks of two different ages. *Journal of Applied Poultry Research* 8, 139–145.

Whitehead, C.C. (1991) Nutrition of the breeding bird and developing embryo. In Tullett, S.G. (ed.) *Avian Incubation*. Butterworth-Heinemann, Rushden, UK, pp. 227–238.

Willemsen, H., Everaert, N., Witters, A., de Smit, L., Verschuere, F., Garian, P., Berckmans, D., Decuypere, E. and Bruggeman, V. (2008) Critical assessment of chick quality measurements as indicator of posthatch performance. *Poultry Science* 87, 2358–2366.

Wilson, H.R. (1997) Effects of maternal nutrition on hatchability. *Poultry Science* 76, 134–143.

Wolanski, N.J., Luiten, E.J., Meijerhof, R. and Vereijken, A.L.J. (2003) Yolk utilisation and chick length as parameters for embryo development. *Avian and Poultry Biology Reviews* 15, 133–134.

Wolanski, N.J., Renema, R.A., Robinson, F.E., Carney, V.L. and Fancher, B.I. (2006) Relationship between chick conformation and quality measures with early growth trials in males of eight selected pure or commercial broiler breeder strains. *Poultry Science* 85, 1490–1497.

Yahav, S. and McMurtry, J.P. (2001) Thermotolerance acquisition in broiler chickens by temperature conditioning early in life – the effect of timing and ambient temperature. *Poultry Science* 80, 1662–1666.

Yahav, S., Sasson Rath, R. and Shinder, D. (2004a) The effect of thermal manipulations during embryogenesis of broiler chicks (*Gallus domesticus*) on hatchability, body weight and thermoregulation after hatch. *Journal of Thermal Biology* 29, 245–250.

Yahav, S., Collin, A., Shinder, D. and Picard, M. (2004b) Thermal manipulations during broiler chick embryogenesis: effects of timing and temperature. *Poultry Science* 83, 1959–1963.

PART VI
Managing the Environment

CHAPTER 14
Photoperiod and Control of Breeding

P.D. Lewis

Department of Animal and Poultry Science, University of KwaZulu-Natal, South Africa

ABSTRACT

Photoperiod is the most important environmental manipulator of sexual development and reproductive activity, and changes in photoperiod exert far greater effects on sexual maturation than constant photoperiods. The critical daylength to initiate sexual maturation appears to be common for all species of poultry at between 10 and 11 h, with a saturation daylength of between 13 and 14 h.

The photoperiodic response varies with the type of poultry and is principally dependent on whether or not the bird exhibits photorefractoriness and if so in which form. Photorefractoriness is a condition that, in its extreme, prevents gonadal maturation in the year of hatch and limits breeding to spring and early summer. Birds are hatched photorefractory and need a period of short days to dissipate the condition, the time requirement for which varies with photoperiod and the degree of feed restriction. Reproductivity is subsequently ended by the onset of adult photorefractoriness.

Geese and game birds exhibit absolute photorefractoriness and have a breeding season limited to about 4 months. Their sexual maturity is indefinitely delayed when reared on long days. Turkeys, and probably Muscovy ducks, exhibit relative photorefractoriness, become sexually mature in their first year and suffer no more than a 3- to 4-week delay in maturity when not reared on short days (below critical daylength). Turkeys do not normally have their feed intake restricted and so require only about 2 months of short days to become photosensitive. Broiler breeders also exhibit relative photorefractoriness, but because their feed intake during rearing is severely restricted, they need 4–5 months of short days to become photosensitive. Meat-type ducks do not exhibit photorefractoriness and rearing on short photoperiods is not essential.

INTRODUCTION

Photoperiod is the length of the light phase within a light/dark (L/D) cycle (Greek, *photos* – light, *periodos* – circuit) and can be regarded as being synonymous with daylength in most contexts. For example, non-stimulatory photoperiods are often referred to as short days and stimulatory photoperiods as long days. Photoperiod is not a synonym for lighting in general, and so this chapter reviews only the involvement of photoperiod in the response of meat-type poultry breeding stock; it does not discuss the effects of other photic components, such as illuminance, wavelength and light source. The review will also be restricted to the effects of photoperiod within a 24 h L/D cycle.

The first demonstration of the benefits of using artificial photoperiods to improve performance in poultry is thought to have been a series of three experiments conducted in the USA between 1889 and 1893. Dr Waldorf, a general practitioner in New York State, reported improvements in egg production, fertility and hatchability in domestic hens from using gas lamps to provide constant 16.5 h photoperiods during the short days of winter (Waldorf, 1920). However, artificial photoperiods have been used to manipulate avian photoperiodic responses for at least four centuries. Dutch bird-netters in the 17th century kept wild birds on short days during spring and summer months and induced bird song in the autumn by transferring them to long days at the end of summer so that they could be used as decoys to facilitate the netting of autumn migrants (Hoos, 1937).

Photoperiod is without doubt the most important element of a lighting programme and, unequivocally, the most potent of all environmental manipulators of sexual development and reproductive activity. Although rearing on different constant photoperiods results in different rates of sexual development, changes in photoperiod exert far greater effects on sexual maturation (Morris, 1967). However, the response to a particular photoperiod or change in photoperiod is not the same for all types of poultry, and the effects of photoperiod on reproductive performance will be discussed separately for broiler breeders, turkeys and waterfowl.

PHOTORECEPTION AND ENDOCRINE RESPONSES

Most studies of avian photoreception and the photoperiodic response have involved either egg-laying hybrids or non-domesticated species. Nevertheless, it is likely that the processes of photoreception that induce a sexual response are common to most avian species.

Light reaches the suprachiasmatic nucleus of the hypothalamus via two main routes. The first of these involves absorption by photopigments in the retina (iodopsin in cones and rhodopsin in rods) and, after conversion to an electrochemical signal, transmission through neural pathways, and the second is the more direct route through feathers, skull and cranial tissues. The retina is the primary photoreceptor site in mammals (e.g. Wurtman, 1974), whereas ocular reception has been shown to be non-essential for photoinduction at normal levels of illuminance in several avian species (e.g. house sparrow – Menaker and Keatts, 1968; domestic fowl – Siopes and Wilson, 1980a; Japanese quail – Siopes and Wilson, 1980b), although it may be more important at low light intensities (Homma and Sakakibara, 1971; Siopes and Wilson, 1980a,b). Further evidence that retinal reception is not essential for photostimulation is the earlier maturity and superior egg production of genetically blind chickens compared with normally sighted ones (Ali and Cheng, 1985).

Photic information is used by the hypothalamus to modify the size and timing of gonadotrophin-releasing (GnRH) and gonadotrophin-inhibiting (GnIH) hormone secretions (Ciccone *et al.*, 2004), which, in turn, control the production of luteinizing (LH) and follicle-stimulating (FSH) hormones by the anterior pituitary gland. There is evidence that the enzyme that catalyses the conversion of thyroxine into triiodothyronine within the hypothalamus, type 2 iodothyronine deiodinase, may also be involved in the photoperiodic control of sexual maturation (Yoshimua *et al.*, 2003).

PHOTOREFRACTORINESS

Photorefractoriness is the inability to respond sexually to an otherwise stimulatory photoperiod and is the mechanism by which reproduction is limited to a time of the year when environmental conditions are most favourable for successfully rearing offspring.

Juvenile photorefractoriness

Seasonal-breeding birds are photorefractory when they hatch, and this juvenile form of the condition prevents truly seasonal-breeding species becoming sexually mature in their first year, even though they may be somatically mature. In species that do not exhibit the extreme form of the condition, such as domesticated turkeys and broiler breeders, sexual maturation can occur at any time of the year in response to a combination of age, body weight and body composition (Lopez and Leeson, 1992; Vandenberghe *et al.*, 1999; Lewis *et al.*, 2005b). Dissipation of the condition occurs naturally in *ad libitum*-fed birds after about 2 months of short days (Follett, 1991) but takes much longer when birds have their feed intake restricted. Typically, broiler breeders take 18–20 weeks to achieve photosensitivity (Lewis *et al.*, 2003). It was reported that juvenile photorefractoriness in birds given constant daylengths dissipated at a rate inversely proportional to photoperiod (e.g. Farner and Follett, 1966; Robinson and Follett, 1982). However, it is not a simple linear effect in broiler breeders because the earliest maturity occurs when they are held on 10 h photoperiods, the latest, indicating the slowest rate of dissipation, between 13 and 14 h, with a progressively quicker rate of dissipation for photoperiods ≥ 15 h (Fig. 14.1). This suggests that the rate of dissipation is not dictated by the duration of the photoperiod but how stimulatory it is (Lewis *et al.*, 2004b). The similarity of the photoperiodic response curve to that reported by Dunn and Sharp (1990) for

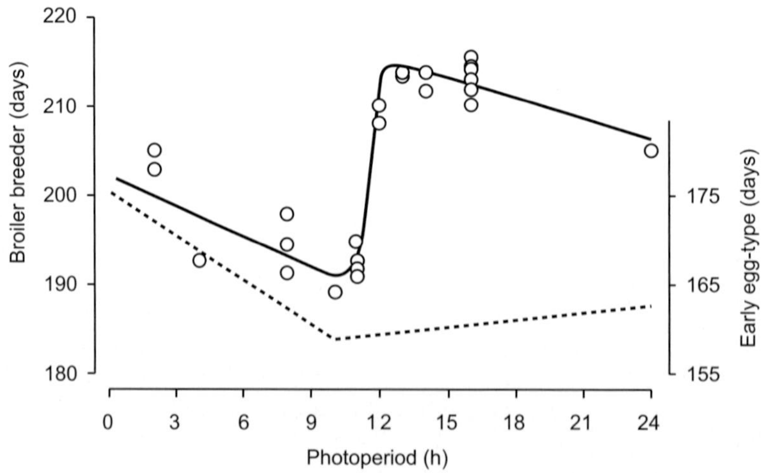

Fig. 14.1. Mean age at sexual maturity for broiler breeder pullets maintained on various photoperiods from 2 days (○ and solid regression, Lewis, 2006) and early strains of egg-type pullets (broken regression, Lewis and Morris, 2005), based on Figure 1 of Lewis and Morris (2006).

changes in plasma LH concentration following transfers from 8 h to various photoperiods between 10.5 and 20 h supports this suggestion (Lewis and Gous, 2006a). The non-linear response by broiler breeders to being maintained on stimulatory photoperiods is in marked contrast to that of non-photorefractory egg-type pullets (Lewis and Morris, 2005), whose maturity is minimally affected by photoperiods greater than 10 h (Fig. 14.1).

Sexual maturity is delayed in photorefractory birds when they are either not given a period of short days during the rearing period or transferred to long days before they have dissipated juvenile photorefractoriness. For example, broiler breeders with normal commercial body weights maintained on photoperiods of between 13 and 17 h (Gous and Cherry, 2004; Lewis *et al.*, 2004b; Lewis and Gous, 2006b) or transferred to 16 h before 10 weeks of age (Lewis *et al.*, 2003, 2007a) matured up to 4 weeks later than birds reared on short days. In the extreme, truly seasonal-breeding species such as partridges may take years to mature if they never experience short days (Siopes and Wilson, 1978; Woodard *et al.*, 1980). When a flock of broiler breeders or turkeys is photostimulated after the first bird becomes photoresponsive but before all birds have achieved photosensitivity, individual ages at first egg will have a bimodal distribution, with some birds having their sexual development accelerated and others retarded. This results in a wide range of ages at first egg, a low peak rate of lay, reduced total egg numbers and variable egg weights.

Adult photorefractoriness

After prolonged exposure to stimulatory photoperiods and 3–4 months of reproduction, truly seasonal-breeding species become photorefractory, undergo gonadal regression and remain infertile until the following spring (Nicholls *et al.*, 1988). However, the mechanism involved in the development of adult photorefractoriness will have been triggered much earlier, because both it and the mechanism that induces sexual maturation are believed to be initiated simultaneously by the initial transfer to long days (Dawson, 2001). Domesticated turkeys and broiler breeders, because they show only a relative form of photorefractoriness, often pause egg laying and resume without any change in daylength (Lewis *et al.*, 2005a; Siopes, 2005). Unsubstantiated evidence from commercial flocks of broiler breeders suggests that a similar scenario also occurs in males, while Muscovy drakes have been observed to experience gonadal regression and spontaneous recrudescence without any change in photoperiod (Jacquet, 1997). In contrast, modern egg-laying hens (Morris *et al.*, 1995) and Pekin ducks (Cherry, 1993), which do not exhibit photorefractoriness, continue to have a high rate of lay throughout the laying period. Figure 14.2 compares typical rates of egg production for truly seasonal-breeding (Emden geese, Chukar partridges), relatively photorefractory (turkeys, broiler breeders and Muscovy ducks) and non-photorefractory (egg-laying hens and Pekin ducks) types of poultry. However, mean rates of lay for flocks of birds mask individual variations in photoresponsiveness. Data for turkey hens have shown that there are three types of bird within a flock towards the end of the laying cycle: those that have developed photorefractoriness and have ceased laying, those that are in lay but have previously had a pause, and those that are in lay and have continued to lay throughout the cycle (Guemene, 1990; Siopes, 2001; Proudman and Siopes, 2004).

Adult photorefractoriness can be terminated artificially by temporarily returning birds to short days (e.g. Leighton *et al.*, 1971), by markedly reducing

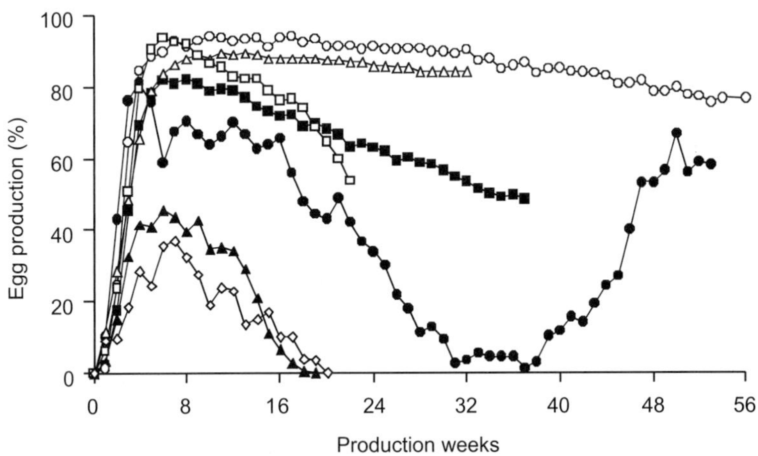

Fig. 14.2. Mean weekly rates of egg production for brown-egg hybrids (○ commercial data), Pekin ducks (△ Cherry, 1993), Muscovies (□ commercial data), broiler breeders (■ unpublished data, University of KwaZulu-Natal), domestic turkeys (● Siopes, 2005), Emden geese (▲ Gillette, 1976), and Chukar partridges (◊ Siopes and Wilson, 1978).

illuminance without altering the daylength (Siopes, 1984) and, because photorefractoriness is in some way facilitated by thyroxine (Follett and Nicholls, 1984; Dawson *et al.*, 2001; Proudman and Siopes, 2002), by pharmacologically inducing hypothyroidism (Siopes, 1997).

CRITICAL AND SATURATION PHOTOPERIODS FOR A PHOTOSEXUAL RESPONSE

Poultry are generally reared on a constant neutral (non-stimulatory) photoperiod and transferred to a longer daylength to initiate sexual maturation. The shortest final photoperiod that results in photosexual induction is called the critical photoperiod and the shortest that achieves the maximum response is called the saturation daylength. Dunn and Sharp (1990) reported the critical daylength to stimulate a rise in plasma LH concentration in dwarf broiler breeders to be between 10.5 and 12.75 h, and the saturation daylength to be between 12.75 and 15.25 h; unpublished data for normal-size broiler breeders supported these findings. Data for the number of birds starting egg production (Wilson *et al.*, 1962) and age at 50% egg production (Siopes, 1994) suggested that the critical daylength for turkeys is between 10 and 12 h. Interestingly, naturally illuminated turkeys start egg laying when the daylength (sunrise to sunset at 35°N) is about 11.5 h (Lien and Siopes, 1993), which means that rapid gonadal development commences when daylength is between 10.5 and 11 h. Siopes (1998) considered that the saturation daylength for turkeys, as indicated by the change in rate of lay when birds were transferred at the end of a laying cycle from various daylengths to a 20 h photoperiod, was at least 12.5 h. Changes in plasma LH concentration indicated that the critical daylength for photoinduction of gonadal development in Muscovy males (Jacquet and Sauveur, 1995) and females (Sauveur and de Carville, 1990) was between 11 and 13 h, and geese exposed to natural lighting begin egg laying when the daylength is between 10 and 12 h (Rousselot-Pailley and Sellier, 1990; Wang *et al.*, 2002).

Thus, it appears that the critical and saturation daylengths for photoinduction are similar for all meat-type strains of poultry, and independent of the exhibition of photorefractoriness (Fig. 14.3).

BROILER BREEDERS

The most important aspect of a broiler breeder's photoperiodic response is photorefractoriness. Although it is only a mild expression of the condition, hatching egg output is only maximized by providing a non-stimulatory photoperiod during rearing and transferring to a photoperiod that is above the critical, but below the saturation, daylength after all individuals in the flock have become photosensitive. The rate of dissipation of juvenile photorefractoriness and the timing of sexual maturation in broiler breeders, whether in response to a constant or a changing photoperiod, is strongly influenced by rate of growth.

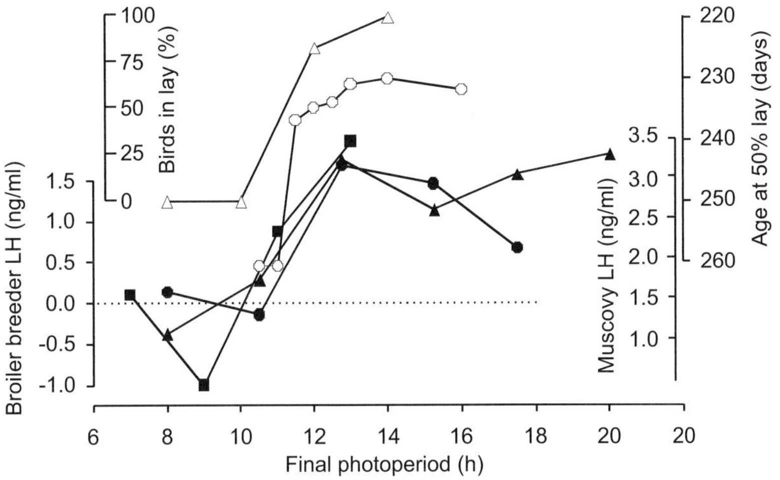

Fig. 14.3. Changes in plasma LH concentration in female normal-size broiler breeders (● unpublished) and dwarf broiler breeders (▲ Dunn and Sharp, 1990) following a transfer from 8 h to various photoperiods at 8 weeks, in Muscovy drakes following a transfer from 6 h to various photoperiods (■ Jacquet and Sauveur, 1995), number of turkeys in lay following a transfer from 6 h to various photoperiods at 33 weeks (Δ Wilson *et al.*, 1962), and age at 50% egg production in turkeys transferred from 8 h to various photoperiods at 30 weeks (○ Siopes, 1994).

Sexual maturation

Sexual maturity advances linearly, up to 10 h, when broiler breeders are maintained on the same photoperiod from soon after hatch (Fig. 14.1). Whilst this response is similar to that of early genotypes of egg-type pullets (Lewis and Morris, 2005), responses to photoperiods > 10 h are very different. Egg-laying pullets experience a delay in maturity of less than 1 day for each 3 h extension of the photoperiod between 10 and 24 h, but broiler breeders maintained on 13 h mature about 25 days later than birds held on 10 h, then commence egg laying progressively earlier when the photoperiod exceeds 13 h (Lewis *et al.*, 2004b). The retarding effects of stimulatory daylengths have also been observed in broiler breeder males (Tyler and Gous, 2006).

Whereas the effect of the initial daylength is minimal when photosensitive birds are transferred abruptly from a neutral to a stimulatory photoperiod (Lewis and Gous, 2006b), the final photoperiod has a significant effect on the rate of sexual development, with the earliest maturity being induced by a transfer to between 14 and 16 h, a response similar to that of egg-type hybrids.

The timing of sexual maturation in broiler breeders is strongly influenced by the degree of feed control applied during the rearing period, and, in the extreme, birds which have had their nutrient intake restricted to the point that it stops them growing remain sexually immature (Dunn and Sharp, 1992). Although body weight modifies the amplitude, rather than the profile, of a photoperiodic response, sexual development is advanced to a significantly greater degree when broiler breeders are maintained on (Lewis *et al.*, 2004b), or transferred to (Lewis and Gous, 2006a), a stimulatory photoperiod than when they are exposed to a neutral or mildly stimulatory one.

Another important factor controlling avian sexual development is the age at which they are photostimulated (Lewis and Morris, 1998). Although the general effect of age at photostimulation is similar for broiler breeder and egg-laying strains, there are significant differences between the two types of fowl in the ages at which photosensitivity is achieved and spontaneous gonadal development commences. Transferring broiler breeders from a neutral to a stimulatory photoperiod before 10 weeks of age delays maturity when they are grown to a body weight of 2.1 kg at 20 weeks. This is because controlled feeding reduces the rate of dissipation of juvenile photorefractoriness and prevents the birds from responding to the incremental increase in photoperiod at stimulation (Lewis *et al.*, 2007a). Photostimulating at 18–20 weeks, when all individuals in a flock are photosensitive, produces the earliest sexual maturity, but photostimulations thereafter result in a progressively smaller advance in gonadal maturation to the point where it occurs spontaneously under the influence of the initial photoperiod (Fig. 14.4).

The age at which photosensitive broiler breeders mature in response to either a constant or a changing photoperiod is strongly influenced by their body weight, decreasing by 2 days for each 100 g change in 20-week body weight up to 2.56 kg (Lewis *et al.*, 2007b). However, because increasing or decreasing body weight also accelerates or delays, respectively, the acquisition of photosensitivity, the response to photostimulation between 10 and 20 weeks varies

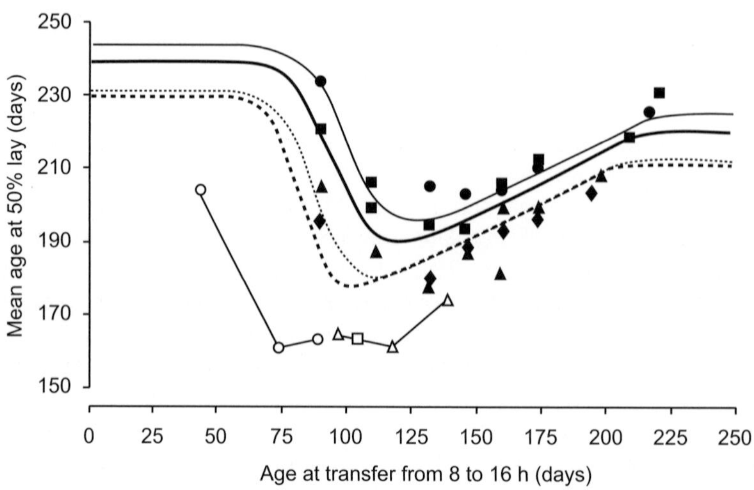

Fig. 14.4. Effect of age at transfer from 8 to 16 h and body weight at 20 weeks on age at 50% egg production in broiler breeders grown to a mean 20-week body weight of 1.85 kg (thin solid line), 2.10 kg (thick solid line), 2.50 kg (dotted line) or 2.80 kg (broken line). Data from the base trial (Lewis *et al.*, 2007a, Experiment 2) are shown for comparison (● 1.85 kg, ■ 2.10 kg, ▲ 2.50 kg, ♦ 2.80 kg). Raw data for broiler breeders grown to a 20-week body weight of between 3.13 and 3.77 kg: ○ (Lewis *et al.*, 2007a, Experiment 3); Δ (Yuan *et al.*, 1994); and □ (unpublished from University of KwaZulu-Natal).

with body weight. An analysis of data from experiments conducted at the University of KwaZulu-Natal showed that increasing 20-week body weight lowers the youngest age for successfully photostimulating all members of a flock by 2 days and reduces the earliest attainable age at sexual maturity by 3.5 days (Lewis and Morris, 2006). The model produced by Lewis *et al.* (2007b) to predict age at 50% egg production following a single increase in photoperiod shows that faster growth produces a bigger advance in maturity and the prospect of photostimulating at a younger age (Fig. 14.4). The raw data in Fig. 14.4 for birds that had been allowed maximum or near maximum growth indicate that the lower limit for uniformly photoinducing sexual development in a flock is probably about 10 weeks of age. This is in good agreement with the 2 months required to dissipate juvenile photorefractoriness in *ad libitum*-fed exotic seasonal-breeding birds.

There are few data for the response of males to photoperiod, but those that exist suggest that males respond similarly to females (Renden *et al.*, 1991; Tyler and Gous, 2006).

Egg production

Total egg production in the laying cycle is negatively correlated with age at sexual maturity, decreasing by about 4 eggs for each 10-day delay in maturity (Lewis *et al.*, 2003; Lewis, 2006; Lewis and Morris, 2006). However, age at sexual maturity is not linearly related to photoperiod (Fig. 14.1), and so egg numbers are not directly related to photoperiod. Lewis (2006) reported that maximum egg numbers for broiler breeders given constant photoperiods are achieved on 10 h, and reduce by 8 eggs/h below, and by 3 to 4 eggs/h above, 10 h. However, broiler breeders are not normally given constant photoperiods, and the relationships between egg production, sexual maturity and daylength in lay are not so clear when they are reared on short days prior to being transferred to long days. Indeed, broiler breeders transferred to a 16 h photoperiod have inferior egg production, especially towards the end of the laying cycle, compared with birds transferred to 11 or 12 h photoperiods, irrespective of the age at which they are transferred (Lewis, 2006). This is because the benefit to egg production of the advance in maturity and earlier peak rate of lay achieved by transferring to a fully stimulatory photoperiod is more than negated by an earlier onset of adult photorefrac toriness and a higher daily energy expenditure that effectively increases the degree of feed restriction. Eleven and 12 h daylengths are not fully stimulatory and so photosensitive birds transferred to them from a neutral photoperiod mature slightly later than if abruptly changed to a 16 h photoperiod, but they have better persistency (Lewis, 2006). Reduced testicular weights at 69 weeks of age for males maintained on 14, 16 or 18 h, compared with those held on 8, 10 or 12 h, suggest that males also become photofractory at an earlier age when they are exposed to more stimulatory photoperiods (Tyler and Gous, 2006).

Sharp (1993) suggested that the initial increase should be to a photoperiod that is between the critical and saturation daylength and that further increments should then be given when required to balance the age-related decline in LH release that is associated with the development of photorefractoriness. However, this hypothesis was subsequently disproved by the finding that persistency was inferior in birds given a final photoperiod of 16 h compared with those that received no further increase after the initial transfer to 11 or 12 h at 20 weeks (Lewis *et al.*, 2007c).

Egg weight is principally a consequence of age and body weight at first egg, but photoperiod also exerts an effect. A broken-stick analysis of data (adjusted for differences in age and body weight at maturity) from trials conducted at the University of KwaZulu-Natal showed that mean egg weight to 60 weeks increased by about 1 g/h, but only up to 13 h (Lewis, 2006). Shell weight, as in egg-type hens, decreases linearly by 30 mg for each 1 h increase in photoperiod (Backhouse *et al.*, 2005). Although it is conjectural as to whether this deterioration in shell quality leads to any reduction in hatchability, shell integrity is positively correlated with hatchability (e.g. Roque and Soares, 1994). The superior shell quality of birds exposed to shorter photoperiods is more likely a consequence of a longer scotoperiod, because concentrations of calcitonin (Klandorf *et al.*, 1997) and parathyroid hormone (Ieda *et al.*, 2000), the two hormones involved in the mobilization of calcium for the shell, peak during the dark period.

Events in the ovulatory cycle are phase-set by the prevailing photoperiod and not by lighting history. Mean oviposition time in broiler breeders is delayed by 0.5 h for each 1 h extension of the photoperiod (Lewis *et al.*, 2004a), which is similar to the rate of change observed in egg-laying hens. However, the actual time of lay for broiler breeders on a given photoperiod is 1 h later than for white-egg hybrids and 2.5 h later than for brown-egg hybrids, and so few eggs are laid before the start of the photoperiod when it is > 12 h.

TURKEYS

Turkeys exhibit photorefractoriness more strongly than broiler breeders, though it is still only the relative form of the condition. Therefore, like broiler breeders, they require a period of short days to hasten its dissipation. Turkeys are fed *ad libitum* and, like exotic seasonal-breeding birds, they acquire photosensitivity more rapidly than broiler breeders. Despite an interaction between the daylength and the length of time needed to dissipate photorefractoriness (for example, 3 weeks of 4 h were reported to equate to 4 weeks of 6 h or 5 weeks of 8 h by Wilson *et al.*, 1962), the modern turkey is usually given 10–11 weeks of 6 or 7 h photoperiods from about 18 weeks in commercial practice.

Sexual maturation

Figure 14.5 shows that photosensitive turkeys are more responsive (8 days later maturity for each 10-day delay in photostimulation) than broiler breeders

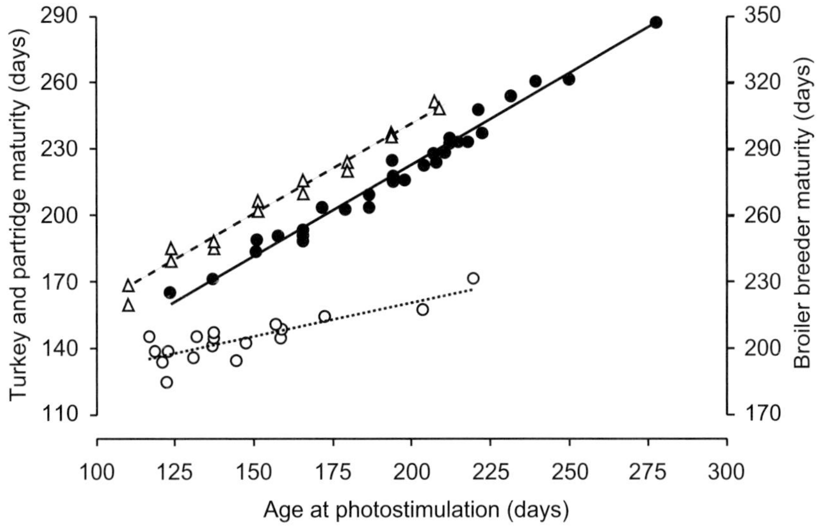

Fig. 14.5. Age at sexual maturity following a transfer from short to long days at various ages in turkeys (•, Lewis and Morris, 1998) and partridges (Δ, Woodard and Snyder, 1978) fed *ad libitum*, and in broiler breeders (○, Lewis *et al.*, 2007b) grown to a body weight of 2.1 kg at 20 weeks.

(3 days per 10 days), but similar to partridges, in their response to a transfer from short to long days. There are insufficient scientific data to evaluate the effect of different photoperiods during the rearing period on sexual development in turkeys.

Egg production

Egg production is strongly influenced by both age at photostimulation and final photoperiod. A hinge analysis indicated that mean rate of lay initially increased by 1.5% for each 10-day delay in photostimulation and was maximized by an increment given at 31 or 32 weeks (typically recommended age for photostimulation), and decreased thereafter by 0.5% per 10-day later transfer to the final photoperiod. Maximum egg numbers are achieved by providing a 14 h final photoperiod, but there is minimal difference between transferring to it abruptly and giving a series of smaller increments (McCartney *et al.*, 1961; Marsden *et al.*, 1962; Siopes, 1998). The lower egg production when daylength provided exceeds 14 h, as in broiler breeders, is a consequence of some individuals becoming photorefractory before the end of the laying cycle (Siopes, 2001). However, turkeys exposed to any stimulatory daylength, especially > 12 h, will eventually show adult photorefractoriness (Siopes, 1994), with the mean onset for birds transferred to 16 h occurring after 24–25 weeks of photostimulation in flocks stimulated in winter and after 16–17 weeks in flocks photostimulated in the spring (Siopes, 2002).

MEAT-TYPE PEKIN DUCKS

There are few data in the literature for the photoperiodic response of ducks. However, in contrast to domestic fowl both the meat and egg-laying strains of duck appear to respond similarly. Meat-type Pekin ducks do not exhibit photorefractoriness (Fig. 14.2), and so neither require a period of short days during rearing to acquire photosensitivity nor mature later when reared on long days (Cherry, 1993). However, in contrast to Muscovy ducks (see below) but similar to egg-type chickens, Pekin ducks do respond differently to increasing and decreasing photoperiods during rearing, with sexual maturity being advanced by an increase and retarded by a decrease in photoperiod (Hill, 1984; Cherry, 1993). Cherry (1993) reported that summer- or autumn-hatched ducks kept in non-lightproof housing and given a step-up lighting from 18 weeks regularly suffer a dip in production after peak (Fig. 14.6), and so it is recommended that they be maintained on 17 h photoperiods from hatch to the end of the laying cycle to avoid the adverse effects of varying natural daylengths. Egg production for spring-hatched ducks given an increasing lighting regimen is similar to that of birds held on 17 h daylengths (Fig. 14.6).

MUSCOVY DUCKS

The photoperiodic nature of Muscovy ducks is somewhat obscure. Egg production data from commercial flocks of Muscovy breeders suggest that they, like broiler breeders and turkeys, exhibit relative photorefractoriness. Peak rate of lay commonly exceeds 90%, but declines rapidly thereafter to reach 50% within

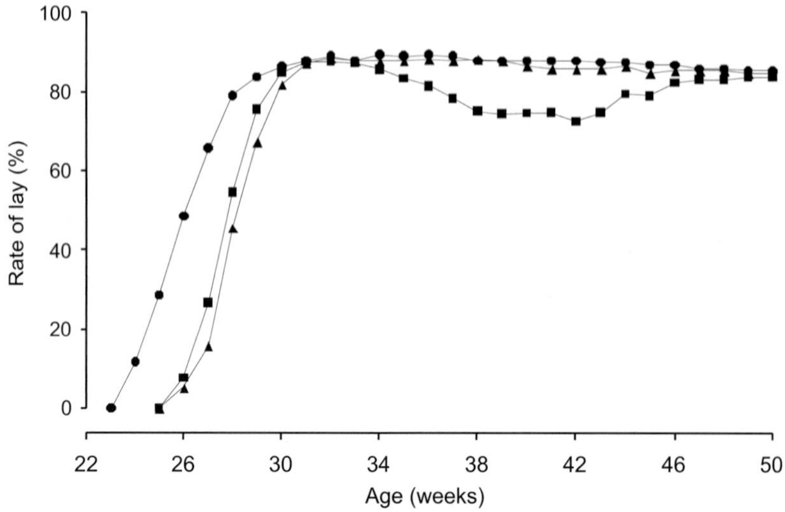

Fig. 14.6. Egg production for Pekin ducks reared under natural daylengths (52°N) and given daylength increases from 18 weeks to reach 17 h at 26 weeks (▲ spring hatched, ■ summer or autumn hatched) or maintained on 17 h throughout (•) (Cherry, 1993).

6 months (Fig. 14.2). Muscovy males have also been observed to undertake gonadal regression and moult within 6 months of being transferred from short to long days, but to then spontaneously regrow their testes and increase plasma LH and testosterone concentrations without any change in photoperiod (Jacquet, 1997). This recrudescence is similar to that seen in turkey hens (Siopes, 2005) and is again indicative of relative photorefractoriness. However, Sauveur and de Carville (1990) reported that birds given 16 h photoperiods during rearing matured 6 weeks earlier than others given an increasing regimen from 8 to 14 h, that birds reared on 13 h and increased to 15 h at 20 weeks reached 50% lay 5 weeks before constant 8 h controls, and that birds given a step-down lighting schedule from 24 to 14 h at 26 weeks matured at a similar time to others reared on an increasing regimen from 8.5 to 14 h. These findings are inconsistent with the response of a species that exhibits photorefractoriness and, despite Muscovies being responsive to a transfer from short to long days (Sauveur and de Carville, 1990), suggest that they are not particularly photoperiodic. It is possible that, in their tropical origins, non-photic environmental cues exerted a larger influence over reproductive activity than light.

GEESE

The common domesticated breeds of geese, Emden, Toulouse, and Roman, exhibit the absolute form of photorefractoriness and are truly seasonal breeding. Under natural lighting, egg laying starts in early spring, peaks between 40 and 50%, and lasts no longer than 4 months (Fig. 14.2). However, Fig. 14.7 shows that, as in other photorefractory species, the provision of a daylength that is

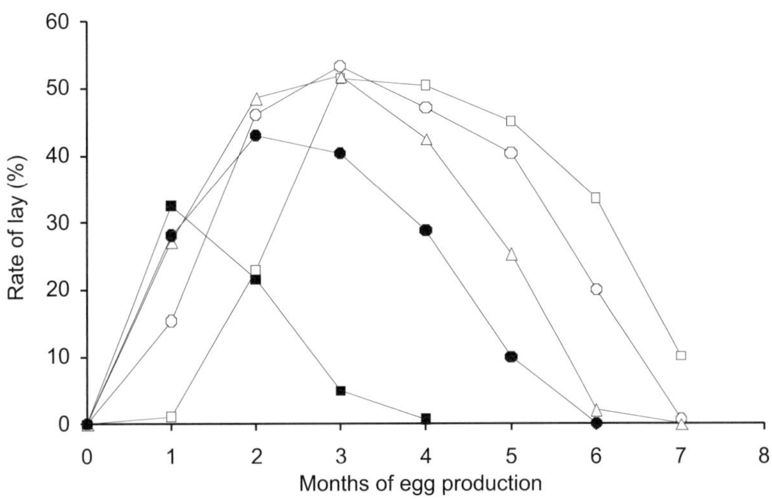

Fig. 14.7. Egg production for Roman geese maintained on 8 (□), 10 (○), 11 (•), 12 (Δ) or 14 h (■) photoperiods (Rousselot-Pailley and Sellier, 1990; Elminowska-Wenda and Rosiński, 1993).

only mildly stimulatory extends the egg-laying period, presumably by delaying the onset of adult photorefractoriness (Rousselot-Pailley and Sellier, 1990; Elminowska-Wenda and Rosiński, 1993).

It is possible, by alternating 10 weeks of 16 h with 10 weeks of 8 h photoperiods, to recycle pheasants (Woodard and Snyder, 1978) and partridge (Woodard *et al.*, 1970), and a modification of this lighting schedule can be used to get more than one laying cycle per year from geese, even when they are kept in open-sided housing. Artificial light is used to supplement natural lighting during winter to create a 20 h day and then removed at the end of the cycle to reduce the daylength, dissipate photorefractoriness and allow the birds to recommence egg laying in response to the naturally increasing daylengths of spring (Wang *et al.*, 2005).

Although most avian species naturally reproduce in the spring, Magang geese, a breed indigenous to southern China, are short-day breeders and have a breeding season that starts soon after the longest day and ends when the daylength exceeds 12 h in the following spring (Huang *et al.*, 2007). However, like the more common breeds of geese, Magang can be encouraged, through photoperiodic manipulation, to lay out of season. Gonadal regression is achieved during the short days of winter by supplementing natural with artificial lighting to create a long day (e.g. 18 h), whilst sexual recrudescence is regained by transferring the birds back to 'short' days (e.g. 11 h) during the summer months.

REFERENCES

Ali, A. and Cheng, K.M. (1985) Early egg production in genetically blind (*rc/rc*) chickens in comparison with sighted (*RC*$^+$/*rc*) controls. *Poultry Science* 64, 789–794.

Backhouse, D., Lewis, P.D. and Gous, R.M. (2005) Constant photoperiods and eggshell quality in broiler breeders. *British Poultry Science* 46, 211–213.

Cherry, P. (1993) Sexual maturity in the domestic duck. PhD thesis, The University of Reading, UK.

Ciccone, N.A., Dunn, I.C., Boswell, T., Tsutsuni, K., Ubuka, T., Ukena, K. and Sharp, P.J. (2004) Gonadotrophin inhibitory hormone depresses gonadotrophin α and follicle stimulating hormone β subunit expression in the pituitary of the domestic chicken. *Journal of Neuroendocrinology* 16, 999–1006.

Dawson, A. (2001) The effects of a single long photoperiod on induction and dissipation of reproductive photorefractoriness in European starlings. *General and Comparative Endocrinology* 121, 316–324.

Dawson, A., King, V.M., Bentley, G.E. and Ball, G.F. (2001) Photoperiodic control of seasonality in birds. *Journal of Biological Rhythms* 16, 365–380.

Dunn, I.C. and Sharp, P.J. (1990) Photoperiodic requirements for LH release in juvenile broiler and egg-laying strains of domestic chickens fed *ad libitum* or restricted diets. *Journal of Reproduction and Fertility* 90, 329–335.

Dunn, I.C. and Sharp, P.J. (1992) The effect of photoperiodic history on egg laying in dwarf broiler hens. *Poultry Science* 71, 2090–2098.

Elminowska-Wenda, G. and Rosinski, A. (1993) Wpyw długości dnia świetlnego na wyniki rozrodu gęsi białych włoskich. *Roczńiki Naukowe Zoothechniki* 20, 339–351.

Farner, D.S. and Follett, B.K. (1966) Light and other environmental factors affecting avian reproduction. *Journal of Animal Science* 25 (Suppl.), 90–115.

Follett, B.K. (1991) The physiology of puberty in seasonally breeding birds. In: Hunzicker, M. and Schwartz, N.B. (eds) *Follicle Stimulating Hormone: Regulation of Secretion and Molecular Mechanisms of Action*. Springer-Verlag, New York, pp. 54–65.

Follett, B.K. and Nicholls, T.J. (1984) Photorefractoriness in Japanese quail: possible involvement of the thyroid gland. *The Journal of Experimental Zoology* 232, 573–580.

Gillette, D.D. (1976) Laying patterns of geese in the Mid West. *Poultry Science* 55, 1143–1146.

Gous, R.M. and Cherry, P. (2004) Effects of body weight at, and lighting regimen and growth curve to, 20 weeks on laying performance in broiler breeders. *British Poultry Science* 45, 445–452.

Guemene, D. (1990) Effect of changing lighting programs at various ages on laying performance in turkey hens (*Meleagris gallopavo*) housed in cages or in floor pen. In: Brillard, J. (ed.) *Control of Fertility in Domestic Birds*. INRA, Paris, pp. 233–241.

Hill, D.A. (1984) Laying date, clutch size and egg size of the mallard, *Anas platyrrhyncos*, and the tufted duck, *Aythya fuligula*. *Ibis* 126, 484–495.

Homma, K. and Sakakibara, Y. (1971) Encephalic photoreceptors and their significance in photoperiodic control of sexual activity in Japanese quail. In: Menaker, M. (ed.) *Biochemistry*. National Academy of Sciences, Washington, DC, pp. 333–341.

Hoos, D. (1937) De Vinkenbaan. Hoe het er toe ging en wat er mee in verband stond. *Ardea* 26, 173–202.

Huang, Y.M., Shi, Z.D., Liu, Z., Liu, Y. and Li, X.W. (2007) Endocrine regulations of reproductive seasonality, follicular development and incubation in Magang geese. *Animal Reproduction Science* 104, 344–358.

Ieda, T., Takahashi, T., Saito, N., Yasuoka, T., Kawashima, M. and Shimada, K. (2000) Changes in parathyroid hormone-related peptide receptor binding in the shell gland of laying hens (*Gallus domesticus*) during the oviposition cycle. *General and Comparative Endocrinology* 17, 182–188.

Jacquet, J.M. (1997) Photorefractory period of the Muscovy duck (*Cairina moschata*): endocrine and neuroendocrine responses to day length after a full reproductive cycle. *British Poultry Science* 38, 209–216.

Jacquet, J.M. and Sauveur, B. (1995) Photoperiodic control of sexual maturation in Muscovy drakes. *Domestic Animal Endocrinology* 12, 189–195.

Klandorf, H., Boyce, C.S., Killefer, J., McGowan, J.A., Petersen, R.A., Valent, M. and Deaver, D.R. (1997) The effect of photoperiod and food intake on daily changes in plasma calcitonin in broiler breeder hens. *General and Comparative Endocrinology* 107, 327–340.

Leighton, A.T., van Kreay, H.P., Moyer, D.D. and Potter, L.M. (1971) Reproductive performance of force-molted turkey breeder hens. *Poultry Science* 50, 119–126.

Lewis, P.D. (2006) A review of lighting for broiler breeders. *British Poultry Science* 47, 393–404.

Lewis, P.D. and Gous, R.M. (2006a) Effect of final photoperiod and twenty-week body weight on sexual maturity and egg production in broiler breeders. *Poultry Science* 85, 377–383.

Lewis, P.D. and Gous, R.M. (2006b) Various photoperiods and *Biomittent*™ lighting during rearing for broiler breeders subsequently transferred to open-sided housing at 20 weeks. *British Poultry Science* 47, 24–29.

Lewis, P.D. and Morris, T.R. (1998) A comparison of the effects of age at photostimulation on sexual maturity and egg production in domestic fowl, turkeys, partridges and quail. *World's Poultry Science Journal* 54, 119–128.

Lewis, P.D. and Morris, T.R. (2005) Change in the effect of constant photoperiods on the rate of sexual maturation in modern genotypes of domestic pullet. *British Poultry Science* 46, 584–586.

Lewis, P.D. and Morris, T.R. (2006) Photoperiod: conventional programmes. In: *Poultry Lighting, the Theory and Practice*. Northcot, Andover, UK, pp. 23–55.

Lewis, P.D., Ciacciariello, M. and Gous, R.M. (2003) Photorefractoriness in broiler breeders: sexual maturity and egg production evidence. *British Poultry Science* 44, 634–642.

Lewis, P.D., Backhouse, D. and Gous, R.M. (2004a) Photoperiod and oviposition time in broiler breeders. *British Poultry Science* 45, 561–564.

Lewis, P.D., Backhouse, D. and Gous, R.M. (2004b) Constant photoperiods and sexual maturity in broiler breeder pullets. *British Poultry Science* 45, 557–560.

Lewis, P.D., Backhouse, D. and Gous, R.M. (2005a) Effects of constant photoperiods on the laying performance of broiler breeders allowed conventional or accelerated growth. *Journal of Agricultural Science, Cambridge* 143, 97–108.

Lewis, P.D., Ciacciariello, M., Nonis, M. and Gous, R.M. (2005b) Simulated natural lighting and constant 14 hour photoperiods for broiler breeders during the rearing period, and interactions of lighting with body weight. *South African Journal of Animal Science* 35, 1–12.

Lewis, P.D., Ciacciariello, M., Backhouse, D. and Gous, R.M. (2007a) Effect of age and growth rate on the photosexual response of broiler breeders to an increase in photoperiod from 8 to 16 h. *British Poultry Science* 48, 601–608.

Lewis, P.D., Gous, R.M. and Morris, T.R. (2007b) A model to predict sexual maturity in broiler breeders given a single increment in photoperiod. *British Poultry Science* 48, 625–634.

Lewis, P.D., Gous, R.M., Ghebremariam, W. and Sharp, P.J. (2007c) Broiler breeders do not respond positively to photoperiodic increments given during the laying period. *British Poultry Science* 48, 245–252.

Lien, R.J. and Siopes, T.D. (1993) The relationship of plasma thyroid hormone and prolactin concentrations to egg laying, incubation behaviour, and molting by female turkeys exposed to a one-year natural daylength cycle. *General and Comparative Endocrinology* 90, 205–213.

Lopez, G. and Leeson, S. (1992) Manage broiler breeder pullets to reduce seasonal variations. *Poultry Digest* September, 31–35.

Marsden, S.J., Cowen, N.S. and Lucas, L.M. (1962) Effect of gradual and abrupt lengthening of photoperiod on reproductive response of turkeys. *Poultry Science* 41, 1864–1868.

McCartney, M.G., Sanger, V.L., Brown, K.I. and Chamberlin, V.D. (1961) Photoperiodism as a factor in the reproduction of the turkey. *Poultry Science* 40, 368–376.

Menaker, M. and Keatts, H. (1968) Extraretinal light perception in the sparrow. II. Photoperiodic stimulation of testis growth. *Proceedings of the National Academy of Sciences of the USA* 60, 146–151.

Morris, T.R. (1967) Light requirements of the fowl. In: Carter, T.C. (ed.) *Environmental Control in Poultry Production*. Oliver and Boyd, Edinburgh, pp. 15–39.

Morris, T.R., Sharp, P.J. and Butler, E.A. (1995) A test for photorefractoriness in high-producing stocks of laying pullets. *British Poultry Science* 36, 763–769.

Nicholls, T.J., Goldsmith, A.R. and Dawson, A. (1988) Photorefractoriness in birds and comparison with mammals. *Physiological Reviews* 68, 133–176.

Proudman, J.A. and Siopes, T.D. (2002) Relative and absolute photorefractoriness in turkey hens: profiles of prolactin, thyroxine, and triiodothyronine early in the reproductive cycle. *Poultry Science* 81, 1218–1223.

Proudman, J.A. and Siopes, T.D. (2004) Effect of a single short-term reduction in photoperiod on photorefractoriness in turkey hens. *Poultry Science* 83, 1199–1202.

Renden, J.A., Oates, S.S. and West, M.S. (1991) Performance of two male broiler breeder strains raised and maintained on various constant photoschedules. *Poultry Science* 70, 1602–1609.

Robinson, J.E. and Follett, B.K. (1982) Photoperiodism in Japanese quail: the termination of seasonal breeding by photorefractoriness. *Proceedings of the Royal Society, London* 215, 95–116.

Roque, L. and Soares, M.C. (1994) Effects of eggshell quality and broiler breeder age on hatchability. *Poultry Science* 73, 1838–1845.

Rousselot-Pailley, D.T. and Sellier, N. (1990) Influence de quelqués facteurs zootechniques sur la fertilitie des oies. In: Brillard, J. (ed.) *Control of Fertility in Domestic Birds*. INRA, Paris, pp. 145–154.

Sauveur, B. and de Carville, H. (1990) Contrôle de la précocité sexuelle de la cane de Barbarie par la photopériode. In: Brillard, J. (ed.) *Control of Fertility in Domestic Birds*. INRA, Paris, pp. 197–203.

Sharp, P.J. (1993) Photoperiodic control of reproduction in the domestic hen. *Poultry Science* 72, 897–905.

Siopes, T.D. (1984) Recycling turkey hens with low light intensity. *Poultry Science* 63, 1449–1452.

Siopes, T.D. (1994) Critical day lengths for egg production and photorefractoriness in the domestic turkey. *Poultry Science* 73, 1906–1913.

Siopes, T.D. (1997) Transient hypothyroidism reinstates egg laying in turkey breeder hens: termination of photorefractoriness by propylthiouracil. *Poultry Science* 76, 1776–1782.

Siopes, T.D. (1998) Absence of a seasonal effect on the critical day length for photorefractoriness in turkey breeder hens. *Poultry Science* 77, 145–149.

Siopes, T.D. (2001) Temporal characteristics and incidence of photorefractoriness in turkey hens. *Poultry Science* 80, 95–100.

Siopes, T.D. (2002) Photorefractoriness in turkey breeder hens is affected by age and season. *Poultry Science* 81, 689–694.

Siopes, T.D. (2005) Spontaneous recovery of photosensitivity by turkey breeder hens given prolonged exposure to long day lengths. *Poultry Science* 46, 584–586.

Siopes, T.D. and Wilson, W.O. (1978) The effect of intensity and duration of light on photorefractoriness and subsequent egg production of Chukar partridge. *Biology of Reproduction* 18, 155–159.

Siopes, T.D. and Wilson, W.O. (1980a) Participation of the eyes in the photo-stimulation of chickens. *Poultry Science* 59, 1122–1125.

Siopes, T.D. and Wilson, W.O. (1980b) Participation of the eyes in the photosexual response of Japanese quail (*Coturnix coturnix japonica*). *Biology of Reproduction* 23, 352–357.

Tyler, N.C. and Gous, R.M. (2006) The effect of photoperiod on testis weight and comb area of male broiler breeders. *Proceedings XII European Poultry Conference*, Verona.

Vandenberghe, N., Ceular, A.L. and Moreno, A. (1999) Use of hatch date for broiler breeder production planning. *Poultry Science* 78, 501–504.

Waldorf, E.C. (1920) Ten eggs per week per hen and how it was done. In: Curtis, G.M. (ed.) *Use of Artificial Light to Increase Winter Egg Production*. Reliable Poultry Journal, Quincy, Illinois, pp. 19–20.

Wang, C.M., Kao, J.Y., Lee, S.R. and Chen, L.R. (2005) Effects of artificial supplemental light on reproductive season of geese kept in open houses. *British Poultry Science* 46, 728–732.

Wang, S.D., Jan, D.F., Yeh, L.T., Wu, G.C. and Chen, L.R. (2002) Effect of exposure to long photoperiod during the rearing period on the age at first egg and the subsequent reproductive performance in geese. *Animal Reproduction Science* 73, 227–234.

Wilson, W.O., Ogasawara, F.X. and Asmundson, V.S. (1962) Artificial control of egg production in turkeys by photoperiods. *Poultry Science* 41, 1168–1175.

Woodard, A.E. and Snyder, R.L. (1978) Cycling for egg production in the pheasant. *Poultry Science* 57, 349–352.

Woodard, A.E., Abplanalp, H. and Wilson, W.O. (1970) Induced cycles of egg production in the Chukar partridge. *Poultry Science* 49, 713–717.

Woodard, A.E., Abplanalp, H. and Snyder, R.L. (1980) Photorefractoriness and sexual response in aging partridge kept under constant long- and short-day photoperiods. *Poultry Science* 59, 2145–2150.

Wurtman, R.J. (1974) The effects of light on man and other mammals. *Annual Review of Physiology* 37, 467–482.

Yoshimua, T., Yasuo, S., Watanabe, M., Iigo, M., Yamamura, T., Hirunagi, K. and Ebihara, S. (2003) Light-induced hormone conversion of T4 to T3 regulates photoperiodic response of gonads in birds. *Nature* 426, 178–181.

Yuan, T., Lien, R.J. and McDaniel, G.R. (1994) Effects of increased rearing period body weights and early photostimulation on broiler breeder egg production. *Poultry Science* 73, 792–800.

CHAPTER 15
Behaviour and Environmental Enrichment in Broiler Breeders

I. Estevez
Neiker-Tecnalia, Arkaute's Agrifood Campus, PO Box 46, Araba, Spain

ABSTRACT

Management practices in broiler breeder production are mostly based on years of farming experience and trial-and-error attempts to address problems as they arise. Some of the problems observed during breeding may be a consequence of the discontinuity in environmental conditions from the rearing to the breeding phase. Rearing takes place in a two-dimensional world, while during breeding birds will have to jump on and off and navigate around farm furniture to find and gain access to nest boxes, feeders and drinkers. Without prior experience this may be a difficult task. Environmental enrichment techniques may help to alleviate this problem as birds gain experience early in life in using the three-dimensional space and would be better prepared for the more complex breeder facility. Enrichment provided by perches, cover panels or other forms of visual barriers not only can reduce the stress of the transition among facilities but may also have a significant impact in reducing the incidence of floor eggs, disturbances, aggression and over-mating. In addition, the presence of cover panels during breeding has been shown to have a positive impact on reproductive performance, and the initial cost of implementation can soon be recovered with additional economic profit. It is clear that environmental enrichment programmes have the potential to benefit animals from the health and welfare standpoint and also improve the management and efficiency of breeding farms.

INTRODUCTION

In poultry science much attention has been given to the behaviour and welfare of caged layers for egg production or to broilers raised for meat. In contrast, aside from studies of reproductive behaviour or the welfare implications of feed restriction programmes for broiler breeder management, our knowledge of the behavioural needs and welfare issues affecting the poultry breeding population is rather limited. Appleby *et al.* (1986) mentioned the lack of much-needed systematic studies on the behaviour of breeders under commercial conditions. Unfortunately, 20 years later we do not know much more. The lack of crucial large-scale behavioural and management studies in broiler breeders is not due to a lack of interest from the investigators but relates to the high cost of breeder research and to the tighter limitations regarding access to commercial breeder farms as a result of the increased biosecurity surrounding breeder flocks.

Current management practices in broiler breeder houses are mostly based on years of farming experience and trial-and-error attempts by farmers and flock supervisors to address problems as they arise. A variety of remedial actions are commonly used by flock supervisors and farmers in commercial broiler breeder operations, including netting males and females separately within the breeder facility right after transfer from the rearing farm, with the purpose that they learn the location of their feeders and drinkers. Farm personnel will 'walk the slats' armed with a variety of utensils to literally scare the females off the slats to make them use the litter area or they will eliminate a proportion of males to control the so-called 'aggression problem', but later during the production cycle will introduce new males to 'spike fertility'. These procedures not only are very time-consuming for flock managers, stressful for the birds and detrimental to egg quality if females are overly stressed, but also add to the cost of production. More importantly, some of these practices can seriously compromise flock biosecurity. In addition, although these practices may help alleviate problems (although this has not been scientifically proven) the positive effects can sometimes be a temporary solution, whereas the real source of the problem may often be missed or overlooked.

THE RELEVANCE OF THE EARLY ENVIRONMENT

In broiler breeder management, there is a disconnection between the conditions in which birds are raised at a young age compared with the environment in which they will be housed during breeding. This disconnection has important implications not only for the welfare of the birds, as the transition can be stressful (Hocking *et al.*, 2005) until they are acquainted with the physical and social features of the new environment, but also for the producer, as some of the above-mentioned management problems observed during breeding can relate to these differences between the rearing and breeding environment. For example, whereas resource distribution during rearing is two-dimensional, the breeding environment is arranged in a three-dimensional space. The location and type of feeders and drinkers can be quite different in shape and colour, and birds are given access to new resources, such as dark nest boxes, that they have never seen before. Furthermore, birds are expected to use these resources properly and quickly. Young breeders may be raised under darker conditions of windowless houses but later at the breeding facility receive plenty of natural light (at least in the USA). In addition, naive broilers at reproductive age will have to learn to navigate the much more complex environment of the breeding facility. Although increased complexity of the captive environment is considered beneficial in terms of animal welfare, as it improves behavioural opportunities and opportunities to hide from aggressive conspecifics (Cornetto *et al.*, 2002), birds may be exposed to higher chances of accidental injuries as they interact with new equipment arranged in a three-dimensional space without prior experience. Acquiring good navigation skills may be important to avoid these accidents but will be most relevant for breeding females so that they can quickly move to avoid overactive males, a problem that may be especially serious early in the breeding phase.

Research has shown that early life experiences can have a dramatic effect on behavioural and neuronal development, use of space and learning ability (Rosenzweig and Bennett, 1996; Gunnarsson *et al.*, 2000; Cornetto and Estevez, 2001a,b; Chaloupkova *et al.*, 2007). It has also been shown that experience in a complex environment attenuates anxiety in stressful situations (see Fox *et al.*, 2006 for a review). Although the majority of the literature regarding the effects of early environment on brain development is based on research in mammals (Rampon *et al.*, 2000; Lewis, 2004 for a review), the hippocampus is known to be larger in those avian species that need excellent orientation and navigational skills to retrieve stored food. These differences in hippocampus size appear to be related to early experiences in food storage, and the hippocampus does not develop as large in birds that never had a chance to actually store food (Clayton and Krebs, 1994). Early-life experience also affects brain development in the domestic fowl. Lowndes and Stewart (1994) found that just 24 h of training to peck at a key was sufficient to increase the density of dendritic spines in chickens and suggested that this brain plasticity may be related to processes involved in long-term information storage.

At the behavioural level, the opportunity to interact early in life with a more complex and enriched environment has been shown to:

1. Affect the spatial distribution and aggression of broilers (Cornetto and Estevez, 2001a; Cornetto *et al.*, 2002; Leone and Estevez, 2007).
2. Affect the incidence of feather pecking and feather damage in layers and young broiler breeders (Hocking *et al.*, 2005; McAdie *et al.*, 2005).
3. Have a positive effect on behaviours associated with fearfulness and learning in young layers (Jones and Waddington, 1992; Krause *et al.*, 2006) and broilers (Newberry and Blair, 1993).
4. Affect the level of fear in young breeder females (Brake *et al.*, 1994).

Early-life experience can also determine how breeders will behave later in a production facility. For example, Brake (1987) and Appleby *et al.* (1988) indicated that young female layer and broiler hens may fail to use nest boxes if no perching opportunities are provided early in life. In an experimental setting, Appleby *et al.* (1988) reported that 86% of hens lay the first egg on the floor, whereas for hens reared with access to perches this proportion was only 21%. The effects of perches were also found to be age dependent, with the best results obtained when perches were provided starting at 4 weeks of age.

Early exposure to perches gives poultry an opportunity to learn to use the three-dimensional space, gain balance and become comfortable jumping on and off perches or from other resources that may be provided at a certain height (such as nest boxes, drinkers and feeders). Although not many data are available for broiler breeders, research in layers kept in aviary systems has shown that early exposure to perches before 4 weeks of age significantly reduced the incidence of floor eggs up to 35 weeks of age and also had a positive effect on the incidence of cannibalism, as hens are provided with an opportunity to get away from cannibalistic birds by jumping on the perches (Gunnarsson *et al.*, 1999). The incidence of floor eggs in broiler breeders is usually observed to be highest during the initial stages of laying (Appleby *et al.*, 1988; Estevez, personal observations). As hens gain experience using the nest boxes, even if no perches are provided, over time they get better at accessing them, thereby reducing the differences between birds raised with and without perches (Gunnarsson *et al.*, 1999). The higher proportion of birds that have early access to perches using nest boxes seems to remain for a much longer period of time in broiler breeders compared with layers (Brake, 1987; Appleby *et al.*, 1988). The results of these studies clearly suggest the involvement of a learning component in the utilization of nest boxes and that the impact of this learning may have long-lasting effects on the preferences of poultry.

It is evident that more research is required regarding the effects of the early environment on behavioural and productive performance during adulthood. Nevertheless, initial research suggests that there may be an early critical period for brain development (or 'wiring') for the formation of long-lasting preferences, which can have major effects on navigation skills, use of space and resources, and preferences that may persist through life.

Equipping animals for a changing environment

Preparedness for the new environment, however, does not necessarily require that young breeders be reared under exactly identical conditions to the ones they will experience later in life. Animals in the wild use a variety of cognitive mechanisms to find scarce and scattered resources in the environment. It has been argued that providing animals with the opportunity to be exposed to a more challenging environment in which to develop their cognitive skills can also improve their welfare (Meehan and Mench, 2007). For example, skills required to deal with a more challenging and variable environment, such as the breeding facility, can be acquired when young if animals are provided with opportunities to learn the spatial competence necessary to navigate and use resources properly later in life. These opportunities can include the provision of perches to add a three-dimensional component or cover panels to increase complexity in the rearing facility (Newberry and Shackelton, 1997; Cornetto and Estevez, 2001a), therefore potentially improving the birds' ability to get around obstacles, to jump on and off fixtures and maintain balance. Research has also shown that domestic animals, including poultry, use visual cues not only to locate food resources but also to exploit them more efficiently, especially in changing environments (Vallortigara, 1996; Roper and Marples, 1997; Howery *et al.*, 2000; Croney *et al.*, 2003; Langbein *et al.*, 2006). It should be possible to use this learning capacity to our advantage to ease the transition of breeders from rearing to breeding facilities. Young breeders may easily identify the location of important resources such as food and water in the breeding house, even when feeders and drinkers vary in design, for example by using colour-coded stickers, flags or other visual cues. If these cues are attached to feeders and drinkers during rearing, then the same colour cues could be used later in the new breeding environment. Furthermore, it should be possible to use different colour codes for resources to be used by males and females. This may not only relieve the stress of confronting a new environment for the animals but can also help improve production performance and ease the management duties of farmers and flock supervisors.

Although most animals will enjoy a certain degree of novelty in their environment that encourages exploratory behaviour, too much novelty can elicit fear and stressful responses (Wemelsfelder and Birke, 1997). The dark enclosure of a nest box may be viewed as a very threatening, unpredictable and therefore unattractive environment for a young female of laying age. Early opportunities to explore and interact with a few nest boxes at the litter level may reduce or even eliminate fear of the novelty component, making the females more likely to use nest boxes later during egg production. Although ideally young hens should be provided with a few nest boxes that they can explore and become familiar with, the essential characteristics of a nest box could be simulated by providing access to disposable boxes of a similar size and colour during rearing. In addition, if early experience in a more complex world helps reduce fear responses and adverse reactions to novelty later in life (Jones and Waddington, 1992; Newberry and Blair, 1993; Krause *et al.*, 2006), then broiler breeders raised in a more challenging and complex environment may have an easier, less stressful and more productive transition from the rearing to the breeding facility.

To make advances in the management of breeders it is very important that we understand that rearing and breeding cannot be viewed as two independent 'aspects'. How broiler breeders and other poultry are raised when young will have long-lasting effects that will be reflected in their health and performance during their reproductive life. Investment in adapting (from an ethological perspective) the rearing environment to prepare young breeders for their reproductive life could pay off from both the economic and the welfare standpoint. Although the suggestions presented here for improvement of the rearing environment are based on strong behavioural principles we still need to gather scientific evidence to show whether these (and other) management techniques will work in a commercial setting, and to determine what exactly will be the long-term benefits for both birds and producers. Nevertheless, available research in this area for poultry (Appleby *et al.*, 1986, 1988; Brake, 1987; Newberry and Blair, 1993) as well as livestock (van de Weerd *et al.*, 2006) provides strong initial evidence that increasing the behavioural opportunities of the early environment can pay off in terms of improving welfare and performance, and reduce the required management time for farmers.

THE VALUE OF LIVING IN A COMPLEX WORLD

Animals in the wild find a variety of landscape features, such as rocks, vegetation or shade, that provide them with protection from environmental conditions and predators, and afford them a safe area for feeding and resting. In contrast, most animal production facilities are relatively barren and offer very few opportunities for suitable challenge or protected areas where animals can retreat from aggressive conspecifics or rest undisturbed. Environmental enrichment programmes serve the purpose of improving the biological functioning of captive animals (Newberry, 1995) by increasing behavioural opportunities (Duncan, 1987; Newberry and Estevez, 1997; Newberry, 1999; Mellen and MacPhee, 2001) or by reducing the incidence of problematic behaviours (Shyne, 2006), and are generally presented as changes to the structure and content of enclosed facilities (van de Weerd *et al.*, 2006; Mason *et al.*, 2007). For example, de Passille *et al.* (1993) demonstrated that calves provided with an artificial teat that they could suckle had lower rates of cross-suckling and improved digestive function. In addition, diverse enrichment treatments have been shown to be effective in reducing the incidence of tail biting in pigs (van de Weerd *et al.*, 2006), aggression and feather damage in layers (Gvaryahu *et al.*, 1994; McAdie *et al.*, 2005) and over-mating in broiler breeders (Estevez, 1999).

The effects of providing environmental enrichment to captive animals can also be observed at the level of the brain. Research conducted mostly in mice and rats clearly demonstrates that providing animals with the opportunity to experience an enriched, complex environment that promotes activity can have massive effects at the neurophysiological and cognitive level (Rosenzweig *et al.*, 1968; Pham *et al.*, 2002; Turner *et al.*, 2003), even if the enrichment takes place later in life (Cotman and Berchtold, 2002; Kempermann *et al.*, 2002). Environmental enrichment has been shown to alter neurotrophin levels (Turner and Lewis, 2003) and gene expression in the brain (Rampon *et al.*, 2000) and to promote recovery of cognitive function after a traumatic brain injury (Hamm *et al.*, 1996).

Nevertheless, results of enrichment programmes can be rather variable, depending on the functional relevance of the resources provided to the animals (Newberry, 1995), how enrichment programmes are presented (Jones *et al.*, 2000), or the timing at which the enrichment is introduced (Mason *et al.*, 2007). Hocking *et al.* (2005) demonstrated that simply providing young broiler breeders with litter during rearing was effective in reducing the incidence of pecking and feather damage as compared with birds raised on a slatted floor. Litter has a high biological value for poultry. Not only does it provide a warm and more comfortable substrate for the birds to rest on but it can also stimulate litter scratching and foraging, especially if small food items can be found hidden in it. On the other hand, toys, such as pecking devices, balls, strings, etc., may be initially attractive and encourage activity. However, initial interest is likely to fade, as seen in Hocking and Jones' (2006) study, if the animal does not get a positive outcome (other than the opportunity of manipulating the object) from these interactions. This is in contrast with the sustained interest for strings found in McAdie *et al.* (2005) and others, and clearly evidence of the variability in results of enrichment programmes.

The decline in the use of the enrichment devices, once the novelty factor disappears, is a relatively common problem of many environmental enrichment programmes, but Newberry (1995) also warns against the assumption that the mere investigation of the introduced devices improves welfare. These behavioural changes may simply reflect short-term curiosity without further repercussions for the animals' long-term health and welfare status. Tarou and Bashaw (2007) also point out that effective enrichment programmes should offer animals with extrinsic reinforcements, for example food, cover or access to conspecifics, that have a biological meaning for the animals, making the enrichment more effective with longer-lasting effects. It appears that serious consideration of the behaviour and life history of the wild ancestors of the domestic fowl could be essential when designing good, functional enrichment programmes. For example, adding complexity to the environment and partitioning the space so that animals have a chance to get away from others, to rest or to hide from cannibalistic or aggressive individuals, are a key requirement for a successful enrichment programme for broiler breeders. From a practical standpoint it is essential to bear in mind that even the best enrichment programme will fail if it is too difficult to implement in practical situations because it is labour intensive, requires constant replacement or can compromise biosecurity.

The effect of enrichment on productive performance

One of the main reasons why enrichment has not been used to a greater extent in animal production is the perception that these programmes may be expensive, can reduce productive performance, especially if the enrichment involves increasing activity of the animals, and are time-consuming. Contrary to this assumption, in layers (Appleby *et al.*, 1993) and broilers (Nicol, 1992; Fiscus-le Van *et al.*, 2000; Pettit-Riley and Estevez, 2001), as well as pigs (Ernst *et al.*, 2005), enrichment has been found not to have a negative impact on performance or feed conversion. In other studies the application of enrichment programmes significantly improved weight gain and feed conversion in both poultry and pigs (Jones *et al.*, 1980; Nicol, 1992; Bell and Adams, 1998; Tauson, 1998; van de Weerd *et al.*, 2006) and reduced mortality related to heat stress in young broilers (Estevez *et al.*, 2002). In addition, the assumption that increased activity necessarily leads to reduced feed efficiency appears to be refutable, at least for broiler breeders (Skinner-Noble *et al.*, 2003). In young turkeys, provision of environmental enrichment in the form of perches, bales of straw and pecking substrates had no discernible effects on performance and final body weight but did reduce the incidence of pecking and injuries (Sherwin *et al.*, 1999; Martrenchar *et al.*, 2001). Further, Leone and Estevez (2008) have shown that, far from being costly, some environmental enrichment programmes can have important economic (and public relations) incentives for the poultry industry, as will be discussed in detail later. Therefore, information currently available suggests that providing environmental enrichment will have some type of initial cost of implementation, depending on each specific programme. However, if the enrichment programme is implemented under standard rearing conditions, such as adding a few pecking devices, perches, cover panels, etc., it does not appear to have a significant negative impact on productive performance. From the animal's perspective, the greater possibilities to increase behavioural activity and control over the environment through enrichment will have a positive impact on welfare. Nevertheless, to be able to determine the realistic 'productive cost' or benefits of enrichment programmes, demonstration projects must be undertaken under commercial conditions, as it is uncertain whether results obtained under well-controlled experimental conditions will translate into commercial situations.

ENRICHMENT IN BROILER BREEDERS

In poultry, enrichment programmes are geared towards increasing the complexity of the environment and challenging the birds with choices but also offering them greater possibilities for exploration and control over the environment, with the goal of improving bird health and welfare (Newberry and Estevez, 2006). One of the most common approaches to providing enrichment for poultry consists of the addition of 'toys', such as infant toys, mirrors, balls, rubber stoppers, tubing, beads, balls, thimbles, strings, etc., in an attempt to reduce and control the incidence of problematic behaviours such as feather pecking, cannibalism and aggression (Jones *et al.*, 1980, 2000; Jones and Waddington, 1992; Jones, 2001; Martrenchar *et al.*, 2001; Huff *et al.*, 2003; McAdie *et al.*, 2005). The advantages of this enrichment approach are that the addition of toys usually does not require significant changes in management procedures, is easy to implement and inexpensive, and they can be easily disinfected so as not to pose a biosecurity risk. Perches have also often been used as an enrichment tool for broilers (Newberry and Blair, 1993; Fiscus-le Van *et al.*, 2000; Pettit-Riley and Estevez, 2001; Bizeray *et al.*, 2002a; Estevez *et al.*, 2002; Pettit-Riley *et al.*, 2002; Bokkers and Koene, 2003), young broiler breeders (Brake, 1987; Appleby *et al.*, 1988; Brake *et al.*, 1994), layers (Duncan *et al.*, 1992; Gunnarsson *et al.*, 2000; Newberry *et al.*, 2001; Vits *et al.*, 2005; Heikkilä *et al.*, 2006) and young turkeys (Martrenchar *et al.*, 2001). Providing more opportunities for increased activity by jumping on and off the perches, or by increasing bone load during perching, was thought to help address some of the leg health issues that are commonly observed in broilers and turkeys or osteoporosis in layers.

More recently the biological value of providing poultry with opportunities to hide by increasing the availability of protective cover has been recognized (Newberry and Shackleton, 1997). This concept was later applied as an attempt to address uneven bird distribution in broilers (Cornetto and Estevez, 2001a) and aggressive behaviour in turkeys (Sherwin *et al.*, 1999). Increasing environmental complexity by adding cover panels has provided some exciting results. Sherwin *et al.* (1999) reported a reduction from a total of 185, 71 and 10 wing, tail and head injuries, respectively, in control pens to 49, 10 and 0 values in enrichment pens that contained cover panels. In broilers, the presence of panels was found to be an effective way of reducing the frequency of disturbances by 50% (Cornetto *et al.*, 2002), and to prevent overuse of the peripheral locations within pens (Cornetto and Estevez, 2001a). This concept has also been employed a number of times when providing enrichment to broiler breeders under commercial conditions in the form of bales of wood shavings (Edmond *et al.*, 2005; Hocking and Jones, 2006) or cover panels (Estevez, 1999; Leone and Estevez, 2008). Despite the high biological value of enrichment programmes that include perches or some form of cover for broiler breeders, this approach requires a larger initial economic investment and higher degree of planning and management, in comparison with the simple addition of pecking devices, probably explaining why the enrichment approach has not been used to its full potential. Nevertheless, as is proposed later in this chapter, the initial investment in environmental enrichment can easily pay off and provide additional benefits for farmers.

Benefits of perches for broiler breeders

Perches are among the most popular ways to provide environmental enrichment to poultry. Access to perches appears to have a positive impact on layers and broiler breeders by minimizing floor eggs, but they should be provided at an early age in order to achieve the greatest benefits for breeders (Appleby, 1984; Brake, 1987; Appleby *et al.*, 1988). Early experience in a three-dimensional world enhances spatial learning and the ability to jump on and off perches and maintain balance, potentially improving the necessary skills that enable birds to access nest boxes later (Newberry and Estevez, 2006). Appleby *et al.* (1988) demonstrated in an experimental setting that, when perches were provided for young layers at 4 weeks of age, floor laying was low and varied initially between 0 and 10%. In contrast, if perches were provided later, between 8 and 20 weeks, then the ratio of initial floor eggs varied between 30 and 100% in early lay, falling to 23 and 43% later on. For broiler breeders under commercial conditions the differences were not as large but still important. Appleby *et al.* (1988) provided perches to 8-week-old broiler breeders in large commercial facilities. Floor laying was initially 21%, dropping to 5% by 30 weeks of age for the experimental group, whereas the proportion of floor eggs in the control flock, reared without access to perches, was 27% and 11% respectively. This commercial trial did not include replications and results have to be considered with caution as they could simply reflect differences across farms rather than differences across treatments. However, it is likely that the differences found correspond to a true effect of early access to perches, given the consistency of these findings with previous trials with layers and with results reported by Brake (1987) in two separate experiments. These studies found that the initial proportion of floor eggs for 25-week-old broiler breeder hens not raised with perches was 41 and 19% respectively for the two experiments. This proportion dropped to 17% for the birds raised with perches. As in Brake (1987) and similar to the results reported in other studies with broilers and layers (Appleby *et al.*, 1988; Gunnarsson *et al.*, 1999), the differences in proportion of floor eggs between females raised with or without perches declined with age. Nevertheless, the differences in results between the two experiments conducted by Brake (1987) were substantial, and, while in one experiment the flocks raised with or without perches became indistinguishable regarding the proportion of floor eggs by 45 weeks of age, in a second experiment differences persisted until 60 weeks. Differences across results between experiments were explained as a consequence of seasonal effects.

An additional but as yet unconsidered advantage of providing perches early in life to broiler breeders is the potential benefits that could be gained in terms of long-term leg health for adults. Leg problems are not uncommon in broiler breeders and can quickly impair male mating success and fertility (Hocking, 1994; Estevez, personal observation). Any steps towards maintaining leg health will contribute to higher farm profitability. Most studies on the effects of early provision of perches to young broilers reared for meat production report no beneficial effects on leg condition, which has been linked to a low interest in the use of the perches provided (Fiscus-le Van *et al.*, 2000; Pettit-Riley and Estevez, 2001; Tablante *et al.*, 2003). In contrast, when the 'perches' consist of low barriers that young broilers have to cross to gain access to feeders and drinkers, tibia diameter at the diaphysis level is significantly larger (Bizeray *et al.*, 2002a,b). This improvement in bone diameter is probably a result of the higher frequency of walking around the barriers to gain access to resources but may also be a response to the high degree of perching observed on the barriers. Similarly, access to perches and provision of an environment that offers opportunities to exercise has been shown to improve bone breaking strength in layers (Hughes and Appleby, 1989; Duncan *et al.*, 1992; Newman and Leeson, 1998; Leyendecker *et al.*, 2005). Broiler breeders tend to be more active and likely to perch than young broilers reared for meat production. It is therefore possible that an early provision of perches may also be beneficial to leg health in broiler breeders as well as breeding turkeys, which have the greatest issues regarding leg integrity (Hester, 1994), potentially leading to better flock fertility.

Interestingly, in one of the few studies on the effects of environmental enrichment in broiler breeders, Edmond *et al.* (2005) found improved eggshell quality over time when birds were provided with enrichment consisting of wood-shaving bales. The authors explained these results as a consequence of improved leg conditions due to the opportunity for the birds to jump on and off the bales during rearing, possibly improving bone strength. Supported by the results of Tullett (1987) on the effects of age on calcium metabolism, Edmond *et al.* (2005) hypothesized that any management tactic that increases the amount of available bone mineral at an early age may be sufficient to maintain eggshell quality later in life once calcium metabolism becomes less efficient as the birds age.

Even though the amount of data on the benefits of access to perches for broiler breeders is very limited, initial results clearly suggest that early access to perches (or other devices that allow birds to jump on and off) may not only be a way of improving the welfare of breeders but may also be economically profitable if the beneficial effects on the incidence of floor eggs, leg integrity, fertility and egg quality are experimentally confirmed.

Cover panels and other visual barriers

Broilers, and most farm and wild animals in captivity, tend to distribute themselves unevenly in the space available, overusing areas close to the walls (Newberry and Hall, 1990) despite large central areas, which may remain unoccupied. Uneven bird distribution comes at a high cost to the poultry industry as the relatively higher densities observed along the wall areas may increase the frequency of disturbances and scratches and may limit airflow among the birds, potentially affecting bird health, performance and their ability to cope with heat stress. It has been experimentally demonstrated with young broilers that the use of panels, simulating walls, placed in the centre of the available space substantially improves spatial bird distribution (Cornetto and Estevez, 2001a) and the use of central areas for resting (Cornetto and Estevez, 2001b), and is effective at reducing the incidence of disturbances by 50% (Cornetto *et al.*, 2002). The use of visual barriers, in combination with supplemental ultraviolet (UV) radiation, straw and pecking substrates, was also found to reduce the incidence of injuries in young male turkeys (Sherwin *et al.*, 1999), as indicated earlier in this review. It is likely that similar results could be obtained by using cover panels or other visual barriers in the rearing of young broiler breeders.

The cover panel concept was tested in two commercial flocks of broilers and resulted in a reduction of carcass condemnations from 1.23 to 0.78% in untreated versus treated houses (Estevez, 1999, unpublished). When applied to broiler breeders, cover panels were able to control a severe over-mating problem in a commercial facility (Estevez, 1999; see Fig. 15.1). In later trials, the effects of providing commercial broiler breeder flocks with cover panels in terms of reproductive performance were investigated. This demonstration trial (Leone and Estevez, 2008) was conducted on five commercial broiler breeder farms, each consisting of a control and a panel treatment room with approximately 7000 females and 800 males each. Cover panels (Estevez and Newberry, 2001) were introduced at either 22 or 27 weeks of age and were constructed of a 10 cm × 10 cm wooden frame, measuring 70 cm high by 70 cm wide. Metal and black plastic mesh was stapled to the front of each panel to slightly obscure the view through the panels, which has been previously shown to be most attractive to chickens (Newberry and Shackleton, 1997; Cornetto and Estevez, 2001a). In this study Leone and Estevez (2008) found that broiler breeder hens provided with panels laid 2.1% more eggs than the hens in the control houses, fertility peaked higher (see Fig. 15.2a) and better persistency of lay was maintained over a longer period of time. Hatchability was also superior in the houses provided with enrichment, with differences being more pronounced as birds aged (see Fig. 15.2b). By 60 weeks of age the difference was 2.6% (68.2% hatchability in control flocks versus 70.8% in enriched flocks). We estimated that the overall improvement in reproductive performance in this demonstration trial was 4.5 additional chicks produced per female in enriched houses, translating into 31,500 more chicks per house per cycle. According to our calculations, the provision of panels should result in an additional income of US$6300 per house, for an initial investment of US$400 for panel construction and materials. Preliminary analysis of a second demonstration trial conducted on four commercial farms confirmed our initial results as there was a significant increase in the number of eggs laid per female (Fig. 15.3). However, thus far it is unclear why reproductive performance is improved by the provision of panels. It has been speculated that providing enrichment to poultry may decrease stress levels (Jones *et al.*, 1980; Jones and Waddington, 1992; Nicol, 1992), which have been shown to have negative consequences for reproductive

Fig. 15.1. Broiler breeder flock during an outbreak of over-mating (top), and the same farm 24 h after inclusion of the cover panels (bottom). Note the difference in the number of females in the litter area. (Reprinted from *Poultry Perspectives*, Estevez, 1999.)

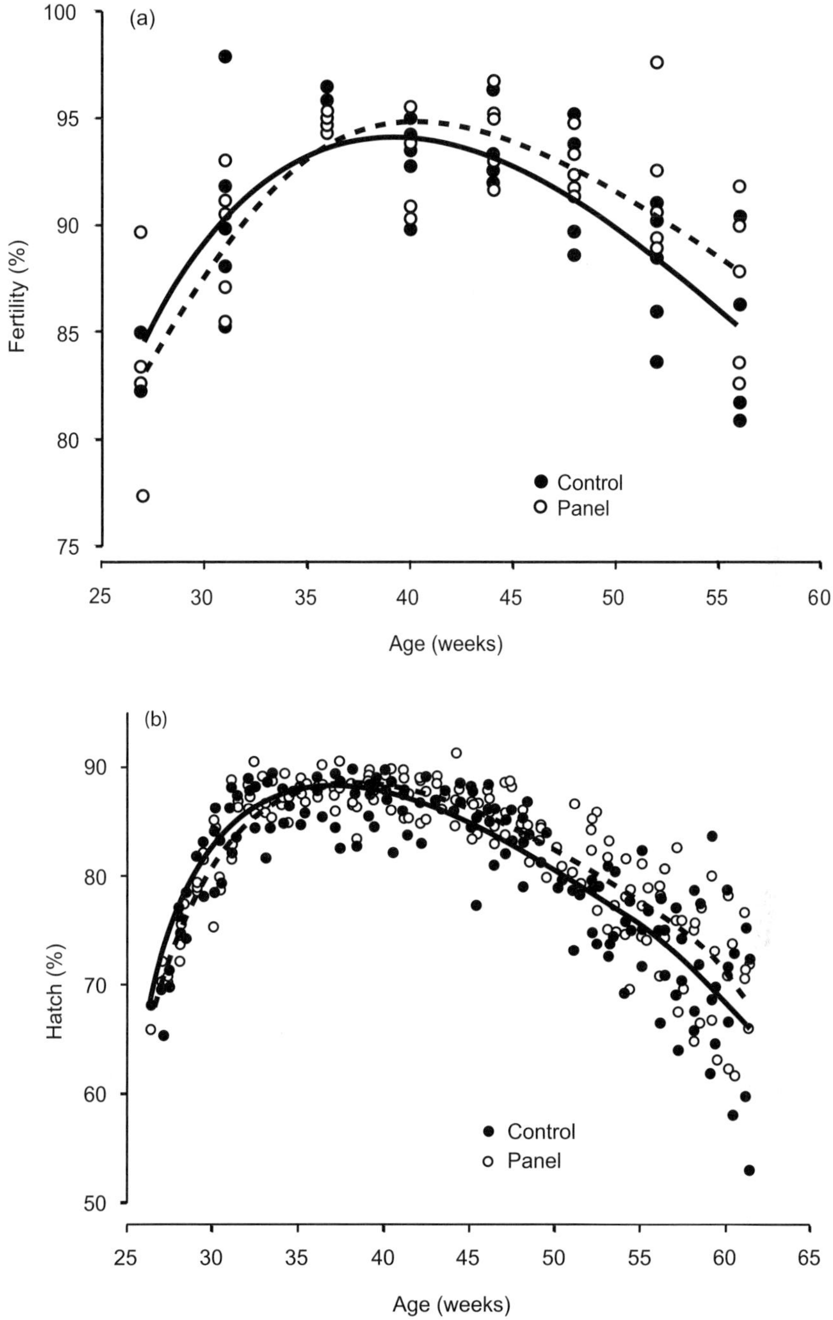

Fig. 15.2. Per cent fertility (a) and hatchability (b) observed across age (in weeks) for control (black circles, solid line) and enriched broiler breeder flocks (open circles, dashed line). The interaction of cover panel and age can be seen in the different slopes of the predicted regression lines for each treatment over time. (From Leone and Estevez, 2008, used by permission of *Poultry Science*.)

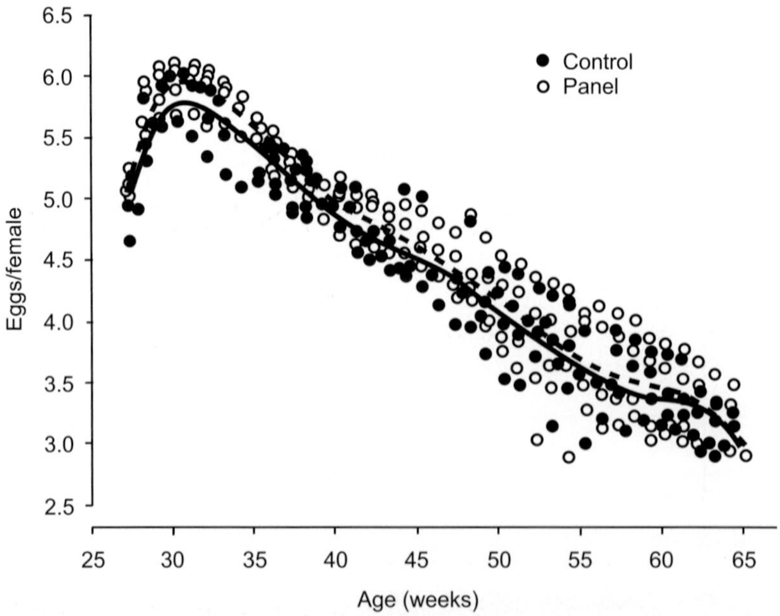

Fig. 15.3. Number of eggs per female per week for control (black circle, solid line) and panel treatment flocks (white circles, dashed lines). There was a significant interaction of treatment by age ($P < 0.001$) and the differences remain significant until 62 weeks.

function in poultry (Jones, 1996). Therefore, a reduction in stress levels related to access to the cover panels may explain the improvements we observed in reproductive performance (Leone and Estevez, 2008). Independent of the mechanism by which reproductive performance is improved, these results clearly demonstrate that environmental enrichment can not only be beneficial for broiler breeder welfare but also be economically advantageous for the farmer: a win–win situation from the standpoint of both animal welfare and performance.

A potential negative side effect of using panels was the possibility of a higher incidence of floor eggs, due to the dark areas created by the panels in the litter area. Contrary to our expectations, we found that after 39 weeks of age hens in enriched houses consistently laid fewer eggs on the floor than controls (1.58 versus 2.12%; Leone and Estevez, 2008). This was explained to be a consequence of the larger number of individuals attracted to the litter area around the panels, possibly making these high-traffic areas unattractive for egg laying (Leone and Estevez, 2008). It is also worth mentioning that in this study, as in other reports (Brake, 1987), we observed a large degree of variability in the incidence of floor eggs across farms, suggesting that management factors play a key role in the incidence of floor eggs. Variability across farms was also observed in hatchability and fertility percentages but to a lesser extent, although still economically relevant, than for the proportion of floor eggs. An epidemiological or risk factor study conducted in commercial conditions could help elucidate the causes of the large variability in performance across farms. It will also be tremendously helpful to determine why the risk factors for floor eggs and fertility and hatchability decline with age in broiler breeder flocks.

In contrast to the positive results of our studies on the effects of cover panels in young and adult breeders and those of Edmond *et al.* (2005) and King (2001) using bales of wood shavings and compact discs (unpublished; reported in Hocking and Jones, 2006), Hocking and Jones (2006) described a lack of effects from providing strings and bales of wood shavings to broiler breeder flocks during rearing. The lack of consistency of results obtained across studies appears to suggest that there could be other factors involved that can affect the outcome of environmental enrichment programmes. These factors may include differences in genetic predisposition or inter-individual differences (Huff *et al.*, 2003; Bell, 2007). Variability in results may also be related to a pronounced variation in the execution of enrichment protocols as used by different investigators (for review see Fox *et al.*, 2006), which may interfere with the results obtained, even if the enrichment devices being used are similar or serve similar purposes.

Cover panels and bird movement

Little information is available on the movement and behaviour of broiler breeders in commercial settings. Young broilers under experimental conditions tend to range across the majority of the space available to them (Hughes *et al.*, 1974; Newberry and Hall, 1990; Estevez *et al.*, 1997). Similar results have been reported for adult male broiler breeders. Although it is speculated that male broilers may establish some kind of territory (McBride and Foenander, 1962; Craig and Guhl, 1969; Oden *et al.*, 2000), our results (Leone and Estevez, 2008), similar to Appleby *et al.* (1985) for broiler breeders and results reported by Oden *et al.* (2004) for adult males of a layer strain, suggest a high degree of overlap between home ranges and a large degree of variability in the size of home ranges (see also Duncan, Chapter 8, this volume). It is unlikely that males in large flocks will try to establish territories (which imply defence of a specific area of the facility) in a situation where territories may be difficult and expensive to maintain due to the large number of competitors (Estevez *et al.*, 1997). Nevertheless, even if exclusive territories are not established, it is possible that differences in the size of home ranges may have repercussions in terms of reproductive success for male poultry. In contrast to the results of Pamment *et al.* (1983), who reported larger use of space for low-ranking males of a layer strain, Oden *et al.* (2004) found larger home ranges in high-rank males (based on frequency of crows). Interestingly, in the report of Leone and Estevez (2008), male home ranges were affected by the provision of cover panels and were significantly larger in enriched as compared with control houses (259 ± 24.4 m^2 versus 184 ± 23.1 m^2). These means represent 22 and 16% of the total house space used by birds in the enriched and control flocks, respectively. In both cases the proportion of area used on average was smaller than the 73% reported by Appleby *et al.* (1985). It is possible that the difference in results between the two studies may be related to the fact that the birds were followed for a shorter period of time, from 27 to 45 weeks of age, whereas Appleby *et al.* (1985) covered from 22 until 55 weeks of age and it is known that the number of observations can affect home range dimensions (Estevez and Christman, 2006).

Despite the fact that differences in use of space when breeders are provided with panels may appear of little relevance, males with larger home ranges may encounter a greater number of females, potentially increasing their chances for mating and fertilization. This may contribute to the better reproductive performance observed in environmentally enriched flocks. Further studies are needed to investigate whether differences in use of space patterns are related to social rank and whether (or not) this translates into a higher frequency of matings and greater reproductive success.

THERMAL CONTROL THROUGH ENVIRONMENTAL ENRICHMENT

In poultry production, three dimensions of the facility are heated, cooled and ventilated while only two dimensions are effectively used for production. A more efficient way of adapting the thermal environment for poultry would focus on providing ideal environmental conditions at the bird level rather than wasting energy cooling or warming large masses of air several feet above the birds with no impact on their health and welfare. A different approach to thermal control of the environment could consist of the addition of enrichment devices that increase environmental complexity and allow birds to use the third dimension while providing them with an ideal thermal environment at the bird level (Estevez, 2006). For example, perches capable of running either hot (during the winter) or cold water (in the summer) could not only be used as a way of providing perching opportunities for young breeders but also create an effective microenvironment at the bird level, where environmental conditions will have the most beneficial effects, while possibly allowing for overall energy savings. Although this concept has never been tested in broiler breeders, access to cool perches is known to have a positive impact on body weight, final carcass weight, feed efficiency and mortalities related to heat stress in young broiler flocks raised for meat production (Reilly *et al.*, 1991; Estevez *et al.*, 2002; Okelo *et al.*, 2003). The use of thermally regulated perches and ramps can be particularly beneficial for adult male broilers as breeder operations in the USA do not provide heat during the winter and cooling is provided mostly by fans. High temperatures have a clear negative impact on male fertility (McDaniel *et al.*, 1996), and it has also been noticed that fertility declines in harsh winters if birds are forced to use feed energy to maintain body temperature (D. Pollock, Princess Ann, 2007, personal communication). Estevez *et al.* (2002) noticed that birds not only perched on cool perches but also obtained some cooling benefits by placing the wing and throat or wattles around the pipes. It is likely that the provision of better cooling or warming opportunities at the bird level by placement of thermal perches or ramps could be a good strategy to provide an enriched and complex environment while possibly improving long-term leg health and maximizing fertility in adult breeders at a low cost. From a welfare standpoint it is likely that in the not so distant future broiler breeder welfare standards will require better control of the thermal environment. Availability of microenvironments has the potential to be an effective strategy to guarantee high welfare and performance levels at low energy costs.

POTENTIAL PROBLEMS ASSOCIATED WITH ENRICHMENT DEVICES

Environmental enrichment can, without question, be a very valuable approach towards improving welfare and performance in animal production. Nevertheless, there are also potential problems that can appear to be associated with the use of enrichment devices and must be carefully considered before implementation. For example, broiler breeders and layers have been found entangled in strings provided for enrichment and some have accidentally hanged themselves (Hocking and Jones, 2006; Estevez, personal observation). Accidental ingestion of pecking devices or accidents from bad landings and collisions have also been observed in enriched facilities provided with perches (Moinard *et al.*, 2004, 2005), and, while it has never been reported, it is possible that falls and bone fractures may occur when breeders are provided with cover panels and bales of wood on which they can climb. Frequent use of perches can increase the chances of keel bone deformities and prevalence of bumble feet, although the incidence will depend on the specific perch design being used (Appleby *et al.*, 1993; Vits *et al.*, 2005). A higher frequency of cracked eggs has been observed for layers in alternative cages (Appleby *et al.*, 1993; Guesdon *et al.*, 2006), but this problem seems to be more related to the nest box and collection system designs rather than access to perches *per se*. Egg laying while perching has only rarely been observed and it is not expected that access to perches by broiler breeders will have an impact on egg quality. However, the incidence of bumble feet has to be considered. Although the availability of perches is supposed to increase the amount of space in the third dimension, some degree of obstruction in the use of space and increased negative effects of high relative densities at the floor level may appear if perches take additional space but are not used at a high enough frequency, as reported for broilers raised for meat production (Pettit-Riley and Estevez, 2001; Heckert *et al.*, 2002). This is unlikely to happen in young or adult breeders as they would be expected to use the available perches at a much higher rate than broilers raised for meat production, because of their greater activity.

When developing an enrichment programme it is also essential to consider that enrichment may occasionally trigger competition and increase aggression (Barnard *et al.*, 1996), especially if the enrichment can only be used by one animal at a time. Huff *et al.* (2003) also reported greater susceptibility to stress-induced opportunistic diseases in a study on environmental enrichment in male turkeys. These results (Huff *et al.*, 2003) may have been more related to stress-induced neophobia, as enrichment devices were changed frequently. Therefore, despite tremendous potential for improving welfare, management and performance for broiler breeders through environmental enrichment, these programmes have to be carefully evaluated prior to implementation because not all enrichment techniques may work equally well. For example, a flat, wide perch may be more appropriate for broiler breeders than a small, rounded perch. Entanglement with string can be prevented by keeping the undone fibres short, while competition and aggression can be minimized by dispersing the enrichment devices throughout the facility, thereby increasing the opportunities for more birds to interact with the device simultaneously.

SUMMARY AND IMPLICATIONS FOR PRODUCTION

In broiler breeder production, there is generally a lack of consideration towards the disconnection between environmental conditions in the rearing and breeding phases. This discontinuity may lead to problems such as a high incidence of floor eggs, lack of interest in nest boxes and inability to find the location of feeders and drinkers or to navigate the more complex broiler breeder house. Providing young broiler breeders with access to a more complex environment that allows them to explore the third dimension of space early on, such as by providing perches, will improve their skills to jump to the nest box for egg laying, or in exploring to find resources. Perches may also improve long-term leg health, perhaps benefiting fertility. The addition of cover panels or other forms of visual barriers increases environmental complexity, affecting not only birds' spatial distribution but also the frequency of aggression and disturbances. In addition, the availability of cover panels during breeding has been shown to have a positive impact on the number of eggs produced per hen, fertility and hatchability. Initial investment in enrichment was overwhelmingly recovered by additional economic profit. All evidence suggests that environmental enrichment programmes can have major benefits for the welfare, health and reproductive performance of broiler breeders.

REFERENCES

Appleby, M.C. (1984) Factors affecting floor laying by domestic hens: a review. *World's Poultry Science Journal* 40, 241–249.

Appleby, M.C., Maguire, S.N. and McRae, H.E. (1985) Movement by domestic fowl in commercial flocks. *Poultry Science* 64, 48–50.

Appleby, M.C., Maguire, S.N. and McRae, H.E. (1986) Nesting and floor laying by domestic hens in a commercial flock. *British Poultry Science* 27, 75–82.

Appleby, M.C., Duncan, I.J.H. and McRae, H.E. (1988) Perching and floor laying by domestic hens: experimental results and their commercial application. *British Poultry Science* 29, 351–357.

Appleby, M.C., Smith, S.F. and Hughes, B.O. (1993) Nesting, dust bathing and perching by laying hens in cages – effects of design on behaviour and welfare. *British Poultry Science* 34, 835–847.

Barnard, C.J., Behnke, J.M. and Sewell, J. (1996) Environmental enrichment, immunocompetence and resistance to *Babesia microti* in male mice. *Physiology and Behaviour* 60, 1223–1231.

Bell, A.M. (2007) Animal personalities. *Nature* 447, 539–540.

Bell, D.D. and Adams, C.J. (1998) Environment enrichment devices for caged laying hens. *Journal of Applied Poultry Research* 7, 19–26.

Bizeray, D., Estevez, I., Leterrier, C. and Faure, J.M. (2002a) Effects of increasing environmental complexity on the physical activity of broiler chickens. *Applied Animal Behaviour Science* 79, 27–41.

Bizeray, D., Estevez, I., Leterrier, C. and Faure, J.M. (2002b). Influence of increased environmental complexity on leg condition, performance and level of fearfulness in broilers. *Poultry Science* 81, 767–773.

Bokkers, E.A.M. and Koene, P. (2003) Behaviour of fast and slow growing broilers to 12 weeks of age and the physical consequences. *Applied Animal Behaviour Science* 81, 59–72.

Brake, J. (1987) Influence of presence of perches during rearing on incidence of floor laying in broiler breeders. *Poultry Science* 66, 1587–1589.

Brake, J., Keeley, T.P. and Jones, R.B. (1994) Effect of age and presence of perches during rearing on tonic immobility fear reactions of broiler breeder pullets. *Poultry Science* 73, 1470–1474.

Chaloupkova, H., Illmann, G., Bartos, L. and Spinka, M. (2007) The effect of pre-weaning housing on the play and agonistic behaviour of domestic pigs. *Applied Animal Behaviour Science* 103, 25–34.

Clayton, N.S. and Krebs, J.R. (1994) Hippocampal growth and attrition in birds affected by experience. *Proceedings of the National Academy of Sciences of the USA* 91, 7410–7414.

Cornetto, T. and Estevez, I. (2001a) Influence of vertical panels on use of space by domestic fowl. *Applied Animal Behaviour Science* 71, 141–153.

Cornetto, T.L. and Estevez, I. (2001b) Behaviour of the domestic fowl in presence of vertical panels. *Poultry Science* 80, 1455–1462.

Cornetto, T., Estevez, I. and Douglass, L.W. (2002) Using artificial cover to reduce aggression and disturbances in domestic fowl. *Applied Animal Behaviour Science* 75, 325–336.

Cotman, C.W. and Berchtold, N.C. (2002) Exercise: a behavioural intervention to enhance brain health and plasticity. *Trends in Neurosciences* 25, 295–301.

Craig, J.V. and Guhl, A.M. (1969) Territorial behaviour and social interactions of pullets kept in large flocks. *Poultry Science* 48, 1622–1628.

Croney, C.C., Adams, K.M., Washington, C.G. and Stricklin, W.R. (2003) A note on visual, olfactory and spatial cue use in foraging behaviour of pigs: indirectly assessing cognitive abilities. *Applied Animal Behaviour Science* 83, 303–308.

de Passille, A.M.B., Christopherson, R. and Rushen, J. (1993) Nonnutritive sucking by the calf and postprandial secretion of insulin, CCK, and gastrin. *Physiology and Behaviour* 54, 1069–1073.

Duncan, E.T., Appleby, M.C. and Hughes, B.O. (1992) Effect of perches in laying cages on welfare and production of hens. *British Poultry Science* 33, 25–35.

Duncan, I.J.H. (1987). The welfare of farm animals: an ethological approach. *Science Progress* 71, 317–326.

Edmond, A., King, L.A., Solomon, S.E. and Bain, M.M. (2005) Effect of environmental enrichment during the rearing phase on subsequent eggshell quality in broiler breeders. *British Poultry Science* 46, 182–189.

Ernst, K., Puppe, B., Schon, P.C. and Manteuffel, G. (2005) A complex automatic feeding system for pigs aimed to induce successful coping by cognitive adaptation. *Applied Animal Behaviour Science* 91, 205–218.

Estevez, I. (1999) Cover panels for chickens: a cheap tool that can help you. *Poultry Perspectives* 1, 4–6.

Estevez, I. (2006) Poultry behaviour, beyond the welfare aspects. *55th Annual National Breeders Roundtable*. 4–5 May, St Louis, MO, pp. 77–85.

Estevez, I. and Christman, M.C. (2006) Analysis of the movement and use of space of animals in confinement: the effect of sampling effort. *Applied Animal Behaviour Science* 97, 221–240.

Estevez, I. and Newberry, R. (2001) Method and device for distributing fowl within an enclosure. U.S. Patent # US 6,289,848, B1 September 18, 2001.

Estevez, I., Newberry, R. and Arias de Reyna, L. (1997) Broiler chickens, a tolerant social system? *Etología* 5, 19–29.

Estevez, I., Tablante, N., Pettit-Riley, R.L. and Carr, L. (2002) Use of cool perches by broiler chickens. *Poultry Science* 81, 62–69.

Fiscus-le Van, N.F., Estevez, I. and Stricklin, W.R. (2000) Use of horizontal and angled perches by broiler chickens. *Applied Animal Behaviour Science* 65, 349–365.

Fox, C., Merali, Z. and Harrison, C. (2006) Therapeutic and protective effect of environmental enrichment against psychogenic and neurogenic stress. *Behavioural Brain Research* 175, 1–8.

Guesdon, V., Ahmed, A.M.H., Mallet, S., Faure, J.M. and Nys, Y. (2006) Effects of beak trimming and cage design on laying hen performance and egg quality. *British Poultry Science* 47, 1–12.

Gunnarsson, S., Keeling, L.J. and Svedberg, J. (1999) Effect of rearing factors on the prevalence of floor eggs, cloacal cannibalism and feather pecking in commercial flocks of loose housed laying hens. *British Poultry Science* 40, 12–18.

Gunnarsson, S., Yngvesson, J., Keeling, L.J. and Forkman, B. (2000) Rearing without early access to perches impairs the spatial skills of laying hens. *Applied Animal Behaviour Science* 67, 217–228.

Gvaryahu, G., Ararat, E., Asaf, F.E., Lev, S.M., Weller, J.I., Robinzon, B. and Snapir, N. (1994) An enrichment object that reduces aggressiveness and mortality in caged laying hens. *Physiology and Behaviour* 55, 313–316.

Hamm, R.J., Temple, M.D., O'Dell, D.M., Pike, B.R. and Lyeth, B.G. (1996) Exposure to environmental complexity promotes recovery of cognitive function after traumatic brain injury. *Journal of Neurotrauma* 13, 41–47.

Heckert, R.A., Estevez, I., Russek-Cohen, E. and Pettit-Riley, R. (2002) Effects of density and perch availability on the immune status of broilers. *Poultry Science* 81, 451–457.

Heikkila, M., Wichman, A., Gunnarsson, S. and Valros, A. (2006) Development of perching in chicks reared in enriched environment. *Applied Animal Behaviour Science* 99, 145–156.

Hester, P.Y. (1994) The role of environment and management on leg abnormalities in meat-type fowl. *Poultry Science* 73, 904–915.

Hocking, P.M. (1994) Assessment of the welfare of food restricted male broiler breeder poultry with musculoskeletal disease. *Research in Veterinary Science* 57, 28–34.

Hocking, P.M. and Jones, E.K.M. (2006) On-farm assessment of environmental enrichment for broiler breeders. *British Poultry Science* 47, 418–425.

Hocking, P.M., Jones, E.K.M. and Picard, M. (2005) Assessing the welfare consequences of providing litter for feed-restricted broiler breeders. *British Poultry Science* 46, 545–552.

Howery, L.D., Bailey, D.W., Ruyle, G.B. and Renken, W.J. (2000) Cattle use visual cues to track food locations. *Applied Animal Behaviour Science* 67, 1–14.

Huff, G.R., Huff, W.E., Balog, J.M. and Rath, N.C. (2003) The effects of behaviour and environmental enrichment on disease resistance of turkeys. *Brain Behaviour and Immunity* 17, 339–349.

Hughes, B.O. and Appleby, M.C. (1989) Increase in bone strength of spent laying hens housed in modified cages with perches. *Veterinary Record* 124, 483–484.

Hughes, B.O., Wood-Gush, D.G. and Jones, R.M. (1974) Spatial organization in flocks of domestic fowl. *Animal Behaviour* 22, 438–445.

Jones, R.B. (1996) Fear and adaptability in poultry: insights, implications and imperatives. *World's Poultry Science* 52, 131–174.

Jones, R.B. (2001) Does occasional movement make pecking devices more attractive to domestic chicks? *British Poultry Science* 42, 43–50.

Jones, R.B. and Waddington, D. (1992) Modification of fear in domestic chicks, *Gallus gallus domesticus*, via regular handling and early environmental enrichment. *Animal Behaviour* 43, 1021–1033.

Jones, R.B., Harvey, S., Hughes, B.O. and Chadwick, A. (1980) Growth and the plasma-concentrations of growth hormone and prolactin in chicks: effects of environmental enrichment, sex and strain. *British Poultry Science* 21, 457–462.

Jones, R.B., Carmichael, N.L. and Rayner, E. (2000) Pecking preferences and pre-dispositions in domestic chicks: implications for the development of environmental enrichment devices. *Applied Animal Behaviour Science* 69, 291–312.

Kempermann, G., Gast, D. and Gage, F.H. (2002) Neuroplasticity in old age: sustained fivefold induction of hippocampal neurogenesis by long-term environmental enrichment. *Annals of Neurology* 52, 135–143.

King, L.A. (2001) Environmental enrichment for broiler breeders. PhD thesis, University of Oxford, Oxford, UK.

Krause, E.T., Naguib, M., Trillmich, F. and Schrader, L. (2006) The effects of short term enrichment on learning in chickens from a laying strain (*Gallus gallus domesticus*). *Applied Animal Behaviour Science* 101, 318–327.

Langbein, J., Nurnberg, G., Puppe, B. and Manteuffel, G. (2006) Self-controlled visual discrimination learning of group-housed dwarf goats (*Capra hircus*): behavioural strategies and effects of relocation on learning and memory. *Journal of Comparative Psychology* 120, 58–66.

Leone, E.H. and Estevez, I. (2007) Patch discrimination as a function of handling time. *Poultry Science* 86, 447–452.

Leone, E.H. and Estevez, I. (2008) Economic and welfare benefits of environmental complexity for broiler breeders. *Poultry Science* 87, 3–13.

Lewis, M.H. (2004) Environmental complexity and central nervous system development and function. *Mental Retardation and Developmental Disabilities Research Reviews* 10, 91–95.

Leyendecker, M., Hamann, H., Hartung, J., Kamphues, J., Neumann, U., Surie, C. and Distl, O. (2005) Keeping laying hens in furnished cages and an aviary housing system enhances their bone stability. *British Poultry Science* 46, 536–544.

Lowndes, M. and Stewart, M.G. (1994) Dendritic spine density in the lobus parolfactorius of the domestic chick is increased 24-h after one-trial passive-avoidance training. *Brain Research* 654, 129–136.

Martrenchar, A., Huonnic, D. and Cotte, J.P. (2001) Influence of environmental enrichment on injurious pecking and perching in young turkeys. *British Poultry Science* 42, 161–170.

Mason, G., Clubb, R., Latham, N. and Vickery, S. (2007) Why and how should we use environmental enrichment to tackle stereotypic behaviour? *Applied Animal Behaviour Science* 102, 163–188.

McAdie, T.M., Keeling, L.J., Blokhuis, H.J. and Jones, R.B. (2005) Reduction in feather pecking and improvement of feather condition with the presentation of a string device to chickens. *Applied Animal Behaviour Science* 93, 67–80.

McBride, G. and Foenander, F. (1962) Territorial behaviour in flocks of domestic fowls. *Nature* 194, 102.

McDaniel, C.D., Bramwell, R.K. and Howarth, B. (1996) The male contribution to broiler breeder heat-induced infertility as determined by sperm–egg penetration and sperm storage within the hen's oviduct. *Poultry Science* 75, 1546–1554.

Meehan, C.L. and Mench, J.A. (2007) The challenge of challenge: can problem solving opportunities enhance animal welfare? *Applied Animal Behaviour Science* 102, 246–261.

Mellen, J. and MacPhee, M.S. (2001) Philosophy of environmental enrichment: past, present, and future. *Zoo Biology* 20, 211–226.

Moinard, C., Statham, P., Haskell, M.J., McCorquodale, C., Jones, R.B. and Green, P.R. (2004) Accuracy of laying hens in jumping upwards and downwards between perches in different light environments. *Applied Animal Behaviour Science* 85, 77–92.

Moinard, C., Rutherford, K.M.D., Haskell, M.J., McCorquodale, C., Jones, R.B. and Green, P.R. (2005) Effects of obstructed take off and landing perches on the flight accuracy of laying hens. *Applied Animal Behaviour Science* 93, 81–95.

Newberry, R.C. (1995) Environmental enrichment – increasing the biological relevance of captive environments. *Applied Animal Behaviour Science* 44, 229–243.

Newberry, R.C. (1999) Exploratory behaviour of young domestic fowl. *Applied Animal Behaviour Science* 63, 311–321.

Newberry, R.C. and Blair, R. (1993) Perches ease handling of broiler chickens. In: Savory, C.J. and Hughes, B.O. (eds) *Proceedings of the 4th European Symposium on Poultry Welfare*. Universities Federation for Animal Welfare, Potters Barr, UK, pp. 270–271.

Newberry, R.C. and Estevez, I. (1997) A dynamic approach to the study of environmental enrichment and animal welfare. *Applied Animal Behaviour Science* 54, 53–57.

Newberry, R.C. and Estevez, I. (2006) Species specific enrichment – poultry. SCAW/FASS Conference: the Humane Care and Use of Agricultural Animals in Research, St. Louis, MO.

Newberry, R.C. and Hall, J.W. (1990) Use of pen space by broiler chickens: effects of age and pen size. *Applied Animal Behaviour Science* 25, 125–136.

Newberry, R.C. and Shackleton, D.M. (1997) Use of visual cover by domestic fowl: a venetian blind effect? *Animal Behaviour* 54, 387–395.

Newberry, R.C., Estevez, I. and Keeling, L.J. (2001) Group size and perching in young domestic fowl. *Applied Animal Behaviour Science* 73, 117–129.

Newman, S. and Leeson, S. (1998) Effect of housing birds in cages or an aviary system on bone characteristics. *Poultry Science* 77, 1492–1496.

Nicol, C.J. (1992) Effects of environmental enrichment and gentle handling on behaviour and fear responses of transported broilers. *Applied Animal Behaviour Science* 33, 367–380.

Oden, K., Vestergaard, K.S. and Algers, B. (2000) Space use and agonistic behaviour in relation to sex composition in large flocks of laying hens. *Applied Animal Behaviour Science* 67, 307–320.

Oden, K., Berg, C., Gunnarsson, S. and Algers, B. (2004) Male rank order, space use and female attachment in large flocks of laying hens. *Applied Animal Behaviour Science* 87, 83–94.

Okelo, R.O., Carr, L.E., Harrison, P.C., Douglass, L.W., Byrd, V.E., Wabeck, C.W., Schreuders, P.D., Wheaton, F.W. and Zimmermann, N.G. (2003) Effectiveness of a novel method to reduce heat stress in broilers: a cool roost system. *Transactions of the ASAE* 46, 1675–1683.

Pamment, P., Foenander, F. and McBride, G. (1983) Social and spatial-organization of male-behaviour in mated domestic fowl. *Applied Animal Ethology* 9, 341–349.

Pettit-Riley, R. and Estevez, I. (2001) Effects of density on perching behaviour of broiler chickens. *Applied Animal Behaviour Science* 71, 127–140.

Pettit-Riley, R., Estevez, I. and Russek-Cohen, E. (2002) Effects of crowding and access to perches on aggressive behaviour in broilers. *Applied Animal Behaviour Science* 79, 11–25.

Pham, T.M., Winblad, B., Granholm, A.C. and Mohammed, A.H. (2002) Environmental influences on brain neurotrophins in rats. *Pharmacology Biochemistry and Behaviour* 73, 167–175.

Rampon, C., Jiang, C.H., Dong, H., Tang, Y.P., Lockhart, D.J., Schultz, P.G., Tsien, J.Z. and Hu, Y.H. (2000) Effects of environmental enrichment on gene expression in the brain. *Proceedings of the National Academy of Sciences of the USA* 97, 12880–12884.

Reilly, W.M., Koelkebeck, K.W. and Harrison, P.C. (1991) Performance evaluation of heat-stressed commercial broilers provided water-cooled floor perches. *Poultry Science* 70, 1699–1703.

Roper, T.J. and Marples, N.M. (1997) Odour and behaviour as cues for taste-avoidance learning in domestic chicks. *Animal Behaviour* 53, 1241–1250.

Rosenzweig, M.R. and Bennett, E.L. (1996) Psychobiology of plasticity: effects of training and experience on brain and behaviour. *Behavioural Brain Research* 78, 57–65.

Rosenzweig, M.R., Love, W. and Bennett, E.L. (1968) Effects of a few hours a day of enriched experience on brain chemistry and brain weights. *Physiology and Behaviour* 3, 819–825.

Sherwin, C.M., Lewis, P.D. and Perry, G.C. (1999) Effects of environmental enrichment, fluorescent and intermittent lighting on injurious pecking amongst male turkey poults. *British Poultry Science* 40, 592–598.

Shyne, A. (2006) Meta-analytic review of the effects of enrichment on stereotypic behaviour in zoo mammals. *Zoo Biology* 25, 317–337.

Skinner-Noble, D.O., Jones, R.B. and Teeter, R.G. (2003) Components of feed efficiency in broiler breeding stock: is improved feed conversion associated with increased docility and lethargy in broilers? *Poultry Science* 82, 532–537.

Tablante, N.L., Estevez, I. and Russek-Cohen, E. (2003) Effect of perches and stocking density on tibial dyschondroplasia and bone mineralization as measured by bone ash in broiler chickens. *Journal of Applied Poultry Research* 12, 53–59.

Tarou, L.R. and Bashaw, M.J. (2007) Maximizing the effectiveness of environmental enrichment: suggestions from the experimental analysis of behaviour. *Applied Animal Behaviour Science* 102, 189–204.

Tauson, R. (1998) Health and production in improved cage designs. *Poultry Science* 77, 1820–1827.

Tullett, S.G. (1987) Egg shell formation and quality. In: Wells, R.G. and Belyavin, C.G. (eds) *Poultry Science Symposium 20, Egg Quality – Current Problems and Recent Advances*. Butterworths, Boston, pp. 123–146.

Turner, C.A. and Lewis, M.H. (2003) Environmental enrichment: effects on stereotyped behaviour and neurotrophin levels. *Physiology and Behaviour* 80, 259–266.

Turner, C.A., Lewis, M.H. and King, M.A. (2003) Environmental enrichment: effects on stereotyped behaviour and dendritic morphology. *Developmental Psychobiology* 43, 20–27.

Vallortigara, G. (1996) Learning of colour and position cues in domestic chicks: males are better at position, females at colour. *Behavioural Processes* 36, 289–296.

van de Weerd, H.A., Docking, C.M., Day, J.E.L., Breuer, K. and Edwards, S.A. (2006) Effects of species-relevant environmental enrichment on the welfare and productivity of finishing pigs. *Applied Animal Behaviour Science* 99, 230–247.

Vits, A., Weitzenburger, D., Hamann, H. and Distl, O. (2005) Production, egg quality, bone strength, claw length, and keel bone deformities of laying hens housed in furnished cages with different group sizes. *Poultry Science* 84, 1511–1519.

Wemelsfelder, F. and Birke, L. (1997) Environmental challenge. In: Appleby, M.C. and Hughes, B.O. (eds) *Animal Welfare*. CAB International, Wallingford, UK, pp. 35–47.

CHAPTER 16
Ratites, Game Birds and Minor Poultry Species

D.C. Deeming

Department of Biological Sciences, University of Lincoln, UK

ABSTRACT

There is considerable emphasis on understanding the breeding biology of domestic fowl, turkeys and duck but not of the other commercially significant species, such as ratites, game birds, guinea fowl and pigeons. In general, our understanding of breeding in these species is poor relative to that of commercial poultry. The only exception is the ostrich, which has been the subject of numerous studies of breeding behaviour, fertility and incubation, both in the wild and in captivity. Other ratite species have received less attention. Despite the economic importance of game birds in Europe and North America, especially pheasant and partridge, studies on reproduction in these species are mainly limited to a few investigations into factors affecting fertility. Improving our understanding of guinea fowl breeding would greatly assist in maximizing production in its native Africa. Of the species described in this chapter, the pigeon is of note because its pattern of reproduction is reliant on long-term pairing of birds and their role in the rearing of the altricial offspring. The paucity of information made it difficult to derive general trends between species in their reproductive biology. However, it was clear that to maximize productivity many of these species of birds needed to be kept as pairs or at least in small groups. This contrasts with industrial poultry production, where group sizes are typically in the thousands. Another aspect of the breeding biology of these minor species is a general lack of breeding technology, e.g. artificial insemination, in egg production. There is considerable scope for further research into these species, which will yield information of both scientific and commercial interest.

INTRODUCTION

Commercial exploitation of birds has typically been associated with a few domesticated species, which usually have a long history (Wood-Gush, 1959;

  Biology of Breeding Poultry (ed. P.M. Hocking)

Appleby *et al.*, 1992). Cooper (1995) reviewed domesticated species in light of sustainable food production and considered those other avian species that are considered as potential sources of human food. Many of these were galliformes or waterfowl, but if they were to be developed then it was considered likely they would remain niche markets in restricted parts of the world. Typically, cultivation of these bird species is localized, small scale and extensive with little control of breeding of individuals (Cooper, 1995). Nevertheless, there are other species of bird that are not widely exploited as food but are (or are potentially) commercially important. Much of this volume is dedicated to breeding in commercial poultry and waterfowl so this chapter focuses on these alternative species, i.e. ratites, various galliformes and pigeons, which are utilized for food, providing meat, eggs or both.

The objective of this chapter is to collate information on breeding biology of these species in an attempt to allow comparison with descriptions of chickens, turkeys and waterfowl included elsewhere in this volume. Unfortunately, the fact that these species do not command considerable financial influence, or that they are relatively recent additions to our plate, means that in most instances relatively little is known about their breeding biology in a commercial environment. However, for some species, e.g. the ostrich, we have a relatively good understanding of the breeding biology. Space precludes a comprehensive review of information about every species but the intention is to provide details of aspects of the breeding biology that are of wider interest, but for some species good scientific studies are rare. It is also an objective to balance out the descriptions of the different species rather than allow any particular species to dominate. This summary will inevitably highlight how little we know and so hopefully stimulate further research into all of these fascinating species.

RATITES

Breeding biology of ratites in the wild is relatively well described (for a general account of all species see Davies, 2002) and especially so for the ostrich (*Struthio camelus*), which is particularly well studied (Bertram, 1992; Deeming, 1999). Ratites have been of commercial importance for around 150 years, although their popularity has fluctuated over that time. The key species are the African ostrich, Australian emu (*Dromaius novaehollandiae*) and the South American greater rhea (*Rhea americana*), although the lesser rhea (*Pterocnemia pennata*) is also gaining prominence in South America. Management of breeding depends on the farming system employed and the location where the birds are being farmed, which is not necessarily in their native lands.

Ostriches

The natural breeding system of the ostrich is well understood and described by Cramp (1977) and Bertram (1992). It is essentially based around males establishing and holding territories, to which they attract females. A major hen will establish a pair bond with the male, but this does not prevent the female from laying eggs in other pairs' nests, or the male copulating with other females that visit his territory. Visiting females can also lay eggs in the male's nest. Incubation is shared and is continuous for 42 days. Chicks are precocial and leave the nest within 48 h of hatching; the formation of a crèche of clutches of chicks is common when pairs of birds escorting chicks encounter each other.

Both the reproductive anatomy and spermatogenesis of the ostrich have been summarized by Soley and Groenewald (1999) and reproductive physiology by Skadhauge and Dawson (1999). Since these excellent reviews there has been little additional information reported in these fields and so they are not described here. Artificial insemination has been developed in the ostrich (Soley and Groenewald, 1999), and modern techniques for assessment of fertility are also used in the ostrich. Assessing fertility by determining the number of holes in the inner perivitelline membrane (see Wishart, Chapter 10, this volume) made by sperm attempting fertilization has proved difficult (Malecki and Martin, 2005). By contrast, assessment of the number of sperm trapped in the outer perivitelline layer (OPVL) has proved useful; duration of sperm storage in sperm storage tubules is estimated to be between 17 and 28 days (Malecki *et al.*, 2004).

Ostrich breeding in a commercial situation attempts to exploit the natural breeding biology in two different ways, based around the group size of the birds. The first approach is to have large numbers of birds living in a large enclosure (hundreds of hectares) where breeding follows a semi-natural pattern. Popular in South Africa and Israel, these enclosures hold 150–200 birds, with an overall ratio of two females per male, and mate selection is natural (Hicks-Alldredge, 1996). Eggs are collected by farm labourers, who search the breeding enclosures. Any eggs found are returned for artificial incubation and can be at any stage of development when placed in an incubator.

The second approach requires much smaller groups, most often trios of two females and one male or pairs of birds. Huchzermeyer (1998) suggests that only breeding in pairs allows for selection of breeding stock, although it is also possible if, on the basis of size and physical characteristics, eggs can be attributed to individual females in a trio (Deeming, 1996). Pairs in South Africa are kept in 0.5 ha enclosures with a shaded nest site and many are left to incubate their own eggs.

The approach of using small groups of birds tends to be very common in areas outside Africa. Despite a pair of birds being favoured in South Africa as the ideal system, during the increased interest in ostrich farming around the world in the 1990s and early 2000s (see Deeming, 1999, 2002) the trio of ostriches (two females and one male) was promoted as being the ideal breeding group to maximize egg and chick production. Pen size is often relatively small (1000–2500 m^2) and males and females are often selected as breeding animals more or less at random. However, the social interaction of birds is very important. Deeming (1996) showed that on a British farm fertility and hatchability of ostrich eggs was not only related to group size but also to how birds were placed together. Two trios of ostriches were known to have selected their own partners whilst in a larger group, and these two pens produced high fertilities (94 and 98% for the two females in one pen, and 89 and 100% in the other). By contrast, another two pens with trios that had not selected their partner and had been randomly placed together had fertilities during the first part of the season of 30 and 0%. During the same laying season, these latter pens converted to pairs of birds by default after one female in each pen died part-way through the season, after which fertility increased to 90 and 56% respectively (Deeming, 1996, 1998).

In general, fertility and hatchability of ostrich eggs in commercial operations are poor (Deeming and Ar, 1999; Navarro and Martella, 2002). Fertility can vary from poor (<50%) to good (85%); reports above 90% (e.g. Deeming, 1996) are uncommon and are associated with particular pairs of birds. Typically, hatch of eggs set is no more than 60%, in part due to low fertility. However, hatch of fertile eggs is typically no more than 70% but is often much less. Such values severely limit the commercial productivity of ostrich farms.

Despite these observations, it was shown by Lambrechts *et al.* (2004) that sex ratio was less important than stocking density. All aspects of both reproductive performance and reproductive behaviour were adversely affected by increases in stocking density (large flocks ranged from 114 to 210 birds/ha and smaller flocks ranged from 9 to 13 birds in 0.13 ha and 0.30 ha camps, respectively). This study found that highest production was from groups of three females with one male, but it was not clear whether this effect was on a per pen or a per female basis. Cloete *et al.* (1998) showed that egg production of pens with pairs was lower than that of pens holding trios, but per female production was lower in the trios.

During a study of the time budget of breeding ostriches on a British farm, McKeegan and Deeming (1997) observed that ostrich behaviour was affected by the presence of people. Observations that gave representative values for the time budget had to be carried out from positions at least 25 m from the birds. This interaction with humans was shown by Bubier *et al.* (1998) to have an important role to play in courtship in ostriches maintained on a British farm. Both male and female ostriches exhibited courtship behaviours towards people standing next to the fence of the birds' enclosure. These behaviours were absent or reduced in frequency when birds were observed from a location where people could not be seen. There was a tendency for those females kept in enclosures with males that exhibited high frequencies of sexual behaviours towards people to produce fewer fertile eggs. Other behavioural problems have been observed in female ostriches on farms but it is unclear whether these affected production (Sambraus, 1994; Deeming and Bubier, 1999). Spray marking one female bird in a trio affects social behaviours, with the male spending more time with the unmarked bird and exhibiting more agonistic behaviour towards the marked bird (Deeming and Bubier, 1999).

Other factors affecting production may have a dietary origin. In South Africa males in pairs that exhibited a history of low productivity consumed more concentrate feed but mated less frequently (Lambrechts and Cloete, 1998). Supplementation of the diet of breeding ostriches with L-carnitine-magnesium (a co-enzyme associated with lipid metabolism) increased courtship activity in both sexes and mating in males (Lambrechts *et al.*, 1998a). This increased egg production and hatchability but only in males greater than 5 years of age (Davis *et al.*, 1998).

Ostrich breeding has involved a degree of improvement of stock from the start of farming operations in the 19th century. Breeding experiments were initially aimed at improving feather quality (see Petitte and Davis, 1999), but the collapse of the feather market around the time of the First World War (see Deeming, 2002) meant that genetic improvement was a low priority. This has changed over recent years, with studies into phenotypic variability increasingly being of interest, particularly at Oudtshoorn, South Africa. Van Schalkwyk *et al.* (1996) found that reproductive traits and body weight of average yearly production of 42–67 mixed-age breeding pairs were highly variable, but this was in part attributed to moderate repeatability of reproduction of pairs from one year to the next. Reproductive traits were moderately repeatable from year to year (from 0.38% for hatchability to 0.51% for embryonic deaths). Since this account was published there have been other reports of studies into factors that should be considered when developing improvements in ostrich breeding (Cloete *et al.*, 1998, 2006; Lambrechts *et al.*, 1998b; Brand *et al.*, 2005; Fair *et al.*, 2005). These reports are all derived from South Africa and are largely based on data collected from many of the same pairs of breeding birds.

It is interesting to note that the age of the breeding female is critical in her reproductive performance, with 2-year-old birds breeding for the first time having significantly lower egg production than older birds (Fair *et al.*, 2005). Moreover, breeding performance increases up to 9 years of age and declines thereafter (Cloete *et al.*, 1998, 2006; Fair *et al.*, 2005). Brand *et al.* (2005) investigated whether parentage affected reproduction in pure-bred, South African, domesticated 'black' ostriches and more wild-type, Zimbabwean 'blue' ostriches. Black ostriches had higher egg production and hatchability irrespective of the sire, although there were no statistically significant effects of strain on fertility. Studies into the genetic improvement of ostriches in parts of the world other than South Africa are lacking. This almost certainly reflects the current level of scientific investment in ostrich farming in countries outside South Africa.

Emus

Our understanding of the breeding biology of emus is largely based on studies of wild birds (see Davies, 2002), but general accounts of emu breeding on farms in the USA are provided by Minnaar and Minnaar (1992) and Brackett (1995). Generally, the management of emus followed the pattern that was adopted by ostrich farmers: keeping birds in pairs or trios and using artificial incubation. Egg production is every 3 days and, as is typical for ratites, incubation is typically for 56–60 days and is conducted solely by the male (see Sales, 2007 for a more detailed description). During the breeding season there is a significant decline in weight of both males and females due to low feed intake during incubation and laying, respectively (Williams *et al.*, 1998). The important aspect of emu breeding is that reproduction is seasonal and only takes place in the winter months, i.e. during short days (Williams *et al.*, 1998). For these reasons the control of reproduction has been relatively well studied over recent years.

Photoperiod controls breeding in this species, with a dissipation of photorefractoriness by short days, leading to an increase in gonadotrophin secretion to levels sufficient to initiate breeding (Malecki *et al.*, 1998; Blache *et al.*, 2001; Sharp *et al.*, 2005). Hence, the process that controls reproduction in the emu is similar to that observed in avian species that breed during long days. Termination of breeding may be associated with a delayed increase in prolactin secretion (Sharp *et al.*, 2005).

Compared with poultry species, the factors affecting the process of fertilization are poorly understood in ratites, but they have been extensively studied in the emu (Malecki *et al.*, 1997a,b, 2000, 2005a,b; Malecki and Martin, 2002a,b; Martin and Malecki, 2004). Assessing fertility by counting the number of sperm holes in the inner perivitelline membrane is possible, but assessment of the number of sperm trapped in the outer perivitelline layer (OPVL) has proved more useful in assessing fertility (Malecki and Martin, 2005). In this species insemination can be on a weekly basis and persistence of sperm in sperm storage tubules averages 16.5 days (Malecki and Martin, 2005). Techniques for artificial insemination have been developed in the emu (Malecki *et al.*, 1997a,b) but it is unclear to what extent they are used in an industrial context.

Rheas

The breeding biology of wild greater rheas has been the subject of recent interest, and various aspects of reproduction and nesting in their natural habitat are well described (Bruning, 1974; Fernández and Reboreda, 1995, 1998, 2000, 2002, 2003, 2007; Fernández and Mermoz, 2003). In general, males control harems of females, who lay their eggs in a single nest. Incubation is around 39 days and chick rearing is by the male (Navarro and Martella, 2002; Sales, 2006). Rhea farming in the USA and Europe is again based on the ostrich model (see above), but in South America other approaches are being investigated, including semi-intensive ranching of birds (Navarro and Martella, 2002). The level of study of the rhea in commercial rearing is much lower than that for the ostrich, but Sales (2006) has undertaken a general review of rhea biology.

Bruning (1973) and Sales *et al.* (2000) have described the reproductive behaviour of rheas in captivity. In a farming environment, reproductive behaviour forms a low proportion of the overall time budget (6.6% for males and 1.6% for females), with most activity taking place during the early afternoon (Sales *et al.*, 2000). In this study the sex ratio was seven females per six males, although higher ratios are reported, e.g. 19 females per four males (Lábaque *et al.*, 2005). There are few data on the appropriate group size or sex ratio in greater rheas, and records of fertility are not common. Lábaque *et al.* (2005)

reported fertility of one batch of eggs as 93%, of which 75% hatched. Egg production in greater rheas was related to management system (Navarro and Martella, 2002). Rheas kept on farms laid more eggs than birds maintained under semi-captive conditions, and eggs laid on farms had a higher hatchability than those from semi-captive or wild birds. Even so, hatchability of rhea eggs was generally low, averaging less than 70% (Navarro and Martella, 2002).

The lesser rhea is a relative newcomer to commercial exploitation, with initial reports dating from 1994 (Sales, 2006), and it is therefore relatively poorly understood. Sarasqueta (2005) described breeding of a small group of this species in a farmed captive environment in Argentina. Breeding was during the period from April to December (southern hemisphere), when birds were kept in pens measuring 20 m × 40 m with three females per male. Sexual maturity was typically achieved between 20 and 24 months and some females were reproductively active at 14 years of age. In Sarasqueta's (2005) study the average period of laying was 137 days (SD = 23.6) and average egg production was 32.6 (SD = 10.6). Males incubate the eggs in natural nests (although other eggs were incubated artificially), when they rely on body reserves rather than feeding. Fertility and hatchability were only reported for two females: at 75 and 63%, fertility was relatively low but comparable to data from other group-breeding ratites (see above). Hatchability of fertile eggs was also low, at 56 and 68%. Reasons for the failure of eggs to hatch were not reported. Chang-Reissig *et al.* (2004) studied fertility and hatchability of lesser rhea eggs on farms in Argentina. Fertility averaged 76% but was highly variable between farms (range of 33–93%). Hatch of fertile eggs was more variable, ranging from 8 to 94%, with a mean of 67%. The reasons for this variation were not clear, but it was suggested that the experience of staff on the farms was an important issue.

A review of productivity of lesser rhea farms by Navarro and Martella (2002) found that both egg production per female and hatchability (of eggs set) was highly variable. A variety of management recommendations were made in light of these findings. First, male density should be lower than three birds per hectare. Second, a 1:1 sex ratio will give the best egg production when natural incubation is practised, but more females per male is possible in small operations or when artificial incubation is used. Other reports of farming of this species are available but sporadic. The group size of breeding birds was variable, ranging from four birds in 2400 m^2 to 78 birds in 19,200 m^2 (Navarro *et al.*, 2003).

There is one report (Delsuc *et al.*, 2007) of hybridization between a male *Rhea* and a female *Pterocnemia*. Two eggs hatched and were confirmed to be molecular hybrids. It is unknown whether the female birds were fertile and capable of breeding with either rhea species.

Incubation

Commercial exploitation of ratites relies on artificial incubation, which is a process that is similar to that employed for poultry and waterfowl with some important differences. Compared with poultry, ratites lay big eggs, weighing

500–600 g for emus and rheas (Deeming, 1997) and an average of 1500 g for ostriches, although the range can be from 1000 to 2000 g (Deeming and Ar, 1999).

Large egg size means that incubators need to be modified to accommodate the eggs, but critically this affects the incubation temperature the eggs require. Commercial experience has shown that in multi-stage incubation these eggs require a lower set-point than that used for poultry eggs. Therefore, ratite eggs are incubated at 36.0–36.5°C rather than the 37.5–37.8°C commonly used for poultry. This does not mean that ratite embryos require a lower incubation temperature but rather indicates that the incubators are unable to remove metabolic heat generated by these large eggs at a sufficient level to ensure that egg temperature does not increase to a suboptimal or lethal level. Incubation at lower temperature than that required for poultry keeps embryonic temperature during the last third of incubation within physiologically acceptable limits. For ostrich eggs, van Schalkwyk *et al.* (1999) showed that at a constant temperature of 37.3°C hatchability was only half of that shown by eggs incubated at 36.0 or 36.5°C (33.8, 63.3 and 60.0% of set, respectively). However, if temperatures can be controlled according to embryonic age, as is the case for single-stage incubation, then incubation at higher temperatures during the first 10 days of incubation has no adverse effect on development (Deeming and Ayres, 1994; Deeming, 1995a). The lower temperatures employed in multi-stage incubation are a compromise that allows sufficient development early in incubation but ensures that egg temperature does not become lethal towards the end of development.

Whilst ratite eggs are subject to most factors that affect hatchability (e.g. dehydration due to excessive weight loss), ostrich eggs are particularly affected by the problem of relatively low porosity eggshells. This means that humidity during incubation has to be low relative to that typically seen for poultry (i.e. <30% RH versus 55–60% RH, respectively). Practically, this has limited ostrich hatchability in those hatcheries that have been unable to maintain a low ambient humidity. Low incubation weight loss reduces hatchability because of shell-induced hypoxia (Tullett and Deeming, 1982; Burton and Tullett, 1983), and the excess water in eggs also causes subcutaneous oedema in near-term embryos, which limits their ability to externally pip in the appropriate place. Deeming (1995b) showed that eggs with lower weight losses pipped nearer the pole of the egg than eggs with higher rates of weight loss.

GALLIFORMES

Pheasants

Originally from the Far East, the pheasant (*Phasianus colchicus*) has been bred for sport for several hundred years in Europe and North America and continues to be commercially important. Poults are generally bred for release into the local environment prior to the shooting season. It is estimated that around 20 million pheasants are released annually in Great Britain alone (Tapper, 1999).

Pheasants are typically kept in large, predominantly female, breeding groups numbering between 50 and 300 birds with a range of 7 to 12 females per male. The size of the pen can vary according to the number of birds to be accommodated but between 4.5 and 6 m^2 should be available for each bird (Game Conservancy Trust, 1993). The choice of the site of the pen is important to minimize long-term disease problems. An alternative approach is to have small fixed or movable pens that accommodate one male and six females. Egg production in these smaller pens is generally better than in large communal pens (Game Conservancy Trust, 1993).

Pheasant production in Great Britain is not very efficient and is highly variable (Table 16.1; Deeming, unpublished observations). Hatchability of eggs set is less than 70%, with losses due to infertility averaging 12.2% (Table 16.1). Losses during the early part of incubation average 4%, which is in line with levels observed in broiler chickens (Deeming and van Middelkoop, 1999), red-legged and grey partridges (Deeming, unpublished observations) and ostriches (Deeming, 1995a). Mid-term mortality (5–20 days) was 4.9%, and late mortality was high, at 10.4% (Table 16.1). The variability in results was related to individual flocks but was also the result of the incubation environment the eggs experienced, which in many cases was far from ideal (Deeming, personal observations). Despite the commercial significance of pheasants and the number of birds released annually, there are very few systematic studies into breeding biology of this species in an attempt to improve these results.

There appears to be little consensus or understanding of the optimal breeding ratio, which can vary from five females per male up to 12 or 16 females per male. Deeming and Wadland (2001) showed that in groups of 250 breeding females a sex ratio of 12 females per male produced consistently and significantly lower fertility than a sex ratio of eight females per male – the average difference was 4% of eggs over a 10-week laying period. It was suggested that the additional males allowed for a higher recruitment and mating of females. More males in a breeding pen increased egg production over the first 5 weeks of laying, and so there was an increase in production of fertile eggs of 10% over the laying season.

Egg quality is critical in determining hatchability in poultry but this is often difficult to quantify. In pheasants, shell colour is typically olive-brown but a

Table 16.1. Summary of productivity of pheasant eggs on British game farms (n = 13) based on dead-in-shell analysis of unhatched eggs of 32 flocks (D.C. Deeming, unpublished data).

	Mean %	SD	Range %
Hatchability of eggs set	67.3	8.2	49.3–87.4
Infertile eggs	12.2	4.6	0.2–20.3
Embryonic mortality			
1–4 days	4.0	1.8	0.1–7.8
5–20 days	4.9	2.1	2.0–10.5
21–25 days	10.4	5.3	2.9–23.9

significant number of eggs are blue in colour (e.g. see Deeming and Wadland, 2001). This colour difference is associated with eggshell morphology: blue shells are thinner and have structural defects (Richards and Deeming, 2001). Low hatchability in these eggs is associated with higher rates of weight loss and there is a significant negative relationship between age of embryonic death and percentage weight loss at 7 days of incubation (Richards and Deeming, 2000).

Partridge

Partridge are reared for game shooting in many countries in Europe, the Middle East and the USA. Three main species are involved: the grey partridge (*Perdix perdix*), the red-legged partridge (*Alectoris rufa*) and the Chukar (*Alectoris chukar*); the rock partridge (*Alectoris graeca*) is also released as a game bird in parts of Europe (Kirikçi *et al.*, 2004). Red-legged and Chukar partridges readily hybridize, producing fertile offspring (Mullin and Boehmer, 1991). However, restrictions on species that can be released within Britain mean that Chukar and hybrids are not found in the UK. Chukar are a popular game bird in areas where they are native (Asia Minor and the Middle East) and in the USA (Mullin and Boehmer, 1991). In each of these species eggs are collected for artificial incubation and chicks are reared intensively prior to release.

Grey partridge can only be kept as pairs, typically in A-framed pens maintained on grass, which have to be moved regularly to minimize disease problems (Game Conservancy Trust, 1993). Battery-type wire cages may also be used. Egg production and fertility are higher if pairs are compatible (Lupo *et al.*, 1990). Forced mating can produce aggression if birds are incompatible. In these cases the female should be removed from the pen and mated with another male (Mullin and Boehmer, 1991).

Red-legged partridge are typically accommodated as pairs in wire-floored pens typically mounted as batteries of six pens (Game Conservancy Trust, 1993). Each pen has covered areas for nesting and to protect the food hopper. Wire floors can reduce disease problems but increase the proportion of cracked eggs if the hen lays outside the nest area. An alternative system is the use of A-frame pens on grass, which can be moved around as the ground becomes soiled. Flock mating of red-legged partridge is limited to relatively small groups of 21 females to 7 males, and maximum flock size should not exceed 40–50 birds. An advantage of flock systems is that it minimizes the adverse effect of incorrectly sexed birds on egg production and fertility, but aggression between individuals needs to be managed by the provision of good-quality cover. Cabezas-Días *et al.* (2005) studied the effect of female age on reproductive performance. Many traits were highly repeatable between the 3 years of the study, although the repeatability of egg mass was relatively low. The oldest birds (third laying season) laid slightly smaller eggs but had larger clutch sizes.

Chukar partridges can be maintained as pairs or trios, or kept in colonies of 500 birds (400 females and 100 males) (Mullin and Boehmer, 1991). However, egg production is lower in colony systems (35–40 eggs per bird per season) than for pair-mated birds (55–60 eggs). Chukar partridges are maintained in wire-bottomed pens similar to those used for red-legged partridge.

Generally, large-scale production of partridges relies on caged pairs of birds. The significance of mate choice in the red-legged partridge was tested by Bottoni *et al.* (1993). Some pairs of birds were allowed to self-select their mates whereas the control birds were placed together in pairs. Males in self-selected pairs had higher levels of testosterone during the breeding season, although there were no significant effects on oestradiol or luteinizing hormone. Calling by males and nest guarding by females were both higher in the self-selected pairs. Egg production was significantly higher (25.9 ± 3.6 versus 16.9 ± 3.3), and wasted scattered eggs significantly lower (1.8 ± 1.0 versus 6.5 ± 2.0), in the self-selected birds. If the birds chose their own mate then in 50% of cases they established two nests and almost all females developed brood patches. In the control birds that were randomly placed together in cages, second nests were not observed and only two out of nine females developed brood patches. Lupo *et al.* (1990) reported similar results for grey partridges, with nesting incubation and hatchability significantly higher in self-selected pairs maintained in semi-natural conditions. Self-selected pairs of birds were also more frequently involved in agonistic behaviours (Beani *et al.*, 1988).

Egg production in red-legged partridge is dependent on pen size (Gaudioso *et al.*, 2002). Pairs of birds kept in 8 m^2 pens laid an average of 27 eggs compared with only 14 eggs for females maintained in pens with an area of 4 m^2. By contrast, there was no effect of rearing environment on the production of eggs. Fertility of red-legged partridge eggs has been shown to be seasonal in southern Spain (González-Redondo, 2006). Fertility was highest from mid-February to mid-March and declined thereafter. Hatch of eggs set was high over the same period but declined thereafter. Hatch of fertile eggs showed less variation over the season and averaged around 83%.

Quail

The two species of quail commonly found in captive conditions are the Japanese quail (*Coturnix coturnix japonica*) and the bobwhite quail (*Colinus virginianus*). The former species have been domesticated in Japan for several centuries and have now been developed into various strains (Ottinger and Rattner, 1999). Bobwhite quail are mainly found on game farms in the USA and are only partially domesticated. Common in the laboratory, where it is used in reproductive studies (Ottinger and Rattner, 1999), there have been few studies of quail reproduction under commercial conditions.

Breeding systems for quail in North America are described by Mullin and Boehmer (1991). Adult Japanese quail are maintained as small groups with two or three females per male or in larger colonies where there are between three and five females per male (Ernst, 1978). For bobwhite quail, three females and one male can be kept in pens measuring 30 cm × 45 cm, but one female paired with one male produces the most eggs per hen. Colony breeding can also be practised in this species, with between two and four females per male.

However, Mullin and Boehmer (1991) did not suggest for either species how birds should be kept in a colony. Under laboratory conditions birds are typically maintained as pairs (Ottinger and Rattner, 1999).

Both Japanese and bobwhite quail are highly photosensitive, and if maintained on long daylengths they rapidly achieve sexually maturity (15 h and 18 h for Japanese and bobwhite quail, respectively; Ottinger and Rattner, 1999). This results in a continuous production of eggs, which may have adverse effects on the health of the laying hen through premature ageing.

Guinea fowl

Guinea fowl (*Numidia meleagris*) has a long history of domestication in Africa, with the first historical record in an Egyptian mural (2400 BC) and representations of the birds on a Greek urn from the 6th century BC. The Romans considered both guinea fowl eggs and birds as delicacies (Appleby *et al.*, 1992). In Africa, the guinea fowl is an abundant, semi-domesticated species and is an important source of animal protein (Ogwuegbu *et al.*, 1988). Commercial exploitation of guinea fowl is geographically limited, with a well-established industry in France and Italy that is absent elsewhere in the world (Butler, 1997). Small-scale breeding is based around a pair or trio of birds and breeding is often by natural incubation (Butler, 1997). In Nigeria the sex ratio is recommended as four females per male, although fertility under laboratory conditions was only 74% of eggs laid (Offiong and Abed, 1980).

Guinea fowl are common in their native Africa, where they are reared extensively on a free-range basis (Cooper, 1995). In their native land this species has many advantages over introduced chickens, including low production costs, premium-quality meat, greater capacity to scavenge for insects and grains, better ability to protect itself against predators and better resistance to common poultry parasites and diseases (Saina, 2005). Unfortunately, Saina (2005) concluded that production of guinea fowl in Zimbabwe was inefficient due to a variety of management issues. In one survey over a 5-month laying season, it was found that egg production was only 42 eggs per hen whereas hatchability was only 71%. Saina (2005) concluded that there was scope for significant improvements in productivity if management practices could be improved. In Nigeria guinea fowl breeding has a distinct season between April and August, when the highest rainfall occurs and it is slightly cooler than the out-of-lay season (Ogwuegbu *et al.*, 1988). However, at the peak of laying, average egg production was only 12 eggs per month per female.

Industrial-scale farming of guinea fowl occurs outside Africa. In France guinea fowl are essentially managed in the same way as poultry (le Coz-Douin, 1994). There are generally three females per male, although this improves to 16 females per male with artificial incubation (J.P. Brillard, 2007, personal communication), the process of which is similar to that used for turkeys (see le Coz-Douin, 1994 for more details).

Social and reproductive behaviour of guinea fowl was studied by Elbin *et al.* (1986) in free-ranging groups at two sites in the USA and a third site in

South Africa. Birds spent time in mixed-sex groups during the non-breeding season but established several short-term male and female associations as the breeding season progressed. After approximately 5 weeks of such associations, a more permanent pair-bond was formed, which involved male courtship, feeding their mates and guarding them during egg laying and incubation.

PIGEONS

The domesticated pigeon, believed to derive from the rock dove (*Columba livia*), is a common laboratory bird but can also be used for meat production as well as for racing. Pigeons were eaten by the Egyptians, Greeks and Romans before poultry were introduced (Barnes, 2003). In Britain, dovecotes maintained by landlords were used to house free-flying birds so that they could be harvested for meat, and farm labourers were prevented from keeping pigeons. More recently, the 20th century has seen the commercial exploitation of pigeons for meat. Although not popular in Great Britain, pigeon meat is popular in the USA and some parts of Europe (Barnes, 2003).

Breeding in this species differs from other birds thus far considered in a variety of ways, mainly based around the altricial nature of the offspring. This means that chicks, or squabs as they are more commonly known, need to be reared by their parents, a situation further complicated by their reliance on 'crop milk' in the initial stages of post-hatching development. Squabs are ready for killing at around 28 days (thereafter, up to 7 weeks of age, the young are called 'squeakers'), when they can attain around 550 g (Aggrey and Cheng, 1993).

Pigeons are monomorphic and sexing of individuals is difficult. Behavioural differences are only obvious during mating, although examination of the cloaca can distinguish the two openings of the vas deferens or the single opening of the oviduct (Nepote, 1999). Typically monogamous, pigeons are capable of mating for life (Hawkins *et al.*, 2001), which has to be accommodated in any management practices. Breeding takes place during the spring and summer months (Nepote, 1999), when the birds are housed in breeding aviaries typically 2 m wide × 2 m long × 2 m high with an area on one wall providing an earthenware or papier mâché nesting pan. Pairings take place any time after moult is completed, typically after December in the northern hemisphere (Parsons, 1996). Males are generally introduced in an aviary first to reduce aggression. Alternatively, males and females can be introduced by placing them in small cages that use a wire mesh to separate the birds prior to release into an aviary. Under captive conditions nesting material has to be provided. Incubation of the two eggs exploits the natural behaviour of the adults, and chicks hatch after 18 days (Nepote, 1999). Pairs are capable of laying a clutch of eggs every 30 days, so there may be two active nests at a time: one with squabs being fed and the other being incubated (Barnes, 2003).

Crop milk is a highly nutritive substance produced by pigeons to feed to their squabs (see review by Sales and Janssens, 2003). It is rich in protein and lipids and is produced during the latter stages of incubation through to the 25th day post-hatching. Crop milk is the sole diet up to day 3 post-hatching, but as the squabs get older the adults mix the milk with grains soaked in the crop. By day 28 after hatching, production of crop milk has ceased and the squabs are encouraged by their parents to feed on grain alone (Aggrey and Cheng, 1993). Growth rates of squabs are maximal between 8 and 14 days post-hatching and maximal size is typically achieved at 28 days, although this can vary between individuals (Aggrey and Cheng, 1993).

Problems with breeding pigeons are described by Parsons (1996). They are mainly associated with infertility, microbial contamination of eggs due to poor nest hygiene and embryonic mortality: potential problems that afflict all species of birds. Lack of eggs may be related to poor sexing of birds, although infertility can be associated with a range of physical, nutritional and behavioural problems. Klint and Enquist (1981) showed that female pigeons allowed to choose their mates laid their first egg earlier, laid more eggs and had a higher fertility. Poor nesting environment with regard to temperature and light can lower fertility, as can excessive human disturbance. Micronutrient deficiencies and poor diet provision are also problematical, while there are potentially a variety of disease situations that reduce fertility.

The reliance on monogamous breeding pairs, small clutch size, natural incubation and feeding with crop milk rather limits commercial pigeon production. Production of 20 squabs per annum would be quite normal (Barnes, 2003), although such levels of production may have adverse effects on quality.

PERSPECTIVES IN BREEDING BIOLOGY OF MINORITY SPECIES

It has been disappointing not to have been able to do more of a comparative analysis of the breeding biology of non-poultry species but this brief review of the available literature has revealed that our understanding of these species is far from complete for any of these birds. Compared with the domestic fowl, the domestic turkey or even the Pekin duck, our knowledge of breeding biology is patchy and dependent on each species concerned. Hence, compared with other non-poultry species we have a relatively good knowledge of many aspects of breeding biology for the ostrich, but for most other species there has been little research. Hence, reports tend to be focused on aspects of the biology of the species that make them different from poultry, for instance mate choice in partridge. Alternatively, the emphasis has been on specific aspects of reproductive biology that have interested particular scientists. This has led to a rather disparate set of information, which yields only a few trends that could perhaps apply to all (or nearly all) of the species concerned. This is not to say that such trends do not exist but rather there is insufficient data for a wide enough variety of species to investigate whether trends exist.

A significant aspect of breeding in the species outlined in this chapter is that to maximize productivity many of the species need to be kept as pairs or at least in small groups. This contrasts with industrial poultry production, where group sizes are typically in the thousands. For ostrich, partridges and pigeons there is good evidence that highest productivity of eggs is achieved in pairs. Moreover, for fertility in particular, there is strong evidence that allowing the breeding birds to select their own mate produces the best results. Evidence of a positive effect of mate selection comes from a variety of physiological and behavioural perspectives in at least four species. The relatively conservative nature of reproduction in birds raises the question whether a causal understanding of the mechanisms involved in mate selection in flock-breeding species would be a profitable venture. In commercial broiler production in floor-mating systems, hatchability declines with flock age, mainly due to lower fertility (see, for instance, Deeming and van Middelkoop, 1999; Duncan, Chapter 8, this volume). Perhaps this is due to changes in mating frequency or changes in behaviour affecting mate selection. That mate choice correlates with high fertility in these minority species perhaps suggests that it would be worthwhile examining more closely mate choice in species that are kept as large flocks.

Another aspect of the breeding biology of the minor species described here is a general lack of breeding technology in egg production. In particular, artificial insemination (AI) has only been applied effectively in a commercial environment in guinea fowl (le Coz-Douin, 1994), although AI techniques have been developed for ostriches (Soley and Groenewald, 1999) and emu (see Martin and Malecki, 2004) as well for a variety of other bird species of little economic significance (see Gee *et al.*, 2004). Otherwise management of breeding populations of these species remains relatively unsophisticated, and lack of financial resources probably means that this will continue to be the case.

Breeding biology of the domestic fowl is probably the best understood for all bird species, yet for many wild species there are excellent accounts of breeding, albeit often limited to behavioural and other non-invasive methods of study. It is likely that the commercial species described in this chapter fall between these extremes, because they do not yield sufficient economic influence, or the domesticated state is not of any great biological interest, to justify significant research effort or funding. This means that production methods are always inefficient and breeding results are poor compared with poultry. However, what we do know about these species is fascinating and there is scope for further research. This would better inform scientists about the breeding biology of birds as a whole and also educate producers, which could have a significant commercial impact. For instance, improvements in the management of breeding guinea fowl could have important economic and social benefits in its native Africa by providing an increased availability of a valuable source of protein. Whether such research will be of interest to scientists and funding bodies remains to be seen.

REFERENCES

Aggrey, S.E. and Cheng, K.M. (1993) Genetic and posthatch parental influences on growth in pigeon squabs. *Journal of Heredity* 84, 184–187.

Appleby, M.C., Hughes, B.O. and Elson, H.A. (1992) *Poultry Production Systems: Behaviour, Management and Welfare*. CAB International, Wallingford, UK.

Barnes, J. (2003) *Commercial Pigeon Breeding and Rearing*. Beech Publishing House, Midhurst, East Sussex, UK.

Beani, L., Cervo, R., Lodi, L., Lupo, C. and Dessi-Fulgheri, F. (1988) Circulating levels of sex steroids and socio-sexual behaviour in the grey partridge (*Perdix perdix* L.). *Monitore Zoologico Italiano (NS)* 22, 145–160.

Bertram, B.C.R. (1992) *The Ostrich Communal Nesting System*. Princeton University Press, Princeton, Massachusetts.

Blache, D., Talbot, R.T., Blackberry, M.A., Williams, K.M., Martin, G.B. and Sharp, P.J. (2001) Photoperiodic control of the concentration of luteinizing hormone, prolactin and testosterone in the male emu (*Dromaius novaehollandiae*), a bird that breeds on short days. *Journal of Neuroendocrinology* 13, 998–1006.

Bottoni, L., Massa, R., Lea, R.W. and Sharp, P.J. (1993) Mate choice and reproductive success in the red-legged partridge (*Alectoris rufa*). *Hormones and Behaviour* 27, 308–317.

Brackett, V.E. (1995) Emu breeder management. In: Drenowatz, C. (ed.) *The Ratite Encyclopedia: Ostrich, Emu and Rhea*. Ratite Records Inc., San Antonio, Texas, pp. 183–188.

Brand, M.M., Cloete, S.W.P., Hoffman, L.C. and Muller, M. (2005) A comparison of live weights, body measurements and reproductive traits in Zimbabwean blue ostriches (*Struthio camelus australis*) and South African black ostriches (*S. camelus* var. *domesticus*). In: Carbajo, E. (ed.) *Proceedings of the 3rd International Ratite Science Symposium of the World's Poultry Science Association*, 14–16 October 2005, Madrid, pp. 73–80.

Bruning, D.F. (1973) Breeding and rearing rheas in captivity. *International Zoo Yearbook* 13, 163–172.

Bruning, D.F. (1974) Social structure and reproductive behavior of the greater rhea. *Living Bird* 13, 251–294.

Bubier, N.E., Paxton, C.M., Bowers, P. and Deeming, D.C. (1998) Courtship behaviour of farmed ostriches in relation to humans. *British Poultry Science* 39, 477–481.

Burton, F.G. and Tullett, S.G. (1983) A comparison of the effects of eggshell porosity on the respiration and growth of domestic fowl, duck and turkey embryos. *Comparative Biochemistry and Physiology* 75A, 167–174.

Butler, J. (1997) *Keeping Guinea Fowl*. Beech Publishing House, Midhurst, East Sussex, UK.

Cabezas-Díaz, S., Virgós, E. and Villafuerte, R. (2005) Reproductive performance changes with age and laying experience in the red-legged partridge *Alectoris rufa*. *Ibis* 147, 316–327.

Chang-Reissig, E., Terzolo, H., Armando, S. and Rogé, A. (2004) Hatching success and embryonic mortality on lesser rhea (*Pterocnemia pennata*) farms in northern Patagonia, Argentina. *British Poultry Science* 45, 471–475.

Cloete, S.W.P., van Schalkwyk, S.J. and Brand, Z. (1998) Ostrich breeding – progress towards a scientifically based strategy. In: *Ratites in a Competitive World. Proceedings of the 2nd International Ratite Congress*, September 1998, Oudtshoorn, South Africa, pp. 55–62.

Cloete, S.W.P., Bunter, K.L., Lambrechts, H., Brand, Z., Swart, D. and Greyling, J.P.C. (2006) Variance components for live weight, body measurements and reproductive traits of pair-mated ostrich females. *British Poultry Science* 47, 147–158.

Cooper, J.E. (1995) The role of birds in sustainable food production. *Biodiversity and Conservation* 4, 266–280.

Cramp, S. (1977) *Handbook of the Birds of Europe, the Middle East and North Africa (Birds of the Western Palearctic), Volume 1, Ostrich to Ducks*. Oxford University Press, Oxford, UK.

Davies, S.J.J.F. (2002) *Ratites and Tinamous*. Oxford University Press, Oxford, UK.

Davis, H.J., Lambrechts, H., Ganzevoort, B., Cloete, S.W.P., King, P.R., Davis, W.H., Blum, R. and Mienie, L.J. (1998) The influence of L-carnitine-magnesium supplement on the egg production and fertility of adult breeding ostriches. In: *Ratites in a Competitive World. Proceedings of the 2nd International Ratite Congress*, September 1998, Oudtshoorn, South Africa, pp. 68–71.

Deeming, D.C. (1995a) Factors affecting hatchability during commercial incubation of ostrich (*Struthio camelus*) eggs. *British Poultry Science* 36, 51–65.
Deeming, D.C. (1995b) The hatching sequence of ostrich (*Struthio camelus*) embryos with notes on development as observed by candling. *British Poultry Science* 36, 67–78.
Deeming, D.C. (1996) Production, fertility and hatchability of ostrich (*Struthio camelus*) eggs on a farm in the United Kingdom. *Animal Science* 63, 329–336.
Deeming, D.C. (1997) *Ratite Egg Incubation – a Practical Guide*. Ratite Conference, High Wycombe, Buckinghamshire, UK.
Deeming, D.C. (1998) Research on the effects of the farming environment on behaviour of adult ostriches in Britain. In: *Ratites in a Competitive World. Proceedings of the 2nd International Ratite Congress*, September 1998, Oudtshoorn, South Africa, pp. 160–166.
Deeming, D.C. (1999) *The Ostrich: Biology, Production and Health*. CAB International, Wallingford.
Deeming, D.C. (2002) The fall and rise of ratite production. *Proceedings of the 11th European Poultry Conference*, Bremen, September 2002. CD-ROM.
Deeming, D.C. and Ar, A. (1999) Factors affecting the success of commercial incubation. In: Deeming, D.C. (ed.) *The Ostrich: Biology, Production and Health*. CAB International, Wallingford, UK, pp. 159–190.
Deeming, D.C. and Ayres, L. (1994) Factors affecting the rate of growth of ostrich (*Struthio camelus*) chicks in captivity. *The Veterinary Record* 135, 617–622.
Deeming, D.C. and Bubier, N.E. (1999) Behaviour in natural and captive environments. In: Deeming, D.C. (ed.) *The Ostrich: Biology, Production and Health*. CAB International, Wallingford, UK, pp. 83–104.
Deeming, D.C. and van Middelkoop, J.H. (1999) Effect of strain and flock age on fertility and early embryonic mortality of broiler breeder eggs. *British Poultry Science* 40, S22–S23.
Deeming, D.C. and Wadland, D. (2001) Influence of mating sex ratio in commercial pheasant flocks on bird health and the production, fertility, and hatchability of eggs. *British Poultry Science* 43, 16–23.
Delsuc, F., Superina, M., Ferraris, G., Tilak, M.K. and Douzery, E.J.P. (2007) Molecular evidence for hybridization between the two living species of South American ratites: potential conservation implications. *Conservation Genetics* 8, 503–507.
Elbin, S.B., Crowe, T.M. and Graves, H.B. (1986) Reproductive behaviour of helmeted guinea fowl (*Numidia meleagris*): mating system and parental care. *Applied Animal Behaviour Science* 16, 179–197.
Ernst, R.A. (1978) Raising and propagating Japanese quail. Division of Agricultural Sciences, University of California, Leaflet 2738.
Fair, M.D., van Wyk, J.B. and Cloete, S.W.P. (2005) Parameter estimates for production traits of ostrich females within breeding seasons. In: Carbajo, E. (ed.) *Proceedings of the 3rd International Ratite Science Symposium of the World's Poultry Science Association*, 14–16 October 2005, Madrid, pp. 21–27.
Fernández, G.J. and Mermoz, M.E. (2003) Group copulation solicitation display among female greater rheas. *Wilson Bulletin* 115, 467–470.
Fernández, G.J. and Reboreda, J.C. (1995) Adjacent nesting and egg stealing between males of the greater rhea *Rhea americana*. *Journal of Avian Biology* 26, 321–324.
Fernández, G.J. and Reboreda, J.C. (1998) Effects of clutch size and timing of breeding on reproductive success of greater rheas. *Auk* 115, 340–348.
Fernández, G.J. and Reboreda, J.C. (2000) Egg losses and nest desertion in greater rheas *Rhea americana*. *Ibis* 142, 29–34.
Fernández, G.J. and Reboreda, J.C. (2002) Nest-site selection by male greater rheas. *Journal of Field Ornithology* 73, 166–173.

Fernández, G.J. and Reboreda, J.C. (2003) Male parental care in greater rheas (*Rhea americana*) in Argentina. *Auk* 120, 418–428.

Fernández, G.J. and Reboreda, J.C. (2007) Costs of large communal clutches for male and female greater rheas *Rhea americana*. *Ibis* 149, 215–222.

Game Conservancy Trust (1993) *Egg Production and Incubation*. The Game Conservancy Trust, Fordingbridge, Hampshire, UK.

Gaudioso, V.R., Alonso, M.E., Robles, R., Garrido, J.A. and Olmedo, J.A. (2002) Effects of housing type and breeding system on the reproductive capacity of the red-legged partridge (*Alectoris rufa*). *Poultry Science* 81, 169–172.

Gee, G.F., Bertschinger, H., Donoghue, A.M., Blanco, J. and Soley, J. (2004) Reproduction in nondomestic birds: physiology, semen collection, artificial insemination and cryopreservation. *Avian and Poultry Biology Reviews* 15, 47–101.

González-Redondo, P. (2006) Influence of the laying date on the fertility and hatchability of red-legged partridge (*Alectoris rufa*) eggs. *Journal of Applied Poultry Research* 15, 579–583.

Hawkins, P., Morton, D.B., Cameron, D., Cuthill, I., Francis, R., Freire, R., Gosler, A., Healy, S., Hudson, A., Inglis, I., Jones, A., Kirkwood, J., Lawton, M., Monaghan, P., Sherwin, C. and Townsend, P. (2001) The pigeon *Columba livia*. *Laboratory Animals* 35, Supplement 1 to issue 4, 97–103.

Hicks-Alldredge, K.D. (1996) Reproduction. In: Tully, T.N. and Shane, S.M. (eds) *Ratite Management, Medicine and Surgery*. Krieger Publishing Company, Malabar, Florida.

Huchzermeyer, F.W. (1998) *Diseases of Ostriches and Other Ratites*. Agricultural Research Council, Onderstepoort Veterinary Institute, Pretoria, South Africa.

Kirikçi, K., Deeming, D.C. and Gunlu, A. (2004) Effects of egg mass and percentage mass loss during incubation on hatchability of eggs of the rock partridge (*Alectoris graeca*). *British Poultry Science* 45, 380–384.

Klint, T. and Enquist, M. (1981) Pair formation and reproductive output in domestic pigeons. *Behavioural Processes* 6, 57–62.

Lábaque, M.C., Maestri, D.M., Hoyos, L., Martella, M.B. and Navarro, J.L. (2005) Do physical and chemical characteristics of greater rhea eggs change throughout the laying season? In: Carbajo, E. (ed.) *Proceedings of the 3rd International Ratite Science Symposium of the World's Poultry Science Association*, 14–16 October 2005, Madrid, pp. 61–65.

Lambrechts, H. and Cloete, S.W.P. (1998) Activity budgets of adult breeding ostriches classified according to annual egg production of the previous breeding season. *Animal Production in Harmony with the Environment. Proceedings of the 36th National Congress of the South African Society of Animal Science*. April 1998, University of Stellenbosch, pp. 93–94.

Lambrechts, H., Cloete, S.W.P. and Davis, H.J. (1998a) Influence of L-carnitine-magnesium supplement on the behaviour of adult breeding ostriches. In: *Ratites in a Competitive World. Proceedings of the 2nd International Ratite Congress*, September 1998, Oudtshoorn, South Africa, pp. 170–172.

Lambrechts, H., Cloete, S.W.P., van Schalkwyk, S.J. and Brand, Z. (1998b) Phenotypic relationships of body measurements with egg production of 2-year-old ostriches. In: *Ratites in a Competitive World. Proceedings of the 2nd International Ratite Congress*, September 1998, Oudtshoorn, South Africa, pp. 72–74.

Lambrechts, L., Swart, D., Cloete, S.W.P., Greyling, J.P.C. and van Schalkwyk, S.J. (2004) The influence of stocking rate and male:female ratio on the production of breeding ostriches (*Struthio camelus* spp.) under commercial farming conditions. *South African Journal of Animal Science* 34, 87–96.

le Coz-Douin, J. (1994) *L'Élevage de la Pintade*. Editions du Point Vétérinaire, Maison-Alfort, France.

Lupo, C., Beani, L., Cervo, R., Lodi, L. and Dessi-Fulgheri, F. (1990) Steroid hormones and reproductive success in the grey partridge (*Perdix perdix* L.). *Bollettino di Zoologia* 57, 247–252.

Malecki, I.A. and Martin, G.B. (2002a) Fertility of male and female emus (*Dromaius novaehollandiae*) as determined by spermatozoa trapped in eggs. *Reproduction, Fertilisation and Development* 14, 495–502.

Malecki, I.A. and Martin, G.B. (2002b) Fertile period and clutch size in the emu (*Dromaius novaehollandiae*). *Emu* 102, 165–170.

Malecki, I.A. and Martin, G.B. (2005) Fertility in ratites. In: Carbajo, E. (ed.) *Proceedings of the 3rd International Ratite Science Symposium of the World's Poultry Science Association*, 14–16 October 2005, Madrid, pp. 45–51.

Malecki, I.A., Martin, G.B. and Lindsay, D.R. (1997a) Semen production in the emu (*Dromaius novaehollandiae*). 1. Methods for semen collection. *Poultry Science* 76, 615–621.

Malecki, I.A., Martin, G.B. and Lindsay, D.R. (1997b) Semen production in the emu (*Dromaius novaehollandiae*). 2. Effect of collection frequency on the production of semen and spermatozoa. *Poultry Science* 76, 622–626.

Malecki, I.A., Martin, G.B., O'Malley, P.J., Meyer, G.T., Talbot, R.T. and Sharp, P.J. (1998) Endocrine and testicular changes in a short-day seasonally breeding bird, the male emu (*Dromaius novaehollandiae*) in south-western Australia. *Animal Reproductive Science* 53, 143–155.

Malecki, I.A., Beesley, J.R.I. and Martin, G.B. (2000) Changes in the characteristics of emu semen with season. *British Poultry Science* 41, S1, 18.

Malecki, I.A., Cloete, S.W.P., Gertenbach, W.D. and Martin, G.B. (2004) Sperm storage and duration of fertility in female ostriches (*Struthio camelus*). *South African Journal of Animal Science* 34, 158–165.

Malecki, I.A., Reed, C.E. and Martin, G.B. (2005a) Hypo-osmotic swelling test for measuring integrity of the emu (*Dromaius novaehollandiae*) sperm membrane. *Avian and Poultry Biology Reviews* 16, 189.

Malecki, I.A., Reed, C.E. and Martin, G.B. (2005b) Membrane integrity and fertilising ability of emu (*Dromaius novaehollandiae*) spermatozoa following storage *in vitro* at 4 and 20 °C for up to 48 hours. *Avian and Poultry Biology Reviews* 16, 190.

Martin, G.B. and Malecki, I.A. (2004) Artificial insemination in the emu (*Dromaius novaehollandiae*): effects of numbers of spermatozoa and time of insemination on the duration of the fertile period. *Animal Science Papers and Reports* 22, 315–323.

McKeegan, D.E.F. and Deeming, D.C. (1997) Effects of gender and group size on the time–activity budgets of adult breeding ostriches (*Struthio camelus*) in a farming environment. *Applied Animal Behaviour Science* 51, 159–177.

Minnaar, P. and Minnaar, M. (1992) *The Emu Farmer's Handbook*. Induna Company, Groveton, Texas.

Mullin, J.M. and Boehmer, P. (1991) *Game Bird Propagation*. Wild Harvest Publications, Inc., Goose Lake, Iowa.

Navarro, J.L. and Martella, M.B. (2002) Reproductivity and raising of greater rhea (*Rhea americana*) and lesser rhea (*Pterocnemia pennata*): a review. *Archiv: für Geflügelkunde* 66, 124–132.

Navarro, J.L., Barri, F.R., Maestri, D.M., Labuckas, D.O. and Martella, M.B. (2003) Physical characteristics and chemical composition of lesser rhea (*Pterocnemia pennata*): eggs from farmed populations. *British Poultry Science* 44, 586–590.

Nepote, K.H. (1999) Pigeons as laboratory animals. *Poultry and Avian Biology Reviews* 10, 109–115.

Offiong, S.A. and Abed, S.M. (1980) Fertility, hatchability and malformations in guinea fowl embryos as affected by dietary manganese. *British Poultry Science* 21, 371–375.

Ogwuegbu, S.O., Aire, T.A. and Adeyemo, O. (1988) Egg-laying pattern of the semi-domesticated helmeted guinea fowl (*Numida meleagris galeata*). *British Poultry Science* 29, 171–174.

Ottinger, M.A. and Rattner, B.A. (1999) Husbandry and care of quail. *Poultry and Avian Biology Reviews* 10, 117–120.

Parsons, D.G. (1996) Breeding problems and neonate diseases. In: Beynon, P.H., Forbes, N.A. and Harcourt-Brown, N.H. (eds) *Manual of Raptors, Pigeons and Waterfowl*. British Small Animal Veterinary Association Limited, Cheltenham, pp. 284–288.

Petitte, J.N. and Davis, G. (1999) Genetics and breeding. In: Deeming, D.C. (ed.) *The Ostrich: Biology, Production and Health*. CAB International, Wallingford, UK, pp. 275–292.

Richards, P.D.G. and Deeming, D.C. (2000) Correlation between shell colour and ultrastructure in pheasant eggs. *Avian and Poultry Biology Reviews* 11, 305–306.

Richards, P.D.G. and Deeming, D.C. (2001) Correlation between shell colour and ultrastructure in pheasant eggs. *British Poultry Science* 42, 338–343.

Saina, H. (2005) Guinea fowl (*Numidia meleagris*) production under smallholder farmer management in Guruve District, Zimbabwe. MPhil thesis, Department of Animal Science, Faculty of Agriculture, University of Zimbabwe.

Sales, J. (2006) The rhea, a ratite native to South America. *Avian and Poultry Biology Reviews* 17, 105–124.

Sales, J. (2007) The emu (*Dromaius novaehollandiae*): a review of its biology and commercial products. *Avian and Poultry Biology Reviews* 18, 1–20.

Sales, J. and Janssens, G.P.J. (2003) Nutrition of the domestic pigeon (*Columba livia domestica*). *World's Poultry Science Journal* 59, 221–232.

Sales, J., Deeming, D.C., van Deventer, P.J.U., Martella, M.B. and Navarro, J.L. (2000) Diurnal time–activity budget of adult greater rheas (*Rhea americana*) in a farming environment. *Archiv für Geflügelkunde* 64, 207–210.

Sambraus, H.H. (1994) Das sexualverhalten des AfrikanischenStaußes (*Struthio camelus*). *Tierärztliche Praxis* 22, 538–541.

Sarasqueta, D.V. (2005) Aspects of rearing, reproduction and hybridisation of Darwin's rhea or choique (*Rhea pennata* syn. *Pterocnemia pennata* spp *pennata*). In: Carbajo, E. (ed.) *Proceedings of the 3rd International Ratite Science Symposium of the World's Poultry Science Association*, 14–16 October 2005, Madrid, pp. 35–44.

Sharp, P.J., van Cleeff, J.K., Martin, G.B. and Blache, D. (2005) Photoperiodic control of seasonal breeding and appetite in the emu. In: Carbajo, E. (ed.) *Proceedings of the 3rd International Ratite Science Symposium of the World's Poultry Science Association*, 14–16 October 2005, Madrid, pp. 53–59.

Skadhauge, E. and Dawson, A. (1999) Physiology. In: Deeming, D.C. (ed.) *The Ostrich: Biology, Production and Health*. CAB International, Wallingford, UK, pp. 51–81

Soley, J.T. and Groenewald, H.B. (1999) Reproduction. In: Deeming, D.C. (ed.) *The Ostrich: Biology, Production and Health*. CAB International, Wallingford, UK, pp. 129–158.

Tapper, S. (1999) *A Question of Balance: Game Animals and Their Role in the British Countryside*. The Game Conservancy Trust, Fordingbridge, UK.

Tullett, S.G. and Deeming, D.C. (1982) The relationship between eggshell porosity and oxygen consumption of the embryo in the domestic fowl. *Comparative Biochemistry and Physiology* 72A, 529–533.

van Schalkwyk, S.J., Cloete, S.W.P. and de Kock, J.A. (1996) Repeatability and phenotypic correlations for body weight and reproduction in commercial ostrich breeding pairs. *British Poultry Science* 37, 953–962.

van Schalkwyk, S.J., Cloete, S.W.P. and Brown, C.R. (1999) The effect of temperature on the hatching performance of ostrich chicks, and its implications for artificial incubation in forced draught wooden incubators. *South African Journal of Animal Science* 29, 92–99.

Williams, K.M., Blache, D., Malecki, I.A., Sharp, P.J., Trigg, T.E., Rigby, R.D.G. and Martin, G.B. (1998) Growth, sexual development and carcass composition in intact and surgically or hormonally gonadectomised male and female emus. In: *Ratites in a Competitive World. Proceedings of the 2nd International Ratite Congress*, September 1998, Oudtshoorn, South Africa, pp. 75–80.

Wood-Gush, D.G.M. (1959) A history of the domestic chicken from antiquity to the 19th century. *Poultry Science* 38, 321–326.

PART VII

Nutrition of Breeding Poultry

CHAPTER 17
Feed Restriction

P.M. Hocking
The Roslin Institute and Royal (Dick) School of Veterinary Studies, University of Edinburgh, Scotland

ABSTRACT

It is common practice to control the feed consumption, and therefore the energy intake, of the parents of commercial broilers and ducks but not turkeys. Genetic selection for high rates of body weight gain results in correlated increases in ovulation rate and disrupted production of eggs with sound shells suitable for hatching. Feed restriction controls multiple ovulation in broiler breeders and ducks but not turkeys. The reasons for these breed and species differences and the biological principles underlying the allocation of feed energy are outlined. A series of experiments provided the basis for a theory of feed restriction in broiler breeders and led to the development of a stochastic model to predict egg production. Research has shown that achieving the target body weight at the onset of lay is critical and that feed supplied after the onset of lay should be adequate to support the peak of egg production. Feed allocation after the peak of egg production has been reached can be gradually decreased to match the requirements for lower egg production and to limit the effects of excessive body weight gain on mortality and productivity. Feed restriction is not practised in turkey breeders because it does not control the ovarian hierarchy and persistency of lay is poor, probably because the birds naturally lose a considerable quantity of body weight after the onset of lay. Conversely, excessive weight gain is undesirable and turkeys are fed low-energy and low-protein diets to limit body weight gain before photostimulation.

The body weight of broiler breeder cocks and drakes is substantially limited during rearing to enhance mating efficiency and fertility. Low-protein or low-energy diets are used to limit excessive body weight gain in stags as this facilitates semen collection and improves health. In males of all three species, body weight is allowed to increase slowly during the mating period to maintain fertility.

The welfare consequences of feed restriction in breeding birds are discussed and alternatives to quantitative feed restriction are outlined. It is concluded that the long-term solution to the need for substantial feed restriction in breeding birds is genetic selection for a decreased propensity of multiple ovulation.

INTRODUCTION

Breeding companies publish detailed and comprehensive manuals about the management of their genetic products, based on empirical evidence, commercial trials and experimental research. These are comprehensive descriptions of best practice for the husbandry and management of breeding birds that are also consistent with legislation protecting the welfare of poultry. Commercial broilers, ducks and turkeys are the offspring of a hybrid hen (a cross of two female lines) and a male line that may or may not be a cross. Recommendations for the parents of different commercial products differ largely in their respective target body weights.

Recommendations for rearing male and female breeding birds consist of target body weight gains that are less than those achieved by the market birds that are their offspring. These targets can be met by a range of different feed ingredients and dietary nutritional values, and, as energy is usually the driver of body weight gains, any discussion of energy intake must be concerned primarily with a description of recommended growth curves. These targets can be met by adjusting the quantity of feed allocated to the flock to meet weekly target body weights and require regular recording of body weight, either by sampling a proportion of the flock or by using automated weighing devices. Body weight control in broiler and duck breeders is substantial, whereas turkeys are fed *ad libitum* for the first 18 weeks, possibly using a relatively low-energy and -protein diet, and body weight targets are only 10–15% less than *ad libitum*. At 15–20 weeks of age, the feed allocated to broiler breeders and ducks is increased to prepare them for laying, when feed intake is in the region of 80% of *ad libitum*. Turkeys are photostimulated at 30 weeks; females are fed *ad libitum* and males 80–90% of *ad libitum*. The reasons for these species differences will be explored in this chapter. The question of feed quality will be addressed in the next two chapters by Fisher and Gous (Chapter 18) and Kidd (Chapter 19).

As there is far more information on broiler breeders than on turkeys and ducks, the theory underlying feed restriction in male and female broiler breeders will be discussed as a basis for the other species. Feed restriction has obvious welfare implications and this will also be explored. Finally, a genetic solution to the problems associated with high degrees of feed restriction will be presented.

OVARIAN FUNCTION IN POULTRY

The reason that feed restriction is practised in female broiler and duck breeders is not to save feed costs but to control excessive ovulation, because this causes poor rates of lay. The ovary of breeding birds (Fig. 17.1) contains a large number of small follicles less than 1 mm diameter, a substantial number of white follicles 1–5 mm diameter and a few yellow follicles 6–8 mm diameter, from which, in the laying hen, one yellow follicle is recruited on successive days to enter a hierarchy of yellow follicles (Gilbert, 1972; Zakaria *et al.*, 1984). The yellow follicles are associated with steroidal competence (Tilly *et al.*, 1991) and proceed to ovulation after 6–7 days, whereas the fate of the vast majority of follicles less than 8 mm diameter is to become atretic. Further consideration of these processes can be found in the respective chapters by Dunn *et al.* and Knight *et al.* (Chapters 6 and 7, this volume).

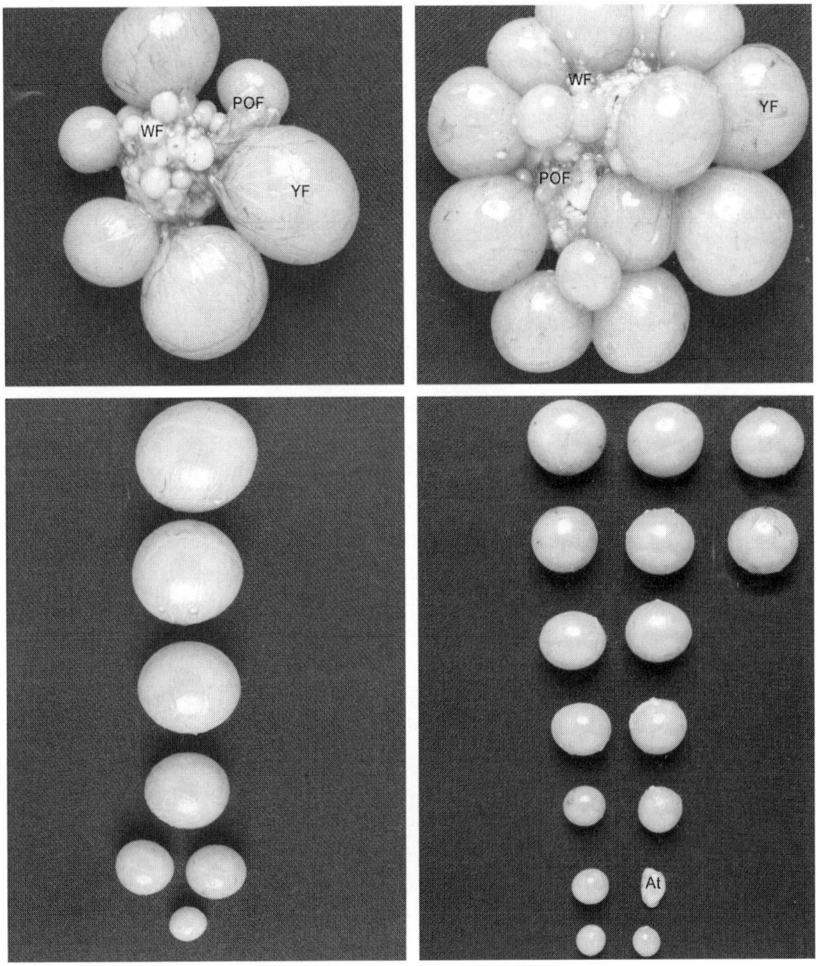

Fig. 17.1. Excised ovaries from broiler breeders at the onset of lay. The ovary on the left is from a bird that was feed restricted and the one on the right was fed *ad libitum*. The lower panels show the yellow yolky follicles in an ordered hierarchy arranged by decreasing maturity as judged by their weight. Note the presence of a single pair of follicles of similar size in the restricted ovary and two sets of triple yellow follicles of similar size, and an atretic follicle, in the yellow hierarchy of the *ad libitum*-fed bird. YF = large yellow follicle; WF = small white follicle; POF = post-ovulatory follicle; At = atretic follicle.

GENETIC SELECTION FOR GROWTH AFFECTS OVARIAN FUNCTION

Genetic selection for higher rates of body weight gain results in larger mature weights and an increase in ovulation rates in a range of animals from fish and mice to cattle and humans (Brien, 1986; Bunger *et al.*, 2005). A positive relationship between adult weight and ovulation rate also exists in poultry, as suggested by comparisons of commercial lines of broiler, duck and turkey breeders with unselected lines (Hocking, 1990a, 1992b; Hocking and Robertson, 2000). Single trait selection experiments for faster rates of gain resulted in larger mature weights and increased numbers of yellow follicles in broilers and turkeys (Udale *et al.*, 1972; Reddy and Siegel, 1976). The converse was also true: Abplanalp *et al.* (1977) selected for a higher incidence of double-yolked eggs in White Leghorn laying hens and reported a prevalence of over 30% multiple-yolked eggs, including a small proportion of eggs containing three to five yolks, after 11 generations of selection compared with only 2% double-yolked eggs in a control flock. Body weight was 1.94 kg in selected and 1.59 kg in the control flock, an increase of 22%. A reduction in both growth and ovarian activity was observed when the dwarfing gene *dw* was introgressed into the selected line (Abplanalp *et al.*, 1987), similarly demonstrating an intimate relationship between high body weight gain and increased ovarian activity during the first laying cycle.

A recent comparison of a selected line with its antecedent that had been maintained without selection for over 25 years (Table 17.1) suggested that genetic selection during this period had not resulted in higher numbers of follicles at the beginning of lay. On the basis of earlier papers (e.g. Jaap and Muir, 1968) the authors suggested that the breeds that formed the basis of current lines of broiler breeders were already characterized by high rates of follicular development. There may be a number of reasons why subsequent genetic selection for broiler traits has not resulted in higher numbers of yellow follicles at the onset of lay. For example, there might currently be an upper genetic physiological limit to the number of hierarchical follicles at the onset of lay or a threshold before an additional follicle is recruited to the hierarchy, which has not yet been achieved. Alternatively, natural or artificial genetic selection for a minimum rate of egg production to reproduce the line may have

Table 17.1. Body weight, abdominal fatness, number of yellow ovarian follicles and age at first egg in a genetically selected male line and a population from the same stock maintained without selection from 1972 fed *ad libitum*. Results for the selected line that was feed restricted are also presented (from Hocking and Robertson, 2005).

Trait	Control 1972 Fed *ad libitum*	Selected 1998 Fed *ad libitum*	Selected 1998 Feed restricted
Follicles (n)	10.6	11.8	7.4
Weight (kg)	3.2	5.6	3.9
Fat (g/kg)	70	51	19
Age (days)	133	140	202

limited further increases in follicle numbers. The resolution of these issues will probably occur only when the DNA changes that control follicular recruitment are elucidated.

FEED RESTRICTION IN FEMALE BROILER BREEDERS

The beneficial effects of feed restriction on productivity in broiler breeders compared with birds fed *ad libitum* are illustrated by the results of an experiment presented in Table 17.2. Female birds fed *ad libitum* experienced high mortality and produced relatively few eggs of poor hatchability compared with feed-restricted birds: feed-restricted birds produced over eight times more chicks per day-old pullet placed in the rearing house (104 versus 13) than those fed *ad libitum*. Hatchability in the *ad libitum*-fed group was half that of the feed-restricted birds. Feed-restricted females were allocated 63, 157 and 151 g/day from hatch to 24, 24–37 and 37–60 weeks of age respectively compared with 163, 192 and 142 g/day in birds fed *ad libitum*. Restricted birds that were fed *ad libitum* after peak rate of lay consumed 203 g/day and were of similar weight at 60 weeks as birds fed *ad libitum* throughout. The rate of lay declined in these birds, producing over 20 fewer eggs to 60 weeks than birds that were allocated a restricted quantity of feed to control body weight. There is an obvious economic advantage to feed restriction in commercial flocks but the major economic factor in broiler breeders is the increase in egg production and improved fertility and hatchability.

The control of body weight has been shown to control ovarian function in a large number of experiments comparing feed-restricted and *ad libitum*-fed broiler breeders (Robbins *et al.*, 1986; Hocking *et al.*, 1987; Yu *et al.*, 1992; Renema *et al.*, 1999). In broiler breeders fed *ad libitum* there are typically two or more ova (follicles) ready to ovulate on any one day (Fig. 17.1). Multiple ovulation results in approximately 50% of normal-shelled eggs because of a number of deleterious consequences. Multiple ovulated ova (yolks) may not be collected by the oviduct and are lost into the body cavity, where they are reabsorbed. Some yolks traverse the oviduct together and produce a multiple- (e.g. double) yolked egg; others will not pass through the oviduct at the same time, leading to the production of eggs with misshapen, malformed, thin shells or membraneous (shell-less) eggs that are either broken or useless for hatching purposes. Double-yolked eggs and eggs with poor shells result in lower hatchability and a higher incidence of embryo defects (Cherms and Wolff,

Table 17.2. Performance at 60 weeks of age of broiler breeder females fed *ad libitum* or feed restricted (from Hocking *et al.*, 2002a).

Trait	*Ad libitum*	Restricted
Body weight (kg)	5.3	3.7
Mortality (%)	46	4
Egg production (n)	58	157
Hatch of eggs set (%)	43	86

1968; Fasenko *et al.*, 2000) and are not incubated. In the experiment of Hocking *et al.* (2002a) shell defects in restricted and *ad libitum*-fed birds were 1.5 and 11.4% of recorded eggs, respectively. Poor shell quality is thought to be the underlying cause of the poor hatchability of apparently normal eggs laid by birds fed *ad libitum*.

Hocking *et al.* (1987) showed that feed restriction controlled the excessive ovulation rate characterized by broiler genotypes fed *ad libitum* (Fig. 17.1). Earlier research by Jaap and Muir (1968) reported the presence of double hierarchies in boiler breeders fed *ad libitum*. The significance of this observation was probably missed because laying hens also produce some double hierarchies at the onset of lay, which results in double-yolked and soft-shelled eggs. In the 1960s feed restriction was introduced to delay the onset of sexual maturity and increase the weight of hatching eggs and also because the control of fatness was thought to enhance egg production by increasing ovulation rate (e.g. Fuller *et al.*, 1969; Chaney and Fuller, 1975). Subsequently, detailed analysis of the relationships between fatness and ovulation rate failed to identify any relationship between measures of fatness and the number of yellow follicles or that genetic selection for broiler traits resulted in excessive fatness (Table 17.1). It was therefore concluded that the role of feed restriction as it affected egg production in broiler breeders was related solely to the control of ovarian function (Hocking and Whitehead, 1990). In retrospect the greater egg productivity of feed-restricted broiler breeders in these early experiments (e.g. Pym and Dillon, 1974) was probably associated with enhanced persistency of lay (see below).

In order to predict egg and chick production and optimize economic feeding (matching energy and protein supply to potential output at least cost) an understanding of the relationship between feed intake and the control of ovarian function is required. The results of a series of experiments to elucidate this relationship in the Ross 308 broiler breeder will be described in the next section. A preliminary model that predicts egg production in feed-restricted broiler breeders based on these experiments will also be described. This model has the potential of integrating information on protein (Chapter 18) and energy (feed intake) responses to optimize the economics of feeding broiler breeders in different economic environments, such as the relative costs of energy and protein, and the effects of different climates or genotypes.

A THEORY OF FEED RESTRICTION IN BROILER BREEDER HENS

Feed restriction during rearing

Hocking *et al.* (1989) transferred groups of birds from restricted to *ad libitum* feeding at 2-week intervals and recorded the numbers of yellow follicles, body weight, abdominal fat weight and age at sexual maturity (the age at which the first egg was laid). Their results are summarized in Fig.17.2 and show that the relationships between the number of yellow follicles and fatness and body weight did not correspond to the effect of feed restriction on the numbers of yellow follicles. These results led to the hypothesis that feed restriction was

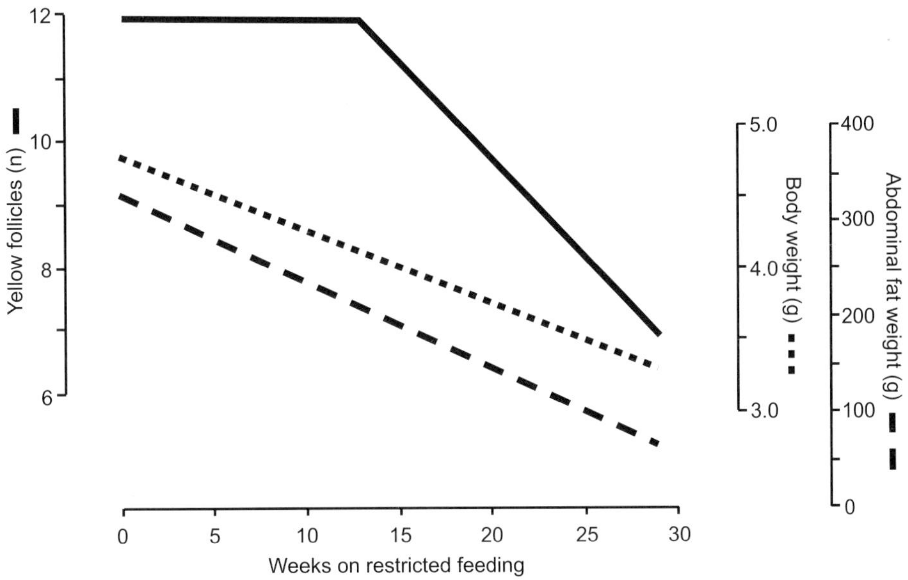

Fig. 17.2. Relationships between the number of weeks on restricted feeding and the numbers of yellow follicles, abdominal fat weight and body weight (from Hocking *et al.*, 1989). The responses for body weight and fatness are not the same as that for the number of yellow follicles.

effective only after 14 weeks of age. This corresponded to the age at which the first signs of ovarian follicular activity were observed in birds fed *ad libitum* throughout, and it was suggested that feed control was effective only over that period corresponding to the stage of maturity in which the ovary became physiologically active. The second conclusion was that the number of yellow follicles should be linearly related to body weight when feed restriction was applied after 14 weeks of age. This conclusion was supported by research that demonstrated a linear relationship between body weight and the number of yellow follicles. A range of body weights at the onset of lay were achieved by rearing groups of broiler breeders to increasing proportions of *ad libitum* body weight (Hocking, 2004b). The group fed to 25% of *ad libitum* did not commence lay and they were eventually fed extra energy, but not protein, and egg production commenced at a mean body weight of 3 kg. This was slightly higher than the mean body weight at the onset of lay in conventionally restricted birds, which were similar in all respects to the breeders reared to 40% of *ad libitum*. The author concluded that the mean minimum body weight that birds of this line commenced lay was 2.8 kg and that energy was the limiting nutrient in the control of ovarian function rather than protein, which is in large theoretical oversupply (Bowmaker and Gous, 1991).

In order to validate the hypothesis that the numbers of yellow follicles at the onset of lay were linearly related to body weight, birds were reared to four different body weights (2.8, 3.3, 3.8 and 4.3 kg) at the onset of photostimulation by a steady increase or decrease in body weight from 14 to 18 or 18 to 22 weeks of age compared with control birds fed *ad libitum* or conventionally feed restricted (Hocking, 1996). The results showed no effect of the two age periods or different degrees of feed restriction and no evidence of a departure from linearity in the relationship between body weight and the numbers of yellow follicles at the onset of lay. The practical application of these experimental results is to suggest that the optimum body weight for the control of yellow follicles at the onset of lay is the minimum body weight that is commensurate with sexual maturity.

Feed restriction after photostimulation

Feed restriction is continued after photostimulation throughout the laying period in commercial flocks to maintain rates of lay. Immediately after photostimulation the feed allocation in commercial flocks is increased and body weight rises until the birds achieve peak rates of lay in excess of 85%. In the experiment of Hocking (1996) body weight and feed intake between photostimulation and the onset of lay were confounded. It is not easy to study the separate roles of feed intake and body weight at any given time after the onset of lay because body weight quickly changes to a new equilibrium where the energetic cost of maintenance and production is balanced by the quantity of feed that is allocated to the birds. Notwithstanding this limitation, the effects of increasing feed allocation in birds of different body weight at photostimulation were examined in a factorial experiment that combined three body weights at photostimulation with three allocations of feed (Table 17.3). The effect of body weight was linear and highly significant but there was also a small positive effect of feed intake on the numbers of yellow follicles. In a second experiment ovarian follicle numbers were recorded 6 weeks after the onset of lay in birds that continued on the restriction programme, fed *ad libitum* for 6 weeks or restricted for 3 weeks before also being fed *ad libitum*. The results of these experiments confirmed that food restriction controlled follicle numbers in proportion to the control of body weight.

Table 17.3. Proportion of multiple yellow follicles and mean body weight at the onset of lay in broiler breeders reared to three body weights by *ad libitum* feeding, conventional restriction or an intermediate body weight and fed three corresponding allocations of feed after photostimulation (from Hocking, 1996).

		After photostimulation		
	Rearing	*Ad libitum*	Intermediate	Restricted
Multiples (%)	*Ad libitum*	87	68	78
	Intermediate	73	53	49
	Restricted	61	34	19
Weight (kg)	*Ad libitum*	5.3	4.9	4.8
	Intermediate	4.5	3.9	3.7
	Restricted	3.9	3.3	3.1

Feed restriction after peak rate of lay

When egg production starts to decline in commercial flocks, the allocation of feed is gradually decreased to maintain relatively small weekly increases in mean body weight. Experimental work supports the commercial reduction in feed allocation after peak rate of lay. Restricted broiler breeders responded to *ad libitum* feeding by rapid increases in body weight and decreased rates of lay, and conversely the rate of lay was maintained and mortality was decreased in birds fed *ad libitum* that were transferred to restricted feeding (Hocking *et al.*, 2002a). The numbers of yellow follicles in this experiment were not determined but the observations suggest that restricted birds quickly take on the characteristics of birds fed *ad libitum* if they are allowed unlimited access to feed. Relatively small short-term experiments typically fail to observe a substantial increase in the numbers of yellow follicles after 40 weeks of age (Robinson *et al.*, 1993; McGovern *et al.*, 1997). Restricted birds that were moulted and fed *ad libitum* or feed restricted showed that birds fed *ad libitum* or restricted during rearing both retained the propensity for producing double ovulations (Hocking, 2004b).

Predictive model of egg production in broiler breeders

A stochastic model (Alvarez and Hocking, 2007) was developed to simulate the egg production of broiler breeders in response to changes in body weight based on the results of the series of experiments described above on the relationships between body weight and ovarian activity. The model was based on a set of equations involving four input parameters. The equations were:

$$\text{YFexp} = \text{YFres} + (\text{YFal} - \text{YFres}) \times (\text{BWexp} - \text{BWres}) / (\text{BWal} - \text{BWres}) \qquad (17.1)$$

where YFexp is the number of yellow follicles and BWexp the body weight of the modelled bird at a given age (days); YFal is the number of yellow follicles and BWal is the body weight in a bird fed *ad libitum*; YFres is the number of yellow follicles and BWres is the body weight in a restricted bird at the same age from experimental data.

YFexp was modified (i.e. set to 0) to account for the presence of 'pause' days and differences in the age at sexual maturity. The length of the sequence of continuous egg production without a pause day declines with age and was calculated from the data of Robinson *et al.* (1990) as:

$$\text{SEQ} = 2.902 + 19.60/(1 + \exp(0.03809 \times (\text{MSEQ}))) \qquad (17.2)$$

where SEQ = sequence length and MSEQ is the age of the flock at the time at which the longest sequence occurred. An intersequence pause day when no ovulation occurred was modelled if a random number from a uniform distribution between 0 and 1 was greater than that predicted from Equation 17.4.

Age at sexual maturity was estimated from the equation:

$$\text{AFE} = 288 - 51 \times \text{BWexp} + 4.6 \times (\text{BWexp})^2 + \text{Rm} \qquad (17.3)$$

where AFE is age at first egg and Rm is a random variable drawn from a normal distribution with mean 0 and standard deviation 7.5 days. YFexp was set to 0 if the modelled bird had not commenced lay as predicted from Equation 17.2.

Egg production is related to the number of yellow follicles in a non-linear manner and was modelled by two equations:

$$\begin{aligned} \text{EGG} &= \text{YFexp}/6 && \text{if YFexp} \le 6 \\ &= 1 - ((\text{YFexp} - 6) \times \text{Re}/6) && \text{if YFexp} > 6 \end{aligned} \quad (17.4)$$

where EGG is the probability of producing an egg on a given day, Re is a random number from a normal distribution with a mean 0.5 and a standard deviation of 0.05, where 0.5 is the proportion of multiple ovulations that result in a single-yolked hatching egg.

Flock size (FLOCK, the probability of a bird housed at 20 weeks being alive and in laying condition) was modelled as:

$$\begin{aligned} \text{FLOCK} &= (1 - \text{MORT} \cdot t) && \text{for age} < 40 \text{ weeks} \\ &= (1 - \text{MORT} \cdot t) \times (1 - \text{NOTLAY} \cdot t) && \text{for age} > 40 \text{ weeks} \end{aligned} \quad (17.5)$$

where MORT and NOTLAY, respectively, are the average daily rate of mortality and the increase in the proportion of birds that are not in lay at time (day) t after photostimulation.

Finally flock rate of lay on the modelled day was calculated as the product of Equations 17.4 (modified as outlined for AFE and SEQ) and 17.5.

In order to drive the iteration over time, four input variables are required: body weight at photostimulation (assumed to be 20 weeks of age in the model), body weight at maturity if fed *ad libitum* (taken as 5 kg) and the daily rates of body weight gain from 20 to 30 and 31 to 70 weeks of age, from which the body weight of the modelled bird at each day of age is predicted. The model accurately predicted the egg production from a commercial trial (185.0 ± 0.76 eggs predicted compared with 187.9 ± 1.64 observed) and in three out of four small-scale experimental treatments (Alvarez and Hocking, 2009). Whereas there are several assumptions and considerable uncertainty about some of the parameters in the model, the model provides a logical basis for future modelling and experimentation. Álvarez and Hocking (2009), for example, investigated the effects on egg production of differences in initial body weight, daily weight gain during lay, age at sexual maturity, variability of body weight and variation in the age at first egg, which would be extremely time consuming and expensive to conduct in experimental conditions. The results of this simulation exercise suggested that total egg production was very sensitive to changes in body weight at photostimulation and body weight gain during the first 10 weeks of lay, whereas changes in body weight gain after peak rate of lay showed only minor effects on egg production. Increasing variability of initial body weight was associated with a linear decrease in the mean and increased variability of egg production. In contrast to body weight, higher variability in age at first egg was associated with increased variability but had little effect on the mean number of eggs produced. These results are in line with experimental data and commercial experience and provide some confidence that the model has a sound basis in the biology of breeding birds. On the other hand, attempts to model genetic changes on ovulation rate were less satisfactory, probably because of a lack of experimental data on the effect of age and body weight on the pattern of multiple ovulation and persistency of lay in birds over 30 weeks of age.

A simulation model of nutritional responses based on an empirical description of body weight and reproduction in broiler breeders and experimental data on efficiencies of nutrient utilization has been outlined by Nonis and Gous (see Fisher and Gous, Chapter 18, this volume). The biological model described above could usefully be combined with the model of nutritional responses to optimize the management of broiler breeders both generally and in specific flocks in a dynamic approach to flock management.

There are many papers in the scientific literature that report the deleterious consequences of providing relatively smaller increments and decrements of feed during rearing than is recommended in breeders' technical manuals (e.g. Leeson and Summers, 1983; Fattori *et al.*, 1991; Robinson *et al.*, 1995, 2007; Wilson *et al.*, 1995). The results of these experiments are generally consistent with the theory of responses to feed restriction in broiler breeders based on the control of ovarian function outlined above and are not reviewed here.

One apparent contradictory result was reported by Bruggeman *et al.* (1999), who alternated *ad libitum* and restricted feeding during different periods and showed that feed restriction from 7 to 15 weeks followed by either *ad libitum* or restricted feeding led to improved reproductive performance, suggesting that long-term feed restriction may not be necessary to attain good reproductive performance. Subsequent research, however, failed to confirm this hypothesis as maximum productivity was only achieved in two experiments with both dwarf and standard broiler breeders by conventional levels of feed restriction applied throughout the rearing period (Bruggeman *et al.*, 2005).

FEED RESTRICTION IN TURKEY HENS

Modern heavy and medium-heavy turkey breeders have multiple hierarchies of ovarian follicles, and the principles upon which broiler breeders are managed, as described in the previous section, suggests that feed restriction should maximize the production of hatching eggs in these birds. Experimental work with turkey breeders has failed to show a consistent improvement in productivity (Whitehead, 1989). One reason for this may lie in the fact that different degrees of feed restriction have frequently been applied during the early rearing period and that during the application of a short photoperiod from 18 weeks to photostimulation body weight was allowed to increase to near *ad libitum* levels. The theory of feed restriction in broiler breeders described above suggests that, to be effective, turkey breeder hens should be severely feed restricted throughout rearing (i.e. 40–50% of *ad libitum* body weight) and early lay. Experimental work, however, has demonstrated that feed restriction does not reliably control the recruitment of two or more follicles or re-establish an orderly hierarchy of yellow follicles in the ovary (Fig. 17.3) and tends to result in lower productivity compared with birds fed *ad libitum*, even if feed restriction during rearing is substantial and continues to early lay. Limiting body weight to 60 and 80% of *ad libitum* in lines of turkeys ranging in mature body weight from 7.6 to 17.0 kg decreased the number of yellow follicles and the proportion of multiples, and the effect was greater in the smaller lines than in the large male lines, where it was relatively ineffective (Table 17.4). Paradoxically these results suggest that feed restriction should be more effective in smaller lines that have relatively few yellow follicles compared with larger male lines that have many. In a medium body weight line housed in cages, feed restriction decreased the number of double-yolked and soft-shelled eggs and increased hatchability and poult production by about 15% (Hocking, 1992a), and this is at the upper end of results of feed-restriction experiments with relatively small turkeys in the

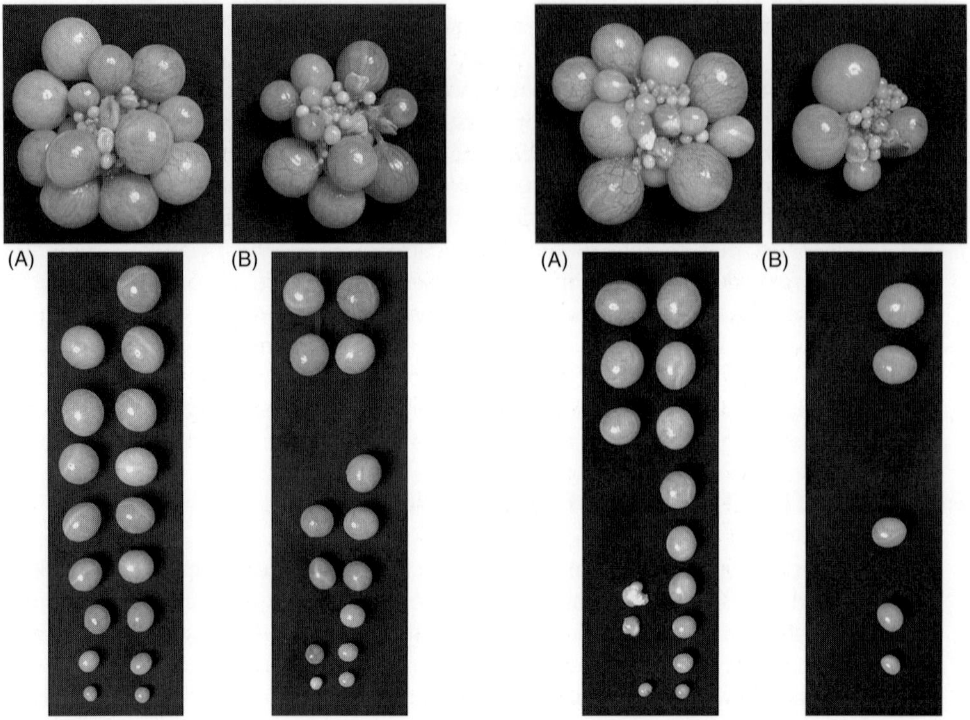

Fig. 17.3. Ovaries and hierarchies of yellow follicles at the onset of lay in a large male line (left) and a small female line (right) of turkeys. The birds were fed *ad libitum* (A) or restricted to 60% of *ad libitum* body weight (B). Note that the yellow follicles require 9–10 days to mature in turkeys compared with 6–7 days in broilers (see Fig. 17.1) and duck breeders and turkeys therefore have more yellow follicles in the ovary. Compared with the broiler breeder (Fig. 17.1) feed restriction is relatively ineffective in controlling multiple ovulation (noted as two follicles of similar weight) and leads to gaps in the hierarchy, which represent days when no yolk will ovulate and no egg will be formed.

Table 17.4. Body weight, yellow follicles and proportion of multiple follicles of three strains of turkeys fed *ad libitum* (AL) or restricted to 60 or 80% of *ad libitum* from 5 weeks of age to the onset of lay (from Hocking, 1992b).

Strain	Restricted (%)	Body weight (kg)	Yellow follicles (n)	Multiples (%)
T5	60	5.2	7.1	4
	80	6.4	8.7	5
	AL	7.4	10.9	34
B6	60	8.3	9.9	17
	80	10.2	11.2	36
	AL	11.9	13.2	50
M6	60	13.3	15.2	65
	80	15.1	14.0	55
	AL	18.9	15.7	68

older (pre-1980) literature. Miles and Leeson (1989) reported a substantial increase from 23 to 32 poults per hen caged in turkey breeders that were feed restricted by only 5% from 2 weeks of age. This is in contrast to the decrease from 40 to 25 eggs in male-line hens respectively fed *ad libitum* or restricted (Hocking and Bernard, 1998). Renema *et al.* (1995) investigated the productivity of male-line turkey breeders in which body weight gain had been restricted either by feeding a low-protein ration or by quantitative feed restriction. A decrease in the proportions of multiple yellow follicles and better shell quality in the restricted treatments were associated with poor persistency and relatively low total egg production. Poor persistency of laying was also observed with birds photostimulated at 18 and 24 weeks of age (Hocking, 1992a) and in the feed-restricted birds of Hocking and Bernard (1998). Crouch *et al.* (2002) reported the effects of body weight reduction of up to 45% during rearing (considerably less than the 60–75% in broiler breeder hens) followed thereafter by rapid feed increases. Initial egg production in the first 5 weeks of lay was higher in turkeys feed restricted from 3 to 16 or 24 weeks of age but persistency was poor. The authors linked this to high seasonal temperatures but it is also consistent with earlier reports of the effects of feed restriction on persistency of lay in modern female turkey breeders in temperate climates. Higher egg production in the turkeys fed *ad libitum* was associated with the loss of proportionally more breast muscle than in the restricted birds, suggesting that sufficient body reserves must exist at the onset of lay to support maximum egg production in turkey hens or the birds will stop laying prematurely. However, it is also possible that the reproductive system of turkey hens of relatively low body weight has not matured sufficiently to respond to short photoperiods and that the inherent photorefractoriness of juvenile birds has not been dissipated and, as a consequence, persistency of lay is poor.

Severely limiting the body weight of male-line turkey hens by feed restriction from 4 weeks of age to the onset of photostimulation failed to control adequately the numbers of yellow follicles and subsequent egg production (Table 17.5). When feed was restricted for 5 weeks post-photostimulation there was still no effective control of yellow follicle numbers (16.4 versus 18.8), even though body weight was decreased to 63% of *ad libitum* (Buchanan *et al.*, 2000).

Table 17.5. Mean number and proportion of yellow ovarian follicles and body weights at different ages in male-line turkey breeder hens fed *ad libitum* or feed restricted from 4 weeks of age (from Hocking and Bernard, 1998).

Trait	Feeding	First egg	44 weeks	54 weeks
Yellow follicles (n)	*Ad libitum*	16.5	13.9	9.5
	Restricted	14.6	6.1	0.0
Multiple follicles (%)	*Ad libitum*	77	72	33
	Restricted	65	45	0.0
Body weight (kg)	*Ad libitum*	15.7	14.9	14.9
	Restricted	11.3	11.3	13.1

Furthermore, about half the birds did not commence lay (47% compared with 87% hens fed *ad libitum*), and those that did were heavier than those that did not (14.0 versus 12.4 kg). An earlier experiment suggested that to achieve a single hierarchy of yellow follicles in this line would require body weight at the onset of lay to be about 7 kg (Hocking *et al.*, 1992), and it is clear that male-line turkeys restricted to this extent, unlike broiler breeders, would not achieve sexual maturity, regardless of any effects on persistency of lay. Research conducted with a medium body weight female line showed that photostimulation at 18 and 24 weeks of age resulted in more yellow follicles than photostimulation at the conventional age of 30 weeks (Hocking, 1992a,b). Male-line hens were therefore photostimulated at 36 rather than 30 weeks of age, but there was no significant difference in the number of yellow follicles at the onset of lay (Buchanan *et al.*, 2000).

Taken together, the results of experiments with different lines of turkey breeders suggest that controlling body weight by a substantial reduction in feed intake may decrease the prevalence of multiple ovulations in commercial female lines and increase shell quality and hatchability during early lay. However, overall productivity in medium and heavy female turkeys is unlikely to be better than conventional *ad libitum*-fed birds because maximum rate and persistency of lay are relatively poor. This fundamental difference between female turkey and broiler breeders may be the result of different physiological systems or different selection practices. Pedigree broiler breeders are feed restricted after selection at a relatively young age and will come into lay in the absence of photostimulation, whereas genetic improvement in turkey breeders is achieved by selection at an age close to maturity, the birds are fed *ad libitum* and need an obligatory dark period before photostimulation (see Lewis, Chapter 14, this volume). Substantial body weight loss after photostimulation occurs in turkey breeders and is associated with low feed intakes and the mobilization of fat and breast muscle (Whitehead, 1989; Bentley, 2006), which are not observed in *ad libitum*-fed broiler breeders. In practice, breeding companies may recommend body weight targets for turkey breeder hens that are slightly less (5–10%) than their maximum potential to maintain fitness and mobility, a practice that also controls feed costs. These objectives are generally achieved by feeding a low-protein diet rather than by limiting feed intake quantitatively. A long-term solution to the poor daily rate of egg production relative to the ovarian potential of turkey hens probably lies in genetic selection against multiple ovulation rather than substantial modifications of the nutritional environment.

FEED RESTRICTION IN DUCKS

Commercial duck breeders are larger and have more paired ovarian follicles than unselected birds (Table 17.6). In contrast to turkey breeders, but like broiler breeders, feed restriction decreases the occurrence of multiple follicles and enhances egg production (Hocking, 1990a). Olver (1995) showed that ducks restricted to 50% of *ad libitum* at 20 weeks and fed *ad libitum* thereafter produced more eggs with higher fertility and lower mortality than ducks fed *ad libitum* throughout (Table 17.7). Less severe feed restriction was associated with lower productivity compared with ducks fed *ad libitum*, and feed restriction from 8 weeks was less effective than restriction from 3 weeks of age (Olver, 1984, 1988). In a heavy line of duck breeders, Cherry (1993) showed optimum productivity was obtained by limiting body weight to a minimum of 65% at sexual maturity followed by an increase to 80% of *ad libitum*, which was easily implemented by limiting the time of access to feed. In spite of the very different growth pattern of ducks compared with chickens, the responses to feed restriction are therefore similar to broiler breeders and very different from turkey breeders.

GROWTH AND FERTILITY OF BROILER BREEDER COCKS

Effective management of the broiler breeder male is primarily about controlling body weight through the allocation of feed to ensure that the birds can mate

Table 17.6. Mean number of yellow follicles and body weight at the onset of lay in selected and unselected ducks fed *ad libitum* or feed restricted (from Hocking, 1990a).

Line	Rearing	Body weight (kg)	Yellow follicles (n)	Multiples (%)
Selected	*Ad libitum*	4.5	9.4	62
	Restricted	2.7	6.6	22
Unselected	*Ad libitum*	3.7	9.1	39
	Restricted	2.4	5.7	1

Table 17.7. Mean body weight, egg production, fertility and mortality in feed-restricted ducks compared with ducks fed *ad libitum* (from Olver, 1995).

Trait	*Ad libitum*	Restricted
Body weight at 20 weeks (kg)	4.0	2.5
Body weight at 60 weeks (kg)	4.3	3.8
Egg production (n/duck)	163	187
Fertility (%)	83	92
Mortality (%)	7.5	2.5

effectively. In an unpublished experiment conducted in the early 1980s, male broiler breeders that weighed 3, 4, 5, 6 and 7 kg at the end of lay were associated with breeder hen fertility of 45, 86, 91, 79 and 22%, respectively. In commercial practice body weight is controlled to provide fit and active males that achieve sexual maturity in the majority of the birds when they are mated, and thereafter the allocated feed is increased to permit a gradual increase in body weight to a maximum of 5 kg at the end of the production period (Hocking and Bernard, 1997; Romero-Sanchez *et al.*, 2007a,b). The fertility of males that are over 5 kg is relatively poor because they are unable to copulate satisfactorily and, conversely, males with low body weight (less that 3.5 kg) have poor fertility associated with small testes (Hocking, 1990b, 1997; Hocking and Bernard, 2000). Male broiler breeders that are not fed increasing amounts of feed to maintain small increases in body weight have relatively poor fertility during the latter half of the breeding period (Hocking and Bernard, 1997). As in female boiler breeders, male body weight is managed to control the variability of body weight (see Duncan, Chapter 8, this volume) because large birds are sexually ineffective and intimidate smaller males, and if body weight is too low, or energy intake is insufficient, cocks become sexually impotent.

BODY WEIGHT CONTROL AND SEMEN PRODUCTION IN TURKEY STAGS

Modern turkeys are reproduced by artificial insemination. This practice was introduced initially to avoid damage to the hens and because of the extreme sexual dimorphism in turkeys – males are about 50% larger than females – and the use of specialized male lines. Turkey male lines are primarily selected on body weight gain and are substantially heavier than males of the female line, to which they are mated. Natural mating is also by a lek system, in which females approach groups of displaying males (see Duncan, Chapter 8, this volume) and is not readily adaptable to modern commercial conditions. Genetic selection has also increased the length and width of the breast muscles of male turkeys and undoubtedly resulted in males that are not able to mate naturally when fed *ad libitum*. Feed restriction decreases body weight and breast size but depresses sexual function. Feed restriction during rearing to less than 70% of *ad libitum*-fed birds was associated with relatively poor semen yields compared with those that were 80 or 90% of *ad libitum* body weight (Hocking, 1988, 1991). However, there has been a greater acceptance of body weight control in turkey breeder males in recent years as they have become even heavier and difficult to manage. Compared with broiler breeder males, the decrease in body weight is relatively mild (less than 20%) and is applied after the birds are 18 weeks of age. The purpose is also rather different, in that it is designed to facilitate ease of semen collection by limiting the development of heavy muscling around the vent, and to increase activity and reduce mortality, particularly in hot environments (Bentley, 2006). As for broiler breeder males, it is important that small increases in body weight are maintained for satisfactory persistency of semen production and to avoid premature onset of a moult.

BODY WEIGHT CONTROL AND FERTILITY IN DRAKES

Cherry (1993) showed that maximum fertility could be achieved by controlling the body weight of drakes in a similar manner as for the ducks described above.

WELFARE CONSEQUENCES OF FEED RESTRICTION

Feed-restricted birds are inevitably hungry, in the sense that they would consume more feed if they had the opportunity to do so. Whereas feed-restricted birds are undoubtedly healthier than birds fed *ad libitum* on conventional nutrient-dense diets, there are also concerns that the degree of feed restriction in broiler breeders is sufficient to compromise the welfare of the birds, at least during rearing. Research in small-scale experiments has shown changes in physiology and behaviour that are characteristic of animals that are subjected to restricted feed intakes. Specifically, broiler breeders are more active (foraging, drinking and pecking with less time resting) and have a higher motivation to feed, an increase in the blood heterophil–lymphocyte ratio and in some studies an increase in the proportion of basophilic cells and plasma corticosterone concentrations; there is no evidence that feed restriction compromised essential bodily functions, as indicated by several enzyme systems, and immune functions were not adversely affected (Kostal *et al.*, 1992; Hocking *et al.*, 1993, 1996, 2001, 2002b; Savory and Maros, 1993; Savory *et al.*, 1993; de Jong *et al.*, 2003; Puterflam *et al.*, 2006). Comparisons of male-line turkey males and females showed similar responses to feed restriction as boiler breeders (Hocking *et al.*, 1998, 1999). Interpretation of these experimental results is not straight-forward: many of the observed behavioural and physiological responses are the expected consequence of limiting feed and an increase in plasma corticosterone concentrations and are not necessarily indicative of compromised welfare (see the review in Hocking, 2004a). Limited evidence from commercial flocks casts doubt on the relevance of some of the observed changes in behaviour, specifically so-called stereotypic pecking, in experimental conditions (Hocking and Jones, 2006). Recently, statistical analysis of behavioural time patterns suggests that changes in behaviour associated with the feeding of greater quantities of feed are quantitative rather than qualitative (Merlet *et al.*, 2005; Hocking *et al.*, 2007). Underlying all these observations is the fact that consistent body weight gains occur in restricted broiler breeders and that mechanisms to cope with food scarcity exist in nature. In contrast to feed-restricted birds it is unquestionably true that the welfare of broiler breeders fed *ad libitum* over the course of their lifetime is not as good as birds subjected to at least some degree of feed restriction. This suggests that there exists an optimum degree of feed restriction that maximizes the overall welfare of the birds and that this is greater than the current level of feed restriction but less than *ad libitum* feed consumption. Unfortunately, as outlined above, any increase in feed intake is also associated with a decrease in the production of hatching eggs. Renema *et al.* (2007) analysed commercial broiler and broiler breeder targets and showed that the target body weights for broiler breeders had changed little over a 30-year period, in contrast to those for broilers, which had more than doubled. As a consequence the recommended body weights for different breeds as a proportion of broiler body weight at 6 weeks had declined from 52% to 35% in males and 27% in females. These results suggest that the welfare issues and management of boiler breeders may become more serious in future. There are two approaches to solve this problem: to feed more feed of a lower quality (typically high-fibre, low-protein diets) or to apply genetic selection to decrease the propensities for multiple ovulation so that broiler breeders could be fed sufficient feed to optimize their welfare without compromising the production of hatching eggs. Feeding diets containing chemical ingredients and appetite suppressants to control body weight have been examined but are unlikely to prove acceptable to flock farmers or the public and there are associated environmental concerns (Hocking and Bernard, 1993; Savory *et al.*, 1996).

Improving broiler breeder welfare by feeding low-protein and high-fibre diets

The protein concentration of commercial breeder feed is in excess of the theoretical requirements for growth (see Fisher and Gous, Chapter 18, this volume) and feeding low-protein rations may mean that more feed could be allocated to maintain the same body weight gain. However, feeding low-protein diets by decreasing the concentration of soy in conventional diets did not improve indices of welfare and led to relatively poor productivity (Hocking *et al.*, 2001). Early experiments that used very-low-protein diets to limit body weight gain in broiler breeders resulted in greater mortality and signs of nutrient deficiency (Lee *et al.*, 1971; Wilson *et al.*, 1971; van Wambeke and Okerman, 1976).

Savory *et al.* (1996) fed a range of diets diluted with indigestible ingredients to caged female broiler breeders and concluded that, whereas body weight might be satisfactorily controlled, variability was increased, and in some diets indexes of welfare were actually worse than on conventional feed restriction. Excessive oral behaviour was abolished in some treatments but there was no decrease in the motivation to feed (Savory *et al.*, 1996; Savory and Lariviere, 2000; Kubikova *et al.*, 2001). Zuidhof *et al.* (1995) fed groups of restricted broiler breeders from hatch to 60 weeks of age on a diet diluted with 15 or 30% of oat hulls. Productivity was higher in birds fed on the 15% diet and there was some evidence that welfare might be enhanced in birds fed the diluted diets. Feeding high-fibre, low-density diets to broiler breeders in a large floor experiment was successful in controlling body weight. One of the treatments resulted in higher egg production, egg weight and hatchability than the control diet and had a small but positive effect on indexes of welfare (de Jong *et al.*, 2005; Enting *et al.*, 2007). Heck *et al.* (2004) fed conventional broiler breeders *ad libitum* or restricted quantities of a conventional diet, and dwarf broiler breeders a finely ground high-fibre meal *ad libitum*. The best productivity was achieved by conventional feed restriction followed by the dwarf line fed *ad libitum*. In two comparable experiments, Bruggeman *et al.* (2005) fed dwarf and normal broiler breeders the same low-energy diet *ad libitum,* restricted from 6 to15 weeks or on a conventional feed-restriction programme. Over the first 24 weeks of lay the highest production of eggs was from dwarf breeders restricted only during the rearing period, followed by the conventionally restricted normal birds. Collectively, these results suggest that, whereas dwarf broiler breeders could be fed a low-energy diet *ad libitum*, egg production would be substantially less than in conventionally restricted standard broiler breeders.

Sandilands and co-workers developed a feeding regimen based on increasing quantities of ground oat hulls and calcium propionate, which controlled the body weight of broiler breeders to the required targets with no effect on subsequent laying performance (Sandilands *et al.*, 2005, 2006; Tolkamp *et al.*, 2005). The diet abolished excessive oral behaviours but had no effect on two physiological indexes of welfare. However, a treatment based on calcium propionate alone was abandoned because of the presence of oral lesions. Whereas this may have been caused by the fineness of the feed, it is not known how calcium propionate controls appetite and its use is unlikely to be acceptable by flock owners or the public. Furthermore, the quantity of feed fed to achieve satisfactory control of body weight was relatively high (14.7 versus 8.1 kg/bird), with implications for feed transport costs and environmental pollution, consistent with the earlier conclusions of Savory *et al.* (1996).

Improving broiler breeder welfare by genetic selection

As discussed at the start of this chapter, genetic selection for high rates of body weight gain is associated with a correlated increase in the prevalence of multiple ovulations and results in the need for feed restriction in broiler and duck breeders, whereas in turkey breeders the producer has to accept relatively low productivity compared with ovarian potential, particularly in male lines of turkeys. Genetic selection to decrease the propensity for multiple ovulation would, in the long term, lead to the possibility of increasing target body weights to optimize the welfare of broiler breeders while maximizing the production of hatching eggs. Genetic selection against multiple ovulation is not possible at the present time because egg production does not reliably reflect ovarian activity and birds have to be killed to estimate follicle numbers. Marker-assisted selection using DNA markers or whole-genome selection (Bijma and Bovenhuis, Chapter 3, this volume) would make such selection possible and is the objective of the author's current research. As broilers become larger, and target body weights become smaller as a proportion of the birds' potential weight (Renema *et al.*, 2007), the search for genetic solutions to the welfare issue will become more pressing. Fortunately the modern genetic tools make this approach feasible and will have the added benefit of contributing to higher rates of lay, which may more than offset the costs of extra feed.

REFERENCES

Abplanalp, H., Lowry, D.C. and Van Middelkoop, J.H. (1977) Selection for increased incidence of double-yolked egg in White Leghorn chickens. *British Poultry Science* 18, 585–595.

Abplanalp, H., Napolitano, D. and Tai, C. (1987) The effects of the gene for sex linked dwarfing (*dw*) on egg production and body composition of hens in a line selected for high numbers of multiple yolked eggs. *Journal of Animal Breeding and Genetics* 104, 304–308.

Alvarez, R. and Hocking, P.M. (2007) Stochastic model of egg production in broiler breeders. *Poultry Science* 86, 1445–1452.

Álvarez, R. and Hocking, P.M. (2009) Stochastic modeling of optimum initial body weight, daily weight gains, and effect of genetic changes in ovulation rate and age at sexual maturity on total egg production of broiler breeders. *British Poultry Science*: in press.

Bentley, J. (2006) Management problems in turkey breeding. In: Romboli, I., Flock, D. and Franchini, A. (eds) *Proceedings of 12th European Poultry Conference*, Verona, 10–14 September. Supplement of the *World's Poultry Science Journal* 62, 441–442.

Bowmaker, J.E. and Gous, R.M. (1991) The response of broiler breeder hens to dietary lysine and methionine. *British Poultry Science* 32, 1069–1088.

Brien, F. (1986) A review of the genetic and physiological relationships between growth and reproduction in mammals. *Animal Breeding Abstracts* 54, 975–997.

Bruggeman, V., Onagbesan, O., D'Hondt, E., Buys, N., Safi, M., Vanmontfort, D., Berghman, L., Vandesande, F. and Decuypere, E. (1999) Effects of timing and duration of feed restriction during rearing on reproductive characteristics in broiler breeder females. *Poultry Science* 78, 1424–1434.

Bruggeman, V., Onagbesan, O., Ragot, O., Metayer, S., Cassy, S., Favreau, F., Jego, Y., Trevidy, J.J., Tona, K., Williams, J., Decuypere, E. and Picard, M. (2005) Feed allowance–genotype interactions in broiler breeder hens. *Poultry Science* 84, 298–306.

Buchanan, S., Robertson, G.W. and Hocking, P.M. (2000) Effects of food restriction or delayed photostimulation on ovarian follicle number, plasma oestradiol concentration and vaginal collagen content in male-line turkeys. *British Poultry Science* 41, 501–507.

Bunger, L., Lewis, R.M., Rothschild, M.F., Blasco, A., Renne, U. and Simm, G. (2005) Relationships between quantitative and reproductive fitness traits in animals. *Philosophical Transactions of the Royal Society B – Biological Sciences* 360, 1489–1502.

Chaney, L.W. and Fuller, H.K. (1975) The relation of obesity to egg production in broiler breeders. *Poultry Science* 54, 200–208.

Cherms, F.L. and Wolff, E. (1968) Incidence of physically abnormal turkey hatching eggs and their relationship to hatchability. *Poultry Science* 47, 760–765.

Cherry, P. (1993) Sexual maturity in the domestic duck. PhD thesis, University of Reading, Reading, UK.

Crouch, A.N., Grimes, J.L., Christensen, V.L. and Krueger, K.K. (2002) Effect of physical feed restriction during rearing in large white turkey breeder hens: 2. Reproductive performance. *Poultry Science* 81, 102–111.

de Jong, I.C., van Voorst, A.S. and Blokhuis, H.J. (2003) Parameters for quantification of hunger in broiler breeders. *Physiology and Behavior* 78, 773–783.

de Jong, I.C., Enting, H., van Voorst, A. and Blokhuis, H.J. (2005) Do low-density diets improve broiler breeder welfare during rearing and laying? *Poultry Science* 84, 194–203.

Enting, H., Kruip, T.A.M., Verstegen, M.W.A. and van der Aar, P.J. (2007) The effect of low-density diets on broiler breeder performance during the laying period and on embryonic development of their offspring. *Poultry Science* 86, 850–856.

Fasenko, G.M., Robinson, F.E., Danforth, B.L. and Zelter, I. (2000) An examination of fertility, hatchability, embryo mortality, and chick weight in double versus single-yolked broiler breeder eggs. *Canadian Journal of Animal Science* 80, 489–493.

Fattori, T.R., Wilson, H.R., Harms, R.H. and Miles, R.D. (1991) Response of broiler breeder females to feed restriction below recommended levels. 1. Growth and reproductive-performance. *Poultry Science* 70, 26–36.

Fuller, H.L., Potter, D.K. and Kirkland, W. (1969) Effect of delayed maturity and carcass fat on reproductive performance of broiler breeder pullets. *Poultry Science* 48, 801–809.

Gilbert, A.B. (1972) Activity of the ovary in relation to egg production. In: Freeman, B.M. and Lake, P.E. (eds) *Egg Formation and Production*. British Poultry Science Ltd, Edinburgh, pp. 3–21.

Heck, A., Onagbesan, O., Tona, K., Metayer, S., Putterflam, J., Jego, Y., Trevidy, J.J., Decuypere, E., Williams, J., Picard, M. and Bruggeman, V. (2004) Effects of *ad libitum* feeding on performance of different strains of broiler breeders. *British Poultry Science* 45, 695–703.

Hocking, P.M. (1988) Effects of restricting body weight gain or limiting the photoperiod during rearing on sexual maturity, viability and semen production in Large White turkey males. *British Poultry Science* 29, 531–544.

Hocking, P.M. (1990a) Comparison of the effects of the degree of food restriction during rearing on ovarian function at the onset of lay in unselected ducks (*Anas platyrhynchos*) and in a line selected for improved feed efficiency. *British Poultry Science* 31, 351–359.

Hocking, P.M. (1990b) The relationship between dietary crude protein, bodyweight and fertility in naturally mated broiler breeder males. *British Poultry Science* 31, 743–757.

Hocking, P.M. (1991) Effects of controlling body weight on the semen production of Large White turkey males. *British Poultry Science* 32, 211–218.

Hocking, P.M. (1992a) Effects of photostimulation at 18, 24 and 30 weeks of age on the productivity of female turkeys fed *ad libitum* or restricted until point of lay. *British Poultry Science* 33, 253–269.

Hocking, P.M. (1992b) Genetic and environmental control of ovarian function in turkeys at sexual maturity. *British Poultry Science* 33, 437–448.

Hocking, P.M. (1996) The role of body weight and food intake after photostimulation on ovarian function at first egg in broiler breeder females. *British Poultry Science* 37, 841–851.

Hocking, P.M. (1997) Broiler breeder reproduction, feeding and welfare: a complex interaction. In: *Proceedings of 3rd Annual Broiler and Breeder Conference*, Linton, Cambridge, UK, pp. 10–17.

Hocking, P. (2004a) Measuring and auditing the welfare of broiler breeders. In: Weeks, C. and Butterworth, A. (eds) *Measuring and Auditing Broiler Welfare*. CAB International, Wallingford, UK, pp. 19–35.

Hocking, P.M. (2004b) Roles of body weight and feed intake in ovarian follicular dynamics in broiler breeders at the onset of lay and after a forced molt. *Poultry Science* 83, 2044–2050.

Hocking, P.M. and Bernard, R. (1993) Evaluation of putative appetite suppressants in the domestic fowl (*Gallus domesticus*). *British Poultry Science* 34, 393–404.

Hocking, P.M. and Bernard, R. (1997) Effects of male body weight, strain and dietary protein content on fertility and musculo-skeletal disease in naturally mated broiler breeder males. *British Poultry Science* 38, 29–37.

Hocking, P.M. and Bernard, R. (1998) Comparative development of the ovary and production, fertility and hatchability of eggs from traditional turkeys and a contemporary male-line fed *ad libitum* or restricted. *British Poultry Science* 39, 291–297.

Hocking, P.M. and Bernard, R. (2000) The effects of the age of male and female broiler breeders on fertility and hatchability of eggs. *British Poultry Science* 41, 370–377.

Hocking, P.M. and Jones, E.K.M. (2006) On-farm assessment of environmental enrichment for broiler breeders. *British Poultry Science* 47, 418–425.

Hocking, P.M. and Robertson, G.W. (2000) Ovarian follicular dynamics in selected and control (relaxed selection) male- and female-lines of broiler breeders fed *ad libitum* or on restricted allocations of food. *British Poultry Science* 41, 229–234.

Hocking, P.M. and Robertson, G.W. (2005) Limited effect of intense genetic selection for broiler traits on ovarian function and follicular sensitivity in broiler breeders at the onset of lay. *British Poultry Science* 46, 354–360.

Hocking, P.M. and Whitehead, C.C. (1990) Relationship between body fatness, ovarian structure and reproduction in mature females from lines of genetically lean or fat broilers given different food allowances. *British Poultry Science* 31, 319–330.

Hocking, P.M., Waddington, D., Walker, M.A. and Gilbert, A.B. (1987) Ovarian follicular structure of White Leghorns fed *ad libitum* and dwarf and normal broiler breeders fed *ad libitum* or restricted to point of lay. *British Poultry Science* 28, 493–506.

Hocking, P.M., Waddington, D., Walker, M.A. and Gilbert, A.B. (1989) Control of the development of the ovarian follicular hierarchy in broiler breeder pullets by food restriction during rearing. *British Poultry Science* 30, 161–174.

Hocking, P.M., Waddington, D. and Walker, M.A. (1992) Changes in ovarian function of female turkeys photostimulated at 18, 24 or 30 weeks of age and fed *ad libitum* or restricted until point of lay. *British Poultry Science* 33, 639–647.

Hocking, P.M., Maxwell, M.H. and Mitchell, M.A. (1993) Welfare of broiler breeder and layer females subjected to food and water control during rearing. *British Poultry Science* 34, 443–458.

Hocking, P.M., Maxwell, M.H. and Mitchell, M.A. (1996) Relationships between the degree of food restriction and welfare indices in broiler breeder females. *British Poultry Science* 37, 263–278.

Hocking, P.M., Bernard, R. and Wess, T.J. (1998) Comparative development of antitrochanteric disease in male and female turkeys of a traditional line and a contemporary sire-line fed *ad libitum* or restricted quantities of food. *Research in Veterinary Science* 62, 29–32.

Hocking, P.M., Maxwell, M.H. and Mitchell, M.A. (1999) Welfare of food restricted male and female turkeys. *British Poultry Science* 40, 19–29.

Hocking, P.M., Maxwell, M.H., Robertson, G.W. and Mitchell, M.A. (2001) Welfare assessment of modified rearing programmes for broiler breeders. *British Poultry Science* 42, 424–432.

Hocking, P.M., Bernard, R. and Robertson, G.W. (2002a) Effects of low dietary protein and different allocations of food during rearing and restricted feeding after peak rate of lay on egg production, fertility and hatchability in female broiler breeders. *British Poultry Science* 43, 94–103.

Hocking, P.M., Maxwell, M.H., Robertson, G.W. and Mitchell, M.A. (2002b) Welfare assessment of broiler breeders that are food restricted after peak rate of lay. *British Poultry Science* 43, 5–15.

Hocking, P.M., Rutherford, K.M.D. and Picard, M. (2007) Comparison of time-based frequencies, fractal analysis and t-patterns for assessing behavioural changes in broiler breeders fed on two diets at two levels of feed restriction: a case study. *Applied Animal Behaviour Science* 104, 37–48.

Jaap, R.G. and Muir, F.V. (1968) Erratic oviposition and egg defects in broiler-type pullets. *Poultry Science* 47, 417–423.

Kostal, L., Savory, C.J. and Hughes, B.O. (1992) Diurnal and individual variation in behavior of restricted-fed broiler breeders. *Applied Animal Behaviour Science* 32, 361–374.

Kubikova, L., Vyboh, P. and Kostal, L. (2001) Behavioural, endocrine and metabolic effects of food restriction in broiler breeder hens. *Acta Veterinaria Brno* 70, 247–257.

Lee, P.J.W., Gulliver, A.L. and Morris, T.R. (1971) Restricted feeding of broiler breeder pullets during the rearing period and its effect on productivity and breeding. *British Poultry Science* 12, 499–510.

Leeson, S. and Summers, J.D. (1983) Consequences of increased feed allowance for growing broiler breeder pullets as a means of stimulating early maturity. *Poultry Science* 62, 6–11.

McGovern, R.H., Renema, R.A. and Robinson, F.E. (1997) Increased feed allocation does not stimulate increased ovarian development or increased egg output in 54-week-old broiler breeder hens. *Canadian Journal of Animal Science* 77, 177–179.

Merlet, F., Puterflam, J., Faure, J.M., Hocking, P.M., Magnusson, M.S. and Picard, M. (2005) Detection and comparison of time patterns of behaviours of two broiler breeder genotypes fed *ad libitum* and two levels of feed restriction. *Applied Animal Behaviour Science* 94, 255–271.

Miles, S.A. and Leeson, S. (1989) Effects of feed restriction during the rearing period and age at photostimulation on the reproductive performance of turkey hens. *Poultry Science* 69, 1552–1528.

Olver, M.D. (1984) Quantitative feed restriction of Pekin breeder ducks during the rearing period and its effect on subsequent productivity. *South African Journal of Animal Science* 14, 136–141.

Olver, M.D. (1988) Quantitative feed restriction of Pekin breeder ducks from 3 weeks of age and its effect on subsequent productivity. *South African Journal of Animal Science* 18, 93–96.

Olver, M.D. (1995) Effect of restricted feeding during rearing and a 'forced moult' at 40 weeks of production on the productivity of Pekin breeder ducks. *British Poultry Science* 36, 737–746.

Puterflam, J., Merlet, F., Faure, J.M., Hocking, P.M. and Picard, M. (2006) Effects of genotype and feed restriction on the time-budgets of broiler breeders at different ages. *Applied Animal Behaviour Science* 98, 100–113.

Pym, R.A.E. and Dillon, J.F. (1974) Restricted feed intake and reproductive performance of broiler breeders. *British Poultry Science* 15, 245–259.

Reddy, P.K. and Siegel, P. (1976) Selection for body weight at eight weeks of age. 11. Ovulation and oviposition patterns. *Poultry Science* 55, 1518–1530.

Renema, R.A., Robinson, F.E., Melnychuk, V.L., Hardin, R.T., Bagley, L.G., Emmerson, D.A. and Blackman, J.R. (1995) The use of feed restriction for improving reproductive traits in male-line Large White turkey hens .2. Ovary morphology and laying traits. *Poultry Science* 74, 102–120.

Renema, R.A., Robinson, F.E., Proudman, J.A., Newcombe, M. and McKay, R.I. (1999) Effects of body weight and feed allocation during sexual maturation in broiler breeder hens. 2. Ovarian morphology and plasma hormone profiles. *Poultry Science* 78, 629–639.

Renema, R.A., Rustas, M.E. and Robinson, F.E. (2007) Implications of changes in commercial broiler and broiler breeder body weight targets over the past 30 years. *World's Poultry Science Journal* 63, 457–472.

Robbins, K.R., McGhee, G.C., Osei, P. and Beauchene, R.E. (1986) Effect of feed restriction on growth, body composition and egg production of broiler females through 68 weeks of age. *Poultry Science* 65, 2226–2231.

Robinson, F.E., Hardin, R.T. and Robblee, A.R. (1990) Reproductive senescence in domestic-fowl – effects on egg-production, sequence length and inter-sequence pause length. *British Poultry Science* 31, 871–879.

Robinson, F.E., Yu, M.W., Lupicki, M.E. and Hardin, R.T. (1993) Short-term consequences of a sudden increase in feed allowance in broiler breeder hens. *Canadian Journal of Animal Science* 73, 159–167.

Robinson, F.E., Robinson, N.A., Hardin, R.T. and Wilson, J.L. (1995) The effects of 20-week body weight and feed allocation during early lay on female broiler breeders. *Journal Applied Poultry Research* 4, 203–210.

Robinson, F.E., Zuidhof, M.J. and Renema, R.A. (2007) Reproductive effeciency and metabolism of female broiler breeders as affected by genotype, feed allocation and age at photostimulation. 3. Reproductive efficiency. *Poultry Science* 86, 2278–2286.

Romero-Sanchez, H., Plumstead, P.W. and Brake, J. (2007a) Feeding broiler breeder males. 1. Effect of feeding program and dietary crude protein during rearing on body weight and fertility of broiler breeder males. *Poultry Science* 86, 168–174.

Romero-Sanchez, H., Plumstead, P.W. and Brake, J. (2007b) Feeding broiler breeder males. 3. Effect of feed allocation program from sixteen to twenty-six weeks and subsequent feed increments during the production period on body weight and fertility. *Poultry Science* 86, 775–781.

Sandilands, V., Tolkamp, B.J. and Kyriazakis, I. (2005) Behaviour of food restricted broilers during rearing and lay – effects of an alternative feeding method. *Physiology and Behavior* 85, 115–123.

Sandilands, V., Tolkamp, B.J., Savory, C.J. and Kyriazakis, I. (2006) Behaviour and welfare of broiler breeders fed qualitatively restricted diets during rearing: are there viable alternatives to quantitative restriction? *Applied Animal Behaviour Science* 96, 53–67.

Savory, C.J. and Lariviere, J.M. (2000) Effects of qualitative and quantitative food restriction treatments on feeding motivational state and general activity level of growing broiler breeders. *Applied Animal Behaviour Science* 69, 135–147.

Savory, C.J. and Maros, K. (1993) Influence of degree of food restriction, age and time of day on behavior of broiler breeder chickens. *Behavioural Processes* 29, 179–190.

Savory, C.J., Maros, K. and Rutter, S.M. (1993) Assessment of hunger in growing broiler breeders in relation to a commercial restricted feeding programme. *Animal Welfare* 2, 131–152.

Savory, C.J., Hocking, P.M., Mann, J.S. and Maxwell, M.H. (1996) Is broiler breeder welfare improved by using qualitative rather than quantitative food restriction to limit growth rate? *Animal Welfare* 5, 105–127.

Tilly, J.L., Kowalski, K.I. and Johnson, A.L. (1991) Stage of follicular development associated with the initiation of steroidogenic competence in avian granulosa cells. *Biology of Reproduction* 44, 305–314.

Tolkamp, B.J., Sandilands, V. and Kyriazakis, I. (2005) Effects of qualitative feed restriction during rearing on the performance of broiler breeders during rearing and lay. *Poultry Science* 84, 1286–1293.

Udale, R.W., Siegel, P.B. and van Krey, H.P. (1972) Rates of ovulation and oviposition in growth-selected lines of chickens. *Poultry Science* 51, 2098–2100.

van Wambeke, F. and Okerman, F. (1976) The effect of rearing and qualitative feed restriction during the rearing period on the development and reproductive performances of broiler breeder males. *Archiv für Geflügelkunde* 40, 64–71.

Whitehead, C.C. (1989) Nutrition of turkey breeding stock. In: Nixey, C. and Grey, T.C. (eds) *Recent Advances in Turkey Science*. Poultry Science Symposium. Butterworths, London, pp. 91–118.

Wilson, H.R., Rowland, L.O. and Harms, R.H. (1971) Use of low protein grower diets to delay sexual maturity of broiler breeder males. *British Poultry Science* 12, 157–163.

Wilson, J.L., Robinson, F.E., Robinson, N.A. and Hardin, R.T. (1995) Effects of feed allocation on female broiler breeders. *Journal of Applied Poultry Research* 4, 193–202.

Yu, M.W., Robinson, F.E., Robinson, N.A. and Hardin, R.T. (1992) Effect of feed allowance during rearing and breeding on female broiler breeders. 2. Ovarian morphology and production. *Poultry Science* 71, 1750–1761.

Zakaria, A.H., Sakai, H. and Imai, K. (1984) Time of follicular transformation to the rapid growth-phase in relation to the ovulatory cycle of laying hens. *Poultry Science* 63, 1061–1063.

Zuidhof, M.J., Robinson, F.E., Feddes, J.J.R., Hardin, R.T., Wilson, J.L., McKay, R.I. and Newcombe, M. (1995) The effects of nutrient dilution on the well-being and performance of female broiler breeders. *Poultry Science* 74, 441–456.

CHAPTER 18
Protein and Amino Acid Responses

C. Fisher[1] and R.M. Gous[2]

[1]*Consultant, 20 Longstaff Court, Hebden Bridge, HX7 6AB, UK;*
[2]*University of KwaZulu-Natal, Pietermaritzburg, South Africa*

ABSTRACT

Unlike the provision of feed or dietary energy to breeding birds, the level of dietary protein does not appear to play any major controlling role in the productivity of reproduction. Amongst few exceptions to this are the control of early growth rate in broiler breeders and the control of growth over a longer period in turkey breeders. Excess protein may have some negative effects on fertility and hatchability but these effects are not well described.

Consideration of the programmes of feeding protein to flocks of breeding birds draws attention to the complexity of the system and the difficulty of carrying out trials that can be assessed in terms of economic performance during the whole production cycle. Practical feeding programmes for broiler breeders, in which severe control of feed intake is practised, are contrasted with turkey breeders, in which virtually full feeding is the norm.

Appropriate protein and amino acid levels in feeds can be more effectively determined by calculation rather than by direct experimentation. This requires knowledge of rates of protein deposition and of protein and amino acid utilization. A rather fragmented literature on these topics in breeding birds is reviewed; most of the available evidence refers to broiler breeder chickens. In comparison with data on egg-laying hens, broiler breeder hens appear to have a very low maintenance requirement for lysine and, on the basis of one experiment, for methionine. Amino acid utilization for egg production is also lower in broiler breeder hens compared with egg-laying hens, but this can be accounted for by the more irregular pattern of egg production.

Empirical experimental evidence about the relationship between dietary protein supply and production of breeding birds is very variable in quantity and applicability. The effect of diet on the reproductive performance of males and evaluation of the (potentially negative) effects of excess protein on reproductive performance of both males and females are most significantly evaluated by such experiments.

A preliminary account is given of a model of broiler breeder hen production and nutritional response developed at the University of KwaZulu-Natal (R. Gous and M. Nonis, unpublished data). In the context of this chapter the model can be used to examine questions about maintenance requirements and utilization of amino acids for egg production.

INTRODUCTION

Although we are concerned here with the biological principles of breeder productivity, in the discussion of dietary protein and amino acids it is difficult to isolate principles from the practical problem of designing a feeding programme. The supply of dietary protein and amino acids to breeding birds seldom appears to play any special role in the control of the reproductive cycle. The practical resolution of the principles involved is mainly to supply sufficient protein to meet the needs of the birds for optimum protein deposition at each stage of their production cycle. In principle this is an economic question, determined by the economic response of the production system to different levels of dietary provision. However, this is a very complex analysis, and, since protein is inexpensive relative to the high-value output of viable hatchlings, in practice the economically optimum level of protein is fortunately very close to that required for maximum production levels (Kemp *et al.*, 2001).

A previous review was published by Lopez and Leeson (1994) and some of the material for amino acids was reviewed by Fisher (1998). In this chapter the evidence on the subject is mostly reviewed from a mechanistic point of view to facilitate the calculation of requirements. This leads at the end of the chapter to a brief and preliminary description of a model of production and protein nutrition in broiler breeder hens. Nutrition of males and the effects of maternal nutrition on the embryo and progeny are mostly reviewed from empirical data. Some reference is made to turkey and duck breeders but, of necessity, most of the evidence and discussion refers to broiler breeders.

PROVISION OF PROTEIN TO BREEDING HENS

Broiler breeders

Different breeds of broiler breeder hens will show important differences in commercial traits but are fairly similar from the point of view of protein nutrition. The one exception is the use of genetically dwarf hens (Picard, 2004), which are controlled at lower absolute feed intakes than conventional hens and will thus receive higher levels of dietary protein to provide a similar level of protein intake.

A conventional scheme for supplying dietary crude protein to flocks of broiler breeder hens (not dwarf) is shown in Table 18.1. Whilst in practice there is considerable variation about any general pattern, the principles involved are common to all situations. In early growth (up to 7–21 days) the birds are fed

Table 18.1. Provision of dietary crude protein to flocks of broiler breeder hens.

Status	Early growth *Ad libitum* feeding	Growth controlled by feeding: Rearing period	Growth controlled by feeding: Pre-lay period	Controlled feeding Egg production
Age (days)	0–21	21–105	105–*ca.* 168	168+
Crude protein (g/kg)	180–200	140–160	150–160	150–160

ad libitum and dietary protein is provided to support a level of growth which is close to maximum for all tissues. After this period, growth is controlled by restricted feeding (i.e. by energy intake) and protein is supplied for growth and tissue development as determined by energy supply. Growth at this stage is much lower than the potential which would be displayed under free feeding. From about 105 days of age, growth will include development of sexual organs, and a slightly higher level of protein might be provided; control of feed intake continues. From sexual maturity at a mean age of about 168 days, protein is supplied to support a low level of continuing growth and egg production. Again the intakes of protein and amino acids required are supplied within a fixed allocation of feed. Within the framework of such a scheme, individual amino acids are controlled relative to lysine or methionine in a balance suitable for growth or egg production (Coon, 2004).

The two main variations on this general theme arise from the use of intermittent or skip-a-day feeding in the growth and development stage, as opposed to everyday feeding. With everyday feeding, the lower range of protein levels (140 g/kg) will frequently be used from about 6 weeks of age. When this is done, an increase in protein supply at about 15 weeks of age is required. This sort of pattern is widely used in Europe. When feeding is skipped on some days, the higher range of protein (about 160 g/kg) may be used from 6 weeks and continued without change up to the start of egg production. This sort of pattern is widely used in the USA.

The principle of controlling productivity of broiler breeder hens rests mainly on controlling feed intake to achieve a predetermined body weight growth pattern. Success in this control rests heavily on feeding management and the maintenance of flock uniformity. There is little evidence that the supply of crude protein plays any controlling role, except in the first few weeks of life and unless gross deficiencies are incurred. There is considerable discussion about what the predetermined growth rate should be, but this is always achieved by feed allocation, with feed composition varied to provide a suitable intake of nutrients.

Turkey breeders

Turkey breeds differ widely in growth rates and body size but the protein nutrition of breeding birds is fairly similar for all types. The conventional scheme for feeding breeder turkey hens (Table 18.2) differs considerably from that for broiler breeders. Turkey breeders are fed at, or close to, *ad libitum* intakes, and thus feed composition can potentially play a larger role in influencing

Table 18.2. Provision of dietary crude protein to flocks of turkey breeder hens.

Status	Feed intake and growth controlled to targets but close to maximum (protein used to control body weight)				Light-controlled period	Egg production *Ad libitum* feeding
Age (days)	0–21	21–42	42–77	77–112	112–203	203+
Crude protein (g/kg)	260–280	230–240	200	190	120–140	140–170

growth, development and production. The principle of the scheme outlined in Table 18.2 is to provide sufficient protein to keep body weight close to a predetermined target at all stages of the birds' life. Contrary to the situation in broiler breeders, the target body weight is fairly close to the maximum potential of the stock. Dietary protein is used to some extent to control the development of the desired body weight in individual flocks. Turkeys are also kept more frequently in open housing, especially during lay, and will therefore experience a wider range of seasonal temperatures. The resulting differences in feed intake will occasion adjustment of dietary protein levels on a seasonal basis.

Turkeys display photorefractoriness, which is dissipated by use of light restriction from about 18 to 30 weeks of age. Growth during this period is low and the importance of feed composition is not well understood, but it seems unlikely that protein supply will be important in controlling the system. Turkey hens also lose body weight during the early stages of lay and, again, the significance of this for protein nutrition and utilization remains unknown.

PROTEIN DEPOSITION IN BREEDING HENS

A detailed description of protein deposition in different types of breeding hens has not been found. However, an approximate picture can be built up, at least for broiler breeders and turkeys, from a variety of sources and approximations.

Broiler Breeders

Growth

Bennett and Leeson (1990) report on the body composition of broiler breeder pullets reared in a conventional way to meet the recommended body weight profile. Figure 18.1, calculated from their data, shows that protein accretion is about 3–4 g per day up to 24 weeks of age. When this growth rate is compared with the putative potential somatic protein growth of these birds, it is clear that protein growth is being severely restricted by energy supply between 2 and 15 weeks of age. Between 15 and 22 weeks the observed rate of protein accretion exceeded the potential for somatic growth, and this is assumed to reflect the development of reproductive tissues.

The main functional protein pools represented by this growth will be organs, muscle, feathers and reproductive organs. The quantitative split between these is not known nor, more importantly, is it known whether there

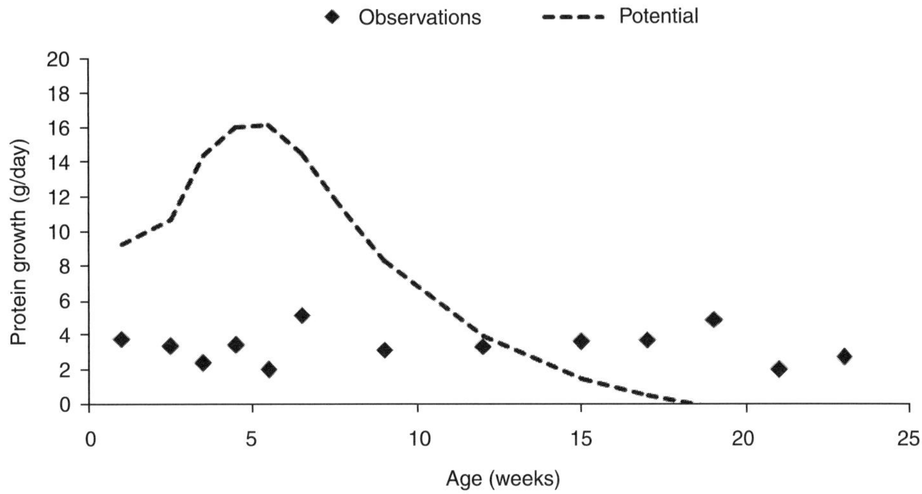

Fig. 18.1. Protein growth rates (body plus feather proteins) in immature broiler breeder hens. Observations from Bennett and Leeson (1990) – standard treatments. The potential growth was estimated using the EFG Software Broiler Growth Model (EFG Software, 2002).

are important partition rules determining the growth of each pool under restricted energy supply. A popular practical suggestion is that feather growth can be encouraged by the supply of additional sulfur amino acids during the growth period, but this is not supported by any description of modified partition rules.

Conformation or fleshing, which is assessed from breast muscle development, has become a practical criterion of body composition in broiler breeders. In particular, the introduction of broiler breeds with enhanced breast muscle development has increased interest in this topic. Birds coming into lay with 'poor' fleshing are likely to show poor persistency of egg production. Whilst this characteristic can undoubtedly be influenced by protein supply at different stages of growth, no quantitative description of these effects has been found.

Once maturity is reached broiler breeders are fed to maintain egg production and to support a continuing, but low, level of body weight gain. Based on practical experience, which is reflected in the recommendations of breeding companies, a body weight gain of 2–3 g per day is targeted. The composition of this gain has not been described in any detail except for the study of Bowmaker and Gous (1989), who describe body (including feather), liver, oviduct and ovary protein growth from 20 to 30 weeks of age. Figure 18.2 shows the rates of growth in these four pools. Body protein growth declines to zero by 30 weeks of age. A simple assumption is that body protein growth does not occur to any significant extent after sexual maturity, as appears to be the case in the laying hen (Martin *et al.*, 1994), but unequivocal data on this point have not been found. It was previously assumed (Fisher, 1998) that this growth had a normal composition (about 150 g/kg protein), based on the work of Pearson and Herron (1981), who showed a protein gain when comparing birds at 21 and 64 weeks. However, the data in Fig. 18.2 show that this period includes the growth occurring as the flock matures and was probably wrongly ascribed to the whole of the laying year. The effects of the assumptions made about body protein growth during lay are trivial so far as net protein needs for growth are concerned, but if maintenance is scaled to body protein mass (see below) then the cumulative effect of small gains can become quite significant.

Egg production

Clearly the mass of eggs produced will be the main determinant of the deposition of egg protein. Variation in the composition of eggs will mainly reflect the proportions of yolk, albumen and shell, and the dry matter contents of these components. As a first approximation it seems reasonable to assume that the egg component dry matter is constant in composition. No body of evidence to support or refute this supposition has been found. It would be expected that egg composition would be highly conserved in the face of random disturbances such as different diets, although some fixed effects, such as breed or age, may have an influence.

In a recent study (M. Nonis, unpublished data) eggs from two important commercial strains of broiler breeder (Ross 308 and Cobb 500, as supplied in South Africa) were used to study variations in egg composition associated with

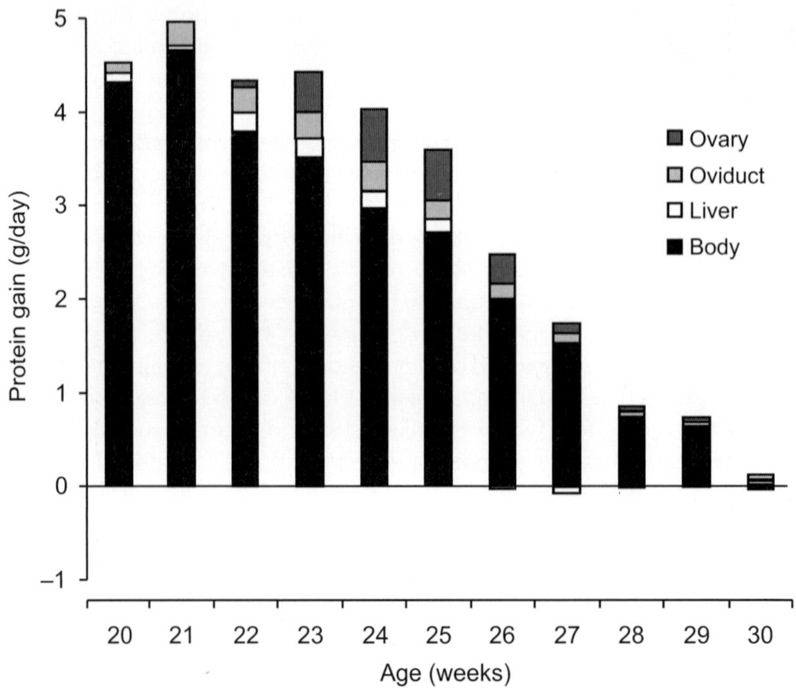

Fig. 18.2. Protein gain in the somatic body (total minus liver, oviduct and ovary), liver, oviduct and ovary in broiler breeder hens from 20 to 30 weeks of age (from Bowmaker and Gous, 1989).

egg weight at a given age and across ages. There were differences in egg composition between strains both at a given age and with changes in age. However, yolk weight (YW) could be predicted from hen age by a single linear by linear function:

$$YW = 28.083 + 34.32/(1 - 0.01836 \times HA)\ (P < 0.001;\ R^2 = 80\%)$$

where HA = hen age in days.

This curve is plotted in Fig. 18.3. To complete the analysis, albumen weight is regressed allometrically on yolk weight and shell weight on egg contents (yolk plus albumen weights), with the results shown in Fig. 18.3. All the allometric equations were highly significant ($P < 0.001$; $R^2 = 58.0$ and 38.7% (albumen, Cobb and Ross respectively); 70.4 and 52.4% (shell, Cobb and Ross respectively)).

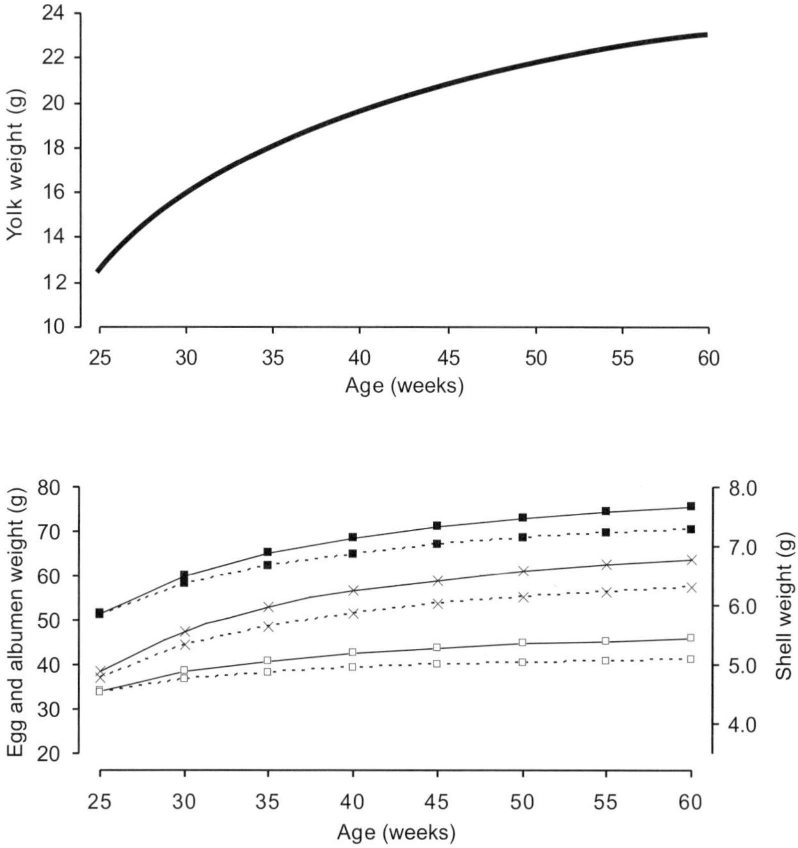

Fig. 18.3. Egg component weights in Ross and Cobb broiler breeder hens at different ages. Upper panel: yolk weight. Lower panel: albumen weight (□) and shell weight (×) predicted from yolk weight and resulting egg weight (■). Solid lines – Cobb; broken lines - - - Ross. See text for details of equations (M. Nonis, unpublished data).

Comparison of these data with similar analyses for three strains of egg-laying hen (Johnston and Gous, 2007b) shows that broiler breeder eggs have larger yolks at a given age, have similar (Cobb) or lower (Ross) rates of albumen increase with yolk size and have heavier shells. From Fig. 18.3 it can be seen that, although the two breeds had very similar yolk weights at a given age, the Cobb eggs have more albumen and slightly heavier shells and are therefore larger at a given yolk weight.

Variations in the dry matter content of egg components with age and other fixed factors have not been described in broiler breeders. Although they may occur this has been ignored here in the calculation of protein deposition rates.

Combining the above calculations with a reasonable assumption about rates of egg production gives the calculated rates of protein and lysine deposition in eggs shown in Fig. 18.4. The peak levels are about 6 g protein per day and 400 mg lysine, falling to about 4 g and 300 mg per day by the end of the laying year. There are small but consistent differences between the two breeds considered here, owing to the higher albumen (and total egg) weight in the Cobb birds used in the experiments.

Turkey breeders

Growth

Unlike broiler breeders, turkeys do not respond positively to feed restriction during the growing period. Whitehead (1989) summarized a number of reports on the subject and concluded that the reduced growth rate resulting from this practice also depresses subsequent reproductive performance. Growth rate to sexual maturity for potential breeders should therefore follow the same path as

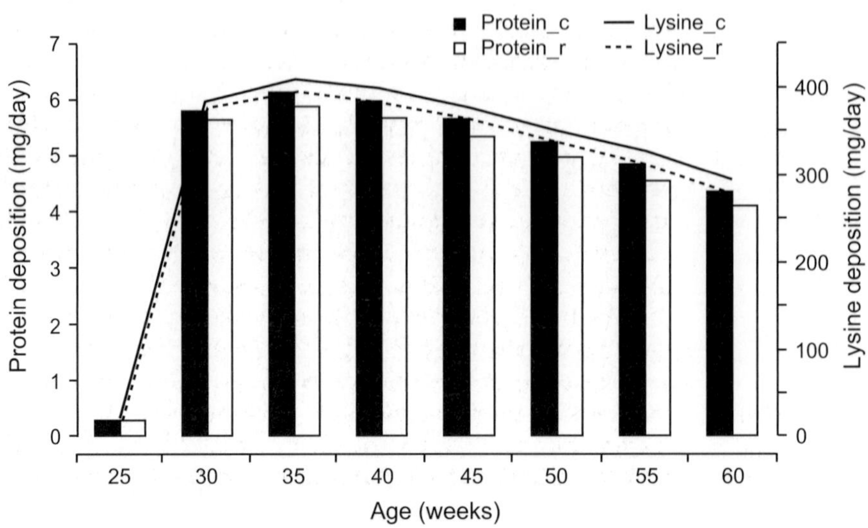

Fig. 18.4. Protein and lysine deposition in eggs of broiler breeder hens of two breeds (Ross – r and Cobb – c) during the laying period. See text for details of the calculations.

that of commercial growing turkeys, and the protein requirements are likely therefore to be similar. Suggested protein contents in feeds for growing turkey breeders are in Cherms *et al.* (1976). Amino acid contents in these feeds should be based on the amino acid content of the body and feather protein being grown and on maintenance requirements, and these can be obtained from Emmans (1989). The question of efficiencies of utilization of amino acids raised in the section on broiler breeders would not apply in the case of full-fed turkeys.

Again, unlike broiler breeders, body weight declines from the time the bird lays its first egg, due to a reduction in food intake. Even though food intake increases soon after egg production has commenced, body weight continues to drop over the period 30–41 weeks, then remains stable for 5 weeks before gradually increasing, but the body weight at 30 weeks is not achieved by 55 weeks of age (Whitehead, 1989). In an experiment reported by Whitehead, body lipid reserves decreased by 580 g during the period 30–43 weeks of age, these presumably being used as an energy source, whereas body protein content remained the same (1275 g/bird). At the end of lay, lipid reserves had increased by 290 g and body protein by 156 g, reflecting the increased food intake during this period. According to Whitehead (1989) these changes in body weight are a normal physiological characteristic of the turkey and hence do not influence reproduction. Attempts to reduce the loss in body weight with the use of high-protein pre-breeder feeds have not been successful (Mather and Harms, 1982; Grimes *et al.*, 1989).

Egg production

The mean daily egg output of turkeys at peak rate of lay (37–43 weeks) is similar to that of broiler breeders, at about 53 g/bird day, but the rate of laying is lower (70 eggs/100 birds) and egg weight higher (76 g/egg). In older hens, egg weight increases to between 80 and 90 g, but rate of laying decreases concomitantly. Consequently the daily amino acid intakes required to sustain such production would appear to be similar to those of broiler breeders, but the maintenance requirements would be considerably higher.

Very little information is available on the amino acid requirements of turkey breeders. A summary of the available data, plus the results of an amino acid response trial, was presented by Whitehead (1989), in which he concluded that the amounts of lysine and methionine required per g of egg output (so-called *a* coefficient, see below) were 10.6 and 5.75 mg/g egg respectively, which are similar to those for laying hens (10.0 and 4.8 mg/g, according to McDonald and Morris, 1985). A turkey female weighs in the region of 10 kg, thus her daily lysine requirement is estimated at 1.1 g, and methionine at 0.57 g, resulting in a protein requirement of about 25 g/day (Whitehead, 1989).This is considerably less than the daily intakes recommended by other authors, which range from 31 g for 8 kg turkeys (Balloun, 1974) to 44 g for 10 kg birds (Krueger *et al.*, 1978), but in agreement with the 25–27 g suggested as adequate by Wilgus (1976).

The relative proportions of yolk and albumen have been reported to be almost identical in turkey and chicken eggs, being 559 and 323 g/kg for albumen and yolk, respectively, in turkeys, and 558 and 319 g/kg, respectively, in chicken eggs (Romanoff and Romanoff, 1949), although these proportions may have changed in modern strains. No reports could be found of the changes that might occur in these proportions as the hens age.

Duck breeders

Growth

Ducks respond positively to feed restriction during the growing period, this causing a delay in age at sexual maturity and an increase in the number of eggs laid to a fixed age (Olver, 1984, 1988; Cherry, 1993). According to these authors, there is a minimum threshold live weight essential for achieving sexual maturity, which appears to be between 50 and 65% of the *ad libitum* live weight. Minimum live weight at sexual maturity to achieve maximum rate of lay is around 75% of *ad libitum* weight (Cherry, 1993). Ducks become sexually mature at around 25 weeks of age, at which time they weigh between 2 and 6 kg (mean 3.2 kg), depending on the strain. An increase in live weight of about 1 kg is expected by 55 weeks.

Cherry (1993) found that there is little, if any, opportunity of manipulating body composition at a specified live weight through manipulation of nutrient intake; hence feed composition appears to have no effect on age at maturity or subsequent performance. Where he reared ducks on feed restriction to 9 weeks and then allowed them *ad libitum* access to feed, they attained their mature body weight by 12 weeks, came into lay at 24 weeks and laid only 15 eggs fewer (164 versus 179) than those restricted through to 18 weeks, even though the *ad libitum* birds were substantially fatter than the restricted birds. Food intake during this repletion phase reached 400 g/day. Evidently, it is not the excess fatness that prevents ducks from achieving a high rate of laying but rather the early sexual maturity that results from full-feeding of ducks during the entire growing period; restriction of growth rate in the first weeks of life prevents ducks from achieving early sexual maturity.

During lay, ducks attempt to achieve a minimum live weight of at least 80% of *ad libitum*-fed mature live weight. Any attempt to reduce this live weight further, by relatively severe feed restriction during the laying period, will result in a decrease in rate of laying (Cherry, 1993).

Feed restriction in ducks is best achieved by reducing the amount of time during which feed is available to the birds. Restricting feeding time to 12 h reduces *ad libitum* feed intake by 2%, whereas giving birds access to feed for only 4 h/day restricts intake by 14% (Fig. 18.5) (Cherry, 1993).

Egg production

The mean number of eggs laid by females to 55 weeks may be as high as 220 (Cherry, 1993), with rate of lay remaining around 93% from 30 to 55 weeks. Mean egg weight is around 82 g; thus the egg output for almost the entire laying period remains at 76 g/bird day. A large proportion of eggs laid by

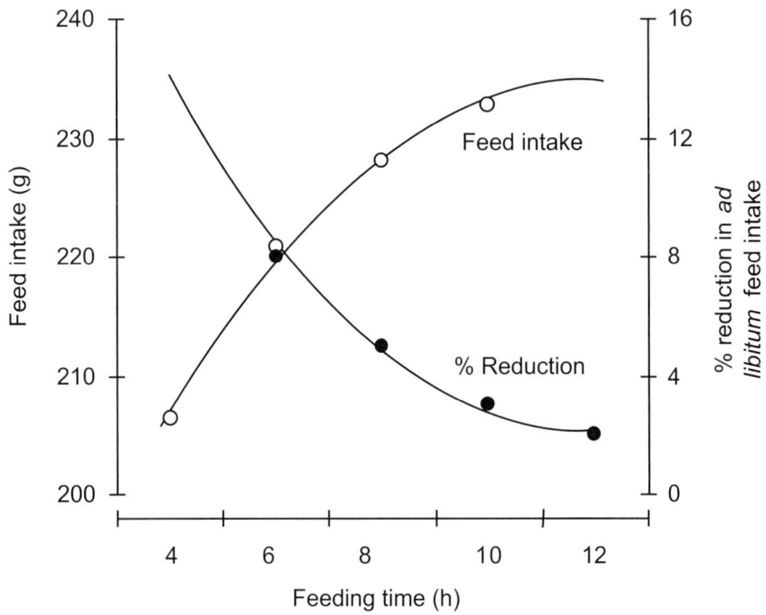

Fig. 18.5. Effect of controlled feeding time on feed intake of Pekin females with a mature weight of 3.6 kg laying about six eggs per female per week and given a diet containing 10.8 MJ ME/kg. Data from unreported trials (P. Cherry, unpublished results, 1989).

commercial ducks before 25 weeks of age weigh less than 70 g, and these eggs are unsuitable for producing commercially viable ducklings; hence there is no advantage in achieving sexual maturity before this age (Cherry, 1993).

Egg weight can be manipulated to an extent by controlled feeding (Fig. 18.6), the range being from 78 to 92 g. Intake of a feed containing 200 g protein/kg required to achieve these weights varied from 170 to 250 g/day (Cherry, 1993) and is most successfully achieved by timed feeding. Hatchability of eggs declines when their weight exceeds 92 g. In the tropics, high temperatures result in lower food intakes and hence lower egg weights than at temperatures below 22°C, but rate of laying remains the same.

RESPONSE TO AMINO ACIDS IN BROILER BREEDER HENS

The economic response to dietary amino acid level is established in laying hens and growing animals by experimental response studies in which different levels of amino acids are fed to groups of birds and the response curve established. Such studies have not been widely used in breeding birds, presumably because the response characteristics of interest are complex and have to be measured over long periods of time. However, such experiments are of interest because they illustrate issues of amino acid utilization and can also form the basis of comparison with a much larger body of evidence available for laying hens.

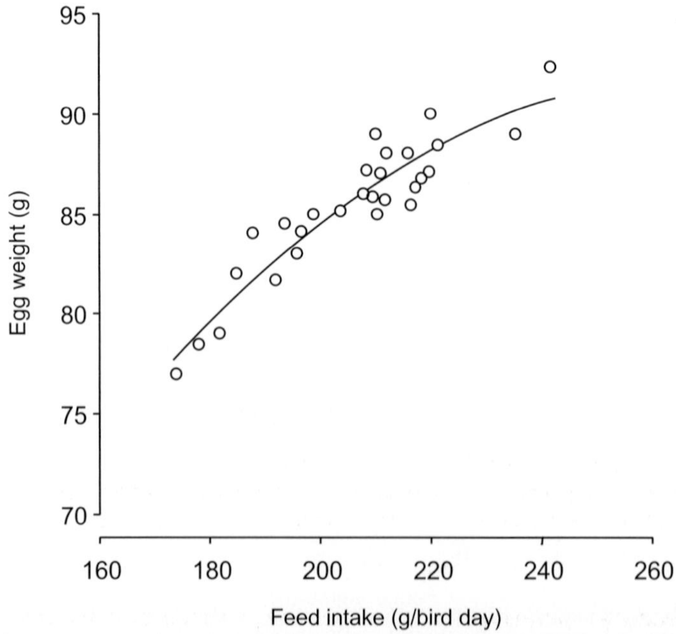

Fig. 18.6. Relationship between controlled feed intake and average egg weight between 34 and 40 weeks of age for Pekin parent stock with a mature weight of about 3.8 kg given a diet containing 10.8 MJ ME/kg and 200 g protein/kg feed. Data from Cherry (1993).

Figure 18.7 shows the response in egg mass to lysine intake from two similar experiments with broiler breeders. Response curves fitted using the Reading Model (Fisher *et al.*, 1973) provide estimates of two parameters, namely a = mg lysine per gram increment in egg output (estimated from the slope of the rising part of the line) and b = mg per kg body weight maintained (estimated by extrapolation of the line to zero output). The estimates of a from the data shown are 13.91 and 15.25 mg/g, and for b are 0.48 and 0.00 mg/kg for the experiments of Fisher *et al.* (2001) and Bowmaker and Gous (1991).

These values differ from those typically reported for egg-laying hens, a being larger and b much lower. For lysine, McDonald and Morris (1985) suggested general values of 10 and 73 mg/kg for a and b respectively from an analysis of several trials. These differences were discussed by Fisher *et al.* (2001). Briefly, the differences in the utilization of lysine for egg production can be associated with the higher incidence of pauses in the egg-laying patterns of broiler breeders. The differences in estimated maintenance requirements remain unexplained, as described in more detail below.

PROTEIN UTILIZATION IN BREEDING HENS

Protein utilization coefficients or protein requirements for specified functions allow protein nutrients in the feed to be related to rates of protein 'use' in the animal. Protein used in the body is specified in exact chemical units but that in

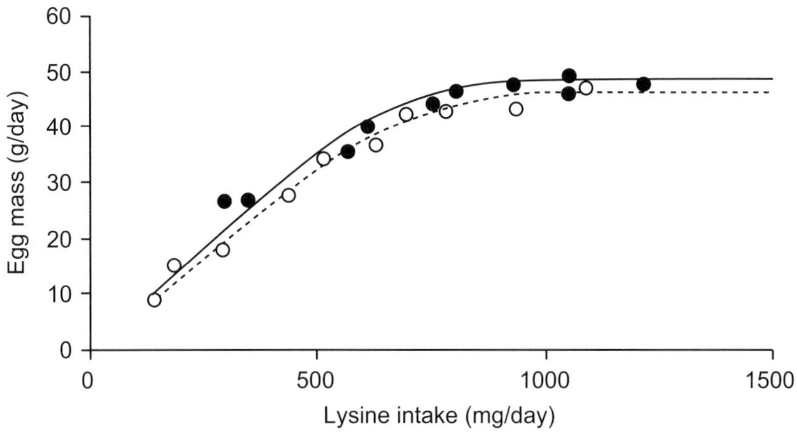

Fig. 18.7. Response in egg mass to lysine intake in two similar experiments. Data from Fisher *et al.* (2001) – solid line and symbols; and from Bowmaker and Gous (1991) – broken line and open symbols. For details of the fitted response curves see text.

the feed may be expressed either in similar chemical units or in units which reflect the digestibility and availability of the nutrient to the animal. Clearly the units used to describe the feeds will be reflected in some way in the utilization coefficients. This is an important practical issue but it is not discussed here. Utilization of protein in the animal is described, with an assumption that the problem of describing feeds and their digestion and availability can be satisfactorily resolved.

All of the information considered relates to broiler breeding hens and the animal functions considered are maintenance, growth and egg production.

Utilization of protein for maintenance

Protein maintenance in poultry continues to be a significant nutritional problem, and in the discussion of amino acid requirements of broiler breeders was identified by Fisher (1998) as a major unresolved issue. The difficulties arise from imprecise definition of what the term maintenance means and especially from uncertainty about experimental methods to determine it. Different scaling rules also bring confusion to the topic.

Emmans and Fisher (1986) first proposed that maintenance requirements for protein and amino acids should be scaled to body protein and, second, proposed a rule which derives from the genetic scaling rule of Taylor (1980):

$$MP = m_p P_m^{0.73} u,$$

where MP = maintenance protein requirement (g/day),
m_p = 0.008 kg ideal protein/unit day,
P_m = mature protein weight (kg),
u = degree of protein maturity (P/P_m).

The composition of ideal protein for maintenance is an issue which remains to be resolved.

These are the most complex scaling rules proposed for dealing with maintenance and are very different from the more usual practice of scaling to body weight to a power of 1, 0.65, 0.73 or 0.75. This remains a controversial area of protein nutrition and it is difficult to devise experiments which will test the various approaches used. The assumptions are important in broiler breeder nutrition because maintenance accounts for about one-third of the amino acid requirement in adult birds and because body weight changes are mostly associated with changes in body fat. For large breeding birds, scaling maintenance to body protein rather than body weight seems to be a logical proposal.

Fisher (1998) summarized data on maintenance lysine requirements from two main sources: N-balance trials with adult roosters and the extrapolation of nutrient response data to zero output. For the purpose of calculation, the equation above was used and ideal protein for maintenance was assumed to have the same composition as body protein. Since that date the following information about maintenance requirements for lysine in broiler breeders has been published.

Fisher *et al.* (2001) reported a lysine dose-response study similar to that of Bowmaker and Gous (1991) and estimated maintenance by extrapolation of the egg mass–lysine intake curve to zero output using a model described by Fisher *et al.* (1973). The estimates in four trials conducted successively during the laying year were indistinguishable from zero. Bowmaker and Gous (1991) had previously reported a value of 11.2 mg lysine per kg body weight.

Sakomura and Coon (2003) reported requirements of 94.2 mg/day/kg $BW^{0.75}$ or 333.3 mg/day/kg body protein. These results were based on N-balance determined by comparative slaughter of 23-week-old immature pullets after feeding purified diets supplemented with single amino acids for 3 weeks. Standard errors on these figures were not reported but the regressions of N-accretion on lysine intakes had r^2 values of about 50%. It seems questionable whether this experimental model will provide an estimate of maintenance which is both accurate and uninfluenced by tissue growth.

Nonis and Gous (2007) estimated maintenance from N-balance data on adult cockerels of an egg-laying strain given, by intubation, different amounts of a lysine-limited feed and free access to an N-free diet for 6 days. The regression of N-retention on lysine intake had an R^2 value of 91.8%. Estimates of lysine maintenance requirement were 39 mg/day/kg BW or approximately 245 mg/day/kg body protein.

The confusion about lysine requirement for maintenance in broiler breeders seems to continue, as it does, but to a lesser extent, for other amino acids. It is important in any application of the factorial method for calculating requirements to make explicit the assumptions about maintenance.

Utilization of protein for egg production

At the level of the flock, measured over time the utilization of amino acids for egg production is lower in broiler breeder than in egg-layer hens. This is assumed to reflect the lower and more irregular pattern of oviposition in broiler breeders. Put very simply, over periods in which hens lay no eggs but continue to eat feed their efficiency is zero, and such periods are a more prominent feature of broiler breeder egg production than of egg-layer production. An attempt to quantify this effect is described by Fisher (1998) and was discussed in the light of one experiment by Fisher *et al.* (2001).

In the model of protein response described below, the sequence of egg laying and pauses, and therefore the time pattern of protein synthesis, is simulated and no assumptions need to be made about the efficiency of amino acid utilization. However, in normal factorial methods, where egg output is averaged over a flock of birds and over a time period, then some system of accounting for the effects on amino acid utilization needs to be included. Methods used for egg-laying hens, which consistently and uniformly lay in closed cycles over long periods, cannot be used directly for broiler breeders, which have more intermittent and variable patterns of oviposition.

The system proposed by Fisher (1998) is summarized by the two graphs in Fig. 18.8. Very simple assumptions are made about the relationship between rate of lay and amino acid utilization for individual birds, and this is combined with the distribution of individual bird laying records to produce a relationship for flocks of birds. This idea and the data on which Fig. 18.8 is based are derived from experiments on egg-laying hens (Fisher, 1994) and, in general, it is not very well supported by the experiment described by Fisher *et al.* (2001). No further work has been carried out or found in the literature on this topic. It is assumed that the ideas in Fig. 18.8 are applicable to calculations made for broiler breeders but that better quantification is required.

Utilization of protein for growth

The questions which arise about the utilization of protein in the grower or developer phase of broiler breeder rearing remain largely unanswered. A simple assumption is that utilization is the same as in full-fed birds at a similar age. However, restriction of growth by energy intake may influence partition rules for energy and protein, and, in particular, it would be expected that utilization is influenced (negatively) by the practice of intermittent or skip-a-day feeding. Skip-a-day feeding involves feeding twice the daily feed allowance on successive days or, in some cases, feeding five-sevenths of the weekly allowance on 5 days with no feed on 2 successive days. In either case the effects of meal feeding on protein utilization come into question. In particular the utilization of crystalline amino acids in the feed is likely to be impaired (Bach Knudsen and Jorgensen, 1986), although the details have not been worked out in broiler breeder birds.

Other factors influencing protein utilization

The utilization of crystalline amino acids in adult broiler breeder hens has been examined recently by Nonis and Gous (2007). Such birds are normally fed

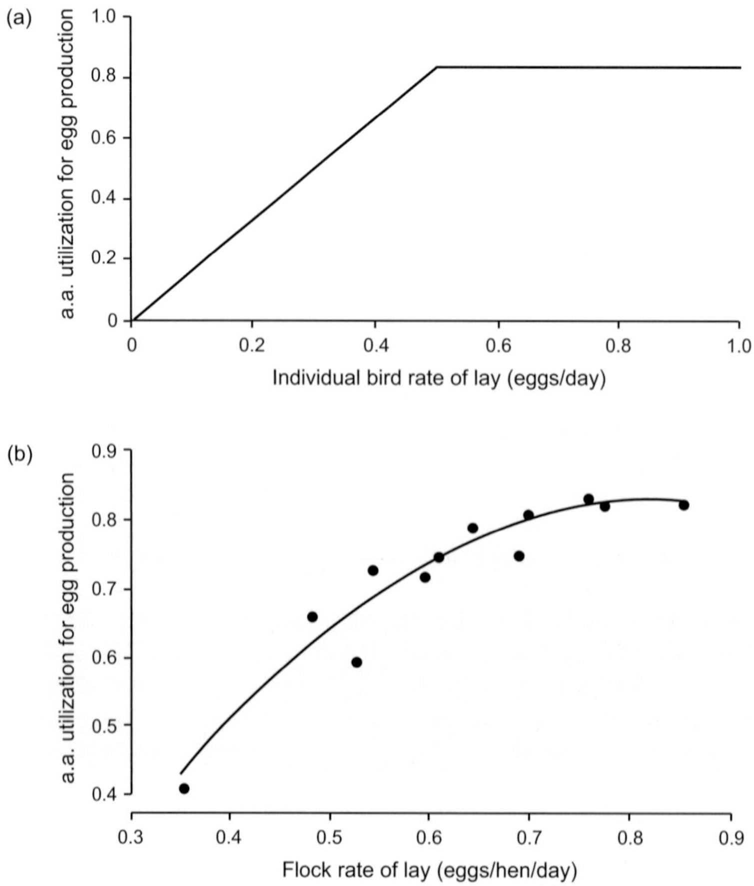

Fig. 18.8. Method used to relate efficiency of amino acid utilization to flock rate of lay (from Fisher, 1998). Figure 18.8a shows the presumed relationship between the instantaneous rate of lay of a single bird and efficiency of amino acid utilization for egg production. Figure 18.8b shows the relationship between flock rate of lay and efficiency. The figure is based on the relationship for an individual bird, as shown in the upper figure, and the distribution of individual bird records at different rates of lay as described by Overfield (1969).

once per day, early in the morning, and may eat their meal in as little as 20–30 minutes. This raises problems about nutrient utilization in comparison with continuously fed birds, and especially of calcium and of amino acids supplied in crystalline form. The general issues involved and evidence from other species have been reviewed by Bach Knudsen and Jorgensen (1986).

Nonis and Gous (2007) fed a series of five diets ranging from a high crude protein level (183 g/kg) and no synthetic amino acids to a low protein (150 g/kg) and a total of 2.3 g/kg synthetic amino acids (lysine and methionine). Methionine appeared to be the first limiting amino acid. Feed, at a fixed level, was given to caged breeder hens either once or twice per day. An outline of the treatments and the results for egg production and protein and methionine utilization are shown in Fig. 18.9.

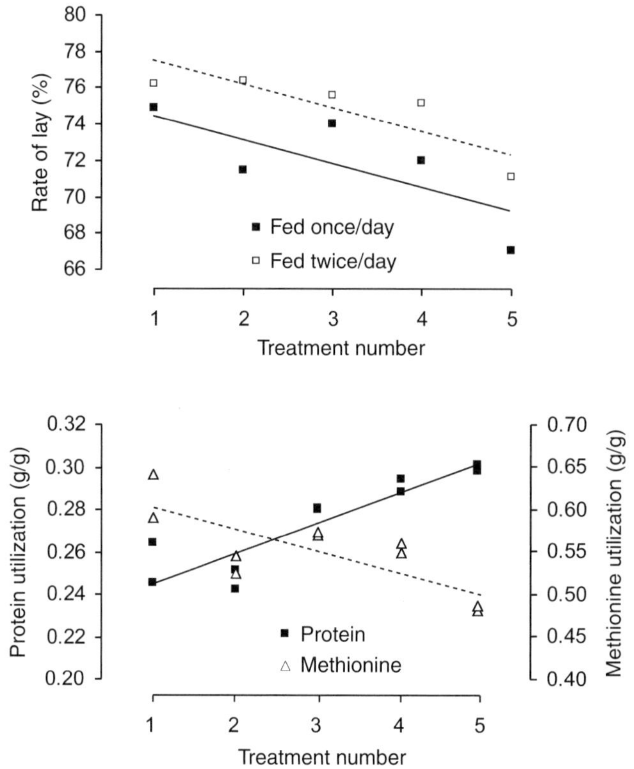

Treatment	1	2	3	4	5
CP, g/kg	183	175	167	158	150
DL-methionine	0	0.15	0.30	0.45	0.60
Synthetic met+lys	0	0.6	1.2	1.7	2.3

Fig. 18.9. Effect of different levels of synthetic amino acids (methionine and lysine) in feeds given once or twice per day (160 g per bird day) to broiler breeder hens. Rate of lay over the last 4 weeks of a 10-week trial (a) and calculated crude protein and methionine utilization (b). The lines are (a) linear regressions of equal slope (*P* different slopes = 0.665) and different intercepts (*P* different intercept = 0.028) and (b) linear regressions ($P < 0.001$). From Nonis and Gous (2007).

As expected, protein utilization increased as synthetic amino acids replaced crude protein, but methionine utilization declined as the synthetic form provided an increasing proportion of the total supply. The reduced supply of 'utilizable' methionine was reflected in lower egg production, but egg weight and feed intake were not significantly affected. Feeding twice each day increased egg production significantly. Methionine utilization was very slightly increased by more frequent feeding (means across all treatments 0.545 and 0.559 for one and two times feeding), but this effect was far from being significant. Data for synthetic amino acid utilization in growing animals, especially lysine in pigs (see Bach Knudsen and Jorgensen, 1986), led to the expectation that reduced utilization would be observed in broiler breeder hens fed once per day. This is confirmed by the results of this trial.

DIETARY PROTEIN SUPPLY AND BREEDING MALES

In the broiler industry, males are typically reared separately from the females so that growth and development can be controlled. After the creation of natural mating groups at sexual maturity (in proportions of approximately ten females to one male), males will be fed separately by the use of various management devices so as to maintain control of average feed intake. The use of separate feeds for males is a very variable practice, and the feeding of males on the same feed as the females is probably most widespread. Thus males frequently receive far more nutrients than required, especially calcium and protein, and the most important question is whether this is detrimental to biological performance.

The rearing of males and the assessment of their lifetime performance are even more difficult than for females, and hence nutritional effects remain very uncertain. Overall there does seem to be a negative effect of higher protein levels, in growth at the later stages of rearing and during the reproductive period, but it is not well quantified. A recent publication by Romero-Sanchez *et al.* (2007) describes a well-executed trial and also illustrates very well the difficulties involved. These authors reared males from 2 to 26 weeks of age on two growth patterns, described as 'concave' and 'sigmoid', using rearing feeds with 140 or 170 g/kg crude protein. Fertility was assessed by natural mating at regular intervals and the results shown in Fig. 18.10 reported. The recovery

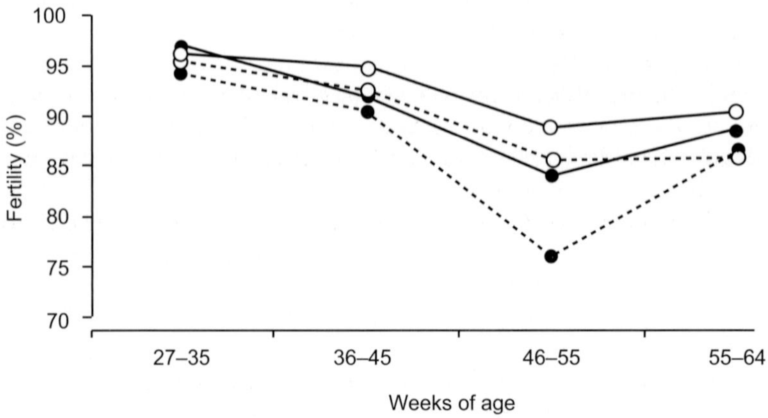

Fig. 18.10. The effect of growth curves and dietary protein level in the rearing period on fertility of males assessed at various ages (from Romero-Sanchez *et al.*, 2007). Males were reared on either a 'concave' (•) or 'sigmoid' (○) growth curve from 2 to 26 weeks using rearing diets containing either 120 (—) or 170 (- - - -) g crude protein/kg.

in fertility between 46 and 56 weeks was associated with an increase in feed allowance and drew attention to the main conclusion from the work, namely the importance of energy supply in the later production period. However, it is clear that the higher level of protein depressed fertility, especially in combination with the concave feeding programme, which involved a higher level of feeding in the later stages of growth.

In early studies of the effect of dietary protein in the feed given during the breeding period, McDaniel (1986) demonstrated that male fertility is improved when low-protein feeds are used, and this led to the practice of separate-sex feeding of breeders. Subsequently several papers have investigated the effect of crude protein (CP) on semen production and fertility in broiler breeder males (Wilson *et al.*, 1987a,b; Hocking, 1989; Revington *et al.*, 1991; Hocking and Bernard, 1997; Zhang *et al.*, 1999). Although the effects are not universal, the general conclusion from this research corroborates the work of McDaniel (1986), showing that low-CP diets have an advantage in semen production over high-CP diets.

DIETARY PROTEIN SUPPLY AND THE EMBRYO

Wiley (1950) reported that embryonic development during the later stages of incubation may be restricted within small eggs laid by young broiler breeders, by the limited space within the eggshell. Bray and Iton (1962) termed the effect of initial egg weight on embryo weight 'a temporary environmental influence which begins after 11 days of incubation', while correlations of near zero have been reported between egg weight and embryo weight up to 14 days of incubation (Hassan and Nordskog, 1971; Al-Murrani, 1978). Growth restriction of embryos within small eggs has been attributed to the protein content of the egg (Al-Murrani, 1978), the decreased oxygen availability (Tullet and Deeming, 1982), the inefficient use of egg nutrients by the embryo (Shanawany, 1984), disturbances in the mechanisms of embryonic metabolism (Noble and Connor, 1987), and the degree of yolk-sac utilization (Wilson, 1991). Tullet and Burton (1982) claimed that the variation in chick weight at hatch not only was accounted for by the size of the egg but was explained by water loss during incubation. Thus, when chicks are incubated in a moist, rather than a dry, environment, the chicks are heavier at hatch.

Subsequent to the above, relatively old, references, when broilers grew more slowly than they now do, McLoughin and Gous (1999) reported that embryonic growth was significantly correlated both with flock age (from 30 to 52 weeks) and with egg size, with growth rate in the smallest eggs falling behind that in larger eggs as early as 8 days after the start of incubation. This 'maternal effect' appears to constrain the growth of broiler embryos to a greater extent now than in the past, this being the result of the improvement in potential growth rate in broilers without a concomitant increase in the size of the egg to accommodate this increased growth rate. Egg size will therefore have an impact on the size of the resultant chick through this maternal effect, and not necessarily because of a nutrient deficiency in the hen's diet.

CARRY-OVER EFFECTS OF MATERNAL PROTEIN SUPPLY

In a summary of a number of trials in which laying hens had been subjected to a range of diets differing in amino acid content, Morris and Gous (1988) showed that egg weight and rate of laying were equally reduced at marginally deficient amino acid intakes, but, as the deficiency became more severe, rate of laying was reduced to a greater extent than was egg weight. The same effect has been observed in broiler breeders (Bowmaker and Gous, 1991), with the maximum decline in egg weight being to 0.8 of the highest egg weight, the latter being achieved on the greatest daily allocation of amino acids. It is unlikely that the amino acid supply to broiler breeders would be restricted to such an extent in commercial practice, but it would be expected that smaller eggs resulting from lower protein intakes would result in smaller chicks at hatching, as studies have shown that the protein levels fed to breeders in production can affect chick body weight and final broiler performance.

Apart from the smaller size of the day-old chick resulting from a smaller egg, there appears to be no other effect of maternal protein supply on chick quality or resultant performance (Kidd, 2003). Where these small chicks are raised together with chicks of a larger size, it is likely that competition for water and food will hinder their growth such that their subsequent performance will be compromised.

EMPIRICAL EVIDENCE ABOUT PROTEIN SUPPLY TO BREEDING BIRDS

Unlike broilers and laying hens, empirical experimental evidence about the protein nutrition of breeding birds has been less successful in establishing reliable guidelines for practical feeding. In addition to such experiments, collective field experience, reflected, for example, in the recommendations of breeding companies, and individual company experience play a large part in guiding commercial decisions. The limitations of empirical evidence arise from the complexity of the systems under study and from the long-term and variable nature of the experiments themselves. In broiler breeders, in particular, the need to use a controlled feeding procedure to achieve a predetermined body weight (see Hocking, Chapter 17, this volume) and the consequences of this for responses to different nutritional treatments make the comparison of different trials and the extension of the results into practice very difficult. The earlier information of this kind was well reviewed by Lopez and Leeson (1994). Selected papers are referred to here but we have not tried to be comprehensive and have not set time limits in the selection of papers.

Broiler breeders

Experiments on early protein nutrition

The general aim of supporting high growth rates in the first few weeks of a broiler breeder hen's life is well recognized. Use of inadequate protein levels in this period will restrict bone growth and frame size, increase flock variability and may lead to lower egg production. The trial by Hudson *et al.* (2000) illustrates the principles involved and provides an update on this topic. Protein levels of 120, 160 and 200 g/kg were fed to 6 weeks of age using a typical commercial feed management programme. Although feeding levels were adjusted to give similar 6-week weights and at later stages the weights and dimensions of all birds were essentially similar, persistent treatment effects on egg production to 33 weeks of age were recorded. An interesting observation was that the laying of eggs on the floor, an undesirable behavioural trait in broiler breeders, was increased by the use of a low-protein diet some 25 weeks previously.

Experiments in the grower or development period

It is generally accepted that feed intake of broiler breeders during the growing period needs to be restricted to maximize egg production and fertility, and to minimize mortality. Whereas a considerable literature on this subject exists, there are few references to the dietary protein requirements during rearing, and among these there is some controversy.

Walsh and Brake (1997) concluded, from a series of studies in which they used different dietary protein contents in the rearing period to rear broiler pullets to the same weight at 20 weeks, that a cumulative crude protein intake of at least 1180 g per pullet to 20 weeks of age was needed to achieve acceptable persistency of fertility in lay. Food intake to and lipid content of these birds at 20 weeks were not reported, but because food intake was adjusted to ensure a similar body weight at 20 weeks it would be expected that considerably more feed would have been allocated on the lowest protein feed to achieve the same body weight gains, as was reported by Lilburn *et al.* (1987) and Hocking *et al.* (2002). As a consequence, differences in body lipid content would have resulted, with pullets on the lower protein feeds exhibiting higher body lipid, and hence lower body protein contents, the former being associated with reproductive dysfunction (Frisch, 1980). This, in turn, is likely to have affected sperm storage function, as suggested by McDaniel *et al.* (1981). It is thus more likely that the reduction in fertility reported on the lowest protein feed was a consequence of the resultant carcass composition at 20 weeks than because the birds had not consumed a minimum cumulative amount of dietary protein. A disturbing feature of the first of the trials reported by Walsh and Brake (1997) was that 22% of the females receiving the lowest protein feed during rearing died as a result of injuries during mating. The surviving females would probably have attempted to avoid mating, and this would have contributed substantially to the low fertility in this group, rather than the cumulative protein intake to 20 weeks. Hocking *et al.* (2002) have subsequently shown no effect of low cumulative protein intakes on fertility or hatchability, nor was there an effect on these two parameters when a conventional rearing feed was diluted with 0.15 or 0.30 of oat hulls (Zuidhof *et al.*, 1995).

Whereas the hypothesis advanced by Walsh and Brake (1997) may be refuted, the use of low-protein feeds during rearing is unwise from the point of view of the development of the chemical body. When low-protein feeds are offered, high food intakes are required to achieve the desired body weight, resulting in a carcass at 20 weeks that is higher in lipid and lower in protein than is appropriate for a breeding bird. Some adjustment of this composition may be possible in the pre-laying period, as has been demonstrated by some authors.

Experiments in the pre-laying period

During the pre-laying period the reproductive organs develop, the liver increases in size (Yu and Marquardt, 1974) and the oestrogen-dependent hepatic biosynthetic pathways are activated (Schjeide *et al.*, 1963). Thus the rate at which the daily food (and protein) allocation should be increased and the age (or stage of development) when this is initiated are of importance.

Cave (1984) and Brake *et al.* (1985) demonstrated improved egg production due to an increased protein intake during the 18- to 24-week pre-breeder period, and they surmised that the additional dietary protein may either elicit greater control over the oviduct, thereby reducing the number of internal layings, or increase reserves of available protein during this period, which would sustain subsequent reproductive performance. The rate of internal laying is likely to be strain-related, which might explain the large difference in the response to additional protein in the two strains used by Cave (1984), the Hubbard strain producing only five additional eggs versus 14 in the Shaver Starbro strain on the high-protein treatment. If this were the reason for the improved performance then the practice of feeding high-protein pre-breeder feeds will be less valuable now, given the improvement in egg production in modern broiler strains.

It is unlikely that a high-protein pre-breeder feed is needed to meet the requirements of the developing ovary and oviduct, as Bowmaker and Gous (1989) calculated the protein requirement by broiler breeder pullets in the period just prior to the onset of laying to be just over 10 g/day, taking account of maintenance requirements and those for the growth of the reproductive organs. A feed containing only 90 g protein/kg would satisfy this requirement if food intake is increased from 100 to 150 g/day between 20 and 26 weeks of age. It is equally unlikely that the high-protein pre-breeder feed builds up reserves of protein in the body (liver), to be available later in lay, as suggested by these authors. Bowmaker and Gous (1989) demonstrated that a range of protein intakes considerably wider apart than those used in the above experiments resulted in differences in liver protein content of only 2.05 g at 30 weeks of age, this difference being insufficient to explain the substantial improvement in egg production noted in the above experiments.

There have been reports in which no advantage has been measured of feeding high-protein pre-breeder feeds. Van Wambeke (1992) fed broiler breeder pullets from 17 to 24 weeks of age on feeds containing either 140 or 175 g crude protein/kg and reported no differences in reproductive traits thereafter. Grimes *et al.* (1989) had previously shown the same lack of effect of dietary protein content (120, 150 and 180 g/kg) in the pre-breeder period on reproductive performance of turkeys. It would appear that a pre-breeder feed high in protein would be advantageous only in cases where pullets have been reared in such a manner that their body composition at 20 weeks has been compromised. The extent to which body composition during this period may be adjusted has not been reported, but it is possible that a high-protein feed will enable the bird to deposit additional protein, given that such a bird would be at a lower degree of maturity than one reared on a higher protein feed, as its potential rate of protein deposition would be higher.

Experiments in the laying period

Measuring the response to dietary protein in the laying period is fraught with difficulties. The rate of laying differs between strains, with age and within individuals. Additionally, broiler breeders tend to exhibit less uniformity in their reproductive cycles than do laying hens, with internal laying and more than one pause day between clutches being relatively common. As will be demonstrated below, such aberrations lead to apparently lower efficiencies of utilization of amino acids and protein for egg production. Experiments that report on the response in reproductive performance of broiler breeders to dietary protein content over the full laying period are therefore relatively unhelpful in determining how to feed these birds. Recommended protein intakes vary from 16.5 g/day, for individually housed breeders (Pearson and Herron, 1982), to 22 g/day (Waldroup *et al.*, 1976), with many values in between (for example, 19.0 g/day by Spratt and Leeson, 1987; 19.5 g/day by Pearson and Herron, 1981; 20.6 g/day by Wilson and Harms, 1984).

Of greater value are experiments in which responses have been measured in individual breeder hens, as this enables calculations of efficiencies of utilization. The relevance of such trials has been questioned on the grounds that broiler breeders are fed in large groups commercially, and, because they are not allowed *ad libitum* access to food, not all birds will consume the same amount of feed daily. This issue would be difficult to resolve experimentally, but some attempt has been made to address the issue through modelling, and this is discussed further below. Setting aside the problem of not being able to guarantee that all birds consume the same amount of food daily, experiments with individually caged breeders have yielded valuable information that may be used to construct models of responses to amino acids. Some of these experiments are discussed below.

Pearson and Herron (1981, 1982) measured responses in individually caged broiler breeders to a range of crude protein and energy contents and provided carcass analysis data as well as production data. They reported no effect of a wide range of crude protein intakes (16.4–27 g/bird day) on reproductive performance, except at the lowest energy intake, where egg production was significantly lower than on the other treatments. Their suggestion was that a lower allowance of crude protein than 16.4 g/day might be adequate, given the high performance achieved on this intake in their trial. Of interest was the increase of 105 g body protein in birds between 21 and 64 weeks of age (Pearson and Herron, 1982): the pullets reached sexual maturity at 27 weeks, during which time the liver, ovary and oviduct would have increased in weight by about 25, 90 and 60 g respectively (Bowmaker and Gous, 1989), this increase accounting for most of the change in protein content, confirming the theory that hens do not deposit body protein once they become sexually mature (Nonis and Gous, unpublished).

Broiler breeder hens were used in an experiment by Bowmaker and Gous (1991) to measure the responses to dietary lysine and methionine over a 10-week period (29–38 weeks of age).The hens were offered 150 g/day of one of 20 dietary treatments, ten being lysine-limiting and ten being methionine-limiting. The feeds were mixed by diluting one of two concentrate (summit) mixes with a protein-free dilution mixture. The lysine-limiting summit feed was designed to supply approximately 1300 mg lysine/bird day, while the other supplied 520 mg methionine/bird day, when fed at 150 g/bird day. The minimum egg weight recorded was 0.8 of the maximum, whereas the rate of lay of birds fed diets with the lowest amino acid concentrations was 0.2 of the maximum. The daily requirement for broiler breeder hens of 3 kg, producing 45 g of egg output per day, was estimated to be 800 mg of lysine and 320 mg of methionine. Birds on the five lowest concentrations of both lysine and methionine did not consume the allotted amount of food, the amount decreasing, in a curvilinear fashion, to approximately 105 g/bird day. Whereas such low-protein feeds are unlikely to occur in commercial practice, this feed intake response raises interesting questions about factors constraining food intake.

In a similar trial, reported by Fisher *et al.* (2001), lysine dose–response experiments were conducted with caged broiler breeder hens at 26, 37, 48 and 60 weeks of age, each for 10 weeks. Requirements, estimated from fitted asymptotic response curves for egg output, were 864, 859, 763 and 687 mg true faecal digestible lysine at respective ages. Measured and calculated estimates of the efficiency of utilization of lysine for egg production were lower than comparable data from egg-laying hens.

INTEGRATION: A MODEL OF PROTEIN UTILIZATION IN FEMALE BROILER BREEDERS

The integrated model of broiler breeder production and nutritional response (M. Nonis, unpublished PhD thesis, University of KwaZulu-Natal and R.M. Gous) predicts the performance of a flock of broiler breeder hens after maturity, taking account of dietary energy and amino acid levels. So far as protein and amino acids are concerned the model integrates the information discussed above in a more flexible way than the usual factorial calculations. The following is a brief description of the model components.

1. A population of birds with appropriate body weights and initial body fat levels is generated. Age at sexual maturity for each bird is predicted from body weight and from information about rearing lighting programmes (Lewis *et al.*, 2008).
2. Egg production, clutch patterns, yolk weights, egg component weights, double-yolk eggs, internal laying and soft-shelled eggs are predicted using the model described by Johnston and Gous (2007a) adapted for broiler breeders. This encompasses a description of the genotype.
3. Body protein growth, including feathers, is assumed to be complete at sexual maturity. Body weight changes after maturity are due entirely to accumulation of fat, with no obligatory requirement for growth.
4. Energy transactions are expressed in units of effective energy (Emmans, 1994).
5. Protein requirements for egg production are calculated from the composition of the next egg to be laid and an assumed fixed chemical composition. Utilization of amino acids for egg production is in the region of 0.80–0.85, as derived for laying hens (Fisher, 1994).
6. Protein requirement for maintenance is calculated from the equation given above (page 343, see Emmans and Fisher, 1986). At present the composition of ideal protein for maintenance is assumed to be the same as body protein. The main implication of this is that the lysine maintenance requirement will be high (see discussion in Fisher, 1998). For calculating maintenance from this equation birds are assumed to be at their mature weight (u = 1). This is in contrast to the assumption by Fisher (1998) that mature body protein = 0.863 kg, which would reflect full feeding of the hens. The consequence of this difference is not very great.
7. Protein partition rules are that maintenance has highest priority, yolk protein deposition second and albumen protein third. Provided that sufficient nutrients are available, yolk protein deposition is assumed to occur continuously to meet the predicted growth rate, unless inter-clutch intervals exceed 2 days, when yolk growth ceases. Nutrients (energy and amino acids) for albumen production accumulate in a pool, which has some controlling influence over ovulation. If the nutrients in the albumen pool will support the predicted development of the next egg then ovulation will occur; otherwise ovulation is delayed until sufficient nutrients are available. All events are timed within a 24-hour day, assuming that feeding is at one time in the morning.
8. Excess energy, including that from unused protein, is stored as body fat within user-controlled limits. Similarly, body fat is always available as an energy source within the bird but subject to a minimum body fat level.
9. A population of birds is created assuming the following factors to vary (the default CV% is shown in parentheses). Initial body weight (10), initial body lipid weight (10), initial body protein weight (3), age at first egg (7), yolk weight constant (10), 'aggressiveness' (10). 'Aggressiveness' is an index producing different feed intakes around the controlled mean. At present there are no covariance terms but these may need to be added to the model. Variation in egg-laying patterns is introduced, as described by Johnston and Gous (2007c).

10. Feeding is described by a daily allowance of a feed of specified composition for up to 280 days. For individual birds this is adjusted according to the aggressiveness index. The desired feed intake is calculated each day according to the supply of the first limiting nutrient (invariably an amino acid) in relation to potential need. Actual feed intake is either the desired intake or the adjusted feed allotment, whichever is the lower.

In order to illustrate the use of the model in the context of this chapter, the experiment described by Fisher *et al.* (2001) was simulated as closely as possible. The experimental period between 30 and 40 weeks of age and data for the higher energy diets only were used. The description of the nutrient levels in the feeds was as given in the original paper.

Figure 18.11 shows the response in egg mass (g/day) to dietary lysine (g/kg), as observed and as simulated in the model. In general the model produces a simulated response essentially similar to the observations, but at low lysine levels the predicted egg output is lower than the observation. The asymptote of the response is similar in the experiment and the simulation. As noted above, the responses observed in the experiment are consistent with maintenance lysine requirements determined by extrapolation, which are very close to zero. In the initial model the composition of protein for maintenance is assumed to be the same as in the body protein, and thus lysine requirement is relatively high. The discrepancy between the model and the simulation is thus not surprising. When maintenance for lysine is set to zero in the model, the agreement between simulation and experiment is much closer (Fig. 18.11).

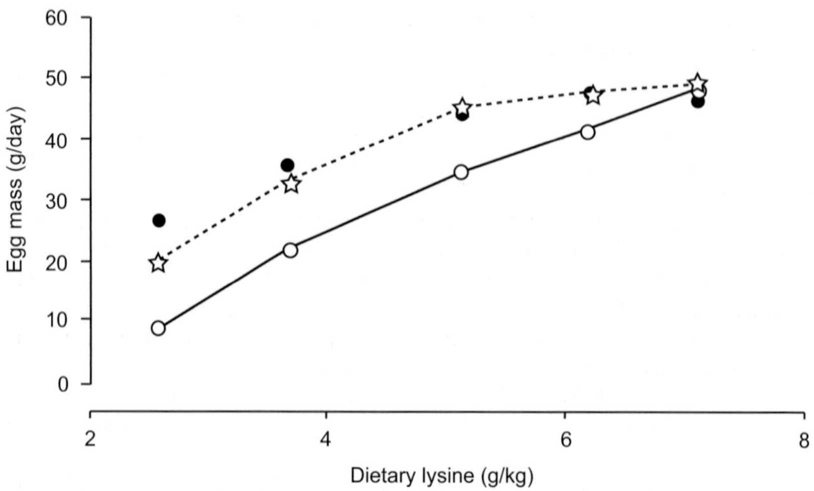

Fig. 18.11. Simulation of the trial described by Fisher *et al.* (2001) using the model developed by Nonis and Gous at the University of KwaZulu-Natal; response in egg mass to lysine intake (final 4-week data from a 9-week assay starting at 26 weeks; data for high-energy treatments only). The experimental data are shown by closed symbols. Simulation points using the first version of the model are shown by open symbols joined by a solid line. Simulation points assuming a zero maintenance requirement for lysine are shown by open stars and joined by a broken line. See text for more details.

This very limited exercise suggests that modelling may provide a way of resolving some of the issues that arise in determining the amino acid requirements of broiler breeder hens in lay. The uncertainty about maintenance is illustrated but is not resolved by the use of the model. More significant is the observation that the slope of the response, which is a measure of average efficiency of lysine utilization for egg production, is very similar in the simulation and in the experiment. In the model the instantaneous efficiency of amino acid utilization is assumed to be 0.85, as observed in high-producing laying hens. The lower efficiency observed in the experiment (about 0.59) is simulated entirely by producing an intermittent pattern of egg production, as observed in broiler breeders. This is a fairly strong test of the hypothesis that this is a correct explanation of the observed lower efficiency in birds laying at a lower average rate.

Early work using modelling or calculation of amino acid requirements in broiler breeder hens was entirely based on the use of factorial equations (Waldroup *et al.*, 1976; Bornstein *et al.*, 1979; Fisher, 1998). More recently Alvarez and Hocking (2007) have described a stochastic model of egg production, but this does not consider nutritional inputs or response. Thus it is premature to attempt any comparison of these different approaches, but the work by Nonis and Gous described briefly here does indicate that simulation modelling may be productive in this class of stock.

REFERENCES

Al-Murrani, W.K. (1978) Maternal effects on embryonic and post-embryonic growth in poultry. *British Poultry Science* 19, 277–281.

Alvarez, R. and Hocking, P.M. (2007) Stochastic model of egg production in broiler breeders. *Poultry Science* 86, 1445–1452.

Bach Knudsen, K.E. and Jorgensen, H. (1986) Use of synthetic amino acids in pig and poultry diets. In: Haresign, W. and Cole, D.J.A. (eds) *Recent Advances in Animal Nutrition – 1986*. Butterworths, London, pp. 215–225.

Balloun, S.L. (1974) Effect on reproductive performance of turkey breeders of nutrient restriction during the development period. *Poultry Science* 53, 625–632.

Bennett, C.D. and Leeson, S. (1990) Body composition of the broiler-breeder pullet. *Poultry Science* 69, 715–720.

Bornstein, S., Hurwitz, S. and Lev, Y. (1979) The amino acid and energy requirements of broiler breeder hens. *Poultry Science* 58, 104–116.

Bowmaker, J.A. and Gous, R.M. (1989) A quantification of reproductive changes and nutrient requirements of broiler breeder pullets at sexual maturity. *British Poultry Science* 30, 663–675.

Bowmaker, J.A. and Gous, R.M. (1991) Response of broiler breeder hens to dietary lysine and methionine. *British Poultry Science* 32, 1069–1088.

Brake, J., Garlich, J.D. and Peebles, E.D. (1985) Effect of protein and energy intake by broiler breeders during the prebreeder transition period on subsequent reproductive performance. *Poultry Science* 64, 2335–2340.

Bray, D.F. and Iton, E.L. (1962) The effect of egg weight on strain differences in embryonic and post-embryonic growth in the domestic fowl. *British Poultry Science* 3, 175–188.

Cave, N.A.G. (1984) Effect of a high-protein diet fed prior to the onset of lay on performance of broiler breeder pullets. *Poultry Science* 63, 1823–1827.

Cherms, F.L., Stoller, M.G. and MacIlraith, J.J. (1976) Reproduction in turkeys as influenced by prebreeder and breeder protein intake and the environment. *Poultry Science* 55, 1678–1690.

Cherry, P. (1993) Sexual maturity in the domestic duck. PhD thesis, University of Reading, Reading, UK.

Coon, C. (2004) The ideal amino acid requirement and profile for broilers, layers and broiler breeders. Proceedings of St. Petersburg Feed and Nutrition Conference, *The American Soybean Association*, pp. 1–34.

EFG Software (2002) Broiler Growth Model, described in www.efgsoftware.net (accessed July 2007).

Emmans, G.C. (1989) The growth of turkeys. In: Nixey, C. and Grey, T.C. (eds) *Recent Advances in Turkey Science*. CAB International, Wallingford, UK, pp. 135–166.

Emmans, G.C. (1994) Effective energy: a concept of energy utilisation applied across species. *British Journal of Nutrition* 71, 801–821.

Emmans, G.C. and Fisher, C. (1986) Problems in nutritional theory. In: Fisher, C. and Boorman, K.N. (eds) *Nutritional Requirements and Nutritional Theory*. Butterworths, London, pp. 9–39.

Fisher, C. (1994) Responses of laying hens to amino acids. In: D'Mello, J.P.F. (ed.) *Amino Acids in Farm Animal Nutrition*. CAB International, Wallingford, UK, pp. 245–279.

Fisher, C. (1998) Amino acid requirements of broiler breeders. *Poultry Science* 77, 124–133.

Fisher, C., Morris, T.R. and Jennings, R.C. (1973) A model for the description and prediction of the response of laying hens to amino acid intake. *British Poultry Science* 14, 469–484.

Fisher, C., Gous, R.M. and Goddard, L. (2001) Lysine requirements of broiler breeder hens. *Proceedings 2001 Australian Poultry Science Symposium*, pp. 183–186.

Frisch, R.E. (1980) Fatness, puberty and fertility. *Natural History* 89, 16–27.

Grimes, J.L., Ort, J.F., Christensen, V.L. and Ball, H.R. Jr (1989) Effects of different protein levels fed during the prebreeder period on performance of turkey breeder hens. *Poultry Science* 68, 1436–1441.

Hassan, G.M. and Nordskog, A.W. (1971) Effects of egg size and heterozygosis on embryonic growth and hatching speed. *Genetics* 67, 279–285.

Hocking, P.M. (1989) Effect of dietary crude protein concentration on semen yield and quality in male broiler fowls. *British Poultry Science* 30, 935–945.

Hocking, P.M. and Bernard, R. (1997) Effects of dietary crude protein content and food intake on the production of semen in two lines of broiler breeder males. *British Poultry Science* 28, 199–202.

Hocking, P.M., Bernard, R. and Robertson, G.W. (2002) Effects of low dietary protein and different allocations of food during rearing and restricted feeding after peak rate of lay on egg production, fertility and hatchability in female broiler breeders. *British Poultry Science* 43, 94–103.

Hudson, B.P., Lien, R.J. and Hess, J.B. (2000) Effects of early protein intake on development and subsequent egg production of broiler breeder hens. *Journal of Applied Poultry Research* 9, 324–333.

Johnston, S.A. and Gous, R.M. (2007a) A mechanistic, stochastic, population model of egg production. *British Poultry Science* 48, 224–232.

Johnston, S.A. and Gous, R.M. (2007b) Modelling the changes in the proportions of the egg components during a laying cycle. *British Poultry Science* 48, 347–353.

Johnston, S.A. and Gous, R.M. (2007c) Extent of variation within a laying flock: attainment of sexual maturity, double-yolked and soft-shelled eggs, sequence lengths and consistency of lay. *British Poultry Science* 48, 609–616.

Kemp, C., Wylie, L.M. and Fisher, C. (2001) Broiler breeder nutrition, nutrient transfer and broiler performance. *Proceedings 13th European Symposium on Poultry Nutrition.* Blankenberghe, Belgium, 7 pp.

Kidd, M.T. (2003) A treatise on chicken dam nutrition that impacts on progeny. *World's Poultry Science Journal* 59, 475–494.

Krueger, K.K., Owen, J.A., Krueger, C.E. and Ferguson, T.M. (1978) Effect of feed and light regimes during the growing period on subsequent reproductive performance of broad breasted white turkeys fed two protein levels. *Poultry Science* 57, 27–37.

Lewis, P.D., Gous, R.M. and Morris, T.R. (2008) Model to predict age at sexual maturity in broiler breeders given a single increment in photoperiod. *British Poultry Science* 48, 625–634.

Lilburn, M.S., Ngiam-Rilling, K. and Smith, J.H. (1987) Relationships between dietary protein, dietary energy, rearing environment, and nutrient utilization by broiler breeder pullets. *Poultry Science* 66, 1111–1118.

Lopez, G. and Leeson, S. (1994) Nutrition and broiler breeder performance: a review with emphasis on response to diet protein. *Journal of Applied Poultry Research* 3, 303–311.

Martin, P.A., Bradford, G.D. and Gous, R.M. (1994) A formal method of determining the dietary amino acid requirements of laying-type pullets during their growing period. *British Poultry Science* 35, 709–724.

Mather, F.B. and Harms, R.H. (1982) Performance of laying turkeys grouped according to daily food intake. *Poultry Science* 61, 1818–1822.

McDaniel, G.R. (1986) Sex separate feeding of broiler parent stock. *Zootechnica International* 1986(11), 54–58.

McDaniel, G.R., Brake, J. and Eckman, M.K. (1981) Factors affecting broiler breeder performance 4. The interrelationship of some reproductive traits. *Poultry Science* 60, 1792–1797.

McDonald, M.W. and Morris, T.R. (1985) Quantitative review of optimum amino acid intakes for young laying pullets. *British Poultry Science* 26, 253–264.

McLoughin, L. and Gous, R.M. (1999) The effect of egg size on pre- and post-natal growth of broiler chickens. *World Poultry* 15, 34–38.

Morris, T.R. and Gous, R.M. (1988) Partitioning of the response to protein between egg number and egg weight. *British Poultry Science* 29, 93–99.

Noble, R.C. and Connor, K. (1987) Lipid metabolism in the chick embryo of the domestic fowl (*Gallus domesticus*). *World's Poultry Science Journal* 40, 114–120.

Nonis, M.K. and Gous, R.M. (2007) Utilisation of synthetic amino acids by broiler breeder hens. *South African Journal of Animal Science* 36, 126–134.

Olver, M.D. (1984) Quantitative feed restriction in Pekin breeder ducks during the rearing period and its effect on subsequent productivity. *South African Journal of Animal Science* 14, 75–78.

Olver, M.D. (1988) Quantitative feed restriction of Pekin ducks from 3 weeks of age and its effect on subsequent productivity. *South African Journal of Animal Science* 18, 93–96.

Overfield, N.D. (1969) Calcium requirements for good egg shell quality. *NAAS Quarterly Review* 86, 84–91.

Pearson, R.A. and Herron, K.M. (1981) Effects of energy and protein allowances during lay on the reproductive performance of broiler breeder hens. *British Poultry Science* 22, 227–239.

Pearson, R.A. and Herron, K.M. (1982) Relationship between energy and protein intakes and laying characteristics in individually-caged broiler breeder hens. *British Poultry Science* 23, 145–159.

Picard, M. (2004) Broiler breeder production: solving a paradox. http://ec.europa.eu/research/agriculture/projects/qlrt_2000_01732_en.htm (accessed June 2007).

Revington, W.H., Moran, E.T. and McDaniel, G.R. (1991) Performance of broiler breeder males given low protein feed. *Poultry Science* 70, 139–145.

Romanoff, A.L. and Romanoff, A.J. (1949) *The Avian Egg*. John Wiley & Sons, Inc., New York.

Romero-Sanchez, H., Plumstead, P.W. and Brake, J. (2007) Feeding broiler breeder males. 1. Effect of feeding program and dietary crude protein during rearing on body weight and fertility of broiler breeder males. *Poultry Science* 86, 168–174.

Sakomura, N.K. and Coon, C. (2003) Amino acid requirements for maintenance of broiler breeder pullets. *Proceedings 14th European Symposium on Poultry Nutrition*, pp. 280–281.

Schjeide, D.A.M., Wilkens, M., Candless, R.G., Munn, R.J., Peterson, M. and Carlsen, E. (1963) Liver synthesis, plasma transport and structural alterations accompanying passage of yolk proteins. *American Zoologist* 3, 167–184.

Shanawany, M.M. (1984) Inter-relationship between egg weight, parental age and embryonic development. *British Poultry Science* 25, 449–455.

Spratt, R.S. and Leeson, S. (1987) Broiler breeder performance in response to diet protein and energy. *Poultry Science* 66, 683–693.

Taylor, St C.S. (1980) Genetic size-scaling rules in animal growth. *Animal Production* 30, 161–165.

Tullet, S.G. and Burton, F.G. (1982) Factors affecting the weight and water status of the chick at hatch. *British Poultry Science* 23, 361–369.

Tullet, S.G. and Deeming, D.C. (1982) The relationship between eggshell porosity and oxygen consumption of the embryo in the domestic fowl. *Comparative Biochemistry and Physiology* 72A, 529–533.

van Wambeke, F. (1992) The effect of a high protein pre-breeder ration on reproductive performance of broiler breeder hens. *Proceedings 19th World's Poultry Congress*, Amsterdam, Netherlands, pp. 260–261.

Waldroup, P.W., Hazen, K.R., Bussell, W.D. and Johnson, Z.B. (1976) Studies on the daily protein and amino acid needs of broiler breeder hens. *Poultry Science* 55, 2342–2347.

Walsh, T.J. and Brake, J. (1997) The effect of nutrient intake during rearing of broiler breeder females on subsequent fertility. *Poultry Science* 76, 297–305.

Whitehead, C.C. (1989) Nutrition of turkey breeding stock. In: Nixey, C. and Grey, T.C. (eds) *Recent Advances in Turkey Science*. CAB International, Wallingford, UK, pp. 91–117.

Wiley, W.H. (1950) The influence of egg weight on prehatching growth rate in the fowl. 1. Egg weight–embryonic development ratios. *Poultry Science* 29, 570–574.

Wilgus, H.S. (1976) Estimation of the amino acid requirements of turkey breeder hens, with comparisons to those of laying chickens. 1. Estimation of requirements. *Feedstuffs, Minneapolis* 48 (47), 38–41.

Wilson, H.R. (1991) Interrelationships of egg size, chick size, posthatching growth and hatchability. *World's Poultry Science Journal* 47, 5–20.

Wilson, H.R, and Harms, R.H. (1984) Evaluation of nutrient specifications for broiler breeders. *Poultry Science* 63, 1400–1406.

Wilson, J.L., McDaniel, G.R. and Sutton, C.D. (1987a) Dietary protein levels for broiler breeder males. *Poultry Science* 66, 237–242.

Wilson, J.L., McDaniel, G.R., Sutton, C.D. and Renden, J.A. (1987b) Semen and carcass evaluation of broiler breeder males fed low protein diets. *Poultry Science* 66, 1533–1540.

Yu, J.Y.L. and Marquardt, R.R. (1974).Hyperplasia and hypertrophy of the chicken oviduct during a reproductive cycle. *Poultry Science* 53, 1096–1105.

Zhang, X., Berry, W.D., McDaniel, G.R., Roland, D.A., Liu, P., Calvert, C. and Wilhite, R. (1999) Body weight and semen production of broiler breeder males as influenced by crude protein levels and feeding regimens during rearing. *Poultry Science* 78, 190–196.

Zuidhof, M.J., Robinson, F.E., Feddes, J.J.R., Hardin, R.T., Wilson, J.L., McKay, R.I. and Newcombe, M. (1995) The effects of nutrient dilution on the well-being and performance of female broiler breeders. *Poultry Science* 74, 441–456.

Chapter 19
Vitamins, Minerals and Micronutrients

M.T. Kidd

Department of Poultry Science, Mississippi State University, USA

ABSTRACT

Breeder hen micronutrient levels must be provided in adequate amounts, as a deficiency in virtually any mineral or vitamin can impair embryonic development and hatchability. As breeder hen experiments are time-consuming and costly, most micronutrients are formulated into premixes with wide safety margins. However, are nutrient levels for hen productivity also adequate for chick health and growth (i.e. chick quality)? Chick quality is an accepted viability index, but its assessment entails numerous measurements that often vary. Increasing some vitamin, mineral and micronutrient levels in hen diets will increase progeny tissue levels of these nutrients, but do these increased maternal transfer levels result in realized benefits? This depends on the nutrient, its source, level and the degree of embryo and chick stress (e.g. physiological, immunological and environmental stressors).

Investigations on hen diet fat-soluble vitamins and trace elements have shown some consistent beneficial effects on progeny performance. Both the fat-soluble vitamins D and E result in carry-over effects that improve chick health, growth and bone quality. It may be that the chick's preferential absorption of glucose over fatty acids limits early fat-soluble vitamin absorption, resulting in the hen's nutrition as the primary source of these vitamins for the young chick. The trace elements zinc, manganese and selenium have a myriad of metabolic needs and have been shown to improve chick quality, especially that of chick health.

INTRODUCTION

A dietary deficiency of hen micronutrients is not common because hen premix specifications contain wide safety margins. Hence, only a handful of studies exist for each vitamin and mineral that assesses the requirements for egg production, embryonic development, hatchability and fertility. A deficiency of any micronutrient in the hen diet has been shown to decrease hatchability and most have been shown to impair some aspect of embryonic development. The reproductive efficiency of breeder males has been shown to be decreased when deficiencies of vitamin B_6, vitamin E, selenium or zinc occur, as mediated through testicular development and semen quality problems. Experimentation utilizing varying levels of micronutrients tends to support the hypothesis that hen dietary levels needed for embryonic development and hatchability are higher than those for good egg production. Moreover, it appears that micronutrient levels needed for optimal chick quality are higher than those for hatchability. As such, this chapter will address micronutrient nutritional needs for hens at their highest demand (i.e. progeny responses).

Embryos from egg-laying species must have vitamins and minerals assimilated in yolk contents at lay, unlike embryos that develop in the womb. As such, hen dietary nutritional adequacies at a consistent level are critical for optimal progeny development. Although many maternal diet deficiencies result in profound embryo abnormalities, maternal diet adequacies for embryo status may vary depending on which embryo or chick measurement is assessed. Indeed, hen diet adequacies or over-adequacies for progeny performance (i.e. health, growth and lean tissue assimilation) are of much interest in today's competitive poultry production climate.

Optimizing chick quality is of utmost importance. Animal welfare programmes require it and competitive broiler production necessitates it. The broiler breeder hen cannot be left out of an effective chick quality programme. Hence, although hatchery management is critical for effective chick quality, shell quality, embryo health and chick viability as affected by hen micronutrient nutrition must not be ignored. Hen nutritional needs that impact hatchability (Beer, 1969; Wilson, 1997) and progeny growth, viability and immunity (Kidd, 2003; see Butter and Walter, Chapter 21, this volume) have been reviewed. This chapter addresses key micronutrient needs that are known to impact progeny viability. Attention is given to fat-soluble vitamins (D and E) and trace metals (manganese, selenium and zinc).

CRITICAL FAT-SOLUBLE VITAMINS

Of the research addressing maternal vitamin transfer, most has concerned fat-soluble vitamins. Hen diet adequacy of fat-soluble vitamins is important for embryo development, as well as supplying these vitamins for the first few days post-hatch through egg yolk. Hence, chick lipase activity and bile secretions are low during the first week of life, which limits fat utilization (Sell, 1996). The utilization of tallow has been shown to increase from 40 to 79% from 1 to 2 weeks post-hatch (Carew *et al.*, 1972). As the chick's digestive tract is physically and functionally immature for the first few days post-hatch (Sell, 1996), increasing dietary fat levels in chick diets does not overcome the inefficiency of fat utilization (Noy and Sklan, 2001). Therefore, heightened yolk stores of fat-soluble vitamins are important to satisfy vitamin needs through the first few days post-hatch, as absorption of fat, and thus fat-soluble vitamins, is limited.

Vitamin D

The primary function of vitamin D in poultry is to increase intestinal absorption of calcium and phosphorus, thereby satisfying bodily needs. For example, vitamin D increases bone deposition and retention of calcium and phosphorus. Broiler breeder hens not only have their calcium and phosphorus needs to manage but also those for internal and external egg contents for offspring.

Modern broiler breeders are housed in curtain- or solid-sided houses to improve efficiency of negative-pressure ventilation. Hen diets are typically based on cereals and oilseeds, making vitamin D a likely candidate for deficiency due to dietary absence, mixing errors, source stability and premix stability. As such, cholecalciferol (vitamin D_3) or 25-hydroxyvitamin D_3 (25-OH-D_3) must be provided in the hen's diet. The 25-OH-D_3 is the only form capable of producing 1,25-dihydroxyvitamin D_3 (1,25(OH_2)-D_3). Production of 1,25(OH_2)-D_3 takes place in the kidney by the action of parathyroid hormone. Parathyroid hormone action is triggered by systemic calcium and phosphorus status. The 1,25(OH_2)-D_3 gives vitamin D its hormonal functionality through its effect on calcium and phosphorus metabolism in intestine and bone.

A dietary deficiency of vitamin D results in rickets. In poultry, classic external and internal characteristics of dietary vitamin D deficiency are rickets and malformed ribs at the spinal column juncture, respectively (Scott *et al.*, 1982). Productive hens fed vitamin D-deficient diets produce thin-shelled eggs, have reduced hatchability and deformed embryos, and eventually develop bone disease (Scott *et al.*, 1982). Hence, the level or metabolite of vitamin D that optimizes chick quality is of utmost importance.

Feeding hens higher vitamin D levels than thought to be adequate has been shown to increase progeny tibia mineralization (Griminger, 1966). This finding was similarly found by Ameenuddin *et al.* (1986), who fed hens 5000 μg/kg vitamin D_3. Their progeny had higher tibia ash, but not tibia calcium levels, than chicks from hens fed lower levels (Ameenuddin *et al.*, 1986). Sunde *et al.* (1978) conducted experiments with White Leghorn hens and evaluated the effect of vitamin D metabolites on embryonic chick development. It was demonstrated that hens fed 1,25(OH_2)-D_3 produced eggs with abnormal embryos, and they concluded that 1,25(OH_2)-D_3 may antagonize the incorporation of vitamin D_3 and 25-OH-D_3 from the hen into the yolk (Sunde *et al.*, 1978). The most noticeable abnormality in embryos produced from hens fed a vitamin D-deficient diet not supplemented with 1,25(OH_2)-D_3 (Sunde *et al.*, 1978) was a shortened beak (underdeveloped upper mandible). The former cited research established the essentiality of vitamin D for the developing chick embryo.

Recent research at the University of Georgia evaluated the effect of vitamin D_3 in hen diets on progeny growth and bone abnormalities (Atencio *et al.*, 2005a,b; Driver *et al.*, 2006) and the efficacy of vitamin D metabolites fed to hens for progeny growth and bone abnormalities (Atencio *et al.*, 2005c). Atencio *et al.* (2005a,b) fed broiler breeder flocks 0, 125, 250, 500, 1000 and 2000 IU of vitamin D_3/kg of diet or the basal diet with a later addition of 4000 IU of vitamin D/kg. Progeny weight gain at 16 days of age was highest (significant linear effect) when hens were fed 2000 or 4000 IU vitamin D_3/kg.

Although responses varied, progeny tibia ash and the lack of tibial dyschondroplasia were improved with higher levels of D_3 in the hen's diet. This work collectively (Atencio *et al.*, 2005a,b) suggests that hens need between 2000 and 4000 IU of vitamin D_3 for optimal chick quality; however, chick growth may be heightened by hen vitamin D_3 levels in excess of 4000 IU/kg. It is noteworthy that the level of vitamin D_3 needed to optimize progeny body weight gain is higher than that previously established for egg production (2800 IU vitamin D_3/kg; Atencio *et al.*, 2004). The results from these studies suggest that the requirement of broiler breeders to produce the heaviest and healthiest chicks (e.g. lower incidence of tibial dyschondroplasia) is consistently higher than the requirement for maximum performance of hens. Additional research (Atencio *et al.*, 2005c) found the relative biological value of 25-OH-D_3 to be higher than vitamin D_3, when lower levels of vitamin D_3 were fed (3125 ng/kg). Although the biological value of 25-OH-D_3 was higher for embryo mortality and body ash of progeny over vitamin D_3, the biological value of 25-OH-D_3 was higher for hens than for progeny. Future research should evaluate combined metabolites (vitamin D_3 and 25-OH-D_3) for hens on progeny growth and subsequent performance.

Vitamin E

Vitamin E consists of a family of tocopherols and tocotrienols known mostly for their roles in antioxidant functions, disease resistance and interactions with selenium in tissue integrity. However, bioactive vitamin E forms are also involved in membrane stability, oxidative reduction reactions, heavy metal detoxification, prostaglandin synthesis, synthesis and metabolism of some vitamins, and metabolism of sulfur-containing amino acids (Scott *et al.*, 1982). Although vitamin E is well known for its antioxidant role, it must be pointed out that other antioxidants can replace vitamin E as an effective antioxidant. It is the antioxidant role, however, that makes vitamin E an important aspect of hen nutrition. Hence, newly hatched chicks have high levels of polyunsaturated fatty acids as a component of their tissue lipids (Cherian and Sim, 1992). In addition to high levels of fatty acids susceptible to free radical damage, as chicks clear the shell they are exposed to atmospheric oxygen, which can further augment oxidative tissue damage. As such, α-tocopherol is the principal tissue antioxidant in animals (Packer, 1992). Breeder hen dietary fortification of a highly bioactive form of vitamin E (i.e. D or DL-α-tocopheryl acetate) is necessary to optimize chick quality status. The form of vitamin E only becomes an antioxidant once hydrolysed in the intestine to α-tocopherol. Because vitamin E is fat soluble, higher dietary concentrations result in concomitant increases in yolk concentration (Surai *et al.*, 1997). Moreover, higher levels of α-tocopherol in yolk are correlated with higher concentrations of α-tocopherol in numerous chick tissues post-hatch (Surai *et al.*, 1997).

In addition to the role of hen vitamin E fortification in elevating vitamin E in yolk and progeny chick tissues, progeny from hens fed diets fortified with fat-soluble vitamin E have decreased susceptibility of peroxidation in chick tissues, resulting in improved health post broiler flock placement in the rearing facility. For example, feeding broiler breeder hens high vitamin E (365 versus 147 μg vitamin E/g feed) has been shown to decrease peroxidation of progeny tissues, especially brain (Surai *et al.*, 1999). Additional dietary means are available to increase vitamin E status of hens. For example, it has been shown that feeding broiler breeder hens organic selenium (0.2 and 0.4 mg/kg of feed) increased yolk vitamin E (Surai, 2000). Moreover, progeny from hens fed the combination of organic selenium and vitamin E had heightened liver, plasma and brain vitamin E post-hatch (Surai, 2000). Vitamin E is the first line of defence against tissue peroxidation, whereas selenium via glutathione peroxidase degradation of free radicals aids the second line of defence (McDowell, 1989). Regarding the hen diet, it is important to understand that selenium and vitamin E are somewhat replaceable, with regard to tissue protection. Hence, hen diet research that delineates vitamin E or selenium needs must make careful reference to the level of each nutrient in the diet. Also, research on the possible interactive effects of hen dietary selenium and vitamin E levels on progeny chick quality is warranted.

As vitamin E has been shown to improve innate and adaptive immunity in poultry, research has also been conducted which has demonstrated that improved adaptive immunity in progeny is mediated by passive transfer of antibodies from the hen (Jackson *et al.*, 1978; Haq *et al.*, 1996; Boa-Amponsem *et al.*, 2001). Chicks depend on passive transfer of antibodies to aid in immunity, as their own immune system is slow to develop (see Butter and Walter, Chapter 21, this volume). As chicks are exposed to a number of pathogens early in life, such improved passive immunity is essential to increase chick quality.

Although the inclusion of vitamin E and other fat-soluble vitamins in hen diets increases the yolk status of such vitamins, care must be taken as fat-soluble vitamins can antagonize each other. For example, excess vitamin A fed to hens has been shown to decrease plasma tocopherols and increase exudative diathesis of progeny (Combs, 1976), and compromise progeny antioxidant status (Surai *et al.*, 1998). As many commercial nutritionists have decreased relative dietary vitamin A levels to increase relative vitamin E levels based on relative costs, experimentation assessing dose titrations of vitamin A in hen diets on vitamin E content of yolk should be assessed to quantify specific antagonistic effects on progeny chick quality.

CRITICAL TRACE ELEMENTS

Hen nutrition considerations

The key inorganic elements in hen nutrition that impact progeny are zinc, manganese and selenium. Other trace elements (e.g. copper) may have important maternal effects, but such roles have not been researched thoroughly. These inorganic elements are commonly referred to as trace elements owing to their small presence in diet and tissues. However, trace elements cannot be synthesized or decomposed *de novo*, and are required for normal metabolism in virtually all living matter. Other than hen diets containing high copper levels, the dietary concentration of manganese is the highest of the trace elements supplemented in mineral premixes, followed by zinc, whereas selenium is supplemented in the smallest quantity. Selenium supplementation to hen diets in the United States is regulated to a maximum supplemental level of 0.3 mg selenium/kg of diet.

The history of feeding inorganic elements began with sodium chloride in the first century AD. Initial discoveries of zinc and manganese essentiality in poultry occurred in the 20th century, and both were found to be required for normal reproduction. Before selenium was found to be essential for tissue repair, research in the early 20th century associated it with toxicity in grazing animals.

Trace element essentiality with respect to maternal needs is poorly understood. Experimental assessments of trace element needs in hens for maternal impact via dose titration methodology are sparse. Moreover, these trace element studies, when using chick quality as the endpoint, are affected by a myriad of factors. In addition, the interrelationship between trace elements with respect to balance is poorly understood. Thus, the level and source of trace elements that result in dietary adequacy for maternal nutrition are of continued concern with breeder nutritionists.

In breeder hen nutrition, trace element level is typically assessed and fed with respect to its level in the diet, without regard to ratios with other minerals or resultant level in hen or chick tissues. If the dietary minimum level of a trace element is questionable, some nutritionists overfeed the element or replace a portion of the element with a highly available source (e.g. an inorganic sulfate form or organic amino acid complex form). The most challenging aspect of this work is quantifying a need for such a small component of the diet while assessing measurements of the resulting offspring, rather than the hen. In addition, difficulty in quantifying the former need is compounded by the hen's tissue reserve of trace elements and the amounts incorporated into egg contents.

Zinc and manganese

Zinc is well known for its role in metalloenzymes, as their representation encompasses all six enzyme classes. A past review on zinc and its impact on avian immunity discussed zinc chemistry with reference to RNA and DNA synthesis, metallothionines and zinc metalloenzymes (Kidd *et al.*, 1996). In the oxidoreductase class, zinc is a key component of numerous dehydrogenases. Of key importance in the oxidoreductase class is superoxide dismutase, a zinc metalloenzyme that protects cells and tissues from oxidative damage (Bannister *et al.*, 1971), especially mononuclear phagocytes during the oxygen burst (Cook-Mills and Fraker, 1993). RNA and DNA polymerase are zinc-dependent enzymes in the transferase class. Hence, cellular metabolism and subsequent tissue accretion rely on zinc adequacy. In the hydrolase class, collagenase is a zinc-dependent enzyme responsible for collagen synthesis and strength (Underwood and Suttle, 2001). In addition to collagen, zinc is required for synthesis of keratin (Underwood and Suttle, 2001), which is of further importance in feather synthesis in avian species in comparison to mammals. Common enzymes in the lyase, isomerase and ligase classes are carbonic anhydrase, phosphomannose isomerase and tRNA synthetase, respectively.

The physiological functions of manganese, like those of zinc, are heavily involved in enzyme activities. Manganese metalloenzymes in mitochondria include superoxide dismutase, arginase, pyruvate carboxylase and glutamine synthetase. In addition to being a structural component of enzymes, its presence activates enzymes in the hydrolase, kinase, decarboxylase and transferase classes. The most common valence states for manganese are +2 and +3, and it has been shown that other cations can spare manganese enzyme activation activity without loss of kinetic capacity (McDowell, 1992). Of the manganese metalloenzymes, superoxide dismutase has received considerable attention for its role in antioxidant protection (Surai, 2005).

Zinc is integral for the functionality of hormones. For example, the zinc-containing hormone thymulin has been shown to be biologically inactive when stripped of zinc via chelation agents (Dardenne *et al.*, 1982). Dietary zinc deficiency has been shown to reduce serum insulin-like growth factor-1 (Roth and Kirchgessner, 1997). These results point to the importance of zinc in normal bone development. Further research that assessed the effect of zinc status on bone growth demonstrated that zinc deficiency directly impacts upon growth plate chondrocyte proliferation, differentiation and apoptosis (Wang *et al.*, 2002). Proliferation was inhibited, whereas differentiation was accelerated, in the presence of activated apoptosis (Wang *et al.*, 2002). Similarly, manganese is essential for proper bone development. The organic matrix of the bone, mucopolysaccharide, is dependent upon manganese availability. Hens fed manganese-deficient diets produced embryos and chicks with reduced bone lengths (Caskey and Norris, 1940). Progeny from this study fed diets fortified with adequate manganese continued to exhibit shortened bones throughout life. This condition was subsequently termed chondrodystrophy and was characterized as chicks from hens fed deficient manganese that exhibited poor growth, bone development and mortality (McDowell, 1992).

Bartsevich *et al.* (2003) conducted research with zinc finger proteins and demonstrated their ability to heighten differentiation of embryonic stem cells. During embryonic development and post-hatch, zinc finger proteins act on DNA to facilitate tissue transcription (Luscombe *et al.*, 2000). Cell turnover has also been shown to be regulated by zinc (Bray *et al.*, 1998). In addition to zinc, manganese is also involved in protein metabolism. Optimum synthesis of proteins in cell-free culture (Hicks and Wallwork, 1987) and in intact cells (Terasaki and Rubin, 1985) is dependent on both zinc and manganese. Progeny performance of chicks from Cobb 500 broiler breeders fed various levels of inorganic and organic zinc and manganese was assessed by Virden *et al.* (2003). Growth and feed utilization of progeny were unaffected by hen diet, but organic sources of zinc and manganese (Availa® zinc and manganese) improved early liveability (Virden *et al.*, 2003). Further statistical analysis of the data in our laboratory (unpublished) revealed that progeny from hens fed organic zinc and manganese had improved overall ($P = 0.06$) liveability over progeny fed the control diet. Of further importance was the analysis (unpublished) of breast meat yields. Progeny from hens fed organic zinc and manganese had numerically greater breast meat yield compared with controls (P = 0.20; 253.0 versus 247.5 g breast meat/kg of body weight). Research on the effect of hen zinc and manganese on progeny protein metabolism, especially in commercial high-yield lines, is warranted.

The mention of improved liveability in the previous paragraph may be attributable to the role of manganese in oxidative stress and zinc's role in immunity. To the author's knowledge, no published work exists on the effect of hen manganese on progeny oxidative stress status. There are numerous published studies on the effect of hen zinc status on progeny immunity. Flinchum *et al.* (1989) fed White Leghorn hens diets containing 100 mg zinc/kg of diet or diets supplemented with 40 or 80 mg zinc/kg of diet in the form of zinc-methionine. Hens fed 180 mg/kg zinc had progeny with improved survival to an *Escherichia coli* challenge. Subsequent studies to compare zinc-methionine's role in progeny immunity were conducted with broiler breeders (Kidd *et al.*, 1992, 1993). In both studies, hens were fed a control level of zinc and the control diet supplemented with an additional 40 mg/kg of zinc in the form of zinc oxide or zinc-methionine. Initially it was found that hens fed supplemental zinc-methionine had progeny with improved innate immunity, whereas differences in adaptive immunity did not occur (Kidd *et al.*, 1992). In the second study (Kidd *et al.*, 1993), supplementation of zinc oxide and zinc-methionine over the control level improved progeny adaptive and innate immunity, respectively. Similarly, Hudson *et al.* (2004) fed hens equal levels of inorganic zinc ($ZnSO_4$) or organic zinc (Availa® zinc) and noted improved adaptive immunity in progeny from hens fed $ZnSO_4$. Differences in innate immunity, however, did not occur. Virden *et al.* (2004) found that some improvements in progeny immunity were higher due to organic zinc and manganese in hen diets over a lower level of zinc and manganese from inorganic sources, but the progeny from hens fed organic zinc and manganese also had improved cardiac function. Although much of the literature suggests that dietary concentrations of zinc, and the combination of zinc and manganese concentrations and source, are important to optimize progeny performance, the optimal dietary concentrations of these elements and the ratio of inorganic to organic forms should be further studied.

Selenium

Vitamin E has been discussed in a previous section and its role is very similar to that of selenium, in that both nutrients protect tissues from oxidative damage. Although adequate levels of antioxidant nutrients are needed in chicks to reduce free radical damage due to high levels of tissue polyunsaturated fatty acids, environmental, physiological and immunological stressors, selenium's antioxidant role in chicks, as mediated by maternal transfer, has not been studied to the extent of other micronutrients. Cantor and Scott (1974) fed Leghorn hens diets containing 0 or 0.03 mg selenium/kg of diet and determined that chicks from hens fed no supplemental selenium had reduced growth. Antioxidant capacity of the former progeny did not differ. It must be pointed out, however, that the maize used in the diet (711 g/kg) was analysed to contain 0.02 mg/kg of selenium (Cantor and Scott, 1974). Bains *et al.* (1975) conducted a large field trial with broiler breeders and found that supplementing the hen's diet with 0.1 mg selenium/kg of diet decreased selenium deficiency symptoms (exudative diathesis) of progeny. After these papers, the occurrence of published work with regard to the importance of maternal selenium, independent of vitamin E research, is somewhat sparse. However, recent research points to the benefits and effectiveness of hen selenium for progeny antioxidant status and tissue selenium status. Pappas *et al.* (2005) fed broiler breeder hens diets containing either 0.03 or 0.42 mg selenium/kg of diet and evaluated the selenium status of progeny up to day 28 post-hatch. Feeding hens high-selenium diets resulted in eggs, chick liver, chick breast muscle and chick whole blood having 7.1, 5.4, 4.3 and 7.7 times more selenium, respectively, than eggs or progeny tissues from hens fed the low-selenium diet. In the presence of low-selenium starter diets, the chicks from hens fed high selenium maintained high tissue selenium and glutathione peroxidase activity up to 28 days post-hatch (Pappas *et al.*, 2005). Subsequent work by Pappas *et al.* (2006) demonstrated that the length of time elevated selenium remained in chick tissues post-hatch from hen administration is 2 weeks. Also, progeny from hens fed elevated selenium had elevated docosahexaenoic acid concentrations (Pappas *et al.*, 2006). Chicks are inevitably exposed to stressors and disease challenges, especially early post-placement, and the elevated levels of selenium in tissues as a result of parental fortification should prove beneficial during such challenges.

AN ADDITIONAL MICRONUTRIENT WORTH CONSIDERING: L-CARNITINE

Researchers in the 1950s published findings that described L-carnitine's essential role as a carrier of activated fatty acids across the mitochondrial membrane (Friedman and Fraenkel, 1955; Fritz, 1955). Its biosynthesis from methylation of lysine and donation of methyl groups from methionine was first demonstrated in chick embryos (Bremer, 1983). Kidd *et al.* (2005) fed broiler breeders diets containing 0 or 25 mg L-carnitine/kg of diet and evaluated progeny status. The rationale for this research was that the developing embryo may need L-carnitine for fatty acid metabolism and its *de novo* synthesis may be limited in embryos, as has been shown in embryos of humans (Rebouche *et al.*, 1989). Further, ingredients rich in L-carnitine are meals derived from tissue of animal origin, which are typically absent from breeder feed. Progeny growth performance was not affected by hen diet, but progeny abdominal fat was reduced in one of three hatches as a result of fortifying hen diet with L-carnitine. In addition, progeny from hens fed L-carnitine and fed on diets high in amino acid density had decreased carcass fat and increased breast meat as compared to progeny from hens fed no L-carnitine and fed lower amino acid density. A limitation of this study is that benefits observed in progeny occurred after hens had been on test feed for 16 weeks, but further hatches were not carried out.

CONCLUSIONS

Egg production, embryonic development, hatchability and fertility are dependent on micronutrient nutrition, and hen diet premix levels should reflect adequacies with non-antagonistic safety margins. Moreover, attention to premix levels for progeny should be considered, as broiler breeder hen micronutrient nutrition impacts progeny metabolism, growth and health. The fat-soluble vitamins (i.e. vitamins D and E) and trace metals (zinc, manganese and selenium) discussed in this chapter clearly influence many attributes of progeny. Although attention was not given to water-soluble vitamin transfer from hen to chick, this area must not be overlooked when establishing vitamin premix specifications, as many water-soluble vitamins have been demonstrated to transfer to progeny and exert some beneficial functions (Kidd, 2003). In addition, micronutrients such as L-carnitine may deliver benefits in progeny. Integrated businesses interested in these potential carry-over nutrient effects should plan hen nutritional strategies with discussions with breeder, hatchery and broiler management, as well as the breeder nutritionist.

Modern broiler breeder hens have higher peak egg production, and the percentage of eggs laid remains higher throughout life than similar genetic stocks of 20 years ago. Moreover, the daily intake of feed in modern hens is considerably less than that of 20 years ago. As such, vitamin and mineral premix levels in hen diets have not increased. The former hen production changes, coupled with the fact that modern broilers accrete muscle tissue with less feed than their predecessors, indicate that micronutrient nutrition of hens for optimal progeny status should be well researched and carefully monitored.

REFERENCES

Ameenuddin, S., Sunde, M.L., DeLuca, H.F. and Cook, M.E. (1986) Excessive cholecalciferol in a layer's diet: decline in some aspects of reproductive performance and increased bone mineralisation of progeny. *British Poultry Science* 27, 671–677.

Atencio, A., Edwards, H.M. Jr and Pesti, G.M. (2004) The D_3 requirement of broiler breeder hens. In: *Proceedings of the XXII World Poultry Congress*. BiLimsel Tavukçuluk Dernegi, Ankara, Turkey CD section N3.

Atencio, A., Edwards, H.M. Jr and Pesti, G.M. (2005a) Effects of vitamin D_3 dietary supplementation of broiler breeder hens on the performance and bone abnormalities of the progeny. *Poultry Science* 84, 1058–1068.

Atencio, A., Edwards, H.M. Jr and Pesti, G.M. (2005b) Effect of the cholecalciferol supplementation of broiler breeder hen diets on the performance and bone abnormalities of the progeny fed diets containing various levels of calcium or 25-hydroxycholecalciferol. *Poultry Science* 84, 1593–1603.

Atencio, A., Pesti, G.M. and Edwards, H.M. Jr (2005c) Twenty-five hydroxycholecalciferol as a cholecalciferol substitute in broiler breeder hen diets and its effect on the performance and general health of the progeny. *Poultry Science* 84, 1277–1285.

Bains, B.S., Margaret, F.A. and McKenzie, R.A. (1975) Selenium-responsive myopathy in broiler breeder hens in Queensland. *Australian Veterinary Journal* 51, 140–145.

Bannister, J., Bannister, W. and Wood, E. (1971) Bovine erythrocyte cupro-zinc protein. *European Journal of Biochemistry* 18, 178–186.

Bartsevich, V.V., Miller, J.C., Case, C.C. and Pabo, C.O. (2003) Engineered zinc finger proteins for controlling stem cell fate. *Stem Cells* 21, 632–637.

Beer, A.E. (1969) A review of the effects of nutritional deficiencies on hatchability. In: Carter, T.C. and Freeman, B.M. (eds) *The Fertility and Hatchability of the Hen's Egg*. Oliver and Boyd, Edinburgh, pp. 93–108.

Boa-Amponsem, K., Price, S.E., Geraert, P.A., Picard, M. and Siegel, P.B. (2001) Antibody responses of hens fed vitamin E and passively acquired antibodies of their chicks. *Avian Diseases* 45, 122–127.

Bray, S.E., Schorl, C. and Hall, P.A. (1998) The challenge of p53: linking biochemistry, biology, and patient management. *Stem Cells* 16, 248–260.

Bremer, F. (1983) L-carnitine metabolism and functions. *Physiological Reviews* 63, 1420–1480.

Cantor, A.H. and Scott, M.L. (1974) The effect of selenium in the hen's diet on egg production, hatchability, performance of progeny and selenium concentration in eggs. *Poultry Science* 53, 1870–1880.

Carew, L.B., Machemer, R.H., Sharp, R.W. and Foss, D.C. (1972) Fat absorption in the very young chick. *Poultry Science* 51, 738–742.

Caskey, C.D. and Norris, L.C. (1940) Micromelia in adult fowl caused by manganese deficiency during embryonic development. *Proceedings of the Society of Experimental Biology and Medicine* 44, 332–335.

Cherian, G. and Sim, J.S. (1992) Preferential accumulation of n-3 fatty acids in the brain of chicks from eggs enriched with n-3 fatty acids. *Poultry Science* 71, 1658–1668.

Combs, G.F. Jr (1976) Differential effects of high dietary levels of vitamin A on the vitamin E-selenium nutrition of young and adult chickens. *Journal of Nutrition* 106, 967–975.

Cook-Mills, J.M. and Fraker, P.J. (1993) The role of metals in the production of toxic oxygen metabolites by mononuclear phagocytes. In: Cunningham-Rundles, S. (ed.) *Nutrient Modulation of the Immune Response*. Marcel Dekker, New York, pp. 127–140.

Dardenne, M., Pleau, J.M., Nabarra, B., Lefancier, P., Derrien, M., Choay, J. and Bach, J.E. (1982) Contribution of zinc and other metals to the biological activity of serum thymic factor. *Proceedings of the National Academy of Sciences of the USA* 79, 5370–5373.

Driver, J.P., Atencio, A., Pesti, G.M., Edwards, H.M. Jr and Bakalli, R.I. (2006) The effect of maternal dietary vitamin D_3 supplementation on performance and tibial dyschondroplasia of broiler chicks. *Poultry Science* 85, 39–47.

Flinchum, J.D., Nockles, C.F. and Moreng, R.E. (1989) Aged hens fed added zinc methionine had chicks with improved performance. *Poultry Science* 68 (Suppl. 1), 55.

Friedman, S. and Fraenkel, G. (1955) Reversible enzymatic acetylation of L-carnitine. *Archives of Biochemistry and Biophysics* 59, 491–501.

Fritz, I.B. (1955) The effect of muscle extracts on the oxidation of palmitic acid by liver slices and homogenates. *Acta Physiologica Scandinavia* 34, 367–385.

Griminger, P. (1966) Influence of maternal vitamin D intake on growth and bone ash of offspring. *Poultry Science* 45, 849–851.

Haq, A., Bailey, C.A. and Chinnah, A.D. (1996) Effect of β-carotene, canthaxanthin, lutein, and vitamin E on neonatal immunity of chicks when supplemented in the broiler breeder diets. *Poultry Science* 75, 1092–1097.

Hicks, S.E. and Wallwork, J.C. (1987) Effect of dietary zinc deficiency on protein synthesis in cell-free systems isolated from rat liver. *Journal of Nutrition* 117, 1234–1240.

Hudson, B.P., Dozier, W.A. III, Fairchild, B.D., Wilson, J.L., Sander, J.E. and Ward, T.L. (2004) Live performance and immune responses of straight-run broilers: influences of zinc source in broiler breeder hen and progeny diets and ambient temperature during the broiler production period. *Journal of Applied Poultry Research* 13, 291–301.

Jackson, D.W., Law, G.R. and Nockels, C.F. (1978) Maternal vitamin E alters passively acquired immunity of chicks. *Poultry Science* 57, 70–73.

Kidd, M.T. (2003) A treatise on chicken dam nutrition that impacts progeny. *World's Poultry Science Journal* 59, 475–494.

Kidd, M.T., Anthony, N.B. and Lee, S.R. (1992) Progeny performance when dams and chicks are fed supplemental zinc. *Poultry Science* 71, 1201–1206.

Kidd, M.T., Anthony, N.B., Newberry, L.A. and Lee, S.R. (1993) Effect of supplemental zinc in either a corn–soybean or a milo corn–soybean meal diet on the performance of young broiler breeders and their progeny. *Poultry Science* 72, 1492–1499.

Kidd, M.T., Ferket, P.R. and Qureshi, M.A. (1996) Zinc metabolism with special reference to its role in immunity. *World's Poultry Science Journal* 52, 309–324.

Kidd, M.T., McDaniel, C.D., Peebles, E.D., Barber, S.J., Corzo, A., Branton, S.L. and Woodworth, J.C. (2005) Breeder hen dietary L-carnitine affects progeny carcass traits. *British Poultry Science* 46, 97–103.

Luscombe, N.E., Austin, S.E., Berman, H.M. and Thornton, J.M. (2000) An overview of the structures of protein–DNA complexes. *Genome Biology* 1(1):REVIEWS001. Epub 2000 Jun 9.

McDowell, L.R. (1989) *Vitamins in Animal Nutrition: Comparative Aspects to Human Nutrition*, 1st edn. Academic Press, Inc., San Diego, California.

McDowell, L.R. (1992) *Minerals in Animal and Human Nutrition*, 1st edn. Academic Press, Inc., San Diego, California.

Noy, Y. and Sklan, D. (2001) Yolk and exogenous feed utilization in the posthatch chick. *Poultry Science* 80, 1490–1495.

Packer, L. (1992) Interactions among antioxidants in health and disease: vitamin E and its redox cycle. *Proceedings of the Society of Experimental Biology and Medicine* 200, 271–276.

Pappas, A.C., Karadas, F., Surai, P.F. and Speake, B.K. (2005) The selenium intake of the female chicken influences the selenium status of her progeny. *Comparative Biochemistry and Physiology B* 142, 465–474.

Pappas, A.C., Acamovic, T., Surai, P.F. and McDevitt, R.M. (2006) Maternal organo-selenium compounds and polyunsaturated fatty acids affect progeny performance and levels of selenium and docosahexaenoic acid in the chick tissues. *Poultry Science* 85, 1610–1620.

Rebouche, C.J., Bosch, E.P., Chenard, C.A., Schabold, K.J. and Nelson, S.E. (1989) Utilization of dietary precursors for L-carnitine synthesis in human adults. *Journal of Nutrition* 119, 1907–1913.

Roth, H.P. and Kirchgessner, M. (1997) Course of concentration changes of growth hormone, IGF-1, insulin and C-peptide in serum, pituitary and liver of zinc-deficient rats. *Journal of Animal Physiology and Animal Nutrition* 77, 91–101.

Scott, M.L., Nesheim, M.C. and Young, R.J. (1982) *Nutrition of the Chicken*, 3rd edn. M.L. Scott and Associates, Ithaca, New York.

Sell, J.L. (1996) Physiological limitations and potential for improvements in gastrointestinal tract function of poultry. *Journal of Applied Poultry Research* 5, 96–101.

Sunde, M.L., Turk, C.M. and DeLuca, H.F. (1978) The essentiality of vitamin D metabolites for embryonic chick development. *Science* 200, 1067–1069.

Surai, P.F. (2000) Effect of selenium and vitamin E content of the maternal diet on the antioxidant system of the yolk and the developing chick. *British Poultry Science* 41, 235–243.

Surai, P.F. (2005) Mineral and anti-oxidants. In: Taylor-Pickard, J.A. and Tucker, L.A. (eds) *Redefining Mineral Nutrition*. Nottingham University Press, Nottingham, UK, pp. 147–177.

Surai, P.F., Gaal, T., Noble, R.C. and Speake, B.K. (1997) The relationship between the α-tocopherol content of the yolk and its accumulation in the tissues of the newly hatched chick. *Journal of the Science of Food and Agriculture* 75, 212–216.

Surai, P.F., Ionov, I.A., Kuklenko, T.V., Kostjuk, I.A., MacPherson, A., Speake, B.K., Noble, R.C. and Sparks, N.H.C. (1998) Effect of supplementing the hen's diet with vitamin A on the accumulation of vitamins A and E, ascorbic acid and carotenoids in the egg yolk and in the embryonic liver. *British Poultry Science* 39, 257–263.

Surai, P.F., Noble, R.C. and Speake, B.K. (1999) Relationship between vitamin E content and susceptibility to lipid peroxidation in tissues of the newly hatched chick. *British Poultry Science* 40, 406–410.

Terasaki, M. and Rubin, H. (1985) Evidence that intracellular magnesium is present in cells at a regulatory concentration for protein synthesis. *Proceedings of the National Academy of Sciences of the USA* 82, 7324–7326.

Underwood, E. and Suttle, N. (2001) *The Mineral Nutrition of Livestock*. CAB International, Wallingford, UK.

Virden, W.S., Yeatman, J.B., Barber, S.J., Zumwalt, C.D., Ward, T.L., Johnson, A.B. and Kidd, M.T. (2003) Hen mineral nutrition impacts progeny livability. *Journal of Applied Poultry Research* 12, 411–416.

Virden, W.S., Yeatman, J.B., Barber, S.J., Willeford, K.O., Ward, T.L., Fakler, T.M., Wideman, R.F. Jr and Kidd, M.T. (2004) Immune system and cardiac functions of progeny chicks from dams fed diets differing in zinc and manganese level and source. *Poultry Science* 83, 344–351.

Wang, X., Fosmire, G.J., Gay, C.V. and Leach, R.M. (2002) Short-term zinc deficiency inhibits chondrocyte proliferation and induces cell apoptosis in the epiphyseal growth plate of young chickens. *Journal of Nutrition* 132, 665–673.

Wilson, H.R. (1997) Effects of maternal nutrition on hatchability. *Poultry Science* 76, 134–143.

PART VIII

Health and Welfare

CHAPTER 20
Vaccination: Theory and Practice

T. Cserep
Intervet UK, Milton Keynes, UK

ABSTRACT

Vaccination plays an essential role in disease control in the modern poultry industry. Without vaccination productivity could not have progressed so successfully and as rapidly as it has since the late 1950s.

Although there are areas where further research is needed to understand the fundamental aspects of immunology, the theoretical basis of avian immune responses to disease challenges and vaccinations is reasonably well understood and this is outlined in the first part of this chapter. Live and inactivated vaccines are used in vaccination programmes for poultry and other avian species to induce specific and long-lasting protection against infectious diseases. The efficacy of these vaccines depends on correct timing, preparation and administration. Details of various mass and individual administration techniques are discussed in the second part of the chapter.

Vaccination in conjunction with biosecurity and management is an essential part of a complex disease strategy to minimize the economic impact of infectious diseases and make poultry production commercially viable.

INTRODUCTION

Many avian infectious diseases are ubiquitous and most are difficult to control, even with very good biosecurity measures; hence vaccination plays an essential role in disease control in the modern poultry industry. Without vaccination productivity could not have progressed so successfully and as rapidly as it has over the last 50 years. However, it should always be remembered that vaccines alone can never provide complete protection. They are only one of the essential measures necessary in a complex disease control strategy and must be used in conjunction with good management and biosecurity practices, as discussed in a later chapter (Collett, Chapter 22, this volume).

THEORETICAL BASIS OF VACCINATION: THE CHICKEN IMMUNE SYSTEM

The avian immune system consists of a multilayered network of physical and chemical barriers, cells and inducible bioactive and signal molecules essential to defend birds from pathogens. The immune system has two main components: innate and adaptive immunity (see Janeway *et al.*, 2004; Delves *et al.*, 2006; Male *et al.*, 2006). The innate system responds rapidly to challenges from pathogens entering the body and limits the spread of infection until the adaptive immune system becomes fully effective and can clear or control the infection. Adaptive responses usually take several days, or even weeks, to develop, as specific clones of cells must be selected, replicate and then differentiate to provide sufficient numbers of mature effector cells. Both the innate and adaptive systems work in concert and both are essential for successful vaccination and survival after field or experimental challenge.

The main representatives of innate immune cells are: specialized epithelial cells, macrophages, dendritic cells, granulocytes, thrombocytes and natural killer (NK) cells. A complete review of these cell types and their respective functions is beyond the scope of this chapter and for more detailed information the reader is referred to Davison *et al.* (2008). However, it is pertinent to point out here that some types of the innate immune cells, such as macrophages and dendritic cells, play an essential role as antigen-presenting cells in adaptive immune responses. Unlike the innate immune system, the adaptive immune system has memory and, once primed by prior exposure to an antigen, cells of the adaptive system respond more rapidly and intensively to that same antigen (secondary response). The adaptive immune system has two distinct arms which complement one another and also work together; these consist of humoral (antibody) responses and cell-mediated immune responses.

Upon exposure to an antigen, clones of bursa (B)-derived lymphocytes that recognize specific antigenic determinants through the B-cell receptor are selected and, with appropriate signals, these clones expand. These lymphoblasts mature into plasma cells which secrete immunoglobulins (Ig). Cell-mediated immune responses are mediated by thymus (T)-derived lymphocytes. Upon stimulation, clonally selected cells replicate and mature to become either cytolytic T lymphocytes (CTL), which kill infected and transformed cells, or regulatory (helper) cells, which modulate both humoral and cell-mediated immune responses by releasing biologically active compounds such as cytokines. For further details on adaptive immune responses the reader is referred to the standard immunology textbooks (Janeway *et al.*, 2004; Delves *et al.*, 2006; Male *et al.*, 2006). More detailed information on avian adaptive immune responses can be found in appropriate chapters of Davison *et al.* (2008).

In order to fully understand how the avian immune system works we need to be familiar with not only the different types of lymphocytes (B cells and T cells) but also the role of the major avian lymphoid organs. The avian immune system begins development in the embryo and only becomes fully functional during the first few weeks after hatching (see Butter and Walter, Chapter 21, this volume). The bursa of Fabricius, a small diverticulum of the avian cloaca, has a crucial role as a primary lymphoid organ that is essential for maturation and differentiation of immunoglobulin (Ig)-expressing B lymphocytes. Through a process of gene conversion, which only occurs in the bursa itself, the B-lymphocyte population differentiates and develops an Ig repertoire enabling production of antibodies to a vast array of different antigens, and hence the ability to respond to diverse invading pathogens (Ratcliffe, 2006). Three isotypes of avian Ig have been identified: the homologues of mammalian IgM and IgA (the latter often referred to as secretory or local antibody) and IgY, the homologue of mammalian IgG (see Davison *et al.*, 2008). As in mammals, IgM is the first Ig isotype to be produced during embryonic development and also after exposure to antigen (primary immune response). After antigenic stimulation, and under the influence of appropriate cytokines, mature IgM-expressing B cells undergo isotype switching – usually in specialized compartments of secondary lymphoid tissues – and switch to secreting either IgA or IgY. After immunization a small, long-lived population of B cells expressing either high-affinity IgA or IgY recirculate through the peripheral lymphoid tissues, providing the memory B-cell population, available to react more rapidly and intensively upon re-exposure to the antigen. These memory cells are essential in vaccination protection.

The isotype IgY is especially important for the neonate as this Ig isotype is secreted by the hen into the yolk of developing ova and provides maternally derived antibody (MDA), which is essential for protecting the hatchling until its own immune system becomes fully functional (see Davison *et al.*, 2008). This usually occurs a few weeks after hatching. MDA is referred to as *passive immunity*, since these antibodies are donated by the hen and consequently only recognize antigens to which she has been exposed.

The other primary lymphoid organ is the thymus, which consists of several lobes located alongside each of the jugular veins in the neck of a bird. The thymus provides a unique environment where T lymphocytes undergo replication, maturation and selection to provide populations of cells that express a T-cell receptor (TCR). T lymphocytes are responsible for cell-mediated immune responses, and two major populations of T cells are present in birds (Göbel, 1996). One population expresses the $\alpha\beta$ receptor (TCR$\alpha\beta$), a heterodimer made up of an α and β molecule, and these lymphocytes are considered to be classical T cells. The TCR$\alpha\beta$ population can be further divided into two subpopulations based on the expression of the β_1 or β_2 molecules, which are identified using specific monoclonal antibodies. These TCR$\alpha\beta$ subpopulations appear to have similar functions but occur at different frequencies in different locations around the body. Although these cells react with antigen through the TCR, their functions are related to the expression of either the CD4 antigen (helper cells) or the CD8 antigen (killer cells) on the cell surface (for reviews see Janeway *et al.*, 2004; Delves *et al.*, 2006; Male *et al.*, 2006). After primary exposure to antigen, such as in vaccination, small populations of CD4 and CD8 T cells remain long-lived and available to replicate and rapidly switch to 'effector' status upon re-exposure (secondary response). The other major avian T-cell population expresses the TCR$\gamma\delta$ receptor on the surface (see Göbel, 1996). Although the role of this population is less well understood, these cells usually express a different form of the CD8 antigen on the surface and are mainly found at, or near, epithelial surfaces so they are considered to be important as a first line of defence against pathogen invasion.

Birds have a number of highly organized secondary lymphoid organs where lymphocytes differentiate and proliferate, although, unlike in the primary lymphoid organs, this proliferation is antigen driven. The spleen plays a particularly important role in systemic immune responses, since birds lack highly compartmentalized lymph nodes like those found in mammals. Lymph nodes have been identified in ducks but these are considered to be rather primitive structures (Higgins, 1996). Apart from the spleen, birds have a number of other specialized secondary lymphoid tissues. Mucosal-associated lymphoid tissues include those in the gut, such as the oesophageal tonsils (Hodges, 1974; Oláh *et al.*, 2002), caecal tonsils, several Peyer's patches in the ileum and Meckel's diverticulum (Oláh *et al.*, 1984). Meckel's diverticulum surrounds the vitelline duct connecting the small intestine with the remnants of the yolk sac and persists for the first 5 weeks after hatching. Lymphoid tissues associated with protecting the eye and nasopharyngeal region include the Harderian gland, a lachrymal gland situated in the eye socket of birds on the nose side, and the conjunctival-associated lymphoid tissue of the lower eyelid. The lymphoid tissue lining the duct of the Harderian gland is particularly important as lymphoblasts from the bursa of Fabricius migrate to this gland and establish themselves in the framework of the tissue. Here, these lymphoblasts undergo maturation and become IgA-producing plasma cells. Once produced, IgA is washed out on to mucosal surfaces of the head and neck, where it adheres, providing local immunity (Davelaar *et al.*, 1982). This local immunity is not influenced by MDA and provides a useful route for day-old vaccination against respiratory viruses. It also provides the opportunity to boost local immunity by repeated spray vaccination, even during the laying period.

IMMUNOLOGY OF VACCINATION

Responses to live vaccines

The value of a live vaccine is that it can be administered by the natural route and multiply in the vaccinated animal, inducing an appropriate immune response(s) against challenge from the pathogen concerned; responses can be humoral or cell-mediated or, more likely, both. Depending on the type of vaccine, the onset and duration of immunity can vary significantly. Administration route (Beard and Easterday, 1967), type and age of flock, the immune status of the birds and a number of other factors influence the immune response. The majority of the live poultry vaccines that are currently used are based on attenuated viruses, which are capable of evoking an immune response without causing an adverse reaction in the host. Live vaccines multiply in the bird and therefore induce a more rapid immune response, which is particularly important against those pathogens that cause an early challenge, for example Marek's disease virus (MDV), infectious bursal disease virus (IBDV or Gumboro (G)) and some respiratory viruses.

The role of the local antibody response against respiratory viruses is very important and has been well documented (Davelaar *et al.*, 1982), while the role of the systemic antibody response is more important for the duration of the immunity (Reynolds and Maraqua, 2000) and the protection of the systemic organs. There is an excellent correlation between a virus-neutralizing antibody response and protection against IBDV, but cell-mediated immunity also plays an important role (van den Berg, 2000).

In contrast, no, or very weak, serum antibody responses can be detected following vaccination with live mycoplasma vaccines. In this case the local antibody response is more likely to play a role in protection (Evans and Hafez, 1992).

Responses to inactivated vaccines

Inactivated vaccines are prepared from bacteria or viruses that have been killed or weakened to such a degree that they are unable to multiply in the vaccinated host. In spite of this, these microorganisms retain their antigenicity, though generally they require an adjuvant to induce a strong enough immune response to provide protection. Currently available poultry vaccines are mainly adjuvanted with water-in-oil emulsions or aluminium hydroxide. Adjuvants form a depot at the site of injection, slowly releasing the antigen and prolonging the time for interaction between antigen and antigen-presenting cells and lymphocytes.

Most vaccination programmes for layer and breeder birds contain one or two different types of inactivated vaccines. These are usually administered towards the end of the rearing period following priming vaccinations with live vaccines. This type of priming and boosting is essential for most vaccines to ensure uniformly high and long-lasting immunity during the laying period. Alternatively, inactivated vaccines are used without live priming, for example against egg drop syndrome (EDS) and salmonella. Two doses of inactivated vaccines are used against the *Salmonella enterica* serovars Enteritidis and Typhimurium. In this situation the first dose acts as primer and the second as booster.

Maternally derived antibodies

Hyperimmunization of laying and breeder hens with live primer and inactivated booster vaccines is used to provide protection throughout the laying period and to provide high levels of antibodies in the eggs. The higher the level of circulating antibody in the hen's blood, the higher the amount of MDA provided for the newly hatched chick. MDA is not long-lived and most of the MDAs have been catabolized by 3 to 4 weeks of age.

Chicks with a high level of the correct MDA can protect against IBDV or *chicken infectious anaemia virus* (CIAV), and MDA can also modulate infections with *Newcastle disease virus* (NDV), *infectious bronchitis virus* (IBV) and reovirus (Calnek and Smith, 1972; van der Heide *et al.*, 1976; Darbyshire and Peters, 1985), although the duration of this type of immunity is relatively short.

MDAs not only modulate the growth of virulent pathogens but also have a considerable effect on replication of live vaccines. The decline of IBDV MDA, for example, is the most important factor determining the optimal timing for vaccination against IBDV. It creates a considerable challenge for the producer to time the first vaccination correctly, before field challenge can infect the flock. Various formulae are in use to 'predict' the optimal age for IBDV vaccination, such as the Kouwenhoven formula (Intervet, 2003) and Deventer formula (de Wit, 2004).

PRACTICAL ASPECTS OF VACCINATION

Objectives of vaccination

In poultry production, the purpose of vaccination is to induce a strong enough immune response in the flock to ensure economically satisfactory protection in the event of natural challenge. It must be remembered, however, that under commercial circumstances vaccination is applied to populations and not to individuals, and vaccination has to be cost effective. In a large poultry flock there are always individuals which, for some reason, do not respond as expected to a vaccine. The other factor that must be remembered is that there is always an interval between the administration of the vaccine and onset of immunity. This period varies with the type of bird and the vaccine concerned, but it usually takes about a week for immunity to develop. If during this lag time the birds are challenged with a field virus, morbidities or even mortalities can occur.

It is generally accepted that it is more cost effective to prevent disease than to treat it. Immunizing birds with different types of vaccines is part of this concept. In a breeder immunization programme there are two primary goals. The first goal is to protect the growing males and females from disease. The objective is to have a healthy and uniform flock that is ready to move to the breeder farm and produce hatching eggs. Once in production, the birds need protection from all diseases that can reduce their reproductive performance. The second goal of immunization is to protect the progeny. Vaccination of the breeder hen is one of the most economical ways to provide early protection to her offspring through MDA. Vaccination also helps to prevent egg transmission (vertical transmission) of a disease from hen to chick, ensuring the chick a better and healthier start in life. The goal of vaccination is not only to reduce the economic losses but also to prevent the mass destruction of large numbers of infected or potentially contagious animals. Vaccines can contribute greatly to the welfare of domestic and also wild (free-living) animals. Finally, the aim of animal vaccination must increasingly be to prevent dissemination of zoonoses, such as salmonellosis, rather than just protect the host itself. This is especially important when the infection is not harmful to the vaccinated animal.

Epidemiologists consider an 85% response in a population to be effective in disease control. In order to achieve this response a great deal of attention to detail is needed, during both the preparation and administration of vaccines.

Poultry vaccines

Currently there are three main types of vaccines available for the poultry producer: conventional live and inactivated vaccines and recently available recombinant vaccines. Some recombinant vaccines are already in use but most have not reached the stage of replacing conventional vaccines.

Conventional live viral and bacterial vaccines contain non-pathogenic field isolates or isolates that have been attenuated so that they do not cause disease after the host has been vaccinated. They can spread laterally within a flock; however, the degree of this spread depends on the type of antigen on which they are based. As already mentioned, in a breeder or layer vaccination programme, live vaccines are frequently used as primers before boosting with inactivated vaccines.

Killed or inactivated vaccines are usually based on whole viruses or bacteria that have been inactivated and require some type of adjuvant. Killed vaccines induce good humoral rather than cellular immune responses. They have to be injected into each individual in the flock because their physical characteristics make them unsuitable for mass application and they do not spread laterally.

Inactivated vaccines are available in monovalent (e.g EDS) or polyvalent (e.g. *turkey rhinotracheitis* (TRT) + IBV + IBVvariant + NDV + G forms).

ADMINISTRATION OF VACCINES

Mass administration of live vaccines

Spray method

When mass vaccination is an option, spray administration of a vaccine is the preferred method. This is particularly true for vaccination against respiratory diseases such as *infectious bronchitis*, *Newcastle disease* and *turkey rhinotracheitis*.

The main reason for this preference is that, with live vaccines, strong, highly effective local immunity can be achieved due to the effect on the mucous membranes of the eyes, nostrils, trachea and, if relevant, deeper parts of the respiratory tract. A live vaccine is usually reconstituted in water (distilled or deionized) and presented to the birds to inhale as droplets in the range of 1–300 μm in diameter. Larger droplets only reach the upper respiratory tract and conjunctiva, whereas fine aerosols or particles (<5 μm) are inhaled into the lower respiratory tract (Baxendale, 1996).

Spray vaccination is carried out with sprayers that break down the vaccine suspension into small droplets, and propel them as a cloud of fine droplets over the birds. The smaller the droplets, the deeper the vaccine virus can penetrate into the lungs and air sacs. The deeper the vaccine virus penetrates, the more rapid the onset of protection. If droplets of certain NDV vaccines sediment too far down in the trachea or the air sacs they can cause vaccination reactions. In general, NDV vaccines are capable of causing more severe reactions in naive birds than IBV vaccines. Ideally a full dose of vaccine should be delivered to the upper respiratory tract of all birds in a population on the same day. Garden sprayers or sprayers based on the spinning disc technology can be adapted for vaccination purposes. Specially designed sprayer cabinets are used in hatcheries for the administration of IBV, NDV, TRT virus or coccidiosis vaccines at 1 day of age. The volume of water used depends on the age of the birds and type of sprayer. For day-old vaccination, 200–400 ml of water per 1000 chicks or poults can be recommended. If garden sprayers are used either in rearing or in laying, 500–2000 ml of water per 1000 birds may be required to achieve a uniform vaccine cover. Spinning disc sprayers use much less water than traditional sprayers and the manufacturer's recommendations should be followed. In closed, artificially lit and ventilated poultry houses, fans should be turned off with the inlets and outlets closed, the lights dimmed and the birds allowed to settle quietly before spraying commences.

Day-old vaccination of chicks with NDV and IBV or a coccidial vaccine can be carried out in specially constructed cabinets or with hand-held sprayers with the chicks confined in hatchery boxes. The recommended droplet size for day-old chicks is 100–200 μm to avoid post-vaccinal reactions. After this age vaccines can be administered with knapsack sprayers, spinning disc sprayers or portable spray generators (Baxendale, 1996).

Drinking water method

This method of application is used for live vaccines that are infectious and that enter the bird via the enteric route. These include IBDV, avian encephalomyelitis (AE) and coccidian vaccines as well as those often given by the spray method but which infect the upper respiratory tract, e.g. IBV or NDV vaccines (Baxendale, 1996).

Drinking water vaccination is the least expensive and perceived as the easiest vaccination method of all. In reality this is the method that requires the most preparation in order to achieve a good, uniform vaccine response in the whole flock. There are quite a few precautions that need to be taken for the vaccination to be effective and reliable.

The basic goal using drinking water vaccination is to distribute the vaccine rapidly throughout the whole house and make it available for a long enough period for all the birds to take at least one protective dose of vaccine. In preparing for water-based vaccination it is essential to know how the drinker system works, what the water consumption pattern of the flock is like and what volume of water is needed for the vaccination period. The volume of water needed depends on the type and age of birds, time of day and ambient temperature. Water meters can provide the most accurate data needed for determining the vaccine volume, but 1 l per thousand birds per day of age is a useful rule of thumb for broilers. There should be enough vaccine in the drinker system to last for 1.5–2.5 h, thereby allowing all birds to drink. Water intake can be encouraged by choosing the morning for vaccination and by withdrawing water for a short period of time. In practice 1–1.5 h has been found to be adequate for commercial broilers when environmental temperatures are comfortable for the birds (RUMA Guidelines, 2006).

The water used for this form of vaccination is usually mains water, which is chlorinated. Live vaccines are susceptible to this and as low as 1 p.p.m. chlorine present in the water can reduce efficacy. Skimmed milk powder or various neutralizing tablets and powders can be added to the water to neutralize the chlorine and hence protect the vaccine. A bonus of doing this is that it provides a colour marker to monitor how well the vaccine is being distributed throughout the house. In addition, some of the dye tablets change the colour of the water so intensely that the tongue of those birds drinking the vaccinated water becomes blue. Based on the proportion of birds with a blue tongue in the house, it is easy to assess the uptake of the vaccine so that corrective measures can be taken if necessary.

Some drinker systems (cage, nipple) can retain substantial volumes of residual chlorinated water, which can not only neutralize vaccine virus but dilute the vaccine and slow down its distribution. In such cases, it is uncertain and unpredictable what proportion of the flock receives a zero dose, an incomplete dose or a maximum dose during the vaccination process. This will lead to a non-uniform immune response after vaccination and a certain proportion of the flock may remain vulnerable to disease. It is therefore essential to allow birds to drink the vaccine only after the drinker system has been drained and then primed with water containing the vaccine. Mock vaccination with dye tablets provides a very good method for understanding how the drinker system works and how to adjust the vaccination method to achieve maximum uptake of a vaccine.

Drinking water application of coccidiosis vaccines

The basic guidance given above is also relevant to coccidiosis vaccines, but they do have some features which set them apart from the other poultry vaccines. These mainly relate to the fact that live coccidiosis vaccines are composed of coccidial oocysts, which are considerably (1000-fold) larger and heavier than bacteria and viruses. The result is that they tend to settle to the bottom of drinker lines and in drinkers. It is certainly perfectly feasible to administer these vaccines into bell drinker systems – the products used in layers and breeders include a component to reduce sedimentation and are usually administered directly into the drinkers by the use of a calibrated syringe. If administering in this way, it is important to swirl the drinker to thoroughly mix it. Nipple systems are more difficult, particularly those in which the outflow from the pipe into the nipple is above the centre of the pipe rather than at the bottom. To get over these problems, alternative methods of application have been devised and tested such as spray-on-feed or day-old spray (RUMA Guidelines, 2006).

Individual vaccination

In ovo *injection*

In ovo injection was developed initially for the administration of Marek's vaccine into broiler chick eggs at transfer (17.5–19 days of incubation). It is, in fact, a form of mass administration but it is carried out mechanically, effectively replicating individual administration. The equipment in its typical configuration moves a tray of incubated eggs into an injection area; the eggs are held in position and punctured over the air cell. The injection needle extends through the puncture into the amniotic compartment, before being removed and sanitized by passing a sanitizer between the punch and needle. The eggs are then transferred into a hatcher basket. The equipment is regularly checked, to ensure that no needles are blocked, and carefully sanitized at the end of each day. As punched holes in the injected eggs are not sealed, hatcheries and egg supply farms must operate to a high standard of hygiene, in particular to avoid problems with aspergillosis (RUMA Guidelines, 2006).

Day-old vaccination in hatcheries against Marek's disease

Commercial layers as well as layer- and broiler breeder-type birds are routinely vaccinated in hatcheries against MDV at 1 day of age. The vaccine is injected by the subcutaneous route into the back of the neck or intramuscularly into a leg. A dose of vaccine is delivered either using a multidose syringe or by various types of automatic vaccinators. The most recent 'robot-type' machines vaccinate the chicks subcutaneously into the neck and treat the beak with ultraviolet light at the same time. Careful preparation and administration of the vaccine is crucial because, under normal circumstances, chicks are not vaccinated again with MD vaccine, and early challenges can result in outbreaks in poorly vaccinated birds.

Subcutaneous or intramuscular injection of inactivated and subunit vaccines

Most commercial layer pullets and breeder flocks are vaccinated with inactivated vaccines towards the end of the rearing period. The vaccination site can vary depending on the type of birds, but most often they are vaccinated either subcutaneously into the neck region or intramuscularly into the leg or breast muscles. When vaccinating into the neck, the skin on the mid to lower back of the neck is pulled up into a flap and the needle is inserted into the space between the skin and underlying muscles. Needles should always be directed away from the head. If the needle is pushed completely through a fold of skin the vaccine will be delivered on to feathers and not subcutaneously, resulting in a missed bird.

Injecting too high, near to the head, or into the neck muscles should be avoided because it can cause pain and excessive swelling. Injection of an oil-emulsion vaccine into the neck muscles can result in loss of ability to raise the head and birds may not be able to eat or drink. The danger of accidental self-injection is greater with this technique than with the intramuscular route.

Intramuscular injection has fewer potential problems than the subcutaneous method, although problems can still occur. The breast muscle area is preferable to the leg muscle for several reasons. Leg muscles are small compared with breast muscles and there are important blood vessels, nerves and tendons above the hock area, which can be easily damaged by a needle, resulting in transient or permanent lameness. If the needle is inserted into this part of the leg it can hit the bone, leading to local inflammation and lameness. It is thought that tenosynovitis and lameness in the early laying period are at least partly due to rough handling and poor leg injection technique.

The size and thickness of the breast muscle offer a relatively safe area for intramuscular vaccination. When injecting into breast muscle, the needle should be placed so as to deposit the vaccine in the thickest part of the muscle. If the needle is inserted too close to the end of the keel or too far to the side it may penetrate the abdominal wall and the vaccine will be deposited in the abdominal cavity. This can result in either mortality, as in the case of a liver puncture, or peritonitis.

Injecting inactivated vaccines into poultry or other birds does not require force. The use of excessive force when inserting a needle into the leg or breast of a bird can be dangerous. The needle can easily hit the leg bone, causing pain for the bird and damage to the bone and the periosteum. In this context it is perhaps not surprising that many layers vaccinated into the leg sit and show general malaise after vaccination. Excessive pressure on the ribcage should also be avoided as the needle can reach the heart or the liver, causing fatal injuries.

All equipment that is used for injection must be properly cleaned, sanitized and maintained to avoid contamination and ensure accurate dosing. If the product to be administered is a live vaccine, it is important to ensure that there is no residue of a chemical sanitizer in the equipment, which can reduce its efficacy.

Inactivated vaccines are very powerful tools in a vaccination programme. However, they do not spread between birds like live vaccines. It is therefore important to be aware that a missed bird is missed for ever! If 20% of a breeder flock is missed then 20% of the potential progeny is also missed.

VACCINATION PROGRAMMES

Chickens

Required vaccinations will vary considerably from region to region and country to country, depending on the local disease situation. Vaccination programmes have been designed to accommodate the required vaccines in a chronological order from day-old until depletion or transfer to laying farms. The order of vaccines is determined by the type of poultry flock, the expected time and type of challenges, presence and level of MDA and possible interactions between vaccines. Veterinary advice should always be taken before deciding which vaccines are required and in what order they should be administered. Table 20.1 shows an example of a vaccination programme for broiler breeders.

Table 20.1. Vaccination programme for broiler breeders (UK).

Age	Vaccine	Route
1 day	Marek's disease[1]	Intramuscular injection
5 days	Coccidiosis	Drinking water
3 weeks	*Newcastle disease* (B1/Clone 30)	Spray/drinking water
	Infectious bronchitis (H120/Ma5)	Spray/drinking water
4 weeks	Infectious bursal disease	Drinking water
6–18 weeks	*Chicken infectious anaemia*	Injection/drinking water
10 weeks	*Newcastle disease* (B1/Clone 30)	Spray/drinking water
	Infectious bronchitis (H120/Ma5)	Spray/drinking water
12 weeks	*Turkey rhinotracheitis*	Spray
	Salmonellosis (SE + ST)[2]	Injection
14 weeks	Infectious avian encephalomyelitis (AE)	Drinking water
18 weeks	Multivalent inactivated vaccine containing various combinations of NDV, IBV, TRTV, EDS, IBDV antigens[3]	Intramuscular injection
	Salmonellosis (SE + ST)	Intramuscular injection
	Reovirus (optional)	Intramuscular injection

[1]Marek's disease bivalent vaccine is usually administered consisting of serotype 1 Rispens vaccine with herpesvirus of turkeys (serotype 3 MDV); [2]SE = *Salmonella enterica* serovar Enteritidis; ST = *Salmonella enterica* serovar Typhimurium; [3]NDV = *Newcastle disease virus*; IBV = *infectious bronchitis virus*; TRTV = *turkey rhinotracheitis virus*; EDS = egg drop syndrome; IBDV = infectious bursal disease virus.

Turkeys

TRT has become endemic in Europe and many other parts of the world, and consequently live TRT vaccines are used to control the disease. In most cases the live vaccine is given at 1 day old in the hatchery by spray and again at about 6 weeks of age, followed by a killed vaccine administered at 14 and 22 weeks. Turkey breeders are immunized against ND and respond better to ND vaccines delivered by spray than when administered in the drinking water. Depending on the epidemiological situation, at least three doses of live ND vaccine may be required. La Sota or Clone 30 type vaccines are used as a primer because they 'take' better than B1 type vaccines. At least two doses of killed vaccine are required. Fowl cholera and haemorrhagic enteritis vaccines are also used in endemic areas. Two doses of killed fowl cholera vaccine are normally given during the rearing period, usually between 10 and 24 weeks. Pox vaccination is often required for turkeys in hotter climates (e.g. California and Australia). Two doses are administered by wing web stab; the timing depends on when challenge is likely to occur.

Vaccines for *rhinotracheitis* are now available in the USA, where this condition is caused by *Bordetella avium*. Paramyxovirus (PMV-3) vaccine is often used to prevent drops in egg production caused by this infection.

Ducks

For breeders and commercial birds, duck hepatitis virus (DHV) and duck viral enteritis (DVE) vaccines are needed. In some countries ducks are also vaccinated against fowl cholera and *S. enterica* serovar Typhimurium.

Game birds

Pheasants and partridges are routinely vaccinated against ND in some European countries. Pheasants can carry viruses such as IB and TRT, which can cause losses in the form of coronavirus-nephritis or respiratory disease, similar to swollen head syndrome of chickens. Vaccination of pheasant flocks against these diseases has proved to be successful in certain game units in the UK.

CONCLUSIONS

Vaccinations will never be able to provide 100% protection against infectious diseases of poultry. They are, however, an essential part of a complex disease control strategy that can minimize the economic impact of diseases and make poultry production commercially viable. In order to achieve maximum benefit from vaccination it is essential to choose the correct vaccine at the appropriate time, and to use it in the proper way.

REFERENCES

Baxendale, W. (1996) Current methods of delivery of poultry vaccines. In: Davison, T.F., Morris, T.R. and Payne, L.N. (eds) *Poultry Immunology*. Poultry Science Symposium Series Volume 24, Carfax Publishing Company, Abingdon, UK, pp. 375–387.

Beard, C.W. and Easterday, B.C. (1967) The influence of the route of administration of Newcastle disease virus on host response. I. Serological and virus isolation studies. *Journal of Infectious Diseases* 117, 55–61.

Calnek, B.W. and Smith, M.W. (1972) Vaccination against Marek's disease with cell-free turkey herpesvirus: interference by maternal antibody. *Avian Diseases* 16, 954–957.

Darbyshire, J.H. and Peters, R.W. (1985) Humoral antibody response and assessment of protection following primary vaccination of chicks with maternally derived antibody against avian infectious bronchitis virus. *Research in Veterinary Science* 38, 14–21.

Davelaar, F.G., Noordzij, A. and van der Donk, J.A. (1982) A study on the synthesis and secretion of immunoglobulins by the Harderian gland of the fowl after eye drop vaccinations against infectious bronchitis at 1 day old. *Avian Pathology* 11, 63–79.

Davison, F., Kaspers, B. and Schat, K.A. (2008) *Avian Immunology*. Elsevier, London.

Delves, P.J., Martin, S., Burton, D. and Roitt, I. (2006) *Roitt's Essential Immunology*, 10th edn. Blackwell Publishing, Oxford, UK.

de Wit, J.J. (2004) Gumboro disease: estimation of optimal time of vaccination by the Deventer formula. http://www.gumboro.com/control/Vaccination/vaccination-timing.asp.

Evans, R.D. and Hafez, Y.S. (1992).Evaluation of a *Mycoplasma gallisepticum* strain exhibiting reduced virulence for prevention and control of poultry mycoplasmosis. *Avian Diseases* 36, 197–201.

Göbel, T.W.F. (1996) The T-dependent immune system. In: Davison, T.F., Morris, T.R. and Payne, L.N. (eds) *Poultry Immunology*. Poultry Science Symposium Series Volume 24, Carfax Publishing Company, Abingdon, UK, pp. 31–45.

Higgins, D.A. (1996) Comparative immunology of avian species. In: Davison, T.F., Morris, T.R. and Payne, L.N. (eds) *Poultry Immunology*. Poultry Science Symposium Series Vol. 24, Carfax Publishing Company, Abingdon, UK, pp. 149–157.

Hodges, R.D. (1974) The circulatory system. In: Hodges, R.D. (ed.) *Histology of the Fowl*. Academic Press, London, pp. 151–241.

Intervet (2003) What is the economical impact of Gumboro disease? *Poultry Focus* October 2003. http://www.intervet.co.uk/binaries/92_60320.pdf.

Janeway, C.A., Travers, P., Walport, M. and Shiomchick, M.J. (2004) *Immunobiology: the Immune System in Health and Disease*, 6th edn. Churchill Livingstone, London.

Male, D., Brostoff, J., Roth, D. and Roitt, I. (2006) *Immunology*, 7th edn. Mosby, London.

Oláh, I., Glick, B. and Taylor, R.L. (1984) Meckel's diverticulum II. A novel lymphepithelial organ in the chicken. *Anatomical Record* 208, 253–263.

Oláh, I., Nagy, N., Magyar, A. and Palya, V. (2002) Esophageal tonsil: a novel gut-associated lymphoid organ. *Poultry Science* 82, 767–770.

Ratcliffe, M.J.H. (2006) Antibodies, immunoglobulin genes and the bursa of Fabricius in chicken B cells development. *Developmental and Comparative Immunology* 30, 101–118.

Reynolds, D.L. and Maraqua, A.D. (2000) Protective immunity against Newcastle disease: the roles of antibodies specific to Newcastle disease virus polypeptides. *Avian Diseases* 44, 138–144.

RUMA Guidelines (2006) Responsible use of vaccines and vaccinations in farm animal production. http://www.ruma.org.uk.

van den Berg, T.P. (2000) Acute infectious bursal disease in poultry: a review. *Avian Pathology* 29, 175–194.

van der Heide, L., Kalbac, M. and Hall, W.C. (1976) Infectious tenosynovitis (viral arthritis): influence of maternal antibodies on the development of tenosynovitis lesions after experimental infection of day old chickens with tenosynovitis virus. *Avian Diseases* 20, 641–648.

CHAPTER 21
Immune Protection of the Hatchling

C. Butter and H.J. Walter
Institute for Animal Health, Compton, Newbury, UK

ABSTRACT

It has become clear that, although their immune responses are impoverished compared with those of adults, neonates are capable of mounting adaptive immune responses, which are dependent on the conditions of antigen exposure. Whilst embryonic vaccination of chicks can successfully induce adaptive immunity, the precise mechanisms whereby this is generated in the face of immune unresponsiveness remain to be elucidated. This chapter reviews existing data and further describes the ontogeny of humoral and Th1 cytokine responses in neonatal chicks and their enhancement through manipulation of the immune system. Neonatal chicks are capable of producing age-related Th1 cytokine profiles following parasitic infection. This contrasts with the mammalian system, in which neonates mount Th2-biased cytokine responses following antigen challenge. However, humoral responses are slow to develop, only being detected in juvenile birds from 18 days post-hatch in response to T-dependent non-replicating antigen. We suggest that the immaturity of this response is due to a defect in the neonatal B-cell population. The manipulation of immune function during the neonatal period, a time when chicks are uniquely vulnerable to infection, offers the prospect of improved methods for prevention and treatment of disease.

PROTECTING THE HATCHLING

Immune protection of the hatchling is one of the most serious challenges for poultry immunologists. In the modern production industry the hatchling is in double jeopardy, being abruptly exposed to a range of pathogens at a time when its immune system is least able to mount protective responses. A relatively recent development in the industry is *in ovo* vaccination, the automated delivery of live viral vaccines to chicks at embryonic day (ED) 18. Although the precise mode of action of *in ovo* vaccination is not yet fully understood, currently more than 90% of US commercial poultry are vaccinated against Marek's disease and infectious bursal (Gumboro) disease using this method. Although only available for a restricted range of pathogens, the protective immune responses induced using *in ovo* vaccination with live attenuated pathogens may be elicited earlier than has been suggested by studies using non-replicating (dead) antigens.

The heavy reliance of the poultry industry on vaccines to prevent infectious disease and the need to maintain poultry health and welfare means that understanding avian immunological mechanisms is a fundamental prerequisite for development of new vaccines and disease control strategies in the future (Davison, 2003). As the majority of poultry vaccinations are administered to very young birds, research focusing on the neonatal immune system should help to establish why *in ovo* vaccination is efficacious, help develop more effective vaccines and contribute to the growing body of acquired knowledge on the development of immunocompetence in neonates of other species.

MAMMALIAN NEONATAL IMMUNOLOGY

The paucity of information regarding the immunology of the avian hatchling contrasts with the substantial body of literature pertaining to its mammalian counterpart.

Mice are classified as neonates from the time of birth until 14 days of age (Klaunberg *et al.*, 2004). Historically, the neonatal immune system was thought to be too immature and prone to tolerogenic signals to be capable of mounting effective immune responses. This idea was pioneered by F.M. Burnet (Burnet and Fenner, 1949), who proposed that, during embryonic development or early life, the immune system learns to distinguish 'self' from 'non-self'. This effectively makes the immune system tolerant to its own tissues whilst retaining the ability to attack invading pathogens. The theory was backed by experiments carried out by P.B. Medawar and colleagues (Billingham *et al.*, 1953), who showed that, while adult mice rejected tissue grafts from genetically different (allogeneic) animals, fetal and newborn (neonatal) mice accepted them and in later life were able to accept transplants from the same donor. Medawar and Burnet (Burnet and Fenner, 1949; Billingham *et al.*, 1953) proposed that this supported the idea that neonatal lymphocytes are uniquely susceptible to the induction of tolerance.

The mechanisms involved in tolerance dominated investigative research of the neonatal immune system for several decades and led to two principal theories. The passive model suggested that neonatal tolerance occurs by negative selection in the same way as natural tolerance (Lederberg and Tatum, 1953; Morrissey *et al.*, 1983). The active model suggested that the neonatal immune system generates suppressor cells that actively protect from self-rejection (Roser, 1989; Kawamura *et al.*, 2002). In 1996, three papers were published which challenged the idea that neonates are inherently susceptible to tolerance induction by showing that, under certain circumstances, neonates are competent to mount adult levels of T-cell responses *in vivo* (Forsthuber *et al.*, 1996; Ridge *et al.*, 1996; Sarzotti *et al.*, 1996).

Ridge *et al.* (1996) showed that the proportion of professional antigen-presenting cells (APC) in Medawar's classical donor inoculum (Billingham *et al.*, 1953) was the factor controlling tolerance or immune activation. Medawar's transfers of splenic or bone marrow cell suspensions contained a large proportion of 'non-professional' APC, which were unable to deliver adequate co-stimulation to naive T cells, resulting in the induction of tolerance. By administering a pure population of 'professional' APC, such as dendritic cells (DC), Ridge *et al.* (1996) showed neonatal T cells could be activated to produce cytotoxic T-lymphocyte (CTL) responses. They also showed that the number of cells administered was important, and it was possible to induce tolerance in adults, provided the number of non-purified splenic cells administered was large enough.

The importance of dose was re-emphasized by Sarzotti *et al.* (1996), who showed that neonatal mice infected with Cas-Br-M *murine leukaemia virus* did not produce the protective CTL and interferon (IFN)-γ responses observed in adult mice, instead producing non-protective T-helper (Th)2 responses. Consequently this led to the development of virus-induced neurological disease. However, if the dose of Cas-Br-M *murine leukaemia virus* was decreased proportionately with the number of T cells in neonatal mice, a switch from Th2 to Th1 occurred and protection was induced.

At this point it is important to describe the Th1–Th2 paradigm, which is a central tenet of modern immunology, and essential for understanding adaptive immune responses. In mammals it is considered that the adaptive immune system is able to polarize functions into the type-1 and type-2 immune pathways, which resolve infections with intracellular or extracellular pathogens, respectively (Janeway, 1992). This polarization of responses is largely regulated by antigen-specific T-helper (Th) cells. Th1 cells drive cell-mediated inflammatory responses and Th2 cells drive humoral (antibody) responses, the latter principally against helminth infections and allergy. Th1 responses typically produce IFN-γ and interleukin (IL)-18. Th2 cells typically produce the cytokines: IL4, IL5, IL9, IL13 and IL19, driven by IL4 and IL13.

The tendency for neonatal immune responses to be Th2-biased has been highlighted in work by Forsthuber *et al.* (1996). By challenging animals with protein antigens in a potent adjuvant (Freund's adjuvant), it was shown that the classic experimental models, which had only investigated lymph nodes for signs of immune induction, had consequently missed the induction of Th2-dominated immune responses in the spleen. Forsthuber *et al.* (1996) went on to show that it was the type of adjuvant administered and not the age at primary challenge which dictated the Th profile of the ensuing immune response. Use of incomplete Freund's adjuvant induced a Th2 profile whereas complete Freund's adjuvant induced a Th1 profile.

Although neonatal immune responses had been described before 1996, collectively these findings engendered resurgence in the investigation of neonatal immunity, with the emphasis on immune responsiveness rather than the lack of response and tolerance. At present it is clear that neonates mount various responses, ranging from deficient to deviant to fully mature, depending on the conditions of antigen exposure. Currently identified differences between adult and neonatal immune systems will now be discussed, with the emphasis on mammalian systems.

Neonatal T cells

Functional differences have been identified between adult and neonatal T cells that are thought to contribute to the limitations observed in neonatal $CD4^+$ Th1 and $CD8^+$ CTL responses following infection or immunization. Quantitative differences, such as the observed lower numbers of immune cells in the periphery of neonates (Ridge *et al.*, 1996; Garcia *et al.*, 2000), and lack of memory cells could affect T-cell responsiveness. Neonatal T cells have been shown to proliferate rapidly (Adkins *et al.*, 2003b), express a highly diverse $CD8^+$ T-cell receptor (TCR) (Wedderburn *et al.*, 2001) and have increased sensitivity to IL-7 ($CD4^+$ and $CD8^+$) and IL-15 ($CD8^+$) (Le Campion *et al.*, 2002; Min *et al.*, 2003; Schonland *et al.*, 2003; Schuler *et al.*, 2004). Neonatal T cells have also been shown to enter the cell cycle more quickly (Adkins *et al.*, 2003b) and develop a memory phenotype faster than adult T cells (Early and Reen, 1999; Min *et al.*, 2003). Heightened cytokine sensitivity and faster memory cell formation have been proposed to allow rapid clonal expansion of recent thymic emigrants, building the T-cell pool, and equipping the neonate with a broad array of memory cells, so increasing their immune potential (Schuler *et al.*, 2004).

Neonatal cytokines

Neonates are viewed as having a Th2 cytokine bias due to the limited $CD4^+$ Th1 and $CD8^+$ CTL responses observed and the finding that, as adults, mice immunized as neonates mount Th2-dominated memory responses following re-exposure to the same antigen (Adkins and Du, 1998; Delespesse *et al.*, 1998; Adkins *et al.*, 2000, 2001, 2002, 2003a; Adkins, 2003; Li *et al.*, 2004). The Th2 bias is less evident in human neonates as low levels of all cytokines can be detected (Cohen *et al.*, 1999), although diminished Th1 cell responses have been observed following certain infections and immunizations (Prescott *et al.*, 1998, 2003; Upham *et al.*, 2002; Protonotariou *et al.*, 2004; Schultz *et al.*, 2004). Several factors which may result in the observed differences in neonatal and adult T-cell cytokine production have been identified, although a conclusive explanation for why these differences have evolved has not yet been reached.

Increased levels of Th2 cytokines (IL-4 and IL-13) and anti-inflammatory cytokines (IL-10 and TGF-β) (Prescott *et al.*, 1998; Rainsford and Reen, 2002; Gasparoni *et al.*, 2003; Schultz *et al.*, 2004) could be related to the increased ratio of CD4:CD8 cells in neonates. Neonatal $CD8^+$ T cells have been found to produce increased levels of IL-13 and IL-4 and increased expression of their receptors, IL-13R and IL-4R, respectively. Increased expression of both these cytokines and their receptors is thought to induce a Th2 bias in primary responses while increasing the disposition of $CD8^+$ cells to apoptosis upon secondary exposure (Ribeiro Do Couto *et al.*, 2001; Li *et al.*, 2004). Th1 cytokine induction may be actively inhibited due to increased levels of cytokines and active suppression by neonatal regulatory cells (Adams *et al.*, 2003; Field *et al.*, 2003). Defective production of IL-12 by professional neonatal APC (macrophages and DC) will also reduce the possibility of a Th1 response being initiated (Upham *et al.*, 2002; Prescott *et al.*, 2003; Goriely *et al.*, 2004; Li *et al.*, 2004). A lack of upregulation of CD40 ligand (L) following activation (Nonoyama *et al.*, 1995; Healy *et al.*, 1997; Jullien *et al.*, 2003; Han *et al.*, 2004) and the influence of the environment (e.g. lymph node or spleen) in which T cells encounter antigen have also been suggested to influence Th2 bias (Adkins *et al.*, 2003a).

Adult-like Th1 and CTL responses have been induced in response to certain infections (Sarzotti *et al.*, 1996; Hermann *et al.*, 2002; Wilson and Morgan, 2002), live viral vaccines (Kovarik *et al.*, 2000, 2001) and DNA vaccines (Martinez *et al.*, 1997), and on exposure to strong Th1-promoting adjuvants (Forsthuber *et al.*, 1996; Chipeta *et al.*, 2000; Weeratna *et al.*, 2001; Martinez *et al.*, 2003). These findings led to the suggestion that factors other than intrinsic T-cell properties may contribute to the variant responses of neonates. The maturational status of neonatal DC was highlighted as the most likely influencing factor.

Neonatal antigen-presenting cells

As with T cells, the functional capacity of neonatal DC is controversial, as results vary considerably due to the differences in the DC population examined, the types of antigen tested and the assay systems used. Some groups have shown neonatal DC to be functionally immature while others find they are capable of producing adult-like responses. It is agreed that the absolute number of DC in neonates is several logs lower compared with adults (Muthukkumar *et al.*, 2000; Dadaglio *et al.*, 2002; Sun *et al.*, 2003). The ratio and distribution of neonatal DC have been shown to differ, with increased numbers of myeloid DC seen in human umbilical cord blood (Navarrete *et al.*, 2003) and increased $CD11c^+$ $CD8^+$ DC found in the spleen of mice (Sun *et al.*, 2003; Dakic *et al.*, 2004). Reduced numbers of Langerhans DC with lower expression of major histocompatibility (MHC) class II antigen and a reduced ability to transport antigen to the lymph nodes have been shown in mice (Dewar *et al.*, 2001). Lower MHC II and co-stimulatory molecule expression (Petty and Hunt, 1998; Muthukkumar *et al.*, 2000; Dewar *et al.*, 2001; Langrish *et al.*, 2002; Simpson *et al.*, 2003), together with an inability to process soluble antigen (Kollmann *et al.*, 2004), has been found to reduce the ability of neonatal DC to stimulate T-cell proliferation. Neonatal DC and macrophages are also suggested to be less efficient at inducing Th1 immune responses and CTL responses due to a reduced ability to produce the IL-12p35 subunit of IL-12 (p70) and IFN-γ (Goriely *et al.*, 2001; Langrish *et al.*, 2002; Upham *et al.*, 2002; 2002; Chelvarajan *et al.*, 2004; Dakic *et al.*, 2004). Collectively these findings accord with the view that there is a Th2 bias in neonates, in which APC are involved. Despite these findings, other groups have shown both murine and human neonatal DC to be capable of producing IL-12 (Karlsson *et al.*, 2002; Salio *et al.*, 2003; Sun *et al.*, 2003), processing and presenting antigen, having adult-like MHC and co-stimulatory expression levels, and stimulating T cells (Dadaglio *et al.*, 2002; Dakic and Wu, 2003; Sun *et al.*, 2003; Dakic *et al.*, 2004).

Neonatal B cells

Reduced humoral responsiveness to both those requiring T-cell help and those that do not – T-dependent (TD) and T-independent (TI) antigens, respectively – is frequently observed in neonates. Neonatal antibody responses are often delayed in onset, reach lower peak levels, are of a shorter duration, have different isotype profiles in mice and humans, possibly due to class switching difficulties, and have lower avidity and reduced heterogeneity due to reduced somatic mutation events (reviewed in Marshall-Clarke *et al.*, 2000a; Siegrist, 2001; Adkins *et al.*, 2004). The neonatal B-cell pool comprises mostly immature ($IgM^{+}IgD^{low/-}$) and B1 B cells (King *et al.*, 1999; Marshall-Clarke *et al.*, 2000b; Benschop *et al.*, 2001). Phenotypic and functional differences between adult and neonatal B cells have been identified, which may affect their ability to respond to antigens directly, or indirectly affect T-cell responses.

Cross-linking of murine neonatal B-cell receptors (BCR) has been found to induce negative signals, inhibiting responses to antigen and making B cells prone to apoptosis (Norvell *et al.*, 1995). Splenic B cells of neonatal mice have shown a selective inability to upregulate the expression of MHC II antigen (Marshall-Clarke *et al.*, 2000a; Muthukkumar *et al.*, 2000) and the co-stimulatory molecule CD86 (King *et al.*, 1999; Muthukkumar *et al.*, 2000; Benschop *et al.*, 2001), the presence of both being essential for effective interactions with T cells. Reduced MHC II antigen expression and a lack of CD86 expression have been suggested to influence the Th2 bias in neonatal T cells and increase the possibility of inducing T- and B-cell anergy (Hosken *et al.*, 1995; Lombardi *et al.*, 1997; Marshall-Clarke *et al.*, 2000a,b). These deficiencies are less pronounced in murine lymph node B cells and have not been detected in neonatal humans.

Lower expression levels of complement receptor 2 (CD21) (Griffioen *et al.*, 1993; Balogh *et al.*, 2001; Tasker and Marshall-Clarke, 2003) and the complement component C3 (Davis *et al.*, 1972) have been detected on neonatal B cells. Reduced CD21 expression results in weaker activation signals being induced after antigen exposure and could explain the finding that neonatal B cells require stronger stimulatory signals to become activated after exposure to pokeweed mitogen, anti-IgM and anti-IgD (Brines and Klaus, 1991; Eisenthal *et al.*, 2003). Reduced expression of these molecules is suggested to contribute to the marked deficiency of neonates to respond to TI antigens. TD responses will also be affected by reduced CD21 expression levels, as binding of CD21L on FDC with the B-cell co-receptor complex (CD21/CD19/CD81) provides the signals necessary for germinal centre induction (Qin *et al.*, 1998; Hase *et al.*, 2004).

The ability of neonatal B cells to enter secondary lymphoid organs and present antigen efficiently has been questioned due to the reduced expression of CD62L and CCR7 (Tasker and Marshall-Clarke, 2003). CD62L is expressed on naive B cells and is required for B-cell entry into secondary lymphoid organs. CCR7 is a B-cell chemokine receptor for CCL19 and CCL21, which allows B cells to home into the T-cell areas of lymphoid organs. On entering the secondary lymphoid organs, B cells are thought to have a major role in initiating structural development through interactions with stromal cells, utilizing lymphotrophic chemokines, various members of the TNF/LT family cytokines and their receptors (Fu and Chaplin, 1999; Mebius, 2003).

Neonatal lymphoid organs

As in chickens, the full development of the lymphoid microarchitecture occurs in mice and humans post-natally. It has not been established if the time taken to develop a mature structure after birth is due to immaturity of B cells, to unresponsiveness of follicular DC (FDC) to maturation signals (Pihlgren *et al.*, 2003) or to as yet unidentified deficiencies. The coincidence of mature lymphoid organ microarchitecture with adult-like antibody responses has led to the suggestion that lack of an appropriate environment in which to present antigen may cause the reduced humoral responses of neonates rather than intrinsic B-cell deficiencies (Mast, 1998). The appearance of marginal-zone B cells in the spleens of neonatal mice and humans coincides with their ability to mount antibody responses against TI antigens (Timens *et al.*, 1989). This has led to the suggestion that TI responses are not only affected by reduced CD21 and C3 expression but also due to a lack of marginal-zone B cells. The ability to mount adult-like TD antibody responses has been found to require fully functional and correctly located B cells, T cells and FDC. This then allows germinal centre formation to be initiated, resulting in adult levels of antibody being induced (Mast, 1998; Pihlgren *et al.*, 2001).

In conclusion, several mechanisms ranging from cell signalling deficits to underdeveloped lymphoid organ structure have been identified that could operate to produce the relative immunodeficiency seen in neonates.

THE AVIAN NEONATAL IMMUNE SYSTEM

Since the late 1950s numerous investigations have been carried out on the humoral responsiveness of chickens. Various antigens, over a range of different doses, have been administered via different routes of immunization to several genetic lines and different ages of chickens. Studies focusing on the development of humoral immunocompetence in chicks, despite differences in experimental designs, have generally concluded that humoral responsiveness in newly hatched birds is less efficient than that of adults. Early work by Wolfe and Dilks (1948) showed that chicks immunized at 1 day after hatching could produce weak (and delayed) antibody responses to bovine serum albumin (BSA) 12 days later. Subsequent work suggested that 4–5 days of age was the earliest age at which detectable humoral responses could be induced (Seto and Henderson, 1968; Soloman, 1968). Four days of age was also shown to be the earliest age at which germinal centre formation could be detected following immunization with BSA at ED18 or 1 day after hatching, although no antibody production was detected (Vainio *et al.*, 1978). Generally, attempts to induce humoral responses to TD antigens in the first week of life fail, even with use of strong adjuvants, such as Freund's complete (Mast and Goddeeris, 1999). The ability of chicks to induce humoral responses increases rapidly after the first week of life, reaching adult levels by 2–3 weeks post-hatching (Blazkovec and Wolfe, 1965; Seto and Henderson, 1968; Soloman, 1968; Vainio *et al.*, 1978; Seto, 1981; Munns and Lamont, 1991; Mast and Goddeeris, 1999).

Before the extensive dissection of the murine neonatal immune system, it had been postulated that a lack of adequate leukocyte numbers (Seto, 1975, 1976, 1980; Fredericksen and Gilmour, 1983; Kai and Franklin, 1984), the presence of neonatal suppresser cells (Droege, 1971; Moticka, 1977) or the lack of a mature secondary lymphoid environment (Szenberg, 1977; Mast and Goddeeris, 1999) may result in the delayed humoral responses of neonatal chicks. Despite advances in the understanding of the mammalian neonatal immune system all three hypotheses are still accepted.

The ontogeny of humoral immune responses in inbred line 6 birds was investigated in our laboratory using the administration of the TD antigen, keyhole limpet haemocyanin (KLH), to chicks at different ages and by different routes (Walter, 2005). Quantification of lymphoid cell numbers in peripheral blood and the appearance of B cells (Bu-1$^+$) in the developing spleen were assessed in order to determine if changes in immunocompetence were related to these characteristics.

Antibody responses were not observed in birds immunized with KLH by the intravenous (i.v.) route before 18 days of age. Birds immunized after 18 days of age were capable of mounting humoral responses. The intensity of the response correlated with the age at immunization, with birds immunized at later times producing more substantial responses. We also demonstrated that the magnitude of the antibody response was affected by the dose of antigen administered by the i.v. route. An increase in the dose did not induce earlier antibody responses in neonatal birds, with only birds immunized at 18 days of age or older producing measurable antibody. The relationship of the dose of immunogen with the magnitude of response has been previously reported in mammals and in various lines of chickens using different challenge antigens (Cerny and Ivanyi, 1966; Kreukniet and van der Zijpp, 1990; Boa-Amponsem *et al.*, 2000). Below a certain threshold dose, most immunogens do not elicit a response. Once the threshold level has been surpassed, the intensity of immune response gradually increases with increasing dose of immunogen until a broad plateau level is reached, followed by a decline at very high doses due to the induction of high-zone tolerance. Although immune deviation has been demonstrated by Sarzotti *et al.* (1996), in our study the absence of immune responses in birds immunized before 18 days of age did not appear to be a result of either this or tolerance induction.

Following subcutaneous immunization with KLH in the proprietary adjuvant Titermax Gold, chicks produced detectable antibody at younger ages than if the antigen was administered by the i.v. route, with low-level responses being detected in birds immunized as early as 1 day of age. In earlier studies adjuvant use was reported to delay the initiation of antibody responses when the protein antigen, human serum albumin, was administered (French *et al.*, 1970; Steinberg *et al.*, 1970). It has also been demonstrated that the use of adjuvant can prolong, and in some cases increase, the intensity of primary antibody responses. This effect was attributed to the slower release rate of antigen (French *et al.*, 1970; Fukanoki *et al.*, 2000) and to the immunostimulatory properties of some adjuvants such as Freund's complete and Titermax Gold, which stimulate cell-mediated (Th1) responses that enhance levels of T-cell help, ultimately improving the humoral immune response (Kreukniet *et al.*, 1992; Forsthuber *et al.*, 1996). It is possible that in our studies the immunostimulatory properties of the Titermax adjuvant may have enhanced antigen presentation and T-cell help (Kreukniet *et al.*, 1992; van Immerseel *et al.*, 2002) sufficiently to allow for the observed induction of antibody in younger birds. It does not appear that these earlier responses were a result of slower antigen release or prolonged antigen exposure, as immune responses were detectable at 10 days of age, 8 days earlier than those observed following i.v. immunization.

The route of inoculation influences the location where immune responses are mounted and can also affect the magnitude of response. When chickens receive i.v. inoculation, antigen is thought to locate predominantly in the spleen, which then becomes the major source of antibody production (White and Henderson, 1975). After subcutaneous (s.c.) or intramuscular (i.m.) inoculation, antigen is thought to associate with lymphoid tissues local to the inoculation site, as well as being presented in the spleen (Donker and Beuving, 1989; Kreukniet *et al.*, 1992). Previous studies with adult chickens have shown that i.v. inoculation with TD antigens elicits greater immune responses than s.c., i.m. and intraperitoneal (i.p.) inoculations (Seto and Henderson, 1968; van der Zijpp *et al.*, 1986; Kreukniet *et al.*, 1992; Boa-Amponsem *et al.*, 2001). Consequently, the spleen was considered to be the most important organ for the generation of humoral immune responses against TD antigens (Mast, 1998). In our work, the neonatal spleen did not develop mature B-cell ($Bu\text{-}1^+$) structures until 14 days of age. Mast (1998) and Mast and Goddeeris (1999) suggested that there was a strong relationship between structural organization and function of lymphoid organs, particularly the spleen, and that the immune function of neonates may be compromised due to the lack of splenic structural development. Indeed, in the present study, antibody responses were only detectable after i.v. immunization (when antigen is thought to locate in the spleen) from 18 days after hatching, when the spleen had developed adult-like structures. This could imply that the initiation of earlier immune responses following s.c. immunization may be attributed to the presentation of antigen in a lymphoid organ other than the spleen, as it has been shown to be possible to produce normal, yet slightly delayed, antibody levels in birds which have been splenectomized (Rosenquist and Wolfe, 1962; Graetzer *et al.*, 1963). For example, KLH antigen may have been processed and presented in peripheral lymphoid organs that acquire immune capability before the spleen (14 days of age), such as the bronchial-associated lymphoid tissue (BALT) or mucosal-associated lymphoid tissue (MALT), both of which have developed by 5 days of age (Jeurissen *et al.*, 1989; Sminia *et al.*, 1989). As the development of peripheral secondary lymphoid organs has been shown to be initiated by antigen exposure, it is possible that the presence of the KLH antigen and adjuvant, or both, could have induced earlier maturation of local lymphoid organs, which, in conjunction with the activation of cellular immune responses due to the adjuvant presence, may have resulted in the induction of immune responses in younger birds.

Analysis of the number and phenotype of peripheral blood lymphocytes (PBL) suggested neonates are leukopenic compared with adults (Fig. 21.1) although this is the first report to show such differences in chickens using a whole blood analysis. In earlier studies (Fredericksen and Gilmour, 1983; Cooper and Chen, 1991) leukocyte numbers were found to be lower in neonatal chicks than adults. It was suggested that the leukocyte deficiency may affect the ability of neonates to respond to challenge. It is difficult to relate neonatal immune functional potential to peripheral cell numbers alone because, as already discussed, the structural development of the lymphoid organs has a major impact on functional ability. In addition, no information on the functional

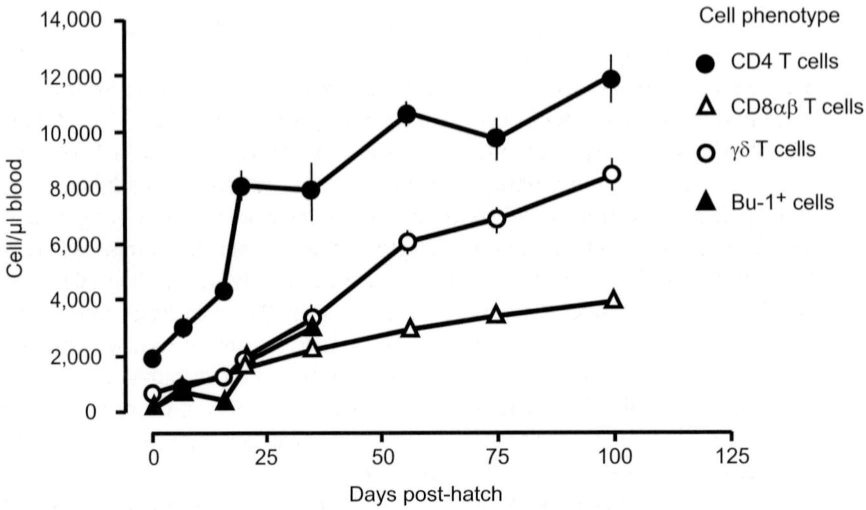

Fig. 21.1. The mean total number of B-cell and T-cell subsets (per ml of blood) as a function of age. All cell subtypes investigated increased over the sampling period. $CD4^+$ T cells were the most numerous lymphocytes present in the circulation but all cell types were at profoundly lower concentrations in the neonate.

maturity of the cells in the periphery can be obtained using this method. Nevertheless, it is evident that numbers of PBL increase substantially in the first 18 days after hatching, after which antibody responses were detectable, irrespective of the immunization protocol.

In summary, chicks appear to cross a threshold at around 18 days post-hatching, after which the functional potential of their immune system rapidly increases, enabling them to respond to primary immunization. This suggests that the acquisition of immune function can be attributed to lymphoid organ structure or cell numbers and, by inference, attainment of cell function. The coincidence of a mature lymphoid structure with antibody production was highlighted twice in our study. First, a mature splenic structure appeared necessary for i.v. immunization to induce antibody responses. Second, s.c. immunizations showed that it may be possible to induce antibody responses before mature spleen structures are developed, possibly because antigen was presented in a mature peripheral lymphoid tissue, such as the bronchial or enteric mucosal-associated lymphoid tissue. The increase in immune response following the acquisition of mature lymphoid structures (14 days of age) may reflect increasing leukocyte numbers in chicks. It remains to be shown if the functional ability of leukocytes directly affects the age at which responses can be induced.

A comparison of the memory humoral responses generated in chicks following secondary immunization with those of adult chickens receiving primary immunization showed no evidence of tolerance induction, irrespective of the route of immunization or dose of antigen administered.

Tolerance has been induced in juvenile and adult chickens following administration of high doses of protein antigens, e.g. human, bovine or duck serum albumin (Ivanyi *et al.*, 1964; Cerny *et al.*, 1965; Cerny and Ivanyi, 1966; Cerny and Viklicky, 1966; Ivanyi and Valentova, 1966; Mast and Goddeeris, 1999), as seen in the mammalian models (Ridge *et al.*, 1996). It has been suggested that the continued presence of antigen is required for sustained tolerance induction in the chicken (Billingham *et al.*, 1953; Ivanyi *et al.*, 1964; Ivanyi and Valentova, 1966). The more general lack of observed tolerance following neonatal immunization is consistent with more recent mammalian literature, which suggests neonates are not inherently susceptible to tolerance induction following antigen challenge as previously suggested by Medawar and Burnet in the 1950s (Burnet and Fenner, 1949; Billingham *et al.*, 1953). These findings are at variance with those of Mast and Goddeeris (1999), who reported tolerance was induced after neonatal chicks were immunized with BSA by the s.c. route. This may be attributed to the different route of immunization, the use of adjuvant or the difference in antigen administered, since different protein antigens have been found to have different immunogenic effects (Ivanyi and Valentova, 1966).

Responses indicative of memory are those which, following secondary immunization, can be induced earlier, more rapidly, become substantially higher or persist longer than those following primary immunization. On this basis we have been able to detect memory responses dependent on the age of the bird at primary inoculation, the use of adjuvant and the dose of antigen. Some memory responses are clearly evident even when no response was detected after primary vaccination.

In summary, these data indicate that neonatal chicks are less able to mount humoral responses than adults, irrespective of the dose of antigen or adjuvant administered, and that immunocompetence develops with age. These deficiencies may be related to the lack of a mature spleen structure and the lower leukocyte numbers in the periphery of neonatal chicks.

THE AVIAN NEONATAL TH1/TH2 CYTOKINE ENVIRONMENT

As previously described, the immune system is capable of functionally polarizing into type 1 and type 2 immune pathways. This polarization is largely regulated by antigen-specific $CD4^+$ cells, although $CD8^+$ T cells, $\gamma\delta$ T cells and DC also play a role in this process (Mosmann and Sad, 1996; Liu *et al.*, 2001). The differentiation into Th1- or Th2-type cells determines the initiation of cell-mediated or humoral immune responses (Mosmann and Sad, 1996). As mentioned earlier, Th1 cells typically produce IFN-γ, driven by the early production of IL-12 and IL-18, and are associated with inflammatory responses necessary for destroying cells infected by viruses and other intracellular microbes. Th2 cells produce IL-4, IL-5 and IL-13, which are associated with antibody production, anti-helminth reactions and IgE production. Consequently, the predisposition of mammalian neonates to a Th2 cytokine profile results in an inability to mount effective Th1 or CTL responses (reviewed in Adkins *et al.*, 2004), leaving them vulnerable to bacterial, viral and intracellular parasitic infections. The recent identification of chicken Th2 cytokine genes (Avery *et al.*, 2004) has now made possible similar studies in chickens.

In order to examine the capacity of neonatal chicks to respond to infection we used the parasite *Eimeria maxima*, known to induce strong Th1-type responses. The intracellular protozoan parasite *Eimeria* spp. infect numerous vertebrate species, including the chicken, and cause coccidiosis, a parasitic disease that has a significant economic impact on the poultry industry (Williams, 1999). Both neonatal and adult birds are susceptible to *Eimeria* infection (Rose, 1967; Hein, 1968; Lillehoj, 1988). Low-dose infection with *E. maxima* results in subclinical coccidiosis, in which chickens show no overt clinical signs but may have depressed growth and poor feed conversion. Infection with small numbers of oocysts is sufficient to induce complete species-specific immunity to reinfection (Rose and Long, 1962; Rose, 1974). Primary infection of adult birds with *E. maxima* results in the induction of Th1-dominated immune responses, with the Th1 cytokine, IFN-γ (Lillehoj and Choi, 1998; Rothwell *et al.*, 2000; Laurent *et al.*, 2001), being produced by T cells (Rose and Hesketh, 1979; Lillehoj, 1994; Smith and Hayday, 2000a,b), and in mice IFN-γ is also produced by NK cells (Schito and Barta, 1997; Schito *et al.*, 1996, 1998a,b). B lymphocytes and macrophages also respond to *Eimeria* infection, although it is not thought these cell types have a role in protection (Talebi and Mulcahy, 1994; Smith and Hayday, 2000a). High-dose infection results in clinical coccidiosis and high levels of mortality.

In our study (Walter, 2005), real-time quantitative polymerase chain reaction (RT-PCR) assays were used to characterize the neonatal cytokine profiles of non-infected and infected chickens. The ability of neonatal chicks to mount protective responses to *E. maxima* was assessed by measuring the levels of Th1 (IL-12α, IL-12β, IL-18, IFN-γ), Th2 (IL-4) and regulatory (IL-10) cytokines over a 10-day period following infection and comparing the levels to those produced by older, more immunologically mature birds.

Overall, the cytokine transcript data demonstrate for the first time that neonatal chickens do not have a Th2 bias and are capable of inducing Th1 responses following challenge with *E. maxima*. The Th1 cytokine, IFN-γ, was constitutively expressed in the gut and spleen of non-infected birds from 2 days of age. This accords with previous research, in which IFN-γ has been detected during embryonic development and in non-infected young chicks (Kogut *et al.*, 2002; Peters *et al.*, 2003). These data suggest that there is no apparent Th2 bias in neonatal chicks and that they are capable of inducing Th1 responses following *Eimeria* infection. Although the intensity of Th1 IFN-γ responses was lower in younger chicks, this appeared to be related to the maturation status of the organs and cells within them or the reduced numbers of parasites accommodated in the gut rather than being due to the active suppression of Th1 responses. It has been suggested that in mammals a Th2 bias is required for successful pregnancy, as it may function to prevent the induction of potentially lethal Th1 and cytotoxic responses in both the fetus and mother, and play a role in the establishment of fetal tolerance (Lin *et al.*, 1993; Wegmann *et al.*, 1993; Chaouat *et al.*, 1997, 2002; Raghupathy, 1997). As the Aves lack the feto-placental interface, such evolutionary adaptation is not necessary, and the avian immune system is consequently able to acquire full Th1 potential earlier than its mammalian counterpart. These results have clear implications for the design of avian vaccines, where targeting to induce cell-mediated responses in the hatchling may be possible.

ENHANCEMENT OF NEONATAL IMMUNE RESPONSIVENESS

Neonatal chicks were found to be relatively lymphopenic in comparison to adults and lacked mature splenic B cell ($Bu\text{-}1^+$) structures until 14 days of age. It has been suggested that reduced leukocyte numbers in neonates affects their ability to respond to antigen challenge (Cooper and Chen, 1991; Ridge *et al.*, 1996). The use of cell transfer methods to increase leukocyte numbers in neonates has been found to improve the levels of responsiveness in both avian (Gilmour *et al.*, 1970; Seto, 1970, 1975, 1977a, 1981, 1983, 1984) and murine (Ridge *et al.*, 1996) models. Work by Seto showed that transfer of bone marrow cells, thymocytes, bursal cells, splenocytes or blood leukocytes into neonatal chicks improves their ability to mount humoral responses (Seto, 1975, 1977a,b, 1981, 1983). Seto (1977a) suggested that the donor cells were able to interact with host cells and could be incorporated into the host organs, possibly enhancing their development (Seto, 1984). Further examples which suggest transferred cells can be integrated into the host system are evident when immune cells are transferred into irradiated hosts. Gilmour *et al.* (1970) reported that the transfer of splenocytes into irradiated chicks restored their ability to respond to TI and TD antigens, whilst transfer of bursacytes restored the ability to respond only to TI antigens. Transfer of splenocytes with bursal cells enhanced immune responses to TD antigens more than transferring splenocytes alone, suggesting that B cells may play a role in improving immune responsiveness.

Our own work (Walter, 2005) has conclusively shown that transfer of adult splenocytes into 1-day-old hatchlings profoundly increases their ability to evoke antibody responses. Furthermore, host immune cells were also induced to participate in these early responses. By depleting the transferred cells of specific immune cell subsets and by bursectomizing donor animals, we were also able to show that B cells ($Bu1^+$ cells) were critical and sufficient for the transfer function.

CONCLUSIONS

Critical protection of the hatchling is only partially achieved by current *in ovo* vaccination strategies. Knowledge of the neonatal immunology of the chick has benefited from mammalian studies, although profound differences, such as the apparent lack of Th1/Th2 bias, offer opportunities for the avian vaccinologist. Continuing advances in the understanding of the avian perinatal immune system should provide immunological explanations as to how, and why, *in ovo* vaccination can be so effective. The ability to experimentally manipulate the neonatal immune system may also help in the development of vaccines that provide greater protection against prevalent diseases at a time when the immune system is otherwise uniquely vulnerable.

REFERENCES

Adams, B., Nagy, N., Paulart, F., Vanderhaeghen, M.L., Goldman, M. and Flamand, V. (2003) CD8(+) T lymphocytes regulating Th2 pathology escape neonatal tolerization. *Journal of Immunology* 171, 5071–5076.

Adkins, B. (2003) Peripheral CD4(+) lymphocytes derived from fetal versus adult thymic precursors differ phenotypically and functionally. *Journal of Immunology* 171, 5157–5164.

Adkins, B. and Du, R.Q. (1998) Newborn mice develop balanced Th1/Th2 primary effector responses *in vivo* but are biased to Th2 secondary responses. *Journal of Immunology* 160, 4217–4224.

Adkins, B., Bu, Y.R., Cepero, E. and Perez, R. (2000) Exclusive Th2 primary effector function in spleens but mixed Th1/Th2 function in lymph nodes of murine neonates. *Journal of Immunology* 164, 2347–2353.

Adkins, P., Bu, Y.R. and Guevara, P. (2001) The generation of the memory in neonates versus adults: prolonged primary Th2 effector function and impaired development of Th1 memory effector function in murine neonates. *Journal of Immunology* 166, 918–925.

Adkins, B., Bu, Y.R. and Guevara, P. (2002) Murine neonatal CD4(+) lymph node cells are highly deficient in the development of antigen-specific Th1 function in adoptive adult hosts. *Journal of Immunology* 169, 4998–5004.

Adkins, B., Bu, Y., Vincek, V. and Guevara, P. (2003a) The primary responses of murine neonatal lymph node CD4+ cells are Th2-skewed and are sufficient for the development of Th2-biased memory. *Clinical and Developmental Immunology* 10, 43–51.

Adkins, B., Williamson, T., Guevara, P. and Bu, Y. (2003b) Murine neonatal lymphocytes show rapid early cell cycle entry and cell division. *Journal of Immunology* 170, 4548–4556.

Adkins, B., Leclerc, C. and Marshall-Clarke, S. (2004) Neonatal adaptive immunity comes of age. *Nature Reviews Immunology* 4, 553–564.

Avery, S., Rothwell, L., Degen, W.D.J., Schijns, V., Young, J., Kaufman, J. and Kaiser, P. (2004) Characterization of the first nonmammalian T2 cytokine gene cluster: the cluster contains functional single-copy genes for IL-3, IL-4, IL-13, and GM-CSF, a gene for IL-5 that appears to be a pseudogene, and a gene encoding another cytokine-like transcript, KK34. *Journal of Interferon and Cytokine Research* 24, 600–610.

Balogh, P., Aydar, Y., Tew, J.G. and Szakal, A.K. (2001) Ontogeny of the follicular dendritic cell phenotype and function in the postnatal murine spleen. *Cellular Immunology* 214, 45–53.

Benschop, R.J., Brandl, E., Chan, A.C. and Cambier, J.C. (2001) Unique signaling properties of B cell antigen receptor in mature and immature B cells: implications for tolerance and activation. *Journal of Immunology* 167, 4172–4179.

Billingham, R.E., Brent, L. and Medawar, P.B. (1953) Actively acquired tolerance of foreign cells. *Nature* 172, 603–606.

Blazkovec, A.A. and Wolfe, H.R. (1965) Factors affecting the primary and secondary responses to bovine serum albumin in chickens. *International Archives of Allergy and Applied Immunology* 26, 80–95.

Boa-Amponsem, K., Dunnington, E.A., Pierson, F.W., Larsen, C.T. and Siegel, P.B. (2000) Antibody responses to different dosages of sheep red blood cells in lines of chickens selected for high and low antibody response to sheep red blood cells. *Poultry Science* 79, 159–162.

Boa-Amponsem, K., Price, S.E.H., Dunnington, E.A. and Siegel, P.B. (2001) Effect of route of inoculation on humoral immune response of White Leghorn chickens selected for high or low antibody response to sheep red blood cells. *Poultry Science* 80, 1073–1078.

Brines, R.D. and Klaus, G.G. (1991) Effects of anti-immunoglobulin antibodies, interleukin-4 and second messenger agonists on B cells from neonatal mice. *International Immunology* 3, 461–466.

Burnet, F.M. and Fenner, F. (1949) *The Production of Antibodies*. Macmillan, Melbourne, Australia.

Cerny, J. and Ivanyi, J. (1966) The dose of antigen required for the suppression of the IgM and IgG rsponse in chickens. II. Studies at the cellular level. *Folia Biologica (Praha)* 12, 343–354.

Cerny, J. and Viklicky, V. (1966) The effect of antigen dose on the changes in the spleen: tolerance versus immunity. In: Cottier, H., Odartchenko, N., Schindler R. and Congdon, C.C. (eds) *Germinal Centers in Immune Responses*. Springer-Verlag, Berlin, pp. 319–328.

Cerny, J., Ivanyi, J., Madar, J. and Hraba, T. (1965) Nature of delay in immune response after administration of large doses of protein antigen in chicks. *Folia Biologica (Praha)* 11, 402–405.

Chaouat, G., Tranchot Diallo, J., Volumenie, J.L., Menu, E., Gras, G., Delage, G. and Mognetti, B. (1997) Immune suppression and Th1/Th2 balance in pregnancy revisited: a (very) personal tribute to Tom Wegmann. *American Journal of Reproductive Immunology* 37, 427–434.

Chaouat, G., Zourbas, S., Ostojic, S., Lappree-Delage, G., Dubanchet, S., Ledee, N. and Martal, J. (2002) A brief review of recent data on some cytokine expressions at the materno-foetal interface which might challenge the classical Th1/Th2 dichotomy. *Journal of Reproductive Immunology* 53, 241–256.

Chelvarajan, R.L., Collins, S.M., Doubinskaia, I.E., Goes, S., Van Willigen, J., Flanagan, D., de Villiers, W.J.S., Bryson, J.S. and Bondada, S. (2004) Defective macrophage function in neonates and its impact on unresponsiveness of neonates to polysaccharide antigens. *Journal of Leukocyte Biology* 75, 982–994.

Chipeta, J., Komada, Y., Zhang, X.L., Azuma, E., Yamamoto, H. and Sakurai, M. (2000) Neonatal (cord blood) T cells can competently raise type 1 and 2 immune responses upon polyclonal activation. *Cellular Immunology* 205, 110–119.

Cohen, S.B.A., Perez-Cruz, I., Fallen, P., Gluckman, E. and Madrigal, J.A. (1999) Analysis of the cytokine production by cord and adult blood. *Human Immunology* 60, 331–336.

Cooper, M.D. and Chen, C.H. (1991) Avian T cell ontogeny. *Advances in Immunology* 50, 87–117.

Dadaglio, G., Sun, C.M., Lo-Man, R., Siegrist, C.A. and Leclerc, C. (2002) Efficient in vivo priming of specific cytotoxic T cell responses by neonatal dendritic cells. *Journal of Immunology* 168, 2219–2224.

Dakic, A. and Wu, L. (2003) Hemopoietic precursors and development of dendritic cell populations. *Leukocyte Lymphoma* 44, 1469–1475.

Dakic, A., Shao, Q.X., D'amico, A., O'Keeffe, M., Chen, W.F., Shortman, K. and Wu, L. (2004) Development of the dendritic cell system during mouse ontogeny. *Journal of Immunology* 172, 1018–1027.

Davis, N.C., West, C.D. and Ho, M. (1972) Effect of aging of serum on quantitation of complement component C3. *Clinical Chemistry* 18, 1485–1487.

Davison, T.F. (2003) The immunologists' debt to the chicken. *British Poultry Science* 44, 6–21.

Delespesse, G., Yang, L.P., Ohshima, Y., Demeure, C., Shu, U., Byun, D.G. and Sarfati, M. (1998) Maturation of human neonatal CD4(+) and CD8(+) T lymphocytes into Th1/Th2 effectors. *Vaccine* 16, 1415–1419.

Dewar, A.L., Doherty, K.V., Woods, G.M., Lyons, A.B. and Muller, H.K. (2001) Acquisition of immune function during the development of the Langerhans cell network in neonatal mice. *Immunology* 103, 61–69.

Donker, R.A. and Beuving, G. (1989) Effect of corticosterone infusion on plasma corticosterone concentration, antibody production, circulating leukocytes and growth in chicken lines selected for humoral immune responsiveness. *British Poultry Science* 30, 361–369.

Droege, W. (1971) Amplifying and suppressive effect of thymus cells. *Nature* 234, 549–551.

Early, E. and Reen, D.J. (1999) Rapid conversion of naive to effector T cell function counteracts diminished primary human newborn T cell responses. *Clinical and Experimental Immunology* 116, 527–533.

Eisenthal, A., Hassner, A., Shenav, M., Baron, S. and Lifschitz-Mercer, B. (2003) Phenotype and function of lymphocytes from the neonatal umbilical cord compared to paired maternal peripheral blood cells isolated during delivery. *Experimental and Molecular Pathology* 75, 45–52.

Field, A.C., Caccavelli, L., Bloch, M.F. and Bellon, B. (2003) Regulatory CD8+ T cells control neonatal tolerance to a Th2-mediated autoimmunity. *Journal of Immunology* 170, 2508–2515.

Forsthuber, T., Yip, H.C. and Lehmann, P.V. (1996) Induction of T(H)1 and T(H)2 immunity in neonatal mice. *Science* 271, 1728–1730.

Fredericksen, T.L. and Gilmour, D.G. (1983) Ontogeny of con A and PHA responses of chicken blood cells in MHC-compatible lines 6(3) and 7(2). *Journal of Immunology* 130, 2528–2533.

French, V.I., Stark, J.M. and White, R.G. (1970) The influence of adjuvants on the immunological response of the chicken. II. Effects of Freund's complete adjuvant on later antibody production after a single injection of immunogen. *Immunology* 18, 645–655.

Fu, Y.X. and Chaplin, D.D. (1999) Development and maturation of secondary lymphoid tissues. *Annual Review of Immunology* 17, 399–433.

Fukanoki, S., Matsumoto, K., Mori, H. and Takeda, R. (2000) Relationship between antigen release and antibody response of infectious coryza water-in-oil-in-water emulsion vaccines. *Avian Diseases* 44, 869–873.

Garcia, A.M., Fadel, S.A., Cao, S. and Sarzotti, M. (2000) T cell immunity in neonates. *Immunologic Research* 22, 177–190.

Gasparoni, A., Ciardelli, L., Avanzini, A., Castellazzi, A.M., Carini, R., Rondini, G. and Chirico, G. (2003) Age-related changes in intracellular Th1/Th2 cytokine production, immunoproliferative T lymphocyte response and natural killer cell activity in newborns, children and adults. *Biology of the Neonate* 84, 297–303.

Gilmour, D.G., Theis, G.A. and Thorbecke, G.J. (1970) Transfer of antibody production with cells from bursa of Fabricius. *Journal of Experimental Medicine* 132, 134–147.

Goriely, S., Vincart, B., Stordeur, P., Vekemans, J., Willems, F., Goldman, M. and de Wit, D. (2001) Deficient IL-12(p35) gene expression by dendritic cells derived from neonatal monocytes. *Journal of Immunology* 166, 2141–2146.

Goriely, S., Van Lint, C., Dadkhah, R., Libin, M., de Wit, D., Demonte, D., Willems, F. and Goldman, M. (2004) A defect in nucleosome remodeling prevents IL-12(p35) gene transcription in neonatal dendritic cells. *Journal of Experimental Medicine* 199, 1011–1016.

Graetzer, M.A., Aspinall, R.L., Meyer, R.K. and Wolfe, H.R. (1963) Effect of thymectomy and bursectomy on precipitin and natural hemagglutinin production in chicken. *Journal of Immunology* 90, 878–885.

Griffioen, A.W., Franklin, S.W., Zegers, B.J. and Rijkers, G.T. (1993) Expression and functional characteristics of the complement receptor type 2 on adult and neonatal B lymphocytes. *Clinical Immunology and Immunopathology* 69, 1–8.

Han, P., Mcdonald, T. and Hodge, G. (2004) Potential immaturity of the T-cell and antigen-presenting cell interaction in cord blood with particular emphasis on the CD40–CD40 ligand costimulatory pathway. *Immunology* 113, 26–34.

Hase, H., Kanno, Y., Kojima, M., Hasegawa, K., Sakurai, D., Kojima, H., Tsuchiya, N., Tokunaga, K., Masawa, N., Azuma, M., Okumura, K. and Kobata, T. (2004) BAFF/BLyS can potentiate B-cell selection with the B-cell coreceptor complex. *Blood* 103, 2257–2265.

Healy, J.I., Dolmetsch, R.E., Timmerman, L.A., Cyster, J.G., Thomas, M.L., Crabtree, G.R., Lewis, R.S. and Goodnow, C.C. (1997) Different nuclear signals are activated by the B cell receptor during positive versus negative signaling. *Immunity* 6, 419–428.

Hein, H. (1968) Resistance in young chicks to reinfection by immunization with two doses of oocysts of *Eimeria acervulina*. *Experimental Parasitology* 22, 12–18.

Hermann, E., Truyens, C., Alonso-Vega, C., Even, J., Rodriguez, P., Berthe, A., Gonzalez-Merino, E., Torrico, F. and Carlier, Y. (2002) Human fetuses are able to mount an adultlike CD8 T-cell response. *Blood* 100, 2153–2158.

Hosken, N.A., Shibuya, K., Heath, A.W., Murphy, K.M. and Ogarra, A. (1995) The effect of antigen dose on CD4(+) T-helper cell phenotype development in a T-cell receptor-alpha-beta-transgenic model. *Journal of Experimental Medicine* 182, 1579–1584.

Ivanyi, J. and Valentova, V. (1966) The immunological significance of taxonomic origin of protein antigen in chickens. *Folia Biologica (Praha)* 12, 36–48.

Ivanyi, J., Hraba, T. and Cerny, J. (1964) Immunological tolerance to human serum albumin (HSA) in chickens of adult and juvenile age. *Folia Biologica (Praha)* 10, 198–205.

Janeway, C.A. Jr. (1992) The immune system evolved to discriminate infectious non-self from non-infectious self. *Immunology Today* 13, 11–16.

Jeurissen, S.H.M., Janse, E.M., Koch, G. and Deboer, G.F. (1989) Postnatal-development of mucosa-associated lymphoid-tissues in chickens. *Cell and Tissue Research* 258, 119–124.

Jullien, P., Cron, R.Q., Dabbagh, K., Cleary, A., Chen, L., Tran, P., Stepick-Biek, P. and Lewis, D.B. (2003) Decreased CD154 expression by neonatal CD4(+) T cells is due to limitations in both proximal and distal events of T cell activation. *International Immunology* 15, 1461–1472.

Kai, O. and Franklin, R.M. (1984) The age-dependent size distribution of chicken peripheral blood cells analyzed by a novel method utilizing a Coulter counter. *Developmental and Comparative Immunology* 8, 375–386.

Karlsson, H., Hessle, C. and Rudin, A. (2002) Innate immune responses of human neonatal cells to bacteria from the normal gastrointestinal flora. *Infection and Immunity* 70, 6688–6696.

Kawamura, H., Kameyama, H., Kosaka, T., Kuwahara, O., Bannai, M., Kawamura, T., Watanabe, H. and Abo, T. (2002) Association of CD8(+) natural killer T cells in the liver with neonatal tolerance phenomenon. *Transplantation* 73, 978–984.

King, L.B., Norvell, A. and Monroe, J.G. (1999) Antigen receptor-induced signal transduction imbalances associated with the negative selection of immature B cells. *Journal of Immunology* 162, 2655–2662.

Klaunberg, B.A., O'Malley, J., Clark, T. and Davis, J.A. (2004) Euthanasia of mouse fetuses and neonates. *Contemporary Topics in Laboratory Animal Science* 43, 29–34.

Kogut, M., Rothwell, L. and Kaiser, P. (2002) Differential effects of age on chicken heterophil functional activation by recombinant chicken interleukin-2. *Developmental and Comparative Immunology* 26, 817–830.

Kollmann, T.R., Way, S.S., Harowicz, H.L., Hajjar, A.M. and Wilson, C.B. (2004) Deficient MHC class I cross-presentation of soluble antigen by murine neonatal dendritic cells. *Blood* 103, 4240–4242.

Kovarik, J., Martinez, X., Pihlgren, M., Bozzotti, P., Tao, M.H., Kipps, T.J., Wild, T.F., Lambert, P.H. and Siegrist, C.A. (2000) Limitations of *in vivo* IL-12 supplementation strategies to induce Th1 early life responses to model viral and bacterial vaccine antigens. *Virology* 268, 122–131.

Kovarik, J., Gaillard, M., Martinez, X., Bozzotti, P., Lambert, P.H., Wild, T.F. and Siegrist, C.A. (2001) Induction of adult-like antibody, Th1, and CTL responses to measles hemagglutinin by early life murine immunization with an attenuated vaccinia-derived NYVAC(K1L) viral vector. *Virology* 285, 12–20.

Kreukniet, M.B. and van der Zijpp, A.J. (1990) Effects of different doses of sheep erythrocytes on the humoral immune-response of chicken lines selected for high or low antibody-production. *Poultry Science* 69, 608–614.

Kreukniet, M.B., van der Zijpp, A.J. and Nieuwland, M.G.B. (1992) Effects of route of immunization, adjuvant and unrelated antigens on the humoral immune-response in lines of chickens selected for antibody-production against sheep erythrocytes. *Veterinary Immunology and Immunopathology* 33, 115–127.

Langrish, C.L., Buddle, J.C., Thrasher, A.J. and Goldblatt, D. (2002) Neonatal dendritic cells are intrinsically biased against Th-1 immune responses. *Clinical and Experimental Immunology* 128, 118–123.

Laurent, F., Mancassola, R., Lacroix, S., Menezes, R. and Naciri, M. (2001) Analysis of chicken mucosal immune response to *Eimeria tenella* and *Eimeria maxima* infection by quantitative reverse transcription-PCR. *Infection and Immunity* 69, 2527–2534.

Le Campion, A., Bourgeois, C., Lambolez, F., Martin, B., Leaument, S., Dautigny, N., Tanchot, C., Penit, C. and Lucas, B. (2002) Naive T cells proliferate strongly in neonatal mice in response to self-peptide/self-MHC complexes. *Proceedings of the National Academy of Sciences of the USA* 99, 4538–4543.

Lederberg, J. and Tatum, E.L. (1953) Sex in bacteria; genetic studies: 1945–1952. *Science* 118, 169–175.

Li, L., Lee, H.H., Bell, J.J., Gregg, R.K., Ellis, J.S., Gessner, A. and Zaghouani, H. (2004) IL-4 utilizes an alternative receptor to drive apoptosis of Th1 cells and skews neonatal immunity toward Th2. *Immunity* 20, 429–440.

Lillehoj, H.S. (1988) Influence of inoculation dose, inoculation schedule, chicken age, and host genetics on disease susceptibility and development of resistance to *Eimeria tenella* infection. *Avian Diseases* 32, 437–444.

Lillehoj, H.S. (1994) Analysis of *Eimeria-acervulina*-induced changes in the intestinal T-lymphocyte subpopulations in 2 chicken strains showing different levels of susceptibility to coccidiosis. *Research in Veterinary Science* 56, 1–7.

Lillehoj, H.S. and Choi, K.D. (1998) Recombinant chicken interferon-gamma-mediated inhibition of *Eimeria tenella* development in vitro and reduction of oocyst production and body weight loss following *Eimeria acervulina* challenge infection. *Avian Diseases* 42, 307–314.

Lin, H., Mosmann, T.R., Guilbert, L., Tuntipopipat, S. and Wegmann, T.G. (1993) Synthesis of T helper 2-type cytokines at the maternal–fetal interface. *Journal of Immunology* 151, 4562–4573.

Liu, Y., Chiriva-Internati, M., Grizzi, F., Salati, E., Roman, J.J., Lim, S. and Hermonat, P.L. (2001) Rapid induction of cytotoxic T-cell response against cervical cancer cells by human papillomavirus type 16 E6 antigen gene delivery into human dendritic cells by an adeno-associated virus vector. *Cancer Gene Therapy* 8, 948–957.

Lombardi, G., Arnold, K., Uren, J., Marelliberg, F., Hargreaves, R., Imami, N., Weetman, A. and Lechler, R. (1997) Antigen presentation by interferon-gamma-treated thyroid follicular cells inhibits interleukin-2 (IL-2) and supports IL-4 production by B7-dependent human T cells. *European Journal of Immunology* 27, 62–71.

Marshall-Clarke, S., Reen, D., Tasker, L. and Hassan, J. (2000a) Neonatal immunity: how well has it grown up? *Immunology Today* 21, 35–41.

Marshall-Clarke, S., Tasker, L. and Parkhouse, R.M. (2000b) Immature B lymphocytes from adult bone marrow exhibit a selective defect in induced hyperexpression of major histocompatibility complex class II and fail to show B7.2 induction. *Immunology* 100, 141–151.

Martinez, X., Brandt, C., Saddallah, F., Tougne, C., Barrios, C., Wild, F., Dougan, G., Lambert, P.H. and Siegrist, C.A. (1997) DNA immunization circumvents deficient induction of T helper type 1 and cytotoxic T lymphocyte responses in neonates and during early life. *Proceedings of the National Academy of Sciences of the USA* 94, 8726–8731.

Martinez, X., Regner, M., Kovarik, J., Zarei, S., Hauser, C., Lambert, P.H., Leclerc, C. and Siegrist, C.A. (2003) CD4-independent protective cytotoxic T cells induced in early life by a non-replicative delivery system based on virus-like particles. *Virology* 305, 428–435.

Mast, J.A.G. (1998) Immunohistochemical analysis of the development of the structural organisation of chicken spleen. *Vlaams Diergeneskundig Tijdschrift* 67, 36–44.

Mast, J. and Goddeeris, B.M. (1999) Development of immunocompetence of broiler chickens. *Veterinary Immunology and Immunopathology* 70, 245–256.

Mebius, R.E. (2003) Organogenesis of lymphoid tissues. *Nature Reviews Immunology* 3, 292–303.

Min, B., McHugh, R., Sempowski, G.D., Mackall, C., Foucras, G. and Paul, W.E. (2003) Neonates support lymphopenia-induced proliferation. *Immunity* 18, 131–140.

Morrissey, P.J., Bradley, D., Sharrow, S.O. and Singer, A. (1983) T-cell tolerance to non-H-2-encoded stimulatory alloantigens is induced intrathymically but not prethymically. *Journal of Experimental Medicine* 158, 365–377.

Mosmann, T.R. and Sad, S. (1996) The expanding universe of T-cell subsets: Th1, Th2 and more. *Immunology Today* 17, 138–146.

Moticka, E.J. (1977) Presence of immunoregulatory cells in chicken thymus – function in B-cell and T-cell responses. *Journal of Immunology* 119, 987–992.

Munns, P.L. and Lamont, S.J. (1991) Research note – effects of age and immunization interval on the anamnestic response to T-cell-dependent and T-cell-independent antigens in chickens. *Poultry Science* 70, 2371–2374.

Muthukkumar, S., Goldstein, J. and Stein, K.E. (2000) The ability of B cells and dendritic cells to present antigen increases during ontogeny. *Journal of Immunology* 165, 4803–4813.

Navarrete, C., Gomez, J. and Borras, F.E. (2003) Cord blood dendritic cells: subsets, functional characteristics and *in vitro* generation. *Leukemia and Lymphoma* 44, 923–928.

Nonoyama, S., Penix, L.A., Edwards, C.P., Lewis, D.B., Ito, S., Aruffo, A., Wilson, C.B. and Ochs, H.D. (1995) Diminished expression of CD40 ligand by activated neonatal T cells. *Journal of Clinical Investigation* 95, 66–75.

Norvell, A., Mandik, L. and Monroe, J.G. (1995) Engagement of the antigen-receptor on immature murine B-lymphocytes results in death by apoptosis. *Journal of Immunology* 154, 4404–4413.

Peters, M.A., Browning, G.F., Washington, E.A., Crabb, B.S. and Kaiser, P. (2003) Embryonic age influences the capacity for cytokine induction in chicken thymocytes. *Immunology* 110, 358–367.

Petty, R.E. and Hunt, D.W. (1998) Neonatal dendritic cells. *Vaccine* 16, 1378–1382.

Pihlgren, M., Schallert, N., Tougne, C., Bozzotti, P., Kovarik, J., Fulurija, A., Kosco-Vilbois, M., Lambert, P.H. and Siegrist, C.A. (2001) Delayed and deficient establishment of the long-term bone marrow plasma cell pool during early life. *European Journal of Immunology* 31, 939–946.

Pihlgren, M., Tougne, C., Bozzotti, P., Fulurija, A., Duchosal, M.A., Lambert, P.H. and Siegrist, C.A. (2003) Unresponsiveness to lymphoid-mediated signals at the neonatal follicular dendritic cell precursor level contributes to delayed germinal center induction and limitations of neonatal antibody responses to T-dependent antigens. *Journal of Immunology* 170, 2824–2832.

Prescott, S.L., Macaubas, C., Holt, B.J., Smallacombe, T.B., Loh, R., Sly, P.D. and Holt, P.G. (1998) Transplacental priming of the human immune system to environmental allergens: universal skewing of initial T cell responses toward the Th2 cytokine profile. *Journal of Immunology* 160, 4730–4737.

Prescott, S.L., Taylor, A., King, B., Dunstan, J., Upham, J.W., Thornton, C.A. and Holt, P.G. (2003) Neonatal interleukin-12 capacity is associated with variations in allergen-specific immune responses in the neonatal and postnatal periods. *Clinical and Experimental Allergy* 33, 566–572.

Protonotariou, E., Malamitsi-Puchner, A., Rizos, D., Papagianni, B., Moira, E., Sarandakou, A. and Botsis, D. (2004) Age-related differentiation of Th1/Th2 cytokines in newborn infants. *Mediators of Inflammation* 13, 89–92.

Qin, D., Wu, J., Carroll, M.C., Burton, G.F., Szakal, A.K. and Tew, J.G. (1998) Evidence for an important interaction between a complement-derived CD21 ligand on follicular dendritic cells and CD21 on B cells in the initiation of IgG responses. *Journal of Immunology* 161, 4549–4554.

Raghupathy, R. (1997) Th1-type immunity is incompatible with successful pregnancy. *Immunology Today* 18, 478–482.

Rainsford, E. and Reen, D.J. (2002) Interleukin 10, produced in abundance by human newborn T cells, may be the regulator of increased tolerance associated with cord blood stem cell transplantation. *British Journal of Haematology* 116, 702–709.

Ribeiro Do Couto, L.M., Boeije, L.C.M., Kroon, J.S., Hooibrink, B., Breur-Vriesendorp, B.S., Aarden, L.A. and Boog, C.J.P. (2001) High IL-13 production by human neonatal T cells: neonate immune system regulator? *European Journal of Immunology* 31, 3394–3402.

Ridge, J.P., Fuchs, E.J. and Matzinger, P. (1996) Neonatal tolerance revisited: turning on newborn T cells with dendritic cells. *Science* 271, 1723–1726.

Rose, M.E. (1967) The influence of age of host on infection with *Eimeria tenella*. *Journal of Parasitology* 53, 924–929.

Rose, M.E. (1974) Immune responses in infections with coccidia: macrophage activity. *Infection and Immunity* 10, 862–871.

Rose, M.E. and Hesketh, P. (1979) Immunity to coccidiosis: T-lymphocyte- or B-lymphocyte-deficient animals. *Infection and Immunity* 26, 630–637.

Rose, M.E. and Long, P.L. (1962) Immunity to 4 species of *Eimeria* in fowls. *Immunology* 5, 79–85.

Rosenquist, G.L. and Wolfe, H.R. (1962) Effect of splenectomy at different ages on precipitin production in chickens. *Immunology* 5, 211–221.

Roser, B.J. (1989) Cellular mechanisms in neonatal and adult tolerance. *Immunological Reviews* 107, 179–202.

Rothwell, L., Muir, W. and Kaiser, P. (2000) Interferon-gamma is expressed in both gut and spleen during *Eimeria tenella* infection. *Avian Pathology* 29, 333–342.

Salio, M., Dulphy, N., Renneson, J., Herbert, M., McMichael, A., Marchant, A. and Cerundolo, V. (2003) Efficient priming of antigen-specific cytotoxic T lymphocytes by human cord blood dendritic cells. *International Immunology* 15, 1265–1273.

Sarzotti, M., Robbins, D.S. and Hoffman, P.M. (1996) Induction of protective CTL responses in newborn mice by a murine retrovirus. *Science* 271, 1726–1728.

Schito, M.L. and Barta, J.R. (1997) Nonspecific immune responses and mechanisms of resistance to *Eimeria papillata* infections in mice. *Infection and Immunity* 65, 3165–3170.

Schito, M.L., Barta, J.R. and Chobotar, B. (1996) Comparison of four murine *Eimeria* species in immunocompetent and immunodeficient mice. *Journal of Parasitology* 82, 255–262.

Schito, M.L., Chobotar, B. and Barta, J.R. (1998a) Cellular dynamics and cytokine responses in BALB/c mice infected with *Eimeria papillata* during primary and secondary infections. *Journal of Parasitology* 84, 328–337.

Schito, M.L., Chobotar, B. and Barta, J.R. (1998b) Major histocompatibility complex class I- and II-deficient knock-out mice are resistant to primary but susceptible to secondary *Eimeria papillata* infections. *Parasitology Research* 84, 394–398.

Schonland, S.O., Zimmer, Z.J., Lopez-Benitez, C.M., Widmann, T., Ramin, K.D., Goronzy, J.J. and Weyand, C.M. (2003) Homeostatic control of T-cell generation in neonates. *Blood* 102, 1428–1434.

Schuler, T., Hammerling, G.J. and Arnold, B. (2004) Cutting edge: IL-7-dependent homeostatic proliferation of CD8(+) T cells in neonatal mice allows the generation of long-lived natural memory T cells. *Journal of Immunology* 172, 15–19.

Schultz, C., Temming, P., Bucsky, P., Gopel, W., Strunk, T. and Hartel, C. (2004) Immature anti-inflammatory response in neonates. *Clinical and Experimental Immunology* 135, 130–136.

Seto, F. (1970) Immunocompetent cells in the blood of immunized chickens. *Poultry Science* 49, 1673–1680.

Seto, F. (1975) Enhancement of chicken hemagglutinin response by thymocytes. *Experimental Hematology* 3, 319–326.

Seto, F. (1976) Antibody-mediated immuno-regulation in chick embryos and baby chicks. *Poultry Science* 55, 172–179.

Seto, F. (1977a) Cell transfer studies with the DK/or inbred chicken lines. *Advances in Experimental and Medical Biology* 88, 275–285.

Seto, F. (1977b) Kinetic analysis of the enhanced immune responsiveness of baby chicks grafted with lymphoid cells from juvenile donors. *Developmental and Comparative Immunology* 1, 145–156.

Seto, F. (1980) Ontogenetic analysis of immune amplification by thymus-derived cells in the chicken. *American Zoologist* 20, 765.

Seto, F. (1981) Early development of the avian immune system. *Poultry Science* 60, 1981–1995.

Seto, F. (1983) Apparent adoptive immune enhancement by bone marrow and bursa of Fabricius cells from juvenile chickens. *Developmental and Comparative Immunology* 7, 313–323.

Seto, F. (1984) Enlargement and immune activity of chick embryo host spleen associated with transfer of B-haplotype-matched donor lymphoid cells. *Experimental Hematology* 12, 53–59.

Seto, F. and Henderson, W.G. (1968) Natural and immune hemagglutinin forming capacity of immature chickens. *Journal of Experimental Zoology* 169, 501–511.

Siegrist, C.A. (2001) Neonatal and early life vaccinology. *Vaccine* 19, 3331–3346.

Simpson, C.C., Woods, G.M. and Muller, H.K. (2003) Impaired CD40-signalling in Langerhans' cells from murine neonatal draining lymph nodes: implications for neonatally induced cutaneous tolerance. *Clinical and Experimental Immunology* 132, 201–208.

Sminia, T., van der Brugge-Gamelkoorn, G.J. and Jeurissen, S.H. (1989) Structure and function of bronchus-associated lymphoid tissue (BALT). *Critical Reviews in Immunology* 9, 119–150.

Smith, A.L. and Hayday, A.C. (2000a) An alpha beta T-cell-independent immunoprotective response towards gut coccidia is supported by gamma delta cells. *Immunology* 101, 325–332.

Smith, A.L. and Hayday, A.C. (2000b) Genetic dissection of primary and secondary responses to a widespread natural pathogen of the gut, *Eimeria vermiformis*. *Infection and Immunity* 68, 6273–6280.

Soloman, J.B. (1968) Ontogeny of cells producing haemolytic antibody or immunocyto-adherence to sheep erythrocytes in the chicken. *Immunology* 14, 611–612.

Steinberg, S.V., Munro, J.A., Fleming, W.A., French, V.I., Stark, J.M. and White, R.G. (1970) The influence of adjuvants on the immunological response of the chicken. *Immunology* 18, 635–640.

Sun, C.M., Fiette, L., Tanguy, M., Leclerc, C. and Lo-Man, R. (2003) Ontogeny and innate properties of neonatal dendritic cells. *Blood* 102, 585–591.

Szenberg, A. (1977) Ontogeny of myelopoietic precursor cells in the chicken embryo. *Advances in Experimental and Medical Biology* 88, 3–11.

Talebi, A. and Mulcahy, G. (1994) High-resolution mapping of B-cell epitopes within an antigenic sequence from *Eimeria tenella*. *Infection and Immunity* 62, 4202–4207.

Tasker, L. and Marshall-Clarke, S. (2003) Functional responses of human neonatal B lymphocytes to antigen receptor cross-linking and CpG DNA. *Clinical and Experimental Immunology* 134, 409–419.

Timens, W., Boes, A., Rozeboomuiterwijk, T. and Poppema, S. (1989) Immaturity of the human splenic marginal zone in infancy – possible contribution to the deficient infant immune-response. *Journal of Immunology* 143, 3200–3206.

Upham, J.W., Lee, P.T., Holt, B.J., Heaton, T., Prescott, S.L., Sharp, M.J., Sly, P.D. and Holt, P.G. (2002) Development of interleukin-12-producing capacity throughout childhood. *Infection and Immunity* 70, 6583–6588.

Vainio, O., Viljanen, M.K. and Toivanen, A. (1978) Early ontogeny of germinal center formation in chicken. *Developmental and Comparative Immunology* 2, 493–504.

van der Zijpp, A.J., Scott, T.R. and Glick, B. (1986) The effect of different routes of antigen administration on the humoral immune response of the chick. *Poultry Science* 65, 809–811.

van Immerseel, F., de Buck, J., de Smet, I., Mast, J., Haesebrouck, F. and Ducatelle, R. (2002) Dynamics of immune cell infiltration in the caecal lamina propria of chickens after neonatal infection with a *Salmonella enteritidis* strain. *Developmental and Comparative Immunology* 26, 355–364.

Walter, H.J. (2005) The development of immunocompetence in the neonatal chick. PhD thesis, University of Oxford, Oxford, UK.

Wedderburn, L.R., Patel, A., Varsani, H. and Woo, P. (2001) The developing human immune system: T-cell receptor repertoire of children and young adults shows a wide discrepancy in the frequency of persistent oligoclonal T-cell expansions. *Immunology* 102, 301–309.

Weeratna, R.D., McCluskie, M.J., Brazolot-Millan, C.L. and Davis, H.L. (2001) CpG ODN can promote Th1 type immune responses and can re-direct pre-established Th2 responses in adult and young mice. *FASEB Journal* 15, A652.

Wegmann, T.G., Lin, H., Guilbert, L. and Mosmann, T.R. (1993) Bidirectional cytokine interactions in the maternal–fetal relationship: is successful pregnancy a TH2 phenomenon? *Immunology Today* 14, 353–356.

White, R.G. and Henderson, D.C. (1975) Localisation of a protein antigen in the spleen. *Immunology* 28, 1–21.

Williams, R.B. (1999) A compartmentalised model for the estimation of the cost of coccidiosis to the world's chicken production industry. *International Journal Parasitology* 29, 1209–1229.

Wilson, A.D. and Morgan, A.J. (2002) Primary immune responses by cord blood CD4(+) T cells and NK cells inhibit Epstein-Barr virus B-cell transformation *in vitro*. *Journal of Virology* 76, 5071–5081.

Wolfe, H.R. and Dilks, E. (1948) Precipitin production in chickens. III. The variation in the antibody response as correlated with the age of the animals. *Journal of Immunology* 58, 245–250.

CHAPTER 22
Managing Current Disease Challenges in Breeders

S.R. Collett

College of Veterinary Medicine, The University of Georgia, USA

ABSTRACT

The need to improve the financial efficiency of the poultry breeder industry through intensification has created new challenges for today's managers. By keeping large numbers of birds in high-density housing and controlling the full range of environmental variables the manager is able to optimize the biological, and hence financial, efficiency of the flock. Unfortunately the psychological stress of close confinement and the physiological stress of performing at capacity predispose these birds to metabolic disease and compromise their capacity to cope with infectious disease challenge.

The biological efficiency of converting feed into disease-free, viable chicks or poults is compromised by disease challenge, and the ultimate success of a breeder operation hinges on optimizing disease status through diligent biosecurity. This involves the coordination of conceptual, structural and operational measures to reduce the risk and consequence of disease. The design of such a system is based on the epidemiological characteristics of the disease and should include an orchestrated mix of immune modulation, bioexclusion, surveillance and biocontainment.

Since the potential impact of disease challenge ranges from catastrophic to inconsequential it is necessary to use economic analysis in designing a disease management programme to ensure optimal resource allocation. Eradication programmes are appropriate for diseases that carry significant public health risk such as *Salmonella enterica* serovar Enteritidis, those that cause a catastrophic drop in breeder productivity such as fowl plague (avian influenza), or those that, through vertical transmission, have a catastrophic effect on offspring productivity. With other less-devastating disease challenges, where eradication is not justified, the focus of disease management is on limiting the risk of challenge through bioexclusion and the consequence of disease challenge through immunization or medication.

There are very few sound epidemiological data on the specifics of disease risk and prevalence in breeder facilities around the world. The collection and comparative analysis of such data would go a long way to narrowing the global variance in breeder liveability.

INTRODUCTION

Narrowing profit margins have driven the process of intensification to make improvements in resource utilization and greater use of mechanization. The resulting large-scale breeder operations have created new disease challenges. The mere scale of these operations has increased the financial risk of disease, and economic pressure has expanded the focus to include less-well-defined, multifactorial diseases that manifest as a decrease in production efficiency. Furthermore, both consumer concern for bird welfare and economic pressure to ensure convergence of genotype and phenotype have expanded the definition of 'disease' to include perceived stress, which might be more appropriately referred to as 'dis-ease'.

Methods of disease control have evolved with the intensification of the poultry industry. While initially focused on diseases of a catastrophic nature, focus has rapidly shifted from defined clinical disease at the individual farm level, to less-well-defined subclinical disease and bird welfare. Similarly, cost justification in decisions on whether or not to implement control measures has become more complex, requiring the aid of formal economic appraisal.

To maximize breeder flock efficiency, disease challenge management requires a carefully designed, multitiered approach, which includes consideration of elements ranging from breeder flock health and productivity through to chick or poult viability. In addition, the term efficiency implies the all important economic component. Since the production system is profit driven, decisions regarding disease challenge can rarely be made solely on biological grounds. Disease management intervention also requires sound economic justification, which begins with clearly defining the financial risk associated with disease.

Unless a disease poses a specific risk to human health or animal welfare, its mere presence in a breeder flock is insignificant unless it is economically advantageous to take action against it. Intervention strategies are consequently chosen based on both their economic and their biological efficiency. This process requires a dynamic integrated combination of epidemiologic and economic analysis to determine and quantify the production effect of, first, the disease challenge and, second, the proposed intervention strategy. Such integrated analysis has become far more significant in today's intensive production systems because the outcome of disease challenge is so markedly influenced by the environmental conditions.

The economic impact of disease is difficult to assess. This is particularly so in an intensive breeder production system, in which the economic return is governed not only by flock productivity but also by product quality and viability. In addition, the consequential loss from disease challenge is, at best, only partially recoverable. Using the cost of disease to justify intervention overemphasizes the consequence of inaction and it is only useful in justifying intervention strategies directed at preventing disease challenge. As the process of economic analysis has evolved, so too the focus has shifted from the cost of disease to the benefit derived from control strategies (Morris and Meek, 1980).

CURRENT DISEASE CHALLENGES

In addressing disease challenge in modern intensive production systems it is important to consider the nature of the production chain. Breeders are kept for a long period of time, which increases the opportunity for exposure to infectious disease, and the production system predisposes these animals to physiological stress and metabolic disease. In addition, the intermittent feeding patterns used to restrict intake make the maintenance of a steady-state gastrointestinal microbiome impossible, thus predisposing broiler breeders to colonization with unfavourable microbial species.

There are very few published data that describe the prevalence of disease in breeders. An extensive survey of broiler breeder mortality during the laying period was conducted in 1978 on three large laying flocks in the UK (Jones *et al.*, 1978). This study was based on an analysis of all mortality occurring during the laying cycle and therefore provides very useful data in terms of the causes of mortality in normal flocks. The most common causes of death in females were reproductive disorders (25%), cellulitis and injury (24%), kidney lesions (10%), liver haemorrhage (7%), Marek's disease (5%) and synovitis and tenosynovitis (4%). Mortality in males was due mainly to synovitis and tenosynovitis (34%) and acute heart failure (15%).

An analysis of laboratory submissions in response to increased mortality from a broiler company in South Africa, although statistically biased by sampling method, provides interesting information as to cause of mortality (Table 22.1). These data show quite clearly the impact of J strain avian leukosis virus (ALV-J) infection in the period between 1996 and 1997. The initial appearance of this disease could possibly have been misdiagnosed as Marek's disease, since the prevalence of Marek's was higher than expected in the period 1993 to 1995. As with the UK study, reproductive disorders and peritonitis accounted for the majority of deaths. Injury appears to be far less common in the South African data, but this is probably due to sampling bias since birds showing clear signs of injury would not be submitted to the laboratory for diagnosis.

A more recent analysis (2000–2003) of submissions to the State Laboratory in Georgia, USA, although also statistically biased by sampling method, provides insight into some of the more important disease entities of broiler breeders (Zavala *et al.*, 2003). Unlike the UK study, these data are derived from laboratory submissions and therefore overemphasize the importance of infectious disease, because cases of environmentally induced, nutritional and metabolic disease are not normally submitted to the diagnostic laboratory. Zavala *et al.* (2003) reported that the most commonly diagnosed conditions were bacterial and parasitic diseases. *Escherichia coli* septicaemia, fowl cholera (*Pasteurella multocida*) and staphylococcal arthritis and septicaemia were the

Table 22.1. Cause of mortality in broiler breeders and proportion of total submitted to the laboratory for post-mortem examination in response to a mortality increase on the farms concerned (5011 samples; University of Georgia, Athens, USA).

Diagnosis	Year			
	1993–1995	1996–1997	1998–1999	1993–1999
Reproductive disorders/peritonitis	20.0	22.0	22.4	22.0
Prolapse and vent cannibalism	5.0	6.7	2.0	5.0
Respiratory disease	4.0	3.5	9.8	5.0
Kidney lesions	8.2	2.2	2.5	4.6
Liver haemorrhage/pathology	5.7	8.0	9.0	5.0
Acute heart failure	13.2	6.0	8.0	10.0
Marek's disease	10.5	3.0	0.9	5.7
Leukosis	0.2	16.0	2.2	7.5
Arthritis	0.03	0.1	4.6	1.0
Septicaemia	0.6	1.4	4.0	1.4
Injury	11.8	3.6	2.0	4.0
Non-specific	8.4	7.6	9.0	8.2
Other	12.37	19.9	23.6	20.6

most common bacterial diseases. Although many of the *P. multocida* isolates were different from the vaccine strains, the majority correlated with those present in the commonly used live vaccines. Some of these isolates could perhaps be vaccine-derived since they have the potential to be mildly pathogenic. Although it is not statistically sound to compare these three sets of data, staphylococcal arthritis appears to be far more common in the USA than in either the UK or South Africa.

Exogenous ALV-J infection rate in broiler breeders in Zavala's US study was, at 40% (14/35), higher than expected, since this part of the analysis covered the period from 2002 to 2003, when ALV-J was considered to have been cleared from most breeder flocks.

Typical broiler breeder pullet losses to 20 weeks of age are 3–5%, and losses in the first 2 weeks account for 30–50% of total mortality. Hen losses from transfer at 20–21 weeks to depletion at 60–65 weeks are in the region of 5–14%, with 30% of the mortality occurring in the 25- to 35-week period. It is this period, from onset of lay to shortly after attaining peak production, in which most breeder flock mortality occurs, and much of this mortality is the result of physiological and metabolic stress. Male mortality is, on average, around three to four times that of the females. Interestingly the parent stock mortality rates in Japan are in the low range, while those in the USA are in the high range, with the UK somewhere in between.

Assuming that mortality rate is a fair reflection of disease challenge, the biggest window of opportunity for improving disease management is between 25 and 35 weeks of age, when approximately 30% of total mortality occurs (Jones *et al.*, 1978). During this period, most deaths are the result of nutritional- and metabolic-induced disease, including prolapse, peritonitis and sudden death syndrome, so it is not surprising that nutritional intervention can markedly change mortality patterns (Gustin, 2007). Data from the UK demonstrate that carefully controlling body weight at this stage of the production cycle has both immediate and long-term benefit in terms of liveability (Fig. 22.1).

Overweight breeders are more susceptible to several disease conditions. Excess loading of the gastrocnemius tendon predisposes it to rupture, while fat deposition in the pelvic area predisposes the oviduct to partial prolapse following oviposition. Secondary bacterial infection, *E. coli* peritonitis, cannibalism and complete prolapse are all common sequelae. Excess fat mobilization and metabolism predispose the liver to fatty degeneration, compromise liver function and make it more susceptible to traumatic rupture. Since the liver is the metabolic powerhouse of the reproductive effort, compromised function during a metabolically stressful period makes sudden death syndrome far more likely.

IMPACT OF DISEASE

The goal of a breeder operation is to convert feed into viable, disease-free chicks or poults as efficiently as possible. The efficiency of this feed conversion is governed primarily by intrinsic or genetic determinants, but under intensive production systems the extrinsic determinants are under management control. Consequently, capital investment is pivotal to success, because both the design characteristics and the management of these modern intensive housing systems alter the extrinsic determinants of disease, and thereby ultimately determine the efficiency of the breeder operation in both biological and financial terms.

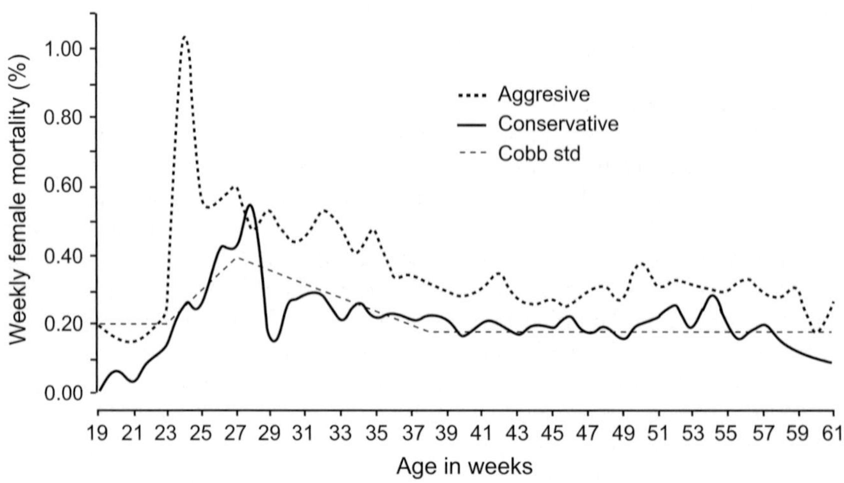

Fig. 22.1. Plots of weekly mortality of two groups of 12 flocks from a European company with different feeding programmes after light stimulation (Gustin, 2007). Mortality of birds fed on standard feed recommendations from the Cobb Breeding Company management manual (Cobb std) is plotted for comparison. Feed allocation at light stimulation, 5% egg production and peak egg production, respectively, for conservative feeding and Cobb standard were 102, 128 and 162 g/bird/day compared with 104, 141 and 170 g/bird/day for aggressive feeding.

Traditional thinking, stimulated by widespread acceptance of Koch's Postulates in the 1900s (Walker *et al.*, 2006), overemphasizes the importance of infectious agents in the disease process. As production systems have evolved, environmental and host disease determinants have played a more obvious role in the disease process, emphasizing the multifactorial nature of disease. The prevalence of specific infectious disease entities has declined as knowledge and control measures have improved. In contrast, the predisposition to, and prevalence of, non-infectious disease has increased with intensification and genetic change. The distinction between infectious and non-infectious disease has become somewhat blurred in intensive agriculture, and a more fully encompassing epidemiological approach to disease diagnosis and control has become necessary.

The breeder house environment, with all its intricacies, is a crucial disease determinant since stress of any kind stimulates a cascade of physiological and biochemical changes, which erode host resistance and productivity (Siegel, 1994). The negative impact that a particular stressor has on performance is directly proportional to the existing stress load, since stress is cumulative and measurably impacts animal performance once the aggregate of individual stresses exceeds the capacity of the coping mechanism (Klasing *et al.*, 1987). Stress level, and hence competence to cope with additional stress, varies with each individual within a flock, thus emphasizing the need to consider the epidemiology of disease within a confined, finite population. Stress lowers the minimum dose of infective agent required for development of infection and increases the risk of infectious or non-infectious challenge developing into clinically detectable disease. The risk and consequence of infectious disease spread within the population are increased by the presence of stress, because susceptible individuals act as amplifiers for the infectious organisms and thus increase the challenge dose to which the pen mates are exposed. While the introduction of a non-infectious disease to a flock may also lower individual resistance, it does not increase the risk of spread (Smith, 1995).

Flock uniformity is crucial to success of the broiler breeder business because, for ease of management, the flock is treated as a single unit. Feeding regimes aimed at restricting body weight are, for example, population based (see Hocking, Chapter 17, this volume). Although the feed allocation is carefully calculated to satisfy the average, most of the birds are either underfed or overfed, and the extent of nutritional stress increases as the uniformity of the flock deteriorates. The physiological stress of producing at, or close to, genetic potential reduces the bird's stress-coping-mechanism reserve. Since stress is cumulative, the impact of additional disease or environmental stress is exaggerated, even in apparently healthy flocks. Although the source of stress may vary, the impact is the same. Metabolic stress in a healthy individual or infectious stress in an unhealthy individual will induce weight loss and result in reduced productivity. Stress- or disease-induced weight loss in breeders is very much more significant than is generally accepted. When challenged, highly stressed individuals will experience a more significant setback than less-stressed individuals, so flock uniformity deteriorates more easily in stressed flocks and magnifies the inadequacies of flock-based intensive management techniques (Humphrey *et al.*, 2003).

Disease causes functional derangement of normal metabolic and homeostatic processes, with a consequential decline in productivity. The economic impact of disease is manifested initially through reduced feed conversion efficiency, either as a consequence of disease-induced anorexia or through specific effects on the physiological processes of nutrient metabolism, respiration and excretion. In breeders, productivity is measured in terms of viable, disease-free chicks per hen, which implies a yield and quality component. Subclinical disease may, however, impact on performance by altering fertility, hatchability or offspring viability without affecting yield or the number of hatching eggs produced. Furthermore, since the breeder operation is part of the food chain, colonization with certain organisms may have far-reaching consequences in regard to offspring viability, without actually causing overt disease. Only in severe cases does disease cause mortality, and yet mortality rate is used universally as a measure of flock health status.

Measures of productivity, such as eggs per hen housed (EHH), are universally used by managers as key performance indicators to evaluate biological and financial efficiency. Flock health status is a little more nebulous and consequently difficult to define or measure. Health and disease are not mutually exclusive. In an intensive breeding production system neither the classic definition of disease (a condition that causes medically significant symptoms) nor that of health (a state of physical and mental well-being) are of value because they are too extreme. Instead, the same parameters used to measure production efficiencies are used to *estimate* flock health. A flock is classed as healthy if it is performing to standard and is free of clinical disease.

Unfortunately, performance indicators are themselves very coarse measures of health. They measure the composite output of a population, which is subject to biological variance and therefore normally changes considerably from day to day. Such measures also give no indication as to the cause (aetiology) or prevalence of disease. Most minor changes in production go undiagnosed, probably because the search for a diagnosis is limited to an infectious cause, when more often than not it is the nutritional, physiological and psychological stress of intensive production systems that creates or predisposes these breeding birds to disease challenge

MANAGING DISEASE CHALLENGE

Biosecurity

Biosecurity includes all procedures implemented to reduce the risks and consequence of introducing an infectious disease into a flock (Gifford *et al.*, 1987). These preventive measures, which are based on applied microbiology and epidemiology, must be practical, enforceable and cost-effective, and thereby form an integral part of the production system. Since the implementation of biosecurity carries a cost, it is necessary to relate this cost to the risk and consequence of infectious disease. Unfortunately, there is no way of accurately defining the relative risk and financial consequence of disease exposure or, for that matter, the effectiveness of preventive measures. Clearly, the development of a cost-effective biosecurity system must entail a calculated estimate of these parameters (Gifford *et al.*, 1987).

A comprehensive biosecurity programme comprises a hierarchy of conceptual, structural and operational components directed at preventing infectious disease transmission. Although physical isolation of the production facility is the most effective means of reducing the risk of introducing an infectious disease into a flock, it is frequently not an option. Most breeder facilities are already built and the capital investment for relocation is prohibitive. Consequently, most biosecurity programmes are focused on structural and operational components

Prevention

Every event in the production process that involves transgression of the house, site, farm or complex boundary creates risk of contact between an infectious organism and the host. Avoidance is the best form of prevention, but, where the event is unavoidable, biosecurity measures need to be implemented to alleviate risk by reducing either the frequency of the transgression or the probability of the event resulting in colonization and infection. Thus risk can be defined as:

Risk of infection = probability of infection × frequency

The probability of infection occurring can be reduced by improving host resistance through immunization, decreasing the challenge dose through cleaning and disinfection, or reducing organism virulence by medication.

Consequences of infection

Any immune response bears a production cost. An appropriate immune response, adequate to contain infectious disease and minimize its impact on productivity, is the cost of health. An inappropriate, excessive or inadequate immune response will depress performance unnecessarily (Klasing *et al.*, 1987; Klasing and Barnes, 1988; Klasing, 1998; Klipper *et al.*, 2000, 2001; Kelly, 2004). Breeding birds are generally intensively vaccinated during the rearing period to ensure high levels of maternal antibody transfer to their offspring (see Butter and Walter, Chapter 21, this volume). This is an advantage in terms of disease protection but costly in terms of immune response.

Leukocyte synthesis in response to antigen stimulation carries a barely detectable nutritional cost (<0.5% of body mass), but the ramifications of the acute-phase (fever) response range from negligible to dramatic (Klasing and Johnstone, 1991; Klasing, 1998). The systemic component of the inflammatory response to vaccination or disease challenge begins with acute-phase protein synthesis in the liver and is followed by several behavioural, hormonal and metabolic responses. Feed intake declines, protein turnover accelerates and birds rapidly undergo transition into negative nitrogen balance (Klasing *et al.*, 1987; Klasing, 1998; Collier *et al.*, 2003). With systemic challenge, most (70%) of the negative impact on growth rate and feed efficiency is attributed to reduced feed intake, while the inefficiencies of catabolism and nutrient absorption account for the rest (30%) (Klasing *et al.*, 1987; Klasing and Barnes, 1988; Klasing, 1998; Klipper *et al.*, 2000; Collier *et al.*, 2003). Aggressive vaccination programmes make it very difficult to maintain good flock uniformity, because the impact of immunological cost on growth rate is proportional to the level of stress (Humphrey *et al.*, 2003).

The practice of maximizing disease resistance (the capacity of a variety of anatomical and physiological systems, including the immune system, to exclude pathogens) has proved to be too costly for the modern broiler breeder industry (Klasing, 1998). An exaggerated or inappropriate immune response is counter-productive in terms of feed efficiency, so emphasis in managing disease challenge has shifted from disease resistance to disease resilience (the capacity to maintain productivity during infectious challenge) (Klasing *et al.*, 1987; Klasing, 1998; Grimble, 2001; Collier *et al.*, 2003).

Just as the cost of an excessive or inappropriate immune response negatively impacts performance, so too does an inadequate immune response. Although usually recognized as an increase in mortality, an inadequate immune response causes negative economic impact through depressed performance long before flock mortality rises. Specific infectious diseases (e.g. infectious bursal disease, chick infectious anaemia, Marek's disease), nutritional deficiencies (e.g. vitamin E, selenium and zinc deficiency), toxicity (mycotoxins) and stress are all factors that can induce sufficient immune suppression to cause an inadequate response (Sword *et al.*, 1991; Ferket and Qureshi, 1992; Siegel, 1994; Qureshi *et al.*, 1998; Ferket *et al.*, 1999; Surai, 2002; Swamy *et al.*, 2002a,b). Any disease control strategy would not be complete without measures to reduce stress.

HEALTH PROGRAMMES

Priority

Cost/benefit analysis is generally used to justify the expense of managing disease challenge, and the primary purpose of economic analysis is to aid in the decision process of resource allocation (Bale *et al.*, 1993). This process begins with grouping diseases according to the extent to which they threaten production system profitability. It is possible to calculate, for comparative purposes, the financial impact that a specific disease challenge has on breeder operation profitability, based on the risk of infection and its effect on performance *per se* (Gifford *et al.*, 1987). Unfortunately this gives no indication of the downstream effect that compromised product quality may have. *Mycoplasma synoviae* challenge may, for example, have a relatively minor effect on breeder performance but the consequential loss resulting from reduced broiler performance through the processing plant and marketing channels may be huge (Gifford *et al.*, 1987). In today's consumer-driven industry the first consideration is the impact of disease on public health and food safety, something that is difficult to justify financially if the disease does not affect breeder productivity.

Disease eradication

Eradication programmes are appropriate when the consequence of the disease is so devastating that it is economically advantageous to implement such drastic control measures. Eradication is only feasible if there is an effective means of detecting infection, containing the infection through clean-out and disinfection, and preventing dissemination of the disease-causing agent (Smith, 1995). There are three categories of disease for which eradication is an appropriate means of control: those that significantly threaten public health, those that have a devastating effect on breeder performance and those that severely compromise product quality. With diseases of this nature, control effort is focused on the complete elimination of the agent from the environment (Smith, 1995). This places the emphasis on preventing contact between the agent and the host, to prevent the disease from occurring, and on early detection of infection.

Of primary concern are those diseases that pose such a threat to public health that they require elimination of the flock. *S. enterica* serovar Enteritidis is an excellent example, since transovarian transmission of this organism can result in contamination of the food chain, and potentially cause a life-threatening, food-borne illness in humans. Avian influenza provides a more recent example, since the currently widely distributed H5N1 virus could severely threaten public health should it gain the ability to spread from human to human (Martinot *et al.*, 2007). The second category of diseases for which eradication programmes are appropriate are those that have a catastrophic impact on breeder productivity. Some of these are host-specific diseases such as those caused by *S. enterica* serovars Gallinarum and Pullorum, while others are diseases that affect a range of avian species, for example velogenic Newcastle disease and fowl plague (highly pathogenic avian influenza). In rare instances eradication policies are appropriate for diseases with a relatively minor effect on breeder performance, for example low pathogenic avian influenza of the H5 or H7 variety, because these viruses have the propensity to mutate and become highly pathogenic. Certain vertically transmitted diseases can have a devastating economic impact on downstream production efficiency and constitute a third category for which eradication programmes are appropriate. *Mycoplasma gallisepticum* (MG), *M. synoviae* (MS) and *Mycoplasma meleagradis* (MM) have for a long time been the primary focus of nationally controlled eradication programmes. Efforts to control MG began in the United States in the 1960s with a voluntary control programme, and MS and MM were added to the programme shortly after (Anonymous, 2000).

The success of an eradication programme hinges on good biosecurity and early detection of disease through the collection, analysis and interpretation of data from the population under consideration. The design of these biosecurity and disease surveillance programmes must be guided by the epidemiological characteristics of the disease-causing agent and its interaction with the host and environment and disease determinants.

In the case of a MG eradication programme, for example, the fact that this is a very fragile (no cell wall), host-adapted (avian only), fastidious (specialized growth requirements) organism which is not able to survive outside the host for more than a few days indicates that the programme focus must be on preventing direct contact with clinically or subclinically infected birds (Kleven, 2002, 2003; Ley, 2003). Special attention needs to be given to backyard flocks, wild birds, commercial layer flocks and previously infected convalescent birds, whether treated or not, as these have been identified as important carriers (Lin and Kleven, 1982b; Glisson and Kleven, 1984; Mohammed *et al.*, 1987a,b; McBride *et al.*, 1991; Ley *et al.*, 1996, 2006).

As with mycoplasma, steps to control food-borne organisms should also start at the parent-flock level. Vertical transmission of gut flora provides the first seeding of the hatchling enteron. Although these individual organisms may not persist throughout the life of the flock, they at the very least create conditions that shape development of the climax flora (Dawson, 2001). Manipulating parent gut flora can have a beneficial effect on offspring resistance to pathogen colonization (Fernandez *et al.*, 2002).

In contrast to mycoplasma species, *S. enterica* serovar Enteritidis is much more resilient, and these organisms can survive in the poultry house between cycles. Contamination of the poultry house environment, if it persists after cleaning and disinfection, plays an important part in the maintenance of salmonella infections, which emphasizes the importance of this as a critical control point (Kraddel and Miller, 1991; Baggesen *et al.*, 1992; Bale *et al.*, 1993). However, the critical control points within the farm ecosystem extend beyond the confines of the building, because litter beetles, rodents and contaminated dust exhausted from the ventilation fans spread the organism beyond the bounds of the house (Lohren, 1994). In breeder operations, the house environment poses a greater risk of salmonella infection than vertical transmission (van de Giessen *et al.*, 1994). In the case of *S. enterica* serovar Enteritidis, healthy adult hens require a relatively large challenge dose (10^3–10^6 organisms) to set up infection, and, since rodent faecal pellets have very high concentrations of salmonella, they are likely to be the primary source of infection for adult flocks (Pivnick and Nurmi, 1982; Opitz, 1993).

Attention to detail is critical in the design and implementation of eradication programmes, since many of the current management practices predispose birds to colonization or infection. Once-a-day feeding techniques will, for example, predispose the gut to colonization with salmonella species. Under normal conditions the flora of the lower gastrointestinal tract spend the majority of their existence in intense competition for a limited source of nutrients (Zinser and Kolter, 2004). Once-a-day feeding techniques, implemented to control body weight in breeders, increase feed passage rate, reduce digestion efficiency and increase nutrient supply to the lower tract. This rapidly changes the downstream gut ecology, allowing unfavourable species such as salmonella to dominate the caecal flora (Abboud *et al.*, 1997). Since shell contamination rates range from 10^3 to 10^8 colony-forming units per egg and these organisms subsequently colonize the hatchling gut, this change in caecal microbiota increases the risk of hazardous eggshell contamination occurring during and after oviposition (Baxter-Jones, 1991; Barrow, 1994; Bruce and Drysdale, 1994; Humphrey, 1994; Jones *et al.*, 2002, 2004).

No eradication programme can be successful without diligent surveillance, since early detection followed by rapid and effective biocontainment is critical to the success of any eradication programme. Highly pathogenic diseases, such as the currently circulating influenza type A virus of the H5N1 subtype, cause acute mortality, which itself attracts rapid veterinary attention, quarantine, diagnosis and control. Less overt disease, such as low pathogenic avian influenza of the H5 or H7 subtype and *S. enterica* serovar Enteritidis infection in adult flocks, may go undetected for long enough to allow widespread dissemination without diligent surveillance.

With a MG surveillance programme for example, in which the aim is to prevent vertical transmission of the disease, breeder flocks must be monitored for MG at regular intervals for early detection of changes in prevalence. A change in prevalence over time indicates a change in incidence, which signals the need for intervention. The procedure used to monitor breeder flocks must be sufficiently specific and sensitive to detect infection within a flock before vertical transmission occurs, or at least before potentially infected eggs hatch. The index case could produce infected eggs within 17 days, but peak shedding occurs when colonization peaks at 3–6 weeks after flock exposure (Glisson and Kleven, 1984, 1985; Sasipreeyajan *et al.*, 1987). After flock exposure to MG, there is a latent phase of 12–21 days in which less than 5% of the flock has a detectable antibody response, so, to prevent vertical transmission with 99% confidence, the monitoring system must be capable of detecting the presence of antibodies at the 5% level (McMartin *et al.*, 1987). The sample size (n) required to estimate this can be derived by calculation as:

$$n = [1 - (1 - 0.99)^{1/d}] \times [N - d/2] + 1$$

where N = the number of birds in the flock and d = the number of positive individuals. Thus, for an average flock of 7000 birds, $n \approx 90$ (Laughlin and Lundy, 1976). To prevent infected eggs from entering the hatchery, it would be necessary to sample flocks every 2 weeks (assuming 100% sensitivity for the test system). The testing interval can be extended by 2 weeks where hatchery tracking systems allow infected egg removal from the setters. This extension also allows sample size to be reduced, since disease prevalence increases to approximately 10% by 13–42 days after exposure, so for a flock of 7000 birds it is only necessary to sample 45 birds, and the sample interval could be extended to 3–4 weeks (DiGiacomo and Koepsell, 1986; McMartin *et al.*, 1987).

Serological methods such as agglutination and haemagglutination-inhibition, and, more recently, commercial ELISA kits, form the basis for mycoplasma control programme monitoring, because they are economical, rapid and have excellent sensitivity and specificity. Serological reactors have traditionally been confirmed by isolation, but recently use of the polymerase chain reaction (PCR) has become a rapid and sensitive alternative. Advancement in PCR technology has gone a long way to enhance early detection and containment/eradication of infection, despite emerging variance in organism virulence and antigenicity (Kleven *et al.*, 1973, 1988; Hnatow *et al.*, 1998; Lauerman, 1998). Ongoing improvements in these molecular diagnostic techniques will help to reduce the risk of management practices such as spiking, which provide obvious biological and financial benefit but at the same time carry significant disease risk.

As with MG, serological monitoring is a useful means of monitoring flocks for invasive salmonella species such as *S. enterica* serovar Enteritidis, since systemic challenge elicits a strong antibody (serological) response. Serological monitoring is, however, not necessarily appropriate for the non-invasive salmonella species. Since they remain part of the gut flora, these species may not elicit a detectable immune response. For non-invasive salmonella species, environmental sample (dust, faecal, drag-swab) collection and culture provide a very sensitive measure of challenge.

For very acute diseases, like highly pathogenic avian influenza, surveillance based on serological monitoring is of limited value, because birds frequently die before mounting an immune response (Suarez *et al.*, 2007). Since disease spread makes biocontainment increasingly difficult, early diagnosis through detection of the disease-causing agent is crucial. While the importance of traditional virus isolation techniques should not be understated, modern molecular diagnostic techniques have been instrumental in facilitating early biocontainment and eradication (Pelzel *et al.*, 2006; Suarez *et al.*, 2007).

Although vaccination can be used to assist with eradication, it should not really be considered part of an eradication programme. First, immunization does not prevent infection; second, a very high level of artificially induced flock immunity is required to eliminate disease; and, thirdly, the presence of circulating antibodies complicates disease surveillance (Smith, 1995). Instead, vaccination programmes are used as an aid in moving towards eradication where the prevalence of the disease is so high that the feasibility and cost of elimination is prohibitive (Capua and Marangon, 2007). With the current highly pathogenic avian influenza outbreak, the 'differentiating of infected from vaccinated animals' (DIVA) vaccination strategy has helped to avoid the complication of compromising surveillance, but there is field evidence to show that inappropriate use of vaccination may lead to the disease becoming endemic (Capua *et al.*, 2003; Capua and Marangon, 2004).

In the case of MG, immunization with a bacterin is effective in reducing the level of colonization, vertical transmission rate, clinical signs of disease and production losses (Hildebrand *et al.*, 1983; Yoder *et al.*, 1984; Kleven, 1985; Yoder and Hopkins, 1985; Karaca and Lam, 1987; Yagihashi *et al.*, 1992). The use of these inactivated vaccines in conjunction with other control measures has been effective in reducing the challenge to a level where eradication becomes feasible. Live vaccines have also been used to displace virulent field strains as a first step to eradication (Kleven *et al.*, 1998). There is a complex relationship between infectivity, pathogenicity and immunogenicity of MG strains. The level of protection induced with a live MG vaccine is correlated to the virulence of the vaccine (Lin and Kleven, 1982a). The more virulent F-strain vaccines induce higher levels of circulating antibodies than the less virulent TS-11 and 6/85 vaccine strains. Protection does not, however, correlate with circulating antibody titre (Lam and Lin, 1984; Talkington and Kleven, 1985; Whithear *et al.*, 1990; Abd-El-Motelib and Kleven, 1993).

Medication is an inappropriate means of eradication because resistance is likely to develop with extended use, and because of the inability of even the modern, highly effective antibiotics to sterilize the infection and prevent the carrier state (Levisohn, 1981; Migaki *et al.*, 1993; Zanella *et al.*, 1998; Wang *et al.*, 2001; Ley, 2003). In the case of MG, several medication strategies have been used successfully to limit vertical transmission, prevent flock infection and avert the full effect of infection on bird performance. Differential pressure egg-dipping in a suitable antibiotic solution, and egg injection at 18 days, either alone or in combination, has been used to reduce vertical transmission (Ghazikhanian *et al.*, 1980; Hodgetts, 1992). These processes, however, adversely affect hatchability and may not totally eliminate vertical transmission.

Disease control

In contrast to eradication, control programmes are aimed at reducing the frequency, and hence economic impact, of disease to a tolerable level. There is a subtle shift in emphasis from prevention, through early detection and elimination, to reducing the consequence or economic impact of the disease, i.e. damage control. Although monitoring and surveillance are still used to gather prevalence data, the primary focus is to measure the level of protection, not the presence of the disease. The principles of prevention through biosecurity still apply, but in a disease control programme the focus shifts to limiting the extent and consequence of exposure. In reality, many of the biosecurity measures taken to eradicate the more devastating diseases provide a solid foundation for the control of the erosive diseases, and immunization is used to bolster host resistance.

Allocating resources to the prevention of diseases that have a major biological and financial impact is relatively easy because, first, the control measures are the cost of doing business and, secondly, as they invariably require eradication, the cost of the disease is usually totally recoverable. In contrast, designing a disease control strategy for diseases that are likely to occur with a high degree of certainty but have less of a financial impact is a great deal more difficult (Gifford *et al.*, 1987). The process begins with clearly defining the estimated cost that the disease presence may incur and the potential benefits that the options for control may provide. Unfortunately there are several unknowns in health-related matters and it is consequently impossible to perform detailed and accurate cost/benefit analysis to ensure optimum resource allocation. Instead, partial farm budgeting is commonly used to compare the economic efficiencies of the various control options, including no action. In such instances immunization and its related biological and financial impact is the cost of health. One of the more difficult but important factors to quantify is the degree of productivity recovery that the control option provides, since disease losses are seldom totally recoverable.

Mesogenic and lentogenic Newcastle disease (ND) and infectious bronchitis (IB) are good examples of diseases that are currently endemic to most areas where broiler breeder flocks are housed around the world. These disease challenges are managed by ensuring birds have a high level of immunity throughout the rearing and laying period, first to prevent drops in productivity as a result of exposure but also to ensure adequate maternal antibody transfer to the offspring. This is achieved by using a combination of live and killed vaccines during the rearing period (McIlroy, 1994). In countries like Mexico and the Middle East, for example, where the threat of velogenic ND challenge is high, repeated vaccination with mesogenic-strain vaccines is necessary (McIlroy, 1994). Flocks are commonly vaccinated every 6–8 weeks during the laying period, to boost immunity and prevent drops in egg production, egg quality and hatchability. These vaccines, usually administered via the drinking water, are in some instances given by aerosol (see Cserep, Chapter 20, this volume). Such practices cause sufficient respiratory tract damage to predispose vaccinates to secondary bacterial infection. Cost/benefit analysis is very difficult since it is impossible accurately to assess and compare the risk and financial consequence of challenge versus vaccination.

The complexity of the disease-causing organism's antigenic structure is inversely proportional to the level of protection afforded by vaccination. While vaccination against viral diseases is a relatively efficient way of limiting the consequence of disease, the same does not apply to bacterial diseases. It would, for example, be relatively easy to justify vaccinating high-risk flocks against infectious coryza (*Avibacterium paragallinarum*), since this bacterial disease causes a very dramatic and sustained drop in egg production. However, although immunization of the pullets with inactivated oil-based vaccines prior to transfer will provide relatively good protection against the clinical signs of disease, it provides relatively little protection against the effect on egg production (Collett, 2005). In contrast, infection with a homologous strain prior to onset of lay will provide solid immunity against both the clinical signs and the effect on productivity. Clearly, an effective control strategy for this disease would involve immunization of the pullet with an inactivated oil-based vaccine, followed by controlled exposure prior to the onset of lay, but such a strategy is not without its risks. Controlled exposure carries the risk of, first, unknowingly introducing potentially pathogenic organisms into the vaccinated population and, second, increasing the number of infected carriers, thereby significantly increasing the risk of infecting adjacent poultry facilities.

Coccidiosis is a very important erosive disease that compromises breeder performance by reducing digestion efficiency and flock uniformity. Coccidiosis control programmes rely on the development of natural immunity through vaccination (attenuated strain), controlled exposure (non-attenuated strains) or suppressed natural exposure (chemoprophylaxis) (McDougald, 2003). While the use of coccidial vaccines in broiler breeders is common practice in some countries, chemoprophylaxis is still widely used in the USA, where the attenuated (precocious strain) vaccines are not registered for use.

The use of chemoprophylaxis in the meat-bird sector and vaccination in the breeder sector has proved to be a logistical problem for some integrated companies. For example, inadvertent delivery of medicated broiler feed to vaccinated breeders during the first 4 weeks can be disastrous, since the coccidial vaccine strains are very susceptible to the effects of the prophylactic agents. Although the compounds used in the control of coccidiosis are relatively safe, they can also have serious effects on breeder productivity if fed at toxic levels. Depending on the dose, ionophores cause transient to permanent paralysis, while nicarbizine causes mottling of the yolks, obvious de-pigmentation of the eggshells and a drop in egg production and hatchability (McDougald, 2003).

As with coccidiosis, chemoprophylaxis has, in the past, been successfully used to control blackhead (*Histomonas melelagridis*). The recent ban of the nitroimidazole compounds (dimetridazole and ipropran), however, has resulted in the re-emergence of this disease, which has required a shift in the method of control to epidemiologically derived intervention strategies (McDougald, 2005).

CONCLUSION

Mortality rates in breeder flocks vary considerably from country to country. This highlights the need to improve disease management, but the identification of specific improvement opportunity is impossible in the absence of reliable epidemiological data. Mortality is a very coarse measure of flock health, and the need for a more sensitive means of assessment increases as breeder performance approaches genetic potential. The epidemiology of disease in an intensive production system is heavily influenced by stress, and therefore the house environment and population dynamics impact performance before overt disease becomes apparent. The diagnosis of a multifactorial disease, before it becomes clinically apparent, is challenging, which makes timely intervention difficult. The collection and comparative analysis of epidemiological data from breeder flocks around the world would facilitate a more efficient allocation of both resources and effort in the management of current disease challenges in breeders.

REFERENCES

Abboud, P., Gallais, A., Janky, E and Gidenne, T. (1997) Caeco-colic digestion in the growing rabbit: impact of nutritional factors and related disturbances. *Livestock Production Science* 51, 73–88.

Abd-El-Motelib, T.Y. and Kleven, S.H. (1993) A comparative study of *Mycoplasma gallisepticum* vaccines in young chickens. *Avian Diseases* 37, 981–987.

Anonymous (2000) *National Poultry Improvement Plan and Auxiliary Provision*, Vol. APHIS-91-55-054. USDA, Washington, DC:.

Baggesen, D.L., Olsen, L.E. and Bisgaard, M. (1992) Plasmid profiles and phage types of *Salmonella typhimurium* isolated from successive flocks of three poultry stock farms. *Avian Pathology* 21, 568–579.

Bale, M.J., Bennett, M., Berringer, J.E. and Hinton, M. (1993) The survival of bacteria exposed to desiccation on surfaces associated with farm buildings. *Journal of Applied Bacteriology* 75, 519–528.

Barrow, P. (1994) The microflora of the alimentary tract and avian pathogens: translocation and vertical transmission. In: Board, R.G. and Fuller, R. (eds) *Microbiology of the Avian Egg*, 1st edn. Chapman and Hall, London, pp. 117–138.

Baxter-Jones, C. (1991) Egg hygiene: microbial contamination, significance and control. In: Tullett, S. (ed.) *Avian Incubation*. Butterworth, London, pp. 269–276.

Bruce, J. and Drysdale, E. (1994) Trans-shell transmission. In: Board, R.G. and Fuller, R. (eds) *Microbiology of the Avian Egg*, 1st edn. Chapman and Hall, London, pp. 63–86.

Capua, I. and Marangon, S. (2004) Vaccination for avian influenza in Asia. *Vaccine* 22, 4137–4138.

Capua, I. and Marangon, S. (2007) The challenge of controlling notifiable avian influenza by means of vaccination. *Avian Diseases* 51, 317–322.

Capua, I., Terregino, C., Cattoli, G., Mutinelli, F. and Rodriguez, J.F. (2003) Development of a DIVA (Differentiating Infected from Vaccinated Animals) strategy using a vaccine containing a heterologous neuraminidase for the control of avian influenza. *Avian Pathology* 32, 47–55.

Collett, S.R. (2005) The evaluation of an inactivated oil based infectious coryza vaccine in broiler breeders. *Proceedings 54th Western Poultry Diseases Conference*. Omnipress, Vancouver.

Collier, C.T., van der Klis, J.D., Deplancke, B., Anderson, D.B. and Gaskins, H.R. (2003) Effects of tylosin on bacterial mucolysis, *Clostridium perfringens* colonization, and intestinal barrier function in a chick model of necrotic enteritis. *Antimicrobial Agents Chemotherapy* 47, 3311–3317.

Dawson, K.C. (2001) Development of an appropriate microflora in the gut. *2nd International Poultry Broiler Nutritionist's Conference: Poultry Beyond 2005*. Poultry Industry Association of New Zealand (Inc.), Rotorua, New Zealand.

DiGiacomo, R.F. and Koepsell, T.D. (1986) Sampling for detection of infection or disease in animal populations. *Journal of American Veterinary Medical Association* 189, 22–23.

Ferket, P.R. and Qureshi, M.A. (1992) Performance and immunity of heat-stressed broilers fed vitamin- and electrolyte-supplemented drinking water. *Poultry Science* 71, 88–97.

Ferket, P.R., Quershi, M.A. and Edens, F.W. (1999) Trace minerals in immunity and stress in poultry. *60th Minnesota Nutrition Conference and Zinpro Technical Symposium*. University of Minnesota Extension Service, Bloomington, Minnesota.

Fernandez, F., Hinton, M. and van Gils, B. (2002) Dietary mannan-oligosaccharides and their effect on chicken caecal microflora in relation to *Salmonella enteritidis* colonization. *Avian Pathology* 31, 49–58.

Ghazikhanian, G.Y., Yamamoto, R., McCapes, R.H., Dungan, W.M. and Ortmayer, H.B. (1980) Combination dip and injection of turkey eggs with antibiotics to eliminate *Mycoplasma meleagridis* infection from a primary breeding stock. *Avian Diseases* 24, 57–70.

Gifford, D.H., Shane, S.M., Hugh-Jones, M. and Weigler, B.J. (1987) Evaluation of biosecurity in broiler breeders. *Avian Diseases* 31, 339–344.

Glisson, J.R. and Kleven, S.H. (1984) *Mycoplasma gallisepticum* vaccination: effects on egg transmission and egg production. *Avian Diseases* 28, 406–415.

Glisson, J.R. and Kleven, S.H. (1985) *Mycoplasma gallisepticum* vaccination: further studies on egg transmission and egg production. *Avian Diseases* 29, 408–415.

Grimble, R.F. (2001) Nutritional modulation of immune function. *Proceedings of Nutrition Society* 60, 389–397.

Gustin, S.J. (2007) New concepts for optimizing liveability in broiler breeders. *The Poultry Informed Professional* 93, 1–4.

Hildebrand, D.G., Page, D.E. and Berg, J.R. (1983) *Mycoplasma gallisepticum* (MG) – laboratory and field studies evaluating the safety and efficacy of an inactivated MG bacterin. *Avian Disease* 27, 792–802.

Hnatow, L.L., Keeler, C.L., Tessmer, J.L., Czymmek, L.K. and Dohms, J.E. (1998) Characterization of MGC2, a *Mycoplasma gallisepticum* cytadhesin with homology to the *Mycoplasma pneumoniae* 30-kilodalton protein P30 and *Mycoplasma genitalium* P32. *Infectious Immunology* 66, 3436–3442.

Hodgetts, B. (1992) Vacuum dipping of broiler hatching eggs. *Zootechnica International* 1992(5), 40–41.

Humphrey, B.D., Koutsos, E.A. and Klasing, K.C. (2003) Requirements and priorities of the immune system for nutrients. In: Lyons, T.P. and Jacques, K.A. (eds) *Nutritional Biotechnology in the Feed and Food Industry*. Nottingham University Press, Nottingham, UK, pp. 69–77.

Humphrey, T.J. (1994) Contamination of egg shell and contents with *Salmonella enteritidis*: a review. *International Journal of Food Microbiology* 21, 31–40.

Jones, D.R., Anderson, K.E., Curtis, P.A. and Jones, F.T. (2002) Microbial contamination in inoculated shell eggs: I. Effects of layer strain and hen age. *Poultry Science* 81, 715–720.

Jones, D.R., Curtis, P.A., Anderson, K.E. and Jones, F.T. (2004) Microbial contamination in inoculated shell eggs: II. Effects of layer strain and egg storage. *Poultry Science* 83, 95–100.

Jones, H.G.R., Randall, C.J. and Mills, C.P.J. (1978) A survey of mortality in three adult broiler breeder flocks. *Avian Pathology* 7, 619–628.

Karaca, K. and Lam, K.M. (1987) Efficacy of commercial *Mycoplasma-gallisepticum* bacterin (Mg-Bac) in preventing air-sac lesions in chickens. *Avian Diseases* 31, 202–203.

Kelly, D. (2004) Regulation of gut function and immunity. In: Tucker, L. and Taylor-Pickard, J. (eds) *Interfacing Immunity, Gut Health and Performance*. Nottingham University Press, Nottingham, UK, pp. 93–105.

Klasing, K.C. (1998) Nutritional modulation of resistance to infectious diseases. *Poultry Science* 77, 1119–1125.

Klasing, K.C. and Barnes, D.M. (1988) Decreased amino acid requirements of growing chicks due to immunologic stress. *Journal of Nutrition* 118, 1158–1164.

Klasing, K.C. and Johnstone, B.J. (1991) Monokines in growth and development. *Poultry Science* 70, 1781–1789.

Klasing, K.C., Laurin, D.E., Peng, R.K. and Fry, D.M. (1987) Immunologically mediated growth depression in chicks: influence of feed intake, corticosterone and interleukin-1. *Journal of Nutrition* 117, 1629–1637.

Kleven, S.H. (1985) Tracheal populations of *Mycoplasma gallisepticum* after challenge of bacterin-vaccinated chickens. *Avian Diseases* 29, 1012–1017.

Kleven, S.H. (2002) Recent developments in *Mycoplasma* diagnosis and control. *Proceedings of the Western Poultry Disease Conference* 51, 109–113.

Kleven, S.H. (2003) Recent developments in diagnosis and control. *World Poultry* 19(10), 8–9.

Kleven, S.H., Fletcher, O.J. and Davis, R.B. (1973) Variation of pathogenicity of isolates of *Mycoplasma synoviae* with respect to development of airsacculitis and synovitis in broilers. *American Journal of Veterinary Research* 163, 1196.

Kleven, S.H., Morrow, C.J. and Whithear, K.G. (1988) Comparison of *Mycoplasma gallisepticum* strains by hemagglutination-inhibition and restriction endonuclease analysis. *Avian Diseases* 32, 731–741.

Kleven, S.H., Fan, H.H. and Turner, K.S. (1998) Pen trial studies on the use of live vaccines to displace virulent *Mycoplasma gallisepticum* in chickens. *Avian Diseases* 42, 300–306.

Klipper, E., Sklan, D. and Friedman, A. (2000) Immune responses of chickens to dietary protein antigens. I. Induction of systemic and intestinal immune responses following oral administration of soluble proteins in the absence of adjuvant. *Veterinary Immunology and Immunopathology* 74, 209–223.

Klipper, E., Sklan, D. and Friedman, A. (2001) Response, tolerance and ignorance following oral exposure to a single dietary protein antigen in *Gallus domesticus*. *Vaccine* 19, 2890–2897.

Kraddel, D.C. and Miller, W.L. (1991) *Salmonella enteritidis* observations on field related problems. *Proceedings of the 40th Western Poultry Disease Conference* 40, 146.

Lam, K.M. and Lin, W.Q. (1984) Further studies on the immunization of chickens with temperature-sensitive *Mycoplasma gallisepticum* mutant. *Avian Diseases* 28, 131–138.

Lauerman, L.H. (1998) Mycoplasma PCR assays. In Lauerman, L.H. (ed.) *Nucleic Amplification Assays for Diagnosis of Animal Diseases*. American Association of Veterinary Laboratory Diagnosticians, Auburn, pp. 41–52.

Laughlin, K.F. and Lundy, H. (1976) The influence of sample size on the choice of method and interpretation of incubation experiments. *British Poultry Science* 17, 53.

Levisohn, S. (1981) Antibiotic sensitivity patterns in field isolates of *Mycoplasma gallisepticum* as a guide to chemotherapy. *Israel Journal of Medical Science* 17, 661–666.

Ley, D.H. (2003) *Mycoplasma gallisepticum* infection. In: Saif, Y.M., Barnes, H.J., Glisson, J.R., Fadly, A.M., McDougald, L.R. and Swayne, D.E. (eds) *Diseases of Poultry*, 11th edn. Iowa State University Press, Ames, pp. 722–744.

Ley, D.H., Berkhoff, J.E. and McLaren, J.M. (1996) *Mycoplasma gallisepticum* isolated from house finches (*Carpodacus mexicanus*) with conjunctivitis. *Avian Diseases* 40, 480–483.

Ley, D.H., Sheaffer, D.S. and Dhondt, A.A. (2006) Further western spread of *Mycoplasma gallisepticum* infection of house finches. *Journal of Wildlife Diseases* 42, 429–431.

Lin, M.Y. and Kleven, S.H. (1982a) Cross-immunity and antigenic relationships among 5 strains of *Mycoplasma gallisepticum* in young Leghorn chickens. *Avian Diseases* 26, 496–507.

Lin, M.Y. and Kleven, S.H. (1982b) Egg transmission of 2 strains of *Mycoplasma gallisepticum* in chickens. *Avian Diseases* 26, 487–495.

Lohren, U. (1994) Measures for disinfection and sterilisation of poultry housing after an outbreak of *S. enteritidis*. *DTW Deutsche Tierarztliche Wochenschrift* 101, 290–292.

Martinot, A., Thomas, J., Theiermann, N. and Dasgupta, N. (2007) Prevention and control of avian influenza: the need for a paradigm shift in pandemic influenza preparedness. *Veterinary Record* 160, 343–345.

McBride, M.D., Hird, D.W., Carpenter, T.E., Snipes, K.P., Danayeelmi, C. and Utterback, W.W. (1991) Health survey of backyard poultry and other avian species located within one mile of commercial California meat-turkey flocks. *Avian Diseases* 35, 403–407.

McDougald, L.R. (2003) Protozoal infections. In: Saif, Y.M., Barnes, H.J., Glisson, J.R., Fadly, A.M., McDougald, L.R. and Swayne, D.E. (eds) *Diseases of Poultry*, 11th edn. Iowa State University Press, Ames, pp. 973–1026.

McDougald, L.R. (2005) Blackhead disease (histomoniasis) in poultry: a critical review. *Avian Diseases* 49, 462–476.

McIlroy, S.G. (1994) The epidemiology and control of economically important diseases of broiler and broiler breeder production. In: Thrusfield, M.V. (ed.) *Proceedings of the Society for Veterinary Epidemiology and Preventive Medicine*. Belfast Meeting, R(D) SVS, Roslin, Vol. 1, pp. 114–127.

McMartin, D.A., Khan, M.I., Farver, T.B. and Christie, G. (1987) Delineation of the lateral spread of *Mycoplasma-gallisepticum* infection in chickens. *Avian Disease* 31, 814–819.

Migaki, T.T., Avakian, A.P., Barnes, H.J., Ley, D.H., Tanner, A.C. and Magonigle, R.A. (1993) Efficacy of danofloxacin and tylosin in the control of mycoplasmosis in chicks infected with tylosin-susceptible or tylosin-resistant field isolates of *Mycoplasma gallisepticum*. *Avian Diseases* 37, 508–514.

Mohammed, H.O., Carpenter, T.E. and Yamamoto, R. (1987a) Economic impact of *Mycoplasma gallisepticum* and *M. synoviae* in commercial layer flocks. *Avian Diseases* 31, 477–482.

Mohammed, H.O., Carpenter, T.E. and Yamamoto, R. (1987b) Evaluation of factors associated with infection of commercial layers with *Mycoplasma gallisepticum* and *Mycoplasma synoviae*. *Avian Diseases* 31, 470–476.

Morris, R.S. and Meek, H.A. (1980) Measurement and evaluation of economic effects of parasitic disease. *Veterinary Parasitology* 6, 165–184.

Opitz, H.M. (1993) Sources of *Salmonella enteritidis* infection for laying hens: a geographically limited study. *WHO Consultation on Control of Salmonella Infection in Animals*. Jena, Germany.

Pelzel, A.M., McCluskey, B.J. and Scott, A.E. (2006) Review of the highly pathogenic avian influenza outbreak in Texas, 2004. *Journal of American Veterinary Medical Association* 228, 1869–1875.

Pivnick, H. and Nurmi, E. (1982) The Nurmi concept and its role in the control of salmonella in poultry. In: Davies, R. (ed.) *Developments in Food Microbiology*. Applied Science Publishers, Barking, UK, pp. 41–70.

Qureshi, M.A., Hussain, I. and Heggen, C.L. (1998) Understanding immunology in disease development and control. *Poultry Science* 77, 1126–1129.

Sasipreeyajan, J., Halvorson, D.A. and Newman, J.A. (1987) Effect of *Mycoplasma gallisepticum* bacterin on egg-transmission and egg production. *Avian Diseases* 31, 776–781.

Siegel, H.S. (1994) Stress and immunity. In: Fisher, C. and Morris, T.R. (eds) *Proceedings of the 9th European Poultry Conference*. WPSA(UK), Andover, II, pp. 122–125.

Smith, R.D. (1995) Cost of disease. *Veterinary Clinical Epidemiology: a Problem Oriented Approach*, 2nd edn. CRC Press, Ann Arbor, pp. 227–246.

Suarez, D.L., Das, A. and Ellis, E. (2007) Review of rapid molecular diagnostic tools for avian influenza virus. *Avian Diseases* 51, 201–208.

Surai, P. (2002) Selenium in poultry nutrition: a new look at an old element. 1 Antioxidant properties, deficiency and toxicity. *World's Poultry Science Journal* 58, 333–347.

Swamy, H.V., Smith, T.K., Cotter, P.F., Boermans, H.J. and Sefton, A.E. (2002a) Effects of feeding blends of grains naturally contaminated with *Fusarium* mycotoxins on production and metabolism in broilers. *Poultry Science* 81, 966–975.

Swamy, H.V., Smith, T.K., MacDonald, E.J., Boermans, H.J. and Squires, E.J. (2002b) Effects of feeding a blend of grains naturally contaminated with *Fusarium* mycotoxins on swine performance, brain regional neurochemistry, and serum chemistry and the efficacy of a polymeric glucomannan mycotoxin adsorbent. *Journal of Animal Science* 80, 3257–3267.

Sword, J.T., Pope, A.L. and Hoekstra, W.G. (1991) Endotoxin and lipid peroxidation *in vivo* in selenium- and vitamin E-deficient and adequate rats. *Journal of Nutrition* 121, 251–257.

Talkington, F.D. and Kleven, S.H. (1985) Evaluation of protection against colonization of the chicken trachea following administration of *Mycoplasma gallisepticum* bacterin. *Avian Diseases* 29, 998–1003.

van de Giessen, A.W., Ament, A.J. and Notermans, S.H. (1994) Intervention strategies for *Salmonella enteritidis* in poultry flocks: a basic approach. *International Journal of Food Microbiology* 21, 145–154.

Walker, L., Levine, H. and Jucker, M. (2006) Koch's postulates and infectious proteins. *Acta Neuropathologica* 112, 1–4.

Wang, C., Ewing, M. and Aarabi, S.Y. (2001) In vitro susceptibility of avian mycoplasmas to enrofloxacin, sarafloxacin, tylosin, and oxytetracycline. *Avian Diseases* 45, 456–460.

Whithear, K.G., Soeripto, Harrigan, K.E. and Ghiocas, E. (1990) Immunogenicity of a temperature sensitive mutant *Mycoplasma gallisepticum* vaccine. *Australian Veterinary Journal* 67, 168–174.

Yagihashi, T., Nunoya, T., Sannai, S. and Tajima, M. (1992) Comparison of immunity induced with a *Mycoplasma gallisepticum* bacterin between high- and low-responder lines of chickens. *Avian Diseases* 36, 125–133.

Yoder, H.W. and Hopkins, S.R. (1985) Efficacy of experimental inactivated *Mycoplasma gallisepticum* oil-emulsion bacterin in egg-layer chickens. *Avian Diseases* 29, 322–334.

Yoder, H.W., Hopkins, S.R. and Mitchell, B.W. (1984) Evaluation of inactivated *Mycoplasma gallisepticum* oil emulsion bacterins for protection against air sacculitis in broilers. *Avian Diseases* 28, 224–234.

Zanella, A., Martino, P.A., Pratelli, A. and Stonfer, M. (1998) Development of antibiotic resistance in *Mycoplasma gallisepticum in vitro*. *Avian Pathology* 27, 591–596.

Zavala, G., Dufour-Zavala, L., Davis, J.F. and Waltman, D. (2003) Broiler breeder health report in Georgia. *Proceedings XIII Congress of the World Veterinary Poultry Association.* AAAP, Athens, Georgia, p. 186.

Zinser, E.R. and Kolter, R. (2004) *Escherichia coli* evolution during stationary phase. *Research Microbiology* 155, 328–336.

PART IX
Abstracts

CHAPTER 23
Poster Abstracts

All contributed abstracts of posters were edited for style and reviewed for the suitability of the material for the symposium by Dr T. Acamovic and Dr P.M. Hocking. The authors alone are responsible for the content and accuracy of the abstracts.

POSTER 1
Major Gene Effect on Semen Quality Characteristics of Nigerian Local Chicken

M.A. Adeleke, S.O. Peters, M.O. Ozoje, O.A. Adebambo and C.O.N. Ikeobi

Department of Animal Breeding and Genetics, University of Agriculture, P.M.B. 2240, Abeokuta, 110001, Ogun State, Nigeria.
Email: maadeleke2003@yahoo.com

The local chicken constitutes between 80 and 90% of the chicken population in Nigeria. These local chickens are mainly naked neck, frizzle feathered and normal feathered. Each of these genes tends to play a significant role in the productive adaptability of the local chicken (Ebozoje and Ikeobi, 1995). This study was aimed at investigating the effects of major genes on semen quality characteristics of Nigerian local chickens.

A total of 18 matured cocks, comprising six naked neck, six frizzle feathered and six normal feathered, maintained at the Poultry Breeding Unit of the University of Agriculture, Abeokuta, Ogun State, Nigeria, were used for this study. The cocks were fed a breeder's ration containing 160 g crude protein/kg, 10.9 MJ/kg metabolizable energy, 32 g calcium/kg and 45 g available phosphorus/kg. Clean water was supplied *ad libitum*. The cocks were trained for artificial semen collection for 3 weeks. Data on semen volume and colour were collected twice weekly in the afternoon for a period of 12 weeks, while data on other traits were taken four times on each cock. Each ejaculate was immediately evaluated for volume, colour, motility, concentration and pH. The colour of the semen was scored using a three-point scale, while motility of the semen sample was expressed as the percentage of cells that were motile under their own power. Semen concentration was determined using a haemocytometer. All data collected were subjected to one-way analysis of variance (SAS, 1999).

All the semen quality characteristics were significantly ($P < 0.01$) affected except pH, colour and proportion active (%). The highest semen volume was recorded in frizzle feathered cocks (0.60 ± 0.02 ml), while the naked neck had highest semen concentration, followed by normal feathered and frizzle feathered, with means and SEs respectively of 4.21 ± 1.45, 4.05 ± 0.65 and 3.40 ± 0.31 10^9/ml. Naked neck strain also had the highest sperm motility of 87.4 ± 0.12%. The three genotypes had creamy white colour with pH ranging between 7.02 ± 0.01 and 7.04 ± 0.02. The semen volume obtained for all the three strains was within the acceptable range for artificial insemination. Variation in semen quality characteristics of Nigerian local chickens as influenced by major genes can be exploited in a poultry breeding programme for genetic improvement.

Ebozoje, M.O. and Ikeobi, C.O.N. (1995) Productive performance and occurrence of major genes in the Nigerian local chicken. *Nigerian Journal of Genetics* 10, 67–77.

SAS (1999) *SAS User's Guide*. Statistical Analysis Institute Inc., Cary, North Carolina, USA.

Poster 2
Validation of a Stochastic Model of Egg Production in Broiler Breeders using Data from Venezuelan Farms

R. Alvarez[1,2] and P.M. Hocking[1]

[1]Roslin Institute and Royal (Dick) School of Veterinary Studies, University of Edinburgh, Roslin, Midlothian, EH25 9PS, UK; [2]Instituto de Producción Animal, Facultad de Agronomía, Universidad Central de Venezuela, Apdo. 4579, Maracay, Venezuela. Email: alvarezr@agr.ucv.ve

A stochastic model was developed to estimate egg production in broiler breeders, using body weight and subsequent daily weight gain as the principal input parameters (Alvarez and Hocking, 2007). The objective of this study was to test the model against a large data set from seven flocks in two commercial farms in Venezuela.

Commercial data consisted of mean weekly egg production, mortality and body weight from 188,177 Ross 308 broiler breeder females from two farms with, respectively, four flocks (in a total of 23 houses) and three flocks (in 21 houses). Initial body weight, subsequent weight gain and mortality for each flock were used as input variables and other parameters were from the published model (Alvarez and Hocking, 2007). For each flock, the data from different houses were combined and total egg production over 42 weeks of lay was simulated 500 times for each set of parameters. Means for each set were determined and analysed by REML using a statistical model with fixed effects of Farm-Flock and Actual-Estimated egg production.

The interaction of Farm-Flock and Actual-Estimated was significant ($P < 0.001$), although the overall means for actual and estimated total egg production were similar (191.1 versus 191.3, SED 0.70 eggs, $P = 0.38$). Inspection of the means showed that total egg production was overestimated by 6.9 eggs and underestimated by 7.1 eggs, respectively, in Farm-Flocks 1–18 and 2–15 (average SED was 1.9 eggs), whereas predictions in the other units were similar (typically less than one egg). In conclusion, this preliminary evaluation suggests that the model can be used to simulate total egg production from commercial flocks of broiler breeders in tropical conditions with sufficient accuracy to serve as a management tool.

RA gratefully acknowledges financial support from the CDCH-Universidad Central de Venezuela and Prof. Cecilia Garcia (Secretaría-UCV). Roslin Institute is supported by the Biotechnology and Biological Sciences Research Council.

Alvarez, R. and Hocking, P.M. (2007) Stochastic model of egg production in broiler breeders. *Poultry Science* 82, 1445–1452.

POSTER 3

The Role of Short-term Feeding Behaviour in Food Intake Control in Two Genetic Lines of Broiler Chickens

J.A. Howie,[1] B.J. Tolkamp,[1] S. Avendano[2] and I. Kyriazakis[1]

[1]*Animal Nutrition and Health Department, SAC, Kings Buildings, West Mains Road, Edinburgh, EH9 3JG, UK;* [2]*Aviagen Ltd, Newbridge, Midlothian, EH28 8SZ, UK. Email: jennifer.howie@sac.ac.uk*

The aim of this study was to characterize normal short-term feeding behaviour (STFB) and its role in feed intake structure and control in broilers. The study used data obtained through computerized records of feeder visits provided by Aviagen Ltd on two lines of broiler chicks aged between 2 and 5 weeks: one selected for higher growth (line A) than the other (line B). The birds were housed in pens of 140 animals (line A = 9 pens; line B = 6 pens), each containing eight feeders linked to scales and computerized recording equipment. Existing methodology was adapted in order to determine a biologically relevant meal criterion, and visits were grouped into meals.

The lines differed significantly in meal size and daily number of meals. The mean (± SE) for lines A and B, respectively, were 12.0 ± 0.02 g and 6.9 ± 0.02 g ($P < 0.001$) for meal size and 9.9 ± 0.03 meals/day versus 14.0 ± 0.05 meals/day ($P < 0.001$) for the number of daily meals. Both lines showed a similar increase in meal size and duration with age. These data show a significant impact of genetic line on the STFB of chicks, as line A consumed more feed in fewer, larger-sized meals.

The correlation between meal size and length of interval after a meal was significantly higher ($r^2 = 23.2\%$) than that between the intermeal interval and following meal size ($r^2 = 8.5\%$, $P = 0.004$) in both lines. This suggests a larger role for hunger than satiety in the control of food intake in both lines. The study shows that genetic selection strongly affects some but not all characteristics of STFB.

Poster 4
Does the Novel cGnRH-Receptor-II Mediate Reproductive Function in the Chicken?

Nerine T. Joseph,[1,2] Kevin Morgan,[2] Robert Millar[2] and Ian C. Dunn[1]

[1] *Roslin Institute and Royal (Dick) School of Veterinary Studies, University of Edinburgh, Roslin, Midlothian, Edinburgh, EH25 9PS, UK;*
[2] *MRC Human Reproductive Sciences Unit, The Queen's Medical Research Institute, Edinburgh, EH16 4TJ, UK. Email: nerine.joseph@roslin.ed.ac.uk*

Reproductive function depends on the stimulation of pituitary gonadotrophin expression and secretion by the hypothalamic peptide gonadotrophin-releasing hormone-I (cGnRH-I). The binding and activation of the GnRH receptor by cGnRH-I is essential for initiating the transcription of gonadotrophin through second messenger signalling cascades. Avians express both cGnRH-I and cGnRH-II ligands, although cGnRH-II is absent from the median eminence. Recent discoveries indicate that two GnRH receptor isoforms with 55% sequence identity are expressed in the chicken: cGnRH-R-I and a novel receptor, cGnRH-R-II.

The aim was to characterize the mRNA distribution and pharmacology of cGnRH-R-II to evaluate if it is the cognate receptor for cGnRH-I, since cGnRH-R-I is ubiquitously expressed in the chicken and has a higher binding affinity and Gq activation with cGnRH-II than I. This served as an *in vitro* test of the hypothesis that the novel cGnRH-R-II is responsible for mediating cGnRH-I's reproductive effects. Expression of cGnRH-R-I and cGnRH-R-II was quantified with real-time PCR and comparative binding affinity and Gq coupling of the receptor isoforms measured by radioligand binding assays and inositol phosphate (IP) assays in transiently transfected Cos-7 cells.

Pituitary cGnRH-R-II mRNA expression was 93 times that in the small intestine. Overall, cGnRH-R-II was 1370 times more abundant than cGnRH-R-I in the pituitary. cGnRH-R-II exhibited higher binding affinity for cGnRH-II than cGnRH-I. Cells expressing cGnRH-R-II had an EC50 for IP production of 5.6 nM for cGnRH-I compared with 0.76 nM for cGnRH-II, indicating that cGnRH-II was more potent.

In conclusion, cGnRH-R-II mRNA is expressed predominantly in the pituitary, and has a higher binding affinity and signal transduction through Gq coupling with cGnRH-II than cGnRH-I. Although neither receptor displays a higher binding affinity for cGnRH-I than cGnRH-II, cGnRH-R-II may be the predominant receptor modulating gonadotrophin synthesis, based on its expression levels in the pituitary, although cellular expression of the receptors remains to be confirmed.

POSTER 5
Genetic Parameters of Different Stages of Embryonic Mortalities in the Ardennaise Chicken Breed

J.-M. Larivière, C. Michaux and P. Leroy

Department of Animal Production, Faculty of Veterinary Medicine, University of Liège, B-4000, Bd de Colonster, 20 B43, Liège, Belgium. Email: jean-marc.lariviere@ulg.ac.be

Genetic factors play a definite role in embryonic viability of hatching eggs. Total embryonic mortality may reach up to 15% of fertile eggs in some poultry species (Krueger, 1990), reducing hatchability of eggs and thus increasing production costs. The purpose of this study is to estimate genetic parameters of embryonic mortality at three stages of incubation (early (EEM), mid-term (MEM) or late (LEM)), as well as fertility (FER) in the Ardennaise, a slow-growing chicken breed.

Heritability and genetic correlations were estimated with the multiple-trait REML VCE package (Groeneveld and Kovac, 1990), taking into account random effects of sires and fixed effects of hatch per year per line. A total of 2040 eggs from six families were recorded over 3 years (902 eggs (year 1), 693 eggs (year 2) and 445 eggs (year 3)). Eggs were identified, incubated, candled at 10 and 18 days, and classified as infertile, hatched, EEM, MEM or LEM.

Estimated heritability of FER, EED, MED and LED were, respectively, 0.10 ± 0.02, 0.03 ± 0.01, 0.11 ± 0.03 and 0.02 ± 0.01. All genetic correlations between FER, EEM, MEM and LEM were negative. Environmental correlations between FER and EEM, MEM, and LEM ranged from −0.03 to 0.05.

The different heritability values of the various stages of embryonic mortality and the negative genetic and environmental correlations between these values suggested that the different stages should be considered as distinct traits. These results are in accordance with those obtained with laying hens (Beaumont *et al.*, 1997).

This work was funded by Région Wallonne (DGA). The authors also thank Ferme de Strée for their assistance.

Beamont, C., Millet, N., Le Bihan Duval E., Kipi, A. and Dupuy, V. (1997) Genetic parameters of survival to the different stages of embryonic death in laying hens. *Poultry Science* 76, 1193–1196.

Groeneveld, E. and Kovac, M. (1990) A generalized computing procedure for setting up and solving mixed linear models. *Journal of Dairy Science* 73, 513–531.

Krueger, K.K. (1990) Fertility of female turkeys: how to manage it? In: Brillard, J.P. (ed.) *Control of Fertility in Domestic Birds*. INRA, Paris, France, pp. 205–212.

Poster 6

Effect of Organic Selenium and 48 h Post-growing Fast on the Pro-oxidant/ Antioxidant System in Heart Muscle of Broiler Chickens

S. Milinković-Tur,[1] J. Piršljin,[1] B. Beer Ljubić,[1] N. Poljičak-Milas,[2] Z. Stojević[1] and T.S. Marenjak[2]

[1]*Faculty of Veterinary Medicine, University of Zagreb, Department of Physiology and Radiobiology, 10000 Zagreb, Heinzelova 55, Croatia;* [2]*Faculty of Veterinary Medicine, University of Zagreb, Department of Pathophysiology, 10000 Zagreb, Heinzelova 55, Croatia. Email: ur@vef.hr*

The objective of the experiment was to determine the effects of organic selenium feed supplements and the 48 h post-growing fast on the antioxidant enzymes activity and lipid peroxidation in the chicken heart muscle.

The experiment was carried out on the Ross 308 broiler chickens. During the growing period one group was fed a standard diet (control), and in another group of chickens 0.3 p.p.m. of organic selenium was added in the standard diet (experimental). After the growing and 48 h fasting period, ten chickens from each group were randomly selected and slaughtered. The activity of glutathione peroxidase (GSH-Px), catalase (CAT), copper zinc superoxide dismutase (Cu,Zn-SOD), manganese superoxide dismutase (Mn-SOD), and reduced glutathione (GSH) and lipid peroxide (TBARS) concentration were determined in the heart muscle from both groups of chickens.

The activity of CAT and Mn-SOD was significantly higher after the fasting period in both groups of chickens ($P < 0.0001$ and $P < 0.01$, respectively, for control and experimental treatments). After the fasting period, GSH-Px activity was significantly higher in chickens with organic selenium supplementation than in the chickens fed the standard diet ($P < 0.02$). There were no statistically significant differences observed in Cu,Zn-SOD activity or in GSH and TBARS concentration between the control and experimental group.

The results show that after the fasting period, chickens with organic selenium supplementation maintained the antioxidant system in the heart muscle more efficiently, with more intense GSH-Px antioxidant defence. These data suggest that organic selenium supplementation may benefit feed-restricted broiler breeders.

POSTER 7
Correlation Between Chick Length and Chick Weight at Hatch and Slaughter Weight and Breast Yield in Broilers

R. Molenaar,[1] I.A.M.Reijrink,[1] R. Meijerhof[2] and H. van den Brand[3]

[1]*HatchTech BV, Gildetrom 25, 3900 AG Veenendaal, The Netherlands;*
[2]*HYBRO BV, Veerstraat 38, 5830 AA, Boxmeer, The Netherlands;*
[3]*Adaptation Physiology Group, Wageningen University, 6700 AH Wageningen, The Netherlands. Email: roos.molenaar@wur.nl*

Chick length at hatch is often used in hatcheries as a tool to measure chick quality. The objective of this experiment was to evaluate the relationship between chick length at hatch and slaughter weight and breast yield in male and female broilers.

At set, 220 eggs of a commercial broiler line were selected on egg weight and ranged between 64 and 67 g. Eggs were incubated at an eggshell temperature of 37.8°C. Chick length and weight of 100 female and 100 male broilers were measured at hatch. Male and female broilers were housed in a commercial broiler house and fed *ad libitum*. Chick weight was measured every week. Slaughter weight and breast yield were measured at 38 days of age.

From 28 days of age, chick weight was higher in males than in females ($P < 0.01$). In males, a positive correlation was found between chick length at hatch and slaughter weight ($r = 0.36$; $\beta = 264.2$; $P < 0.01$), and chick length at hatch and breast yield ($r = 0.25$; $\beta = 44.7$; $P < 0.05$). In males no correlation was found between chick weight at hatch and slaughter weight or breast yield ($P > 0.05$). In females, only a negative correlation was found between chick weight at hatch and breast yield ($r = 0.24$; $\beta = -9.5$; $P < 0.05$).

Based on these results it can be concluded that, in male broilers, chick length at hatch has a predictive value for final body weight and breast meat yield, while chick weight has not. For female broilers, both parameters have no predictive value for final body weight and breast yield.

POSTER 8
Gene Segregation Effects on Fertility and Hatchability of the Nigerian Local Chickens

S.O. Peters, M.O. Ozoje and O.A. Adebambo

Department of Animal Breeding and Genetics, University of Agriculture, P.M.B. 2240, Abeokuta 110001, Ogun State, Nigeria. Email: so_peters@yahoo.com

There is a paucity of information on the effects of sire and the interaction between sire and dam genotypes on fertility and hatchability of artificially inseminated chicken eggs in the humid zone of Nigeria. This study, therefore, was initiated to examine the segregation effect of major genes on the fertility and hatchability of local chicken genotypes raised in south-west Nigeria.

The population of the chickens used in this study consisted of six mature sires and 50 dams per genetic group, classified as normal feathered, naked neck and frizzle feathered. Dams were inseminated twice weekly and eggs were collected twice daily. On days 10 and 18 of incubation, the eggs were candled to identify and remove infertile eggs. The remaining eggs were transferred to the hatching tray for hatching. The numbers of hatched chicks, including the normal, weak, abnormal chicks and dead chicks, were recorded at hatch. All percentage data were transformed to their arcsin$\sqrt{\%}$ values before analysis. All data were analysed using SAS (www.sas.com/technologies/analytics/statistics/), which was also used to pre-adjust the data for the effects of sire age, monthly variation and egg batch.

Sire and dam genotype significantly ($P < 0.01$) affected the number of eggs set, number of eggs fertile, percentage fertility and percentage hatchability ($P < 0.05$). The frizzle feathered sire group produced the highest number of fertile eggs, while the naked neck sire genotype sired the least fertile eggs. However, the normal feathered genetic group had the highest hatchability percentage. The effect of dam genotype was significant ($P < 0.05$), affecting percentage fertility and hatchability. The mean values of hatchability as affected by dam genotype in this study suggest that hatchabilty may not entirely be a function of fertility because of some intrinsic factors associated with the eggs.

In conclusion, the results of this study showed that the variations in the genetic groups of the Nigerian local chicken significantly affected fertility and hatchability of eggs from these chicken types. Fertility and hatchability reduced when sires and dams carrying the naked neck gene were involved in crossing to produce fertile eggs. Similarly, cross-bred mating of naked neck × frizzle feathered birds resulted in about 73.9% dead-in-shell, confirming earlier suggestion of the lethal effect of these genes when combined in a genotype. The normal feathered genetic group was generally superior to other groups.

Poster 9

Reduced Dietary Phosphorus Levels and Phytase Can Decrease Phosphorus Output from Broiler Breeders without Affecting Performance of the Broiler Progeny

P.W. Plumstead,[1] H. Romero-Sanchez,[2] R. Maguire[3] and J. Brake[4]

[1]Danisco Animal Nutrition, PO Box 777, Marlborough, Wiltshire, SN8 1XN, UK; [2]Hybro B.V., PO Box 305830 AA, Boxmeer, The Netherlands; [3]Department of Crop and Soil Environmental Sciences, Virginia Tech, Blacksburg, VA 24061, USA; [4]Department of Poultry Science, North Carolina State University, Raleigh, NC 27695-7608, USA.
Email: peter.plumstead@danisco.com

Effects of phytase and non-phytate phosphorus (NPP) levels in broiler breeder (BB) diets on performance and phosphorus excretion from BB and broiler progeny performance were investigated. Four treatments (A, B, C, D) were assigned to four replicate floor pens of 68 BB pullets and one pen of 50 cockerels. From 10 to 21 weeks, treatments A to D, respectively, contained 3.7, 2.7, 2.7 and 1.7 g/kg NPP, with 300 FTU/kg phytase added to treatments B and D. At 22 weeks, birds were transferred to a 2/3 slat-litter breeder house with 16 pens of 60 pullets and six cockerels each, and NPP in treatments A to D, respectively, were adjusted to 3.7, 2.7, 1.9 and 0.9 g/kg, and phytase increased to 500 FTU/kg feed in treatments B and D. At 55 weeks, four replicate pens of 30 broilers hatched from each BB treatment were assigned to four broiler treatments (I, II, III, IV) with respective NPP levels of 4.5, 4.5, 3.5 or 2.5 g/kg in starter diets (0–2 weeks), 4.5, 3.5, 2.5 or 2.5 g/kg in grower diets (3–4 weeks), and 3.5, 2.5, 2.5 or 2.5 g/kg in finisher diets (5–6 weeks). Hen-day egg production increased ($P < 0.05$) when BB diets contained 0.9 g/kg NPP with phytase, while total and soluble phosphorus in manure was decreased by 42%. Fertility was lower when BB diets contained less than 3.7 g/kg NPP ($P < 0.05$), but there was no effect of BB NPP or phytase on hatchability of fertile eggs, number of chicks produced per hen housed or performance of the broiler progeny.

POSTER 10
Breath Coefficient of Chick Embryos as a Criterion for their Adaptation Possibilities at High Osmotic Pressure

O.I. Stanishevskaya
All-Russian Research Institute of Farm Animal Genetics and Breeding, St Petersburg-Pushkin, Moskovskoye Shosse, 196625, Russia.
Email: olgastan@list.ru

It is well known that increased osmotic pressure (egg weight loss more than 14%) negatively affects embryos' mass, water content in tissues and liveability. Chicken embryos are able to maintain homeostasis by increasing oxidation of yolk lipids. This can be controlled by measurement of the embryo's breath coefficient (BC). Usually it decreases. As a result of experiments carried out on 185 embryos (Cornish and Plymouth Rocks), we found that the degree of change in osmotic pressure varied. In cases where egg weight loss was 14–20%, some embryos (40–60%) sustained their mass and water content in tissues as in the control group (11–13% egg weight loss). The rest of the embryos cannot change their lipid metabolism and suffer from dehydration. For example, in the experiment with the embryos of BAROS meat cross, egg weight was 64.0 ± 0.17 g. Osmotic pressure was induced by day 11 by a decrease in humidity. BC was evaluated at day 15 and embryos were autopsied at day 18. The embryos from the control treatment (n = 18) had BC = 0.753 ± 0.012; mass 32.1 ± 0.5 g, water content in breast muscles 83.8 ± 0.6%. Embryos (n = 12) with the high dehydration rate (14–20%) revealed high BC = 0.786 ± 0.015, decreased mass 31.3 ± 0.7 g and low water content in muscles 79.7 ± 0.9% ($P < 0.001$), i.e. they were not able to regulate their homeostasis. Another group of embryos from eggs with high dehydration (n = 22) had low BC = 0.677 ± 0.003 ($P < 0.001$), mass as in the control group 32.4 ± 0.7 g and water content in muscles 82.2 ± 0.6, i.e. they revealed high adaptation abilities. This phenomenon can be used in chicken selection for dehydration resistance. Dehydration often occurs among neonatal chicks during long-lasting transportation or among hens in hot ambient temperatures.

Poster 11
Analysis of Fertility in Broiler Breeder Flocks – Male-side Approaches

B.Vegi, E. Varadi, Zs. Szoke, K. Liptoi and J. Barna
Institute for Animal Breeding and Nutrition, H-2100 Gödöll , Hungary.
Email: vbarbus@katki.hu

The aim of the study was to clarify the effect of male number and age on the fertility of broiler breeders in the second half of the reproduction cycle. The effect of various mating strategies in seven small flocks of Ross 308 broiler breeders was analysed by weekly monitoring during a whole reproduction cycle.

At the age of 26 weeks, each flock of 80 females and eight males was put in floor pens. The primary breeder's management manual recommendations were followed throughout. In the flocks, various sexual ratios (decreasing, increasing or unchanged number of males), spiking techniques (50–100% replacements at the age of 44 weeks) and exchange of the cockerels between two groups were tested. For fertility determination, GD IPVL hole counting and examinations of candling-rejected eggs were done. Data were analysed by the Mann–Whitney test and regression analysis.

The fertility level (from the average of 95.1 to 90.4%) and the number of holes decreased significantly in each flock in the second half of the cycle, independently from the number and age of males. Flocks with 100% spiking had more holes compared with 50% spiking and flocks with males *unchanged* in number ($P < 0.05$); however, this value was similar to flocks with *increasing*, or even with *decreasing*, male number. However, the 'true' fertility was higher in flocks with young cockerels compared with old ones with the same median values. The relationship between 'true' fertility and median number of holes showed limited correlation ($r = 0.52$).

Increasing the number of males, spiking or exchanging of males could not maintain the earlier level of IPVL holes in the second half of the cycle, which indicates the definitive role of the female genital tract in the decline in fertility with time.

POSTER 12
Effects of Digestible Lysine and Methionine Concentrations on the Economics of Poultry Feeding

I. Vitina, V. Krastina, J. Nudiens and J. Miculis

Research Institute of Biotechnology and Veterinary Medicine 'Sigra' LUA, Instituta Street 1, Sigulda, Latvia. Email: sigra@lis.lv

Dietary concentrations of digestible lysine and methionine essentially determine poultry feed efficiency and cost of production. It is possible to ensure high poultry productivity if the total protein level in poultry feed is decreased relative to normal concentrations if the digestible lysine and methionine contents are increased.

We investigated economically profitable digestible lysine and methionine concentrations in commercial hybrid broiler (Hybro-G, n = 300) and layer (Lohmann Brown, n = 400) feed in diets where the protein concentration was decreased by 10, 20 and 30 g/kg compared with commercial recommendations. Amino acid levels in the feed were maintained by L-lysine HCL and DL-methionine supplementations. In the broiler diet under local conditions, from 0 to 28 days it was economically profitable to decrease protein level by 20 g/kg, and after 28 days of age by 15 g/kg, provided digestible lysine and methionine were increased by 1.2 times their commercial concentrations. It was economically profitable to decrease protein in layer feed by 10 g/kg and increase digestible lysine and methionine content by 1.2 times their commercial concentrations. Lysine and methionine digestibility and utilization in the gastrointestinal tract increased on average by 2.02–2.45%. Broiler live weight and laying intensity in hens had a tendency to increase by 1–2% and feed conversion to decrease by 4.8 and 2.8%, respectively, in broiler chickens and layers in comparison to control diets. Poultry productivity did not decrease if dietary protein level was decreased by 30 g/kg if, correspondingly, synthetic amino acid concentrations were increased appropriately.

POSTER 13
Non-feed Removal Strategies to Induce Moulting in Laying Hens: I. Influence on Hen Performance during 6 Weeks of Moulting

M. Yousaf and A.S. Chaudhry

School of Agriculture, Food and Rural Development, Newcastle University, Newcastle, NE1 7RU, UK. Email: muhammad.yousaf@newcastle.ac.uk

Common methods of moulting laying birds involve feed removal and are unacceptable on welfare grounds. Therefore, we compared different non-food removal strategies to induce moulting in laying hens.

The study involved four treatments: *ad libitum* feed intake (FI) with either 16 h (A) or 8 h (B) photoperiods or 8 h photoperiod with restricted feed (C) or C plus 15 p.p.m. aluminium using Al_2O_3 (D). Ninety-six 70-week-old brown hens were distributed into 32 groups of three hens each (mean ± SE live weight (LW) = 1680 ± 59 g/hen) and housed in 32 two-tier cages under predetermined photoperiods. Each treatment was allocated to eight cages having 24 hens per treatment. All hen groups received a diet (170 g CP/kg) depending upon the treatments. The birds were monitored for health, FI, LW, egg production and moult scores and the data were statistically compared for treatments at $P < 0.05$.

The hens on Treatment B tended to show greater FI and LW than Treatment A for each week (1630–1730 g/hen for B versus 1609–1650 g/hen for A = $P > 0.05$). While egg production decreased with weeks, Treatment B had more eggs in the first 2 weeks and fewer eggs than Treatment A afterwards ($P > 0.05$). The Treatment C and D hens had the same restricted FI, and so the LW loss of 16% was greater in the first week for both treatments than the 6% for Treatment C versus 4% for Treatment D afterwards. While Treatment A and B hens did not moult, Treatment C and D hens showed greater moult scores for D versus C, but D took slightly longer than C to cease egg production. It appeared that feed restriction without or with aluminium at 8 h photoperiod can induce moulting in laying hens.

The Pakistan Higher Education Commission and Yorkshire Agricultural Society provided financial support.

POSTER 14

Non-feed Removal Strategies to Induce Moulting in Laying Hens: II. Influence on Hen Performance during 10 Weeks Post-moult

M. Yousaf and A.S. Chaudhry

School of Agriculture, Food and Rural Development, Newcastle University, Newcastle, NE1 7RU, UK. Email: muhammad.yousaf@newcastle.ac.uk

Yousaf and Chaudhry (2007) (Poster 13, page 452) found that restricted feed (Treatment C) or C plus 15 p.p.m. aluminium using Al_2O_3 (Treatment D) with an 8 h photoperiod induced a moult in Bovan Hyline hens and terminated their egg production within 4 weeks compared with *ad libitum* feeding at 16 h (Treatment A) or 8 h photoperiods (Treatment B). This research reports the effects of those moulting methods on hen performance during 10 weeks post-moult.

All hens received a 16 h photoperiod and *ad libitum* feed (170 g CP/kg) while housed in the same cages. The hens were monitored for food intake (FI), live weight (LW), egg production and moult scores, which were statistically compared for the moulting effects (photoperiod, restricted feed and aluminium) at $P < 0.05$.

The photoperiod effect (Treatments A versus B) was non-significant for FI and LW, but Treatment B maintained greater LW ($P > 0.05$) than Treatment A during post-moult. The hens in Treatments C and D increased FI and were similar to (Treatment C) or greater than (Treatment D) those in Treatment B (restricted-feed effect) within 2 weeks post-moult. The Treatment D hens regained their pre-moult LW a week earlier than Treatment C (aluminium effect). Egg production decreased in the last 4 weeks post-moult for Treatments A and B, but it was greater for Treatment B than Treatment A ($P < 0.05$, photoperiod effect). The hens on Treatment D started egg laying earlier than Treatment C, and the egg production for Treatments C and D surpassed Treatments A and B after 4 weeks and remained higher afterwards (restricted-feed effect). The feather scores were higher for Treatment D than C, but no moult was observed in Treatments A and B. It appeared that food restriction alone or with dietary aluminium supplementation can induce moulting to improve subsequent egg production.

The Pakistan Higher Education Commission and Yorkshire Agricultural Society provided financial support.

Yousaf, M. and Chaudhry, A.S. (2007) Non-feed removal strategies to induce moulting in laying hens: I. Influence on hen performance during 6 weeks of moulting.

Poster 15
Aromatase Pathway Mediates Chicken Sex Change in Each Direction

J.X. Zheng and N. Yang
Department of Animal Genetics and Breeding, China Agricultural University, Beijing 100094, China. Email: katie80881@gmail.com

In the present study, the fertile sex-reversed males were induced from the genetic dwarf females by the treatment of aromatase inhibitor at day 3.5 of incubation.

Five hundred chicken eggs were randomly sorted in two groups of 250 eggs each. Eggs in one group received an injection of 0.1 ml PBS; eggs in another group received 1 mg Fadrozole (4-(5,6,7,8-tetrahydroimidazo(1,5-α) pyridin-5-yl)benzonitrile) monohydrochloride (CGS 16949A, Novartis Pharma AG, Basle, Switzerland). The solution was injected into albumen under the air cell of eggs and the holes were sealed with melted paraffin.

One hundred and twenty-six females and 103 males were obtained from PBS-injected eggs, 104 females and 77 males from those injected with Fadrozole. Among sex-reversed males in the Fadrozole group, two individuals with ID 11636 and 11668 were found to be fertile. They produced sperm and reproduced normal progeny after mating with normal hens by artificial insemination. The fertility for ID 11636 and 11668, respectively, was 22.4 and 5.6% at 30 weeks old. The results of semen quality tests showed that the concentration of sperm for sex reversals, 1.22×10^9/ml, was nearly half of that for the control males, 2.62×10^9/ml. The percentage of abnormal sperm of fertile, sex-reversed males was 74%. Furthermore, it was also demonstrated that W sperm possessed natural fertility and resulted in hatched chicks.

The results suggested that avian sex determination and differentiation are more labile than in mammals and might be manipulated by the hormone milieu and especially by oestrogen. Furthermore, the aromatase pathway might mediate sex changes in both directions in the chicken.

Index